Thin on the Ground

Advances in Human Biology

Series Editors:
Matt Cartmill
Kaye Brown
Boston University

Titles in this Series

Thin on the Ground: Neandertal Biology, Archeology, and Ecology
by Steven E. Churchill

THIN ON THE GROUND

Neandertal Biology, Archeology, and Ecology

STEVEN EMILIO CHURCHILL
Department of Evolutionary Anthropology
Duke University
USA

SERIES EDITORS: MATT CARTMILL AND KAYE BROWN

WILEY Blackwell

Editorial offices: 1606 Golden Aspen Drive, Suites 103 and 104, Ames, Iowa 50010, USA
The Atrium, Southern Gate, Chichester, West Sussex, PO19 8SQ, UK
9600 Garsington Road, Oxford, OX4 2DQ, UK

For details of our global editorial offices, for customer services and for information about how to apply for permission to reuse the copyright material in this book please see our website at www.wiley.com/wiley-blackwell.

Library of Congress Cataloging-in-Publication Data has been applied for

ISBN: 978-1-1185-9087-4 (cloth)

A catalogue record for this book is available from the British Library.

Wiley also publishes its books in a variety of electronic formats. Some content that appears in print may not be available in electronic books.

Cover drawing by Matt Cartmill

Set in 10/12pt Garamond by Aptara Inc., New Delhi, India

1 2014

MIX
Paper from responsible sources
FSC www.fsc.org FSC® C013604

Contents

Series Introduction ix

Preface xi

Acknowledgements xv

Chapter 1 Thin on the Ground: Population Density and Technological Innovation **1**

Note 7

Chapter 2 The Neandertals in Time and Space **9**

2.1 Geographic and Temporal Boundaries 10
2.2 Defining the Neandertals 13
2.3 Neandertal DNA 27
2.4 Neandertal Taxonomy 30
2.5 Regional and Temporal Variation in Neandertal Morphology .. 30

2.6 The Evolutionary History of the Neandertals 32
Notes 39

Chapter 3 Neandertal Material Culture **41**
3.1 Neandertal-associated Lithic Industries 42
3.2 Variation in the Eurasian Middle Paleolithic: Technology as Adaptive Interface 53
3.3 Composite Technology, and the Archeologically Less-visible Component of Technology 59
3.4 Subsistence Technology 61
3.5 Domestic Technology 67
Notes 69

Chapter 4 The Body Neandertal **71**
4.1 Neandertal Body Size: Short but Massive 72
4.2 Body Composition: "Scaled Up" Inuit? 79
4.3 The Cost of Size: Feeding a Large Body and Large Brain 82
4.3.1 Nutrition and Somatic Maintenance: Neandertal Bodies were Energetically Costly 83
4.3.2 Nutrition and Reproduction: Were Neandertal Mothers Like Polar Bears? 92
4.3.3 Dietary Shortfalls: Hard and Lean Winters 97
4.4 The Benefits of Size: Neandertal Body Size in Ecological Context 99
Notes 104

Chapter 5 Surviving the Cold **107**
5.1 How Cold Was It? 108
5.1.1 Cold–Temperate Conditions (MIS 5d-a and 3) 109
5.1.2 Cold Glacial Conditions (MIS 6 and 4) 111
5.2 Human Adaptation to the Cold 114
5.3 Cold Adaptation and Neandertal Morphology 117
5.3.1 The Survival Value of Neandertal Body Form 124
5.3.2 Craniofacial Morphology and Cold Adaptation 129
5.4 Physiological Solutions to Cold Stress 137
5.5 Cold Stress and Neandertal Behavior 141
5.5.1 Activity 141
5.5.2 Extrasomatic Heat Production and Conservation: Fire, Shelter and Clothing 142
5.6 Thermogenic Capacity and Cold Tolerance 146
5.7 The Neandertals *Were* Cold-adapted 148
Notes 149

Chapter 6 The Caloric Economy of Pleistocene Europe **151**
6.1 Issues in the Reconstruction of Past Environments 152
6.2 Pleistocene Biomes of Europe and Western Asia 157

6.2.1 Interglacial Europe: Marine Isotope Stages 7 and 5e.... 159
6.2.2 Productivity and Edible Resources of Interglacial Environments 162
6.2.3 Temperate Interglacial Europe: Marine Isotope Stages 5d–a and 3 169
6.2.4 Cold Steppic Europe: Marine Isotope Stages 6 and 4 ... 173
Notes 175

Chapter 7 Neandertals as Consumers **179**
7.1 Analysis of Food Residues: The Macromammal Component of Neandertal Diet 181
7.2 Analysis of Food Residues: The Small Animal Data 187
7.3 Analysis of Food Residues: Macrobotanical Remains 191
7.4 Dental Wear and Food Residues on Teeth 194
7.5 Stable Isotope and Trace Element Analyses 200
7.6 The Thorny Issue of Cannibalism 214
7.7 The Trophic Ecology of Neandertals 216
Notes 218

Chapter 8 Red in Tooth and Claw: Neandertals as Predators **219**
8.1 Neandertal Morphology and Predation 220
8.1.1 Scapular Glenoid Fossa 222
8.1.2 Elbow Joint Morphology 224
8.1.3 Superior Pubic Ramus Length 226
8.1.4 Humeral Diaphyseal Cross-sectional Geometry 226
8.1.5 Entheseal (Muscle Marking) Morphology 233
8.2 Neandertals as Close-range Predators 236
8.3 Prey Size, Hunting "Pack" Size, and Risk of Injury to Neandertal Hunters 245
8.4 Neandertal Hunting in Ecological Context 246
Notes 249

Chapter 9 In the Company of Killers: Neandertals as Carnivores **251**
9.1 Large-bodied Carnivores of the Eurasian Late Pleistocene 253
9.2 The Members of the Eurasian Pleistocene Large-bodied Carnivore Guild 256
9.2.1 *Homotherium latidens* 256
9.2.2 *Panthera leo spelaea* 258
9.2.3 *Panthera pardus* 259
9.2.4 *Crocuta crocuta spelaea* 260
9.2.5 *Hyaena hyaena* 262
9.2.6 *Canis lupus* 262
9.2.7 *Cuon alpinus* 263
9.3 Competition within the Carnivore Guild 264
9.3.1 Exploitation Competition 265
9.3.2 Interference Competition 268

9.4 Neandertals Were Not the Socially-Dominant Members of the Carnivore Guild 271
9.5 Neandertal Ecology in the Context of Competition within the Carnivore Guild 276
Notes 284

Chapter 10 The Cost of Living in Ice Age Europe **287**
10.1 Subsistence Organization and Mobility 289
10.2 Home Range Size 299
10.2.1 Lithic Raw Material Movement 300
10.2.2 Carnivore Models for Estimating Neandertal Mobility 302
10.3 Paleontological Reflections of Neandertal Mobility 309
10.4 The Energetic Cost of Mobility 316
10.5 The Energetic Cost of Domestic Activities 321
10.6 Neandertal Physical Activity Levels 324
Notes 331

Chapter 11 Neandertal Social Life, Life History, and Demography **333**
11.1 Subsistence Labor Demands, Group Size, and Social Structure 336
11.2 Neandertal Life History 342
11.3 Neandertal Demography 347
Notes 352

Chapter 12 From Thin to Thick: The African MSA **353**
12.1 Tipping the Scales on Population Growth 355
12.2 Culture Change in the Late MSA and Mousterian 360
Note 363

References 365

Index 445

Series Introduction

For us, the experience of reading Steve Churchill's book *Thin on the Ground: Neandertal Biology, Archeology, and Ecology* was like that of reading *The Origin of Species* for the first time. In both Churchill's and Darwin's books, the reader is led carefully and meticulously through a beautifully organized presentation of all the evidence bearing on a vexed and long-standing problem, arriving at a novel answer that resolves many issues all at once. Like Darwin, Churchill makes his case with such a wide-ranging, comprehensive, and judicious presentation that when the overall conclusion is fully laid out in the last chapter, its force is inescapable.

In *Thin on the Ground,* Churchill attempts to answer the overriding question of why the Neandertals became extinct. Over the past 150 years, many answers have been offered to this question. Some have claimed that Neandertals were too dim-witted or inarticulate to compete with the modern humans that began streaming into their European homeland some 40,000 years ago. Others have sought the cause of the Neandertals' demise in disease, or in changing climates that grew too hot or too cold for them, or in genocidal persecution by our own ancestors. Still others have argued that the Neandertals simply evolved into modern Europeans, and never became extinct at all. The search for understanding the disappearance of the Neandertals has seemed both speculative and never-ending.

In this new text, Churchill carefully demonstrates the inadequacy of all these answers. He marshals evidence from a broad range of sciences – genetics, anatomy, archeology, ecology and climatology – to support a complex answer of his own: Neandertals inhabited a ecologically marginal and energetically precarious position in the trophic pyramid of Pleistocene Europe, from which they (and some other large carnivores) were ousted by invaders whose physiology and subsistence strategies gave them an insuperable competitive edge.

We are enormously proud to begin our *Advances in Human Biology* textbook series, aimed at professionals as well as advanced undergraduate students, with this accessible yet magisterial book by Steve Churchill. We firmly believe it will become a landmark in the scientific study of the fossil record of the human lineage.

MATT CARTMILL AND KAYE BROWN

Preface

This is an exciting time to be studying Neandertals. In the last few years we have seen the publication of complete genome sequences of two Neandertals (one of them at very high coverage) which, along with ongoing work on mitochondrial sequences, is fomenting a revolution in our understanding of Neandertal biology. The analysis of ancient DNA has provided a new and powerful tool that has changed the nature of the questions that we can ask about the Neandertals, and added an independent line of evidence to our traditional sources of information: Neandertal fossils and archeological residues. In addition to ancient DNA research, the past decade has seen the addition of important new specimens and archeological sites, improvements in dating methods, and increasingly sophisticated methods of reconstructing Pleistocene environments and Neandertal diet, life history, and behavior. The past year has brought the publication of important new discoveries about Neandertal genetics and physiology, life history, technological innovation, and symbolic behavior, as well as announcements of new fossils of Neandertals or their near ancestors in Europe and the Near East. Late Pleistocene paleoanthropology is a dynamic, rapidly-moving field.

On the other hand, this is probably the worst time to write a book about the Neandertals. Our understanding of these fascinating Ice Age humans is evolving quickly, and new papers reporting important new findings come out almost on a weekly basis.

More than 90 articles about Neandertals or their ancestors, the Middle Paleolithic, or Pleistocene environments came out in major journals in 2013. Any book such as this, which attempts to review the already voluminous and rapidly growing literature on Neandertals, will quickly be out of date (indeed, the process of writing this book has been one of unending revision, in a near-futile attempt to cope with the constant bombardment of new research). Nonetheless, I felt compelled to undertake this project to address a single, perplexing question about these hominins. The Neandertals represent the terminus of a human lineage that survived the waxing and waning of glacial episodes in Europe and western Asia for more than a half million years. To do so, they must have had effective adaptations for dealing with harsh climates and ungenerous ecosystems, and the adaptive flexibility to cope with major climatic and ecological shifts over the millennia. The ancestors of the Neandertals were the first true colonizers of cold–temperate environments outside of the tropics, and the persistence of this lineage in Ice Age Eurasia is a real evolutionary success story. Yet a common refrain in papers on both Middle Paleolithic archeology and archaic human genetics is that the Neandertals lived at low population densities, that is, they were very "thin on the ground." Given their seeming adaptive success, and given that demographic growth in modern humans has been persistent over the last few tens of thousands of years (bringing us to seven billion or so today), what was it about the ecology of the Neandertals that kept the growth of their numbers in check? This book is my attempt to answer that question.

In the pages that follow, I present two mutually-compatible arguments – one concerning Neandertal energy budgets and the other involving carnivore community ecology – that I think provide an answer to the question. Both of these arguments concern the energetic ecology of the Neandertals, and thus this is largely a book about energy flow through the Neandertal world. After framing the central question (Chapter 1) and covering some background material on Neandertal morphology and material culture in Chapters 2 and 3, the book explores the basic energetic demands faced by Neandertals – the energy required for the growth and maintenance of relatively massive bodies, and that needed for staying warm during cold–temperate and glacial periods (Chapters 4 and 5, respectively). Chapter 6 looks at what inferences can be made about the abundance and distribution of humanly-edible calories in the various Ice Age environments of Europe and western Asia, and Chapter 7 reviews what we know about which of those resources the Neandertals were actually eating. The next two chapters explore the capture of what appears to have been the major source of calories for the Neandertals – animal prey – considering first the nature of hunting during the Middle Paleolithic (Chapter 8), and then exploring the dynamics of competition the Neandertals must have experienced with the numerous large carnivore species with which they co-existed (Chapter 9). Chapter 10 looks at overall energy budgets, and Chapter 11 relates those energy budgets to Neandertal social organization, life history, and demography. The final chapter delves into the cultural consequences of Neandertal population density, and addresses the question of how early modern humans might have managed to break free of the demographic constraints that had kept Neandertal populations small.

Although the study of energetics has a deep history in human biology, it has only recently started to become a focus of research in paleoanthropology. But despite this recency, this approach has already provided interesting and important insights into human life history evolution and hominin socioecology, and into the biogeography and site settlement systems of Neandertals. Elsewhere I have argued (Churchill 2006) that an energetics perspective can also help morphologists with the tricky problem of equifinality – which in its broad sense refers simply to the fact that there may exist many different ways to arrive at the same end state. Of importance to paleontologists wishing to infer function or adaptation from morphology is the reality that very different selective pressures may favor the same morphology. The large, capacious chests of Neandertals, for example, may be an adaptation to cold climates (as part of a short and stocky body plan with a relatively low surface area to volume ratio) or to very high activity levels (which demand high ventilatory rates), or both. Likewise, the foreshortened limbs of the Neandertals may also reflect cold adaptation (as a means of reducing skin surface area relative to body volume), or might be an adaptation for increased musculoskeletal leverage (in technologically-unsophisticated hominins who may at times have needed to exert great muscular force on objects in their environments). Similar examples of competing adaptive hypotheses can be cited concerning Neandertal nasal morphology, masticatory biomechanics, overall facial form, pelvic anatomy, and locomotor morphology. Certainly, various adaptive hypotheses need not be mutually exclusive – large chests may have been advantageous for both thermoregulation and respiratory physiology – but our understanding of Neandertal biology is hampered by our inability to discern the major selective agents behind their morphological evolution or, indeed, to even determine if selection was the evolutionary force behind any given set of anatomical features. Without some means of evaluating the costs and benefits of proposed morphological solutions to various adaptive problems, adaptive hypotheses simply accrue over time with little hope of winnowing out the untenable ones. Energetics provides us with a useful tool for unraveling this knot, especially in the areas of thermoregulation and locomotor biomechanics. By allowing us to assess the caloric costs of developing or maintaining some aspect of morphology relative to the energetic benefits which that morphology provides, we can at least begin to evaluate the fitness advantage of proposed adaptations, and the likelihood that they are the product of selection (for example, we might ask how much heat is actually conserved by foreshortening of the limbs, whether it is enough to make much of a difference to survival, and how the attendant increases in locomotor costs compare to potential energy savings in thermoregulation). Energetics is not the only tool at our disposal (experimental work is also a powerful way of testing adaptive hypotheses), but it appears to be equally applicable to questions about the adaptive valence of both the morphology and behavior of earlier hominins, and thus a useful tool with which to explore the nature of constraints that may have operated on natural selection, or on potential behavioral responses to ecological circumstances. Energetics approaches provide us a richer picture of the nature of human adaptation in the Late Pleistocene.

Applying an energetics approach to extinct hominins is, however, fraught with difficulty. It generally requires making abundant simplifying assumptions, and indeed it often demands the blatant over-simplification of complex systems of interactions. It also requires lots and lots of estimations, with plenty of compounded error as estimates of one thing are entered into methods to estimate other things. These estimates occur at every level, from that of the ecosystem down to that of the individual, and estimating the important parameters at all of these levels is difficult. These problems are reflected in the cogent concerns voiced by Alan Turner (1992: 111), who noted that estimates of key ecological parameters for palaeoecosystems "are fraught with difficulties and probably offer no more than a spurious air of precision," and of Steve Kuhn and Mary Stiner (2006: 971) who dismissed assessments of Neandertal energy requirements as tending "to yield highly relative, gross, and oft-revised estimates of daily caloric needs." I cannot argue with these critiques. But I can at least offer a rationale for why this approach is still worth pursuing. Although estimates of Neandertal energetic demand may differ between studies (see Chapter 10), these studies uniformly agree that Neandertals had high caloric expenditures and tight energy budgets, and they thus seem to all be reflecting some fundamental underlying reality about Neandertal ecology. Thus I would argue that, while none of the estimates may be exactly right, in combination they are right enough to provide us with an idea of the energetic ecology of Neandertals, and to appreciate its biological and behavioral consequences (these estimates are, as my colleague Steve Vogel is fond of saying, "right within an order of magnitude"). If nothing else, this kind of exercise is good for forcing us to look at things in new ways, and for generating hypotheses which can potentially be tested with the already rich yet still growing datasets derived from hominin paleontology, Paleolithic archeology, and ancient DNA studies. For these reasons, I make here an earnest effort to follow the calories through the Neandertal world, and I unabashedly make estimation upon estimation in my pursuit of Mousterian caloric economics.

This book is intended to serve as reference reading for graduate-level coursework on the Neandertals. My hope is that, whether or not one accepts my arguments, the material that is reviewed here will provide students (and colleagues) with an expedient entrée to the literature, a good source for stimulating seminar discussion, and a useful reference about Neandertal morphology and behavior. This book diverges from most textbooks in that it does endeavor to advance an argument. And while I have tried my best to review the literature with fairness and objectivity, I have not shied away from expressing my own take on things, or spinning the evidence (within reason) in ways conducive to the arguments that I am trying to develop. I expect that, as more intellectually-mature readers, graduate students will be able to recognize my biases, see both the valid and problematic aspects of my arguments, and read this text with both critical eyes and open minds. My deeper hope is that some readers will find here ideas worth challenging, and that they will be motivated to not only prove me wrong, but to seek their own answers to the outstanding questions about these fascinating archaic humans. Our knowledge of the Neandertals, and our understanding of human evolution, will be much the richer for it.

Steven Churchill
Durham
December, 2013

Acknowledgements

The work presented in this book has been enriched by the research projects and class papers of numerous students over the years. Their energy, creativity, and enthusiasm has truly benefited this work, and I am enormously grateful to all of them: Kristin Ambrosi, Amy Berman, Leon Chang, Dorian Cohen, Julie Daniel, Katie Doswell, Lynn ElHarake, Heather Frieman, Nicole Georgi, Elizabeth Kane, Lily Kimbel, Elissa Krakauer, Hilary McKean-Peraza, Katarina Mucha, Jeanna Novelli, Paul Salem, Mandy Silberman, Adam Schwartz, and Sunny Warren. Many of the ideas expressed here either developed during, or were shaped by, my collaborations with former and current graduate students, especially Michael Black, Bob Cieri, Andrew Froehle, Adam Hartstone-Rose, Patrick Lewis, Laura Shackelford, Chris Walker, and Todd Yokley, and to all of them I am grateful. I also thank Leslie Aeillo, the late Lew Binford, Alison Brooks, Alena Boydon, Bob Franciscus, Trent Holliday, Jean-Jacques Hublin, Steve Kuhn, Joanna Lambert, Bill Leonard, Paul Pettitt, Yoel Rak, Mike Richards, Jill Rhodes, Daniel Schmitt, John Shea, Fred Smith, John Speth, Chris Stringer, Mary Stiner, Erik Trinkaus, Steven Vogel, and Milford Wolpoff for interesting and valuable conversations and email exchanges about the Neandertals and Middle Paleolithic technology over the course of the last decade. Laura Gruss, Adam Hartstone-Rose, Branka Hrvoj, Ryan Long, and Todd Yokley provided critical comments on parts of the manuscript, which is greatly appreciated. Chris Carbone graciously provided unpublished carnivore data, and Lars Werdelin provided useful information about Pleistocene carnivore fossils.

I would especially like to thank Matt Cartmill and Kaye Brown for their efforts as editors of the Advances in Human Biology series at Wiley-Blackwell, and particularly for their valuable ideas on the overall structure of the book and flow of ideas. Thanks also to Matt for the use of his excellent drawings of the La Ferrassie 1 Neandertal fossils, and for producing the wonderful drawing that graces the cover. I'd also like to thank Anna Ehler and Stephanie Dollan at Wiley-Blackwell for their help (and patience) in moving this book towards publication. Finally, I wish to thank the many curators and institutions, too numerous to list here, across Europe, the Near East, and Africa that have allowed me access to fossils of Neandertals, non-Neandertal archaic humans, and early modern humans.

CHAPTER ONE

Thin on the Ground: Population Density and Technological Innovation

The Neandertals are no more. Their ancestors colonized the cold-temperate zones of Ice Age Eurasia, and over the millennia their lineage evolved technological and physical solutions to the adaptive challenges they faced there. The Neandertals persisted for several hundreds of thousands of years in the varied climates and ecosystems of Pleistocene Eurasia, from bitter cold episodes in which much of Europe was carpeted with open steppe and tundra, to warm interglacials when broadleafed forests extended from the Mediterranean to the Baltic Sea. Arguably, the lineage that gave rise to the Neandertals accomplished the first real human colonization of lands outside of the tropics, and their adaptive success over a half million years or more represents a watershed in the human use of technology to deal with the harsh realities of cold, seasonal environments (Foley 1999). But around 50,000 years before present (Ka BP), modern human populations had begun to encroach on the Neandertal homeland, beginning initially in the Near East, but very soon thereafter moving into eastern and central Europe and

Thin on the Ground: Neandertal Biology, Archeology, and Ecology, First Edition. Steven Emilio Churchill.

spreading westward (Hoffecker 2009; Hublin 2012). By about 32 Ka BP the Neandertals were gone, leaving modern humans in sole possession of their former lands.

There is no shortage of hypotheses as to why the Neandertals went extinct. To some, their extinction had nothing to do with the range expansion of modern humans that was occurring at the same time – the two events were only coincidentally related, or perhaps the dying-off of the Neandertals left vacant space into which expanding modern human populations quickly flowed. Thus to some workers the Neandertals went extinct because the climate got too cold for them (Gilligan 2007), or because of a period of climatic instability in which environments and ecosystems changed faster than they could adapt (Finlayson 2005; Finlayson et al. 2004; Finlayson and Carrión 2007; see also Bradtmöller et al. 2012). To others, it was a reduction in carrying capacity as the climate worsened towards the last glacial maximum, leading to the widespread extinction of many elements of the European Pleistocene fauna – Neandertals included (Stewart 2004, 2007; see also Shea 2008). Still others have suggested that they got fried by increased UV-B radiation from a temporary reduction in the ozone layer (Valet and Valladas 2010), or they were unable to cope with an extended "volcanic winter" brought about by a super-eruption (Fedele et al. 2008; Golovanova et al. 2010: but see Lowe et al. 2012), or they baked in their own skin thanks to mitochondria that leaked heat (Hudson et al. 2008), or they did themselves in with transmissible spongiform encephalopathies, perpetrated by their bad habit of eating one another's brains (Chiarelli 2004; Underdown 2008). To others, the modern human diaspora from Africa was an integral part of the demise of the Neandertals, perhaps as a direct result of competition and competitive exclusion (Flores 1998; Banks et al. 2008; see also Svoboda 2005), or due to some combination of climate change and competition with modern humans (d'Errico and Sánchez Goñi 2003, 2004; Stringer et al. 2003; Jiménez-Espejo et al. 2007; Müller et al. 2011). It has also been suggested that modern human populations, expanding into Eurasia from disease- and parasite-rich tropical areas, may have introduced novel diseases into the Neandertal population (Wolff and Greenwood 2010; see also Sørensen 2011). Direct, violent aggression from modern human invaders (Gat 1999; Hortolà and Martínez-Navarro 2013), with some possible intergroup cannibalism (see Ramírez Rozzi et al. 2009), has also been suggested. What with being fried by the sun, frozen by a volcanic winter, driven crazy by "Mad Neandertal" disease and leaking heat from every cell, while being hunted, sickened, and out-done by Neandertal-hungry modern humans, all while the climate wavered and their habitats shrunk – well, one can imagine that the Neandertals might have welcomed extinction when it came!

From anthropological and ecological perspectives, the hypotheses that posit a role of modern humans in the extinction of the Neandertals are perhaps the most interesting. With only one exception (that being the idea that modern humans were vectors of infectious diseases), all of the hypotheses involving modern humans revolve around the concepts of competition and competitive exclusion – either exploitative competition (where both groups were contending for the same resources), or interference competition (direct, aggressive encounters between groups), or both. Modern humans are ecological dominators (Foley 1999; see also Flinn et al. 2005), with a history of progressively monopolizing the productivity of the ecosystems we colonize, altering their landscapes, and growing our populations to densities unknown in other mammals. One component of the expansion of modern humans out of Africa was a noticeable reduction of the large mammal diversity in the areas into which they moved (a trend which, unfortunately, continues today). Loss of biodiversity following modern human colonization is an empirical reality, regardless of one's opinion on whether the two phenomena are causally related (Martin and Klein 1984; Owen-Smith 1987; Klein 1992; Johnson 2002; Brook and Bowman 2004; Pushkina and Raia 2008). Species that were direct resource competitors with modern humans tended not to fare well (Chapter 9: Berger 1999), and thus we might see the extinction of the Neandertals as part of a larger Late Pleistocene, modern human-mediated alteration of the mammalian communities of Eurasia (Stewart et al. 2003; Stewart 2007). Seen in this light, the demise of the Neandertals might best be seen as an integral part of our own story.

If Neandertal extinction was mainly due to competition with modern human invaders, it would imply that the latter had a competitive advantage over the indigenous Neandertals. This in turn raises the interesting question as to how modern human newcomers to the Ice Age environments of Eurasia were able to outcompete a group of hominins that were seemingly well-adapted to those environments (given that they and their ancestors had successfully survived there for a half million years or more), an argument that many find troubling (see for example, Finlayson et al. 2004). Recent human history is certainly replete with cases of invading colonists replacing indigenous peoples, but in these instances infectious disease ecology and technological superiority (coupled with a demographic advantage) are generally invoked to explain the ultimate demise of the native populations (Diamond 1997). Because the post-50 Ka BP incursion of modern humans into the Near East and Europe is attended by the first appearance of Upper Paleolithic assemblages (whereas the local Neandertals were still largely or entirely using Middle Paleolithic toolkits: see Chapter 3.1), it is natural to think that differences between the two groups in technological sophistication may have played an important role in the competitive dynamics between them. Consistent with this idea is the observation that, in artifact assemblages associated with modern humans, rates of technological innovation have steadily accelerated over the last 75,000 years of our evolution (another trend which continues today), whereas the technology of the Neandertals remained largely unchanged (from the perspective of innovation) over hundreds of thousands of years.

Beginning sporadically in the later part of the Middle Stone Age (MSA) and continuing with increasing regularity into the Later Stone Age (LSA) and Upper Paleolithic (UP), modern human-associated assemblages[1] document a rapid florescence of new technologies, including leptolithic and microlithic tools, greater artifact diversity, bone and antler working, heat treatment and pressure flaking of flint, long-range projectile weapons, grindstones, fishing and birding gear, trapping technology, sophisticated pyrotechnology, and possibly watercraft (Valde-Nowak et al. 1987; Mellars 1989a, 1989b; Straus 1991, 1993; Davidson and Noble 1992; Brooks et al. 1995, 2005; Yellen et al. 1995; Henshilwood and Sealy 1997; Ambrose 1998a; Holliday 1998; McBrearty and Brooks 2000; Henshilwood et al. 2001; Shea 2006; d'Errico and Henshilwood 2007; Backwell et al. 2008; Brown et al. 2009; Villa et al. 2009b: Lombard and Phillipson 2010; Mourre et al. 2010). Also during this period we begin to see increasing evidence of symbolic behavior and abstract thought, in the form of pigment processing, personal adornment, incised notational pieces, musical instruments, and mobilary and parietal art (McBrearty and Brooks 2000: Henshilwood et al. 2002, 2004, 2009; Conard 2003, 2009; d'Errico et al. 2005, 2009; Bouzouggar et al. 2007; Marean et al. 2007; Broglio et al. 2009; Conard et al. 2009; Higham et al. 2012). Furthermore, modern human-associated faunal and lithic assemblages from the late MSA onwards provide evidence for expanded diet breadth and innovations in subsistence strategies, expanded social networks, and long-distance exchange (McBrearty and Brooks 2000; Bar-Yosef 2002; Henshilwood and Marean 2003). Together these behaviors – from technological innovation to symbolic expression to niche expansion to enriched social complexity – signal the emergence of what has been called "behavioral modernity" or "fully symbolic *sapiens* behavior" (Henshilwood and Marean 2003; Nowell 2010).

The technological explosion that occurred coincident with the modern human diaspora reflects a notable aspect of our behavior, that being our extraordinary capacity for cumulative technological evolution (CTE), or "cultural ratcheting" (Tennie et al. 2009). This is the process by which multiple actors, who may be well-separated in space and time, contribute innovations towards the development of a single piece of technology or a technological system. Using the development of projectile weapon systems as an example, one individual might have loosely tied feathers to the proximal end of a spearthrower dart to develop the first fletching, and the attendant improvement in flight performance might cause this innovation to catch on and spread. Decades later and miles away, another individual might have devised a better way of binding the feathers to the shaft to further improve flight performance. Cultural ratcheting seems to be a component of modern human technological behavior from Marine Isotope Stage (MIS) 4 onwards, whereas it does not appear to have characterized the Neandertals' relationship to technology. Neandertal-associated Middle Paleolithic material culture, while dynamic and flexible in its own right, seems to lack the regular innovation of tool forms and new ways of using material items for symbolic expression that are seen in modern human-associated assemblages (this is not to say that innovation is totally lacking in Neandertal material culture, just that it is relatively rare: see below). The Neandertals certainly weren't stupid (their brains were every bit as large as ours, and in fact were a little larger on average: Chapter 4), and they and their ancestors had the adaptive wherewithal to survive the rigors of Ice Age Europe for more than 500,000 years. Why, then, this dramatic difference in technological acumen between two closely genetically related, behaviorally-flexible, ecologically-similar human groups?

The apparent technological dichotomy between Neandertals and early modern humans suggests to some that there were important cognitive differences between groups, and that Neandertals may have lacked the capacity for innovation, planning depth, abstract thought, and symbolic behavior that underlies behavioral modernity (see McBrearty and Brooks 2000). Since increased CTE and the geographic expansion of modern humans out of Africa occurred roughly 100–150 Ka after their earliest appearance in the fossil record (White et al. 2003; McDougall et al. 2005), this would suggest that the earliest modern humans likewise lacked the capacity for modern behavior. The persistent expression of symbolic behavior, as well as the marked acceleration of CTE, does not appear to be firmly established until the development of the LSA (in Africa) and UP (in Eurasia) sometime around 50 Ka BP (Klein 2000, 2008; Bar-Yosef 2002; Mellars 2006b; Nowell 2010). This has led some to argue that behavioral modernity resulted from an upgrade in cognitive abilities at around 50 Ka BP, perhaps reflecting a relatively rapid appearance and fixation of new alleles governing neural development (Mellars 2006b). Different cognitive enhancements have been proposed as being key to the emergence of modern behavior, including enhanced working memory (Wynn and Coolidge 2004, 2010; Ambrose 2010), domain-sharing intelligence (Klein 1995), linguistic and symbolic capacities (Mellars 1989b, 2007; Klein 2000, 2003), abstract thinking (Lewis-Williams 2002), "latching" (Amati and Shallice 2007), and the ability to attain higher levels of intentionality (Dunbar 2003).

There are, however, compelling reasons to suspect that Neandertals had cognitive and behavioral capabilities that were on a par with those of the early modern human makers of the LSA and UP. As noted above, their brains were every bit as large as those of early modern humans. There is also growing evidence that Neandertals had the cognitive capacity for substantial technological innovation (Gaudzinski 1999; Villa and d'Errico 2001; Hardy et al. 2013; Soressi et al. 2013), and that they were fully able to engage in symbolic behavior (Zilhão et al. 2010; d'Errico and Stringer 2011; Caron et al. 2011; Peresani et al. 2011). These findings imply that the critical difference between Neandertals and modern humans, and also between modern humans before and after 50 Ka BP, was not in cognitive capacity, but rather in the prevalence and persistence of technological innovation and symbolic behavior.

Increasingly, archeologists and cognitive psychologists are examining the role of cultural, historical, ecological, and demographic factors in both CTE and the persistent expression of symbolic behavior. There is a growing school of thought that holds that the capacity for modern behavior may have emerged at the same time that anatomical modernity did (that is, coincident with the first appearance of modern humans some 200–150 Ka BP), or possibly before (and thus is perhaps a shared primitive trait [symplesiomorphy] that both Neandertals and modern humans inherited from a common ancestor), but that social and demographic factors prevented its consistent expression until the end of the MSA (Chase 2006; Jacobs and Roberts 2009; Nowell 2010; d'Errico and Stringer 2011). Demographic expansion and population density especially are increasingly thought to have played a central role in the expression of behavioral modernity (Jacobs and Roberts 2009; Richerson et al. 2009). Cultural ratcheting requires both cultural innovation and transmission, and variation in these processes is conditioned by the rate of interaction between social learners (Shennan 2001; Henrich 2004). The absolute number of innovators, and rates of technological transmission, are dependent on population size and structure. When population density is low, innovations may arise but their rate of transfer to other groups is generally insufficient to offset the stochastic loss of cultural knowledge that occurs as individuals or local groups die off, and the low relative abundance of technologically-talented individuals combined with the dynamics of social learning may result in a reduction in average technological skill levels in a group over time (Henrich 2004). Computer modeling (Shennan 2001; Powell et al. 2009) and analyses of cultural complexity in small-scale societies (Henrich 2004; Kline and Boyd 2010) both show that populations with either large overall size or high connectedness between subpopulations are more successful in generating, retaining, and diffusing cultural innovations. Thus the sporadic occurrences of symbolic behavior and technological innovation in the earlier part of the MSA (McBrearty and Brooks 2000) may represent geographically-restricted, transient peaks in population density (in which CTE begins to take off) followed by demographic crashes (possibly caused by downturns in climate) (Mellars 2006a; Jacobs and Roberts 2009; Powell et al. 2009; Richerson et al. 2009). Similar transient increases in Neandertal population densities may account for the irregular manifestation of technological innovation and symbolic expression in the European Mousterian. Later expansion of modern human populations in Africa around 80–70 Ka BP, and their subsequent demographic expansion into Eurasia between 60–40 Ka BP (Watson et al. 1997; Excoffier and Schneider 1999; Stiner et al. 1999; Forster 2004; Steele and Klein 2005) seem to have produced population densities sufficient for a high rate of CTE. Thus it may have been demographic factors, rather than cognitive capabilities, that account for the persistent expression of behavioral modernity in the LSA and UP (Shennan 2001; Powell et al. 2009).

When viewed in this light, the technologically-mediated competitive advantage of early modern humans which may have allowed them to supplant native archaic human populations (such as the Neandertals) across the Old World may have been an epiphenomenon of population growth. The Neandertals, on the other hand, appear to have lived at relatively low population densities throughout their tenure. The relatively small metapopulation size of the Neandertals would have impeded CTE and promoted culture loss, while at the same time making the Neandertals vulnerable to extinction even before modern humans reached Eurasia.

The chapters that follow advance two mutually compatible arguments as to why the Neandertals lived at such low population densities – that is, why they were so "thin on the ground." First, consideration of the Neandertals' adaptive solutions to environmental and ecological challenges shows that these solutions were effective but energetically costly (for example, Neandertals had bodies that were good at producing heat: turning up the furnace is an effective way to maintain a constant body temperature in the cold, but it is more costly than adding insulation). The nature of these adaptations resulted in a high energetic overhead to Neandertal life in Ice Age Europe, and the high cost of somatic maintenance and foraging resulted in Neandertal energy budgets that were very tight. This constrained the extent to which Neandertals could store energy as fat (making them "thin on the ground" in another sense), and constrained the amount of energy that could be invested in reproduction. As a consequence, Neandertals likely had a difficult time attaining the levels of fertility needed for population growth. Indeed, given what appears to have been high adult mortality, even reproduction sufficient for the maintenance of stable populations may have been a challenge for the Neandertals at some times. Second, consideration of the behavioral ecology of the Neandertals' resource competitors – members of the Eurasian later Pleistocene large mammal carnivore guild – suggests that Neandertal population growth may have also been ecologically constrained. Consideration of the size, social organization, and behavior of the various carnivore species of Pleistocene Eurasia, as well as what we can infer about Neandertal weapon technology and hunting behavior, suggests that Neandertals were not the socially-dominant members of the carnivore community. Based on recent research on exploitation and interference competition between carnivores, this further suggests that Neandertal population sizes may have been suppressed by a rich and ferocious cast of carnivore competitors. In conjunction, these energetic and ecological factors likely kept Neandertals at low population density throughout their Eurasian tenure. In the chapters that follow, we'll examine the nature of energy flow through the Neandertal world, explore the competive dynamics that surrounded the capture of calories for these Middle Paleolithic foragers, and address the interplay of Neandertal energetic ecology, life history, and demography. In the final chapter, we'll explore the question of how early modern humans of the later MSA were able to overcome similar kinds of ecological challenges and grow in numbers, and why such an outcome was less likely for the Neandertals.

NOTE

1. It is not certain that modern humans (*Homo sapiens*) were entirely responsible for all the innovations that appear in the archeological record of the MSA, since populations of archaic humans (*H. heidelbergensis*) may have still been extant in the late MSA. Evidence for symbolic behavior and technological innovations occurs sporadically in the African MSA after about 100–80 Ka BP. Fossils representing either early modern humans or early modern/archaic human admixed populations occur in east Africa at 195–150 Ka BP (White et al. 2003; McDougall et al. 2005), and in southern Africa by ca. 115–90 Ka BP (Grün and Stringer 1991; Klein 1999). Archaic/modern transitional specimens (such as Florisbad at 259 ± 35 Ka BP: Grün et al. 1996), considered by some to be early modern humans (Lieberman 2011), appear during the early MSA. Thus the hominins responsible for the cultural innovations of the late MSA were either early modern humans or quite nearly so.

CHAPTER TWO

The Neandertals in Time and Space

Any attempt to define the Neandertals and delineate their evolutionary history soon bumps into several thorny issues, including (1) the appropriate choice of species concept when dealing with fossil taxa, (2) the most appropriate methods of taxonomic diagnosis, and (3) how best to reconstruct phylogenetic relationships between paleospecies. One difficulty in defining the Neandertals involves the interdependence of various open questions. For example, the question of whether the Neandertals deserve separate species status (from *H. sapiens*) depends on one's choice of species concept, which in turn guides the choice of traits one uses to define the group, and which presupposes an answer to the question of their evolutionary fate. Hominin paleontology is commonly said to be analogous to solving a large and complicated jigsaw puzzle with most of the pieces missing and without benefit of the box top. In many respects, it is also similar to solving a crossword puzzle, in which one's solutions to intersecting questions both suggest and constrain solutions to the question at hand, and in which a seemingly good yet inaccurate solution to one question may frustrate efforts to resolve

Thin on the Ground: Neandertal Biology, Archeology, and Ecology, First Edition. Steven Emilio Churchill.

adjacent questions. Thus defining the Neandertals, and drawing distinct lines around their temporal and geographic distributions, is difficult.

Yet define the Neandertals we must if we wish to address questions concerning their adaptive strategies and behavioral capabilities. The following represents a tentative attempt to define the group and delineate their extent in time and space, while attempting to avoid overly dogmatic perspectives that, if inaccurate, may further cloud continued research.

2.1 GEOGRAPHIC AND TEMPORAL BOUNDARIES

Delimiting the spatial and temporal distribution of Neandertals presupposes the existence of a clear morphological definition that allows us to draw boundaries between them and their near neighbors geographically, or between them and the human occupants of Europe and western Asia that preceded and followed them. The Neandertals may constitute a distinct, easily recognizable morphological group (Tattersall and Schwartz 2006), yet they have perpetually frustrated attempts to sharply demarcate them from their neighbors in space and time. This is in part because the Neandertals appear to have arisen via a temporally broad series of anagenic changes in ancestral European *H. heidelbergensis* populations (Hublin 1998; Dean et al. 1998; Lebel et al. 2001).[1] Gradual, anagenic speciation events in general do not lend themselves well to clear circumscription of ancestor and descendant species, and the Neandertals are no exception. Additionally, but less problematically, interbreeding between Neandertals and in-migrating early modern humans (Bräuer 1984; Smith et al. 1989; Churchill and Smith 2000a; Trinkaus 2007; Green et al. 2010) may have resulted in the occurrence of some Neandertal autapomorphies (features unique to the Neandertals) in European early modern human specimens. Finally, although they were perhaps largely isolated in peninsular Europe during glacial intervals, the Neandertals were certainly participants in circum-Mediterranean networks of gene flow (Smith et al. 1995; Hutchinson 2000), making a sharp delineation of their geographic range difficult, and raising in some people's mind the question as to whether western Mediterranean Neandertals were in fact Neandertals (Arensburg and Belfer-Cohen 1998; see also Tillier et al. 2008).

Despite these difficulties, we can draw some loose lines around the Neandertal's spatial and temporal realms. Geographically, Neandertal remains have been recovered from the Iberian Peninsula in the west to the Altai Mountains of southern Siberia (Russia) in the east, and from Westphalia (Germany) and the north shore of Wales in the north[2] to Mount Carmel (Israel) in the south (Figure 2.1). However, the bulk of the Neandertal fossil record comes from France, Belgium, Germany, Italy, Croatia, Israel and Iraq. The specimens from Israel (from the caves of Tabun, Kebara and Amud) show slight differences in craniofacial morphology from European "classic" Neandertals (see below), but most likely belong in the same clade (see Section 2.5). The biogeographic distribution of Middle Paleolithic sites and Neandertal fossils has prompted the argument for a Neandertal "core area," located in southern and southwestern Europe, in which they originated and in which they persisted continuously until their extinction (Serangeli and Bolus 2008). Their occasional dispersal out of this area into the north of Europe and Great Britain, or eastward into eastern Europe and Siberia, or southeastward into the Near East, represented temporary population expansions during times

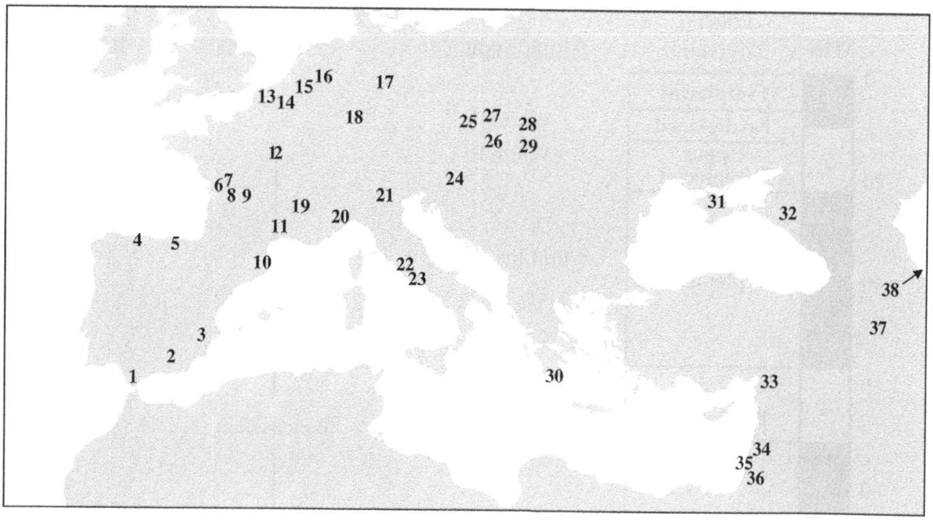

FIGURE 2.1

Approximate locations of major Neandertal fossil sites: (1) Forbes' Quarry, Devil's Tower & Zafarraya; (2) Carigüela; (3) Cova Negra & Las Palomas; (4) El Sidrón; (5) Lezetxiki; (6) St. Césaire; (7) La Quina, Châteauneuf, La Chaise & Marillac; (8) Le Moustier, La Ferrassie, Regourdou, Pech-de-l'Azé, Combe-Grenal & Roc de Marsal; (9) La Chapelle-aux-Saints; (10) Bañolas; (11) Hortus; (12) Arcy-sur-Cure; (13) Biache-St-Vaast; (14) Spy & La Naulette; (15) Engis, Fond-de-Forêt & Scladina; (16) Feldhofer Grotto; (17) Ehringsdorf; (18) Reilingen; (19) Moula Guercy, Payre & Bau de l'Aubesier; (20) Grotta delle Fate; (21) Mezzena; (22) Saccopastore; (23) Grotta Guattari; (24) Vindija & Krapina; (25) Kůlna; (26) Sala; (27) Sipka; (28) Gánovce; (29) Subalyuk & Istállóskö; (30) Apidima & Kalamakia; (31) Kiik-Koba; (32) Mezmaiskaya; (33) Dederiyeh; (34) Amud; (35) Tabun; (36) Kebara; (37) Shanidar Cave; (38) Teshik Tash (located in Uzbekistan, east of the Aral Sea), & Denisova and Okladnikov Caves (located in the Altai Mountains of southern Siberia, Russia).

of favorable climate. Consideration of patterns of genetic and morphological variation within the group (Section 2.5), however, suggests that this model might be overly simplistic.

Fossils that can reasonably be considered Neandertals appear in the late Middle Pleistocene (during Marine Isotope Stage [MIS] 7 and early MIS 6: around 230–180 Ka BP: Figure 2.2) at the sites of Biache-Saint-Vaast, Fontchevade 2, La Chaise Suard, La Chaise Bourgeois-Delaunay and Le Lazaret (France), Ehringsdorf (Germany) and possibly Reilingen (Germany) and Pontnewydd (England) (Hublin 1998; Dean et al. 1998). Neandertals are best known, however, from last interglacial (MIS 5e) and Late Pleistocene (MIS 5d-3) deposits – broadly between 130 and 30 Ka BP – which have produced remains of close to 300 individuals from over 80 sites (cf. Klein 1989a). Neandertals survived in Europe well into the Würm interplenniglacial (MIS 3), seemingly for several millennia after modern humans appeared on the continent (but see Conard and Bolus 2003). In the regions that once saw perhaps the densest occupations

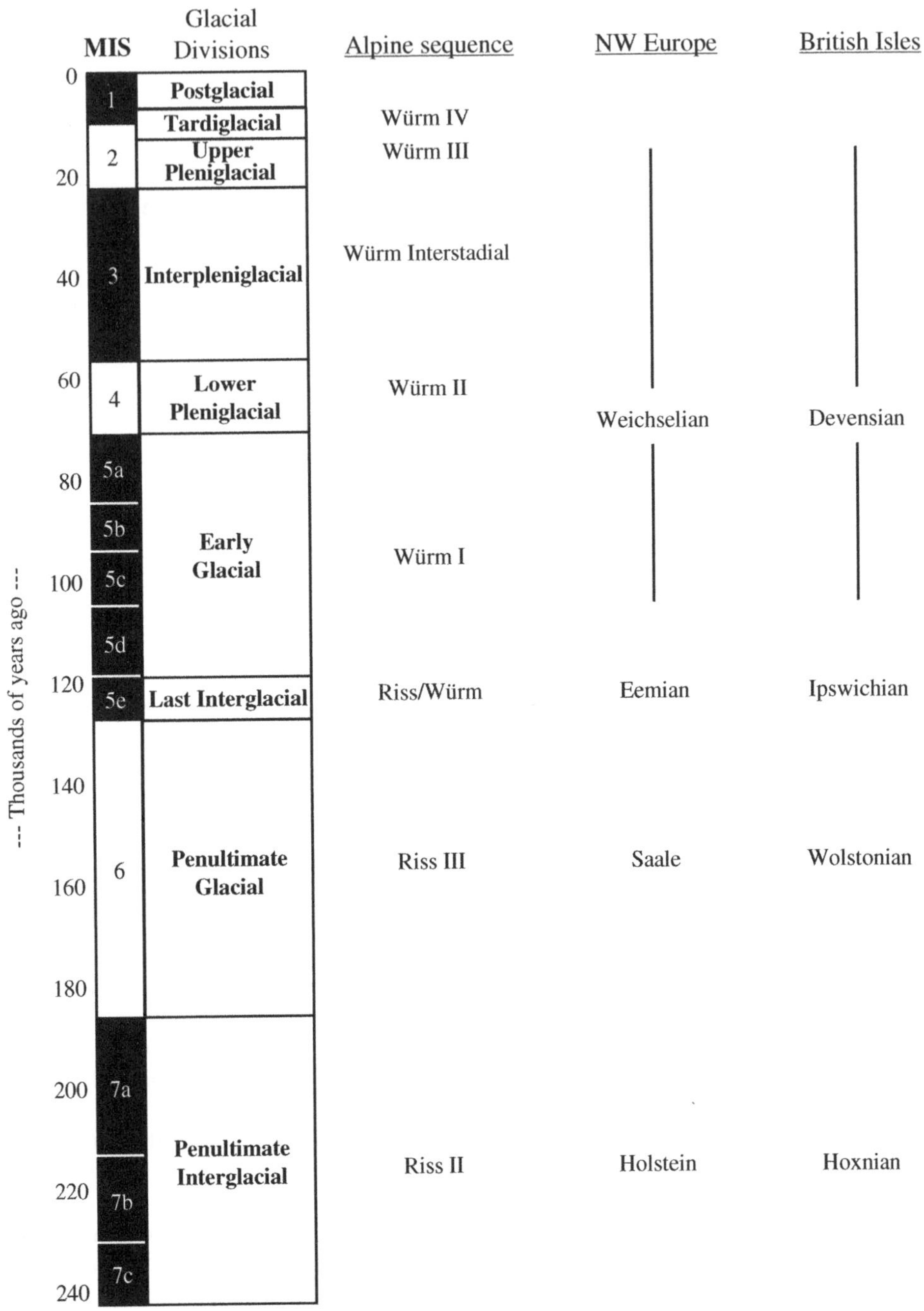

FIGURE 2.2

Chronology of marine oxygen isotope stages (MIS) and their correlation with major named terrestrial glacial and interglacial episodes for western Europe. Dates following Bassinot et al. 1994.

of Neandertals – such as the Dordogne region of France – they seem to have disappeared (perhaps having been displaced by modern humans) by about 33 Ka BP (Pettitt 1998).[3] Similar dates have been obtained for the latest Neandertals in the Dinaric Alps of Croatia (Higham et al. 2006). Claims for the late survival of Neandertals in more peripheral refugia (Finlayson 2008) are controversial. Neandertals may have lived on until about 30–28 Ka BP (corresponding to calendar dates of ca. 35–32.5 Ka BP cal) in the southern Iberian peninsula (Hublin et al. 1995; Finlayson et al. 2006), with some suggestion that they may have persisted there until about 24 Ka BP (Finlayson et al. 2006), although this date has been challenged on stratigraphic grounds (Delson and Harvati 2006). The Caucasus Mountains of the Republic of Georgia have likewise been argued to have sheltered late-surviving Neandertals (at ca. 29 Ka BP: Ovchinnikov et al. 2000), although subsequent dating of local Mousterian (the tool culture generally associated with Neandertals: see Chapter 3) sites suggests that the Neandertals had gone extinct in this area by 37 Ka BP cal (Pinhasi et al. 2011, 2012), supporting the suggestion that in-migrating modern humans may have found large portions of eastern Europe unoccupied (Hoffecker 1999). Finally, Mousterian artifacts from the northern Russian site of Byzovaya might suggest a northern refuge for Neandertals at about 34–31 Ka BP cal (Slimak et al. 2011), although human remains which might confirm that the site was created by Neandertals have not been recovered.

2.2 DEFINING THE NEANDERTALS

Producing a taxonomic diagnosis of the Neandertals (whether one considers them a subspecies of *Homo sapiens* or a distinct species) is made difficult by the nature of their morphological distinctiveness from other members of the genus *Homo*. The Neandertals are problematic because they are better defined by combinations of features (Wolpoff and Caspari 1996) – of which some are primitive and some are derived – than by a list of autapomorphies (uniquely derived traits). Careful consideration of morphological variation among and between groups of Middle and Late Pleistocene humans shows that Neandertals are distinguished by relatively few autapomorphies (Trinkaus 2006b)[4] – roughly half of their derived features are shared (either through parallel evolution or common ancestry) with modern humans. Additionally, some of the derived features that characterize the Neandertals do not occur with 100% frequency in the group, and some may be expressed in low frequencies in non-Neandertals. Neandertals represent the end group of an evolutionarily dynamic lineage of chronospecies – ones that likely exhibited variable amounts of geographic subspecific differentiation (see Fabre et al. 2009) and variable amounts of contact and gene flow with neighboring groups at different times and in different places. Absent a sharp, cladogenic origin of Neandertals (in which all autapomorphies arose full blown and "instantaneously" with respect to geologic time: see Weaver 1980; Crockford 2002, 2003), a definition based on autapomorphic features is likely to be sloppy – that is, a definition that does not fit all possible specimens that belong in the taxa, and one that is plagued by variability in trait frequencies in space and time. Barring the rapid fixation of Neandertal autapomorphies from a brief, intense selective episode or by genetic drift during an extreme population bottleneck event, the use of a taxonomic definition based on cladistic criteria (i.e., autapomorphies) is bound to be problematic at the time scale at which we are operating. This means that a clear definition – one that encompasses all of the specimens generally thought to be Neandertal – may not be attainable.

With respect to taxonomic diagnosis, the most defensible route seems to be to identify, following Hublin (1998) and Trinkaus (2006b), features which appear to be either unique in Neandertals or, if not unique, most frequent in them and rare in other hominins. Most of these features are also established at the time of birth (Gunz et al. 2011), and thus appear to reflect genetic differences from other groups. I have attempted to compile such a list in Table 2.1. In this exercise we must allow that some or all of these features may be variably expressed in the lineage of European archaic humans (*H. heidelbergensis*) that gave rise to the Neandertals, and thus while the overall constellation of these features may reasonably define the Neandertals *sensu stricto*, they will vary in the extent to which they separate Neandertals from other members of their European Middle and Late Pleistocene clade. Thus the features listed in Table 2.1 represent a mixture of derived traits unique to Neandertals and "regional clade features" that characterize the Neandertals but are also expressed, with increasing frequency through time, in earlier members of the clade.

Morphologically, Neandertals are best described by craniofacial features that distinguish them from other members of the genus *Homo*. Postcranially the Neandertals also exhibit what appear to be uniquely derived traits, although continued study of the small but growing sample of early *Homo* postcranial remains may show a number of these traits to be primitive for our genus. The derived features of the Neandertal skeleton occur along with, or sometimes represent modifications of, ancestral (plesiomorphic) features found in earlier humans. In the skull, these primitive features include thick cranial vault bones, platycephalic and low vaulted neurocrania with a low frontal angle, cranial vault superstructures (supraorbital and occipital tori), wide nasal apertures and a projecting external nose, shovel-shaped incisors, and mandibles with broad symphyseal regions but lacking chins (Stringer 1984; Trinkaus 1988; Franciscus and Trinkaus 1988a). Primitive features retained in the Neandertal postcranial skeleton include robust bones with large joint surfaces, ulnae with anteriorly directed trochlear notches, subequal pollical phalangeal proportions, os coxae with elongated, thin superior pubic rami and posteriorly positioned acetabula, and talar articular enlargement (references in Trinkaus 1988). Since some of the primitive traits that enter into these complexes may be shared with modern humans through common descent from a shared Middle Pleistocene ancestor (symplesiomorphies), extreme care must be exercised in assessing the phylogenetic and taxonomic implications of morphological similarities and differences between Neandertals and modern humans.

Neandertal skulls are characterized by two complexes of developmentally and/or functionally integrated features that distinguish them from ancestral *Homo*.[5] The first of these involves accentuated midfacial prognathism, that is, anterior projection of the face along the midline associated with laterally receding ("swept back") zygomatic arches (Figures 2.3 and 2.4) (the second complex, involving the shape of the cranial vault, is discussed below). This configuration apparently derived from a posterior migration of the masticatory muscles and their supporting structures in a face that retained the primitive condition of overall facial and dental prognathism (Trinkaus 1988). Neither of these base features are unique to Neandertals – total facial prognathism is a primitive trait shared with earlier *Homo* (i.e., a symplesiomorphy), while

TABLE 2.1 ■ Neandertal clade features[a]

Feature	Comments	Reference
Face/mandible		
Double-arched supraorbital torus that is arcuately continuous across glabella and that thins laterally (relative to earlier *Homo*), and that is heavily pneumatized in its medial portion by the frontal sinuses. No separation between the *arcus superciliaris* and *arcus supraorbitalis*.	Neandertals are often described as having high, rounded orbits – likely a secondary consequence of their supraorbital morphology.	Cunningham 1908; Hublin 1998; Schwartz and Tattersall 1996b; Stringer and Gamble 1994
Nasal (piriform) aperture that is tall and wide both inferiorly and superiorly (reflected in **anteroinferiorly broad nasal bones**).	Neandertal noses tend to be wide with respect to both piriform aperture height and nasal height.	Franciscus 1995
Prominent and complex inferior nasal sill, with high frequency of bilevel nasal floors.		Franciscus 2003
Medial projection of the lateral wall of the posterior portion of the internal nasal cavity.	Variably present, and likely related to expansion of the maxillary sinus.	Schwartz and Tattersall 1996a; Yokley 1999
Mid-facial prognathism characterized by low nasiofrontal and subspinal angles and a large difference between M^1 alveolus and zygomaxillare radii.		Hublin 1998
Flat or convex infraorbital plates (no canine fossa) enclosing large maxillary sinuses and receding obliquely in alignment with the **anterolaterally flattened surface of the zygomatic**.		Hublin 1998
Posterior rooting of the facial crest.		Hublin 1998
Retreating or vertical mandibular symphysis with a nonprojecting (or entirely absent) mental trigone.		Dobson and Trinkaus 2002
Mandibular mental foramen positioned below M_1.		Hublin 1998; Wolpoff and Caspari 1996
Diagonally-oblique mylohyoid line on mandible.		Rosas 2001

(*continued*)

TABLE 2.1 ■ (*Continued*)

Feature	Comments	Reference
Truncate gonion resulting in rounder posterior border of the mandibular ramus.		Rosas 2001
Horizontal-oval mandibular foramen (lingular bridging of foramen).	Variably present (slightly greater than 50%).	Hublin 1998; Wolpoff and Caspari 1996
Mandibular retromolar space formed by anterior positioning of the dental arcade relative to the ascending ramus.	Distinct from retromolar space formed by pre-angular notch on anterior margin of ascending ramus, as sometimes found in modern humans.	Rak 1998
Shallow mandibular notch with deepest portion adjacent to condyle.		Rak 1998
Laterally placed mandibular condyle.		Stefan and Trinkaus 1998
Lateral tubercle on mandibular condyle.	Formed by the medial position of the intersection of the mandibular notch (incisura) and condyle.	Rak 1998; Hublin 1998
Large medial pterygoid tubercle on mandible.		Rak et al. 1996; Antón 1996
<u>Dentition</u>		
Buccolingually wide anterior teeth.		Hublin 1998; Stringer and Gamble 1994
Maxillary incisors with lingual marginal ridges, lingual tubercles, and labial convexity.		Crummett 1994
High frequency of mesial ridges on maxillary canines		Bailey 2002
Markedly skewed maxillary molars with small metacones and large hypocones, and relatively small occlusal polygons with narrow distal segments.		Bailey 2004
Mandibular second premolars with large mesially-placed metaconid, uninterrupted transverse crest, high frequency of multiple lingual cusps, and asymmetrical occlusal outline.		Bailey 2002; Bailey and Lynch 2005

TABLE 2.1 ■ (*Continued*)

Feature	Comments	Reference
Mandibular molars with well-developed mid-trigonid crest and large anterior fovea, with relatively high frequencies of cusp 6, and with a high frequency of Y-grooves in the M_2.		Bailey 2002; Bailey and Hublin 2006b
Taurodont molars	Characterized by expanded pulp cavities and partially fused roots.	Trinkaus 1978; Hublin 1998
<u>Neurocranium</u>		
Platycephalic cranial vault with a large endocranial capacity.	Endocranial capacity shared homoplastically with modern *H. sapiens*.	Hublin 1998
Cranial vault shape "en bombe" in *norma occipitalis*.	Produced by a constellation of features including, among other things, reduction and medial curving of the mastoid region, maximum cranial breadth situated low on the parietals and indistinct parietal bosses (see Trinkaus 1988; Manzi et al. 2000)	Hublin 1978
Horizontal occipital torus of uniform vertical thickness, limited to the central portion of the occipital and with no or very slight development of an external occipital protuberance.	The median portion of the torus (midsagittal plane) is generally least developed, with maximal development laterally, giving a bipartite appearance to the torus.	Hublin 1978, 1988a; Santa Luca 1978
Well developed, **horizontally extended suprainiac fossa,** triangular in outline with the apex in the midline and the base in contact with the occipital torus, and generally demarked by an uplifted area of bone surrounding the depression.	Usually simple in form, but sometimes divided into upper and lower portions or right and left portions. The fossa, marking the insertion of *Mm. semispinalis capitis*, is generally rugose or porus.	Santa Luca 1978; Hublin 1978, 1988a; Balzeau and Rougier 2010
Convex occipital squama (*planum occipitale*).		Hublin 1978, 1998
Low, symmetrically arched temporal squama, with an anterior margin that is flexed medially, forming, along with inferomedial flexture of the posterior margin of the greater wing of the sphenoid, an infratemporal fossa distinct from the temporal fossa.		Schwartz and Tattersall 1996b; Hublin 1998

(*continued*)

TABLE 2.1 ■ (*Continued*)

Feature	Comments	Reference
Presence of a *tuberculum mastoideum anterius* (mastoid protuberance) behind the external auditory meatus.	Development of this protuberance can be variable, but it is always expressed.	Santa Luca 1978; Hublin 1978, 1988a
Occipitomastoid crest usually larger than the mastoid process, often straddling the occipitomastoid suture, separated from surrounding bone by distinct sulci medially and laterally.	See Schwartz and Tattersall 1996b for a discussion of variability in cresting patterns in this region.	Santa Luca 1978; Hublin 1978, 1988a
Laterally flattened and inferiorly tapering mastoid process with medially inclined apex.		Trinkaus 1988; Hublin 1998; Manzi et al. 2000
Oval, superoposteriorly oriented external auditory meatus, situated horizontally above the roof of the mandibular fossa (and on the same horizontal plane as the temporal zygomatic process).		Trinkaus 1983b, 1988; Condemi 1988; Vinyard and Smith 2002
Small, more circular and inferiorly situated posterior semicircular canal; small, narrow anterior semicircular canal with greater torsion than seen in modern humans; larger lateral semicircular canal.		Hublin et al. 1996; Spoor et al. 2003
Mandibular (glenoid) fossa that are large, shallow and poorly delimited, with an entoglenoid formed entirely from the temporal bone, a weakly developed articular eminence, and with a mediolaterally expanded postglenoid process (posterior zygomatic tubercle) forming a significant portion of the posterior wall of the fossa.		Condemi 1988; Trinkaus 1988; Hublin 1998; Vinyard and Smith 2002
Mediolaterally broad tympanic plate (reflected in a large distance between the stylomastoid foramen and the lateral tympanic wall), with a vaginal process that does not contact the mastoid process.		Trinkaus 1983b; Schwartz and Tattersall 1996b; Vinyard and Smith 2002

TABLE 2.1 ■ (*Continued*)

Feature	Comments	Reference
Anteroposteriorly long, mediolaterally narrow foramen magnum.		Rak et al. 1996
Postcranial skeleton		
Robust lower cervical spinous processes.		Trinkaus 2006b
Foreshortening of the distal extremities (bones of the forearm and leg) reflected in low brachial and crural indices.	Homoplastically shared with modern humans indigenous to cold climates.	Trinkaus 1981; Holliday 1997b
Scapula with narrow glenoid fossa and **dorsal sulcus on axillary border**.	Axillary border morphology variable, but 82% of specimens have true dorsal sulci or bisulcate borders with greater development of the dorsal sulcus: Churchill 1994a	Trinkaus 1977, 1983b; Churchill and Trinkaus 1990; Churchill 1994a; Trinkaus 2008b
Humeri with narrow, two-crested deltoid tuberosities and with mediolaterally wide olecranon fossae bounded by relatively narrow distodorsal pillars.		Churchill 1994a; Churchill and Smith 2000b
Radii with medially oriented biciptal tuberosities and **markedly bowed radial shafts**.		Trinkaus 1983b; Trinkaus and Churchill 1988
***Opponens pollicis* flange on first metacarpal.**		Trinkaus 2006b
Pollex with a flattened carpometacarpal joint and with proximal and distal phalanges of subequal length.	Possibly pleisiomorphic or secondarily related to appendicular robusticity: Trinkaus 1988	Trinkaus 1983b
Os coxae with elongated and dorsoventrally thin superior pubic rami, and with posteriorly positioned acetabula.	Possibly the primitive condition: see Rosenberg 1998; Arsuaga et al. 1999a	Trinkaus 1984

[a] Features considered autapomorphic by Trinkaus (2006b) are denoted by bold type.

posterior retreat of the masticatory region is a derived feature that is shared, presumably through homoplasy, with modern humans. However, only in the Neandertals do these two features occur together, producing one of several cases in which Neandertals are morphologically distinguished from other members of the genus *Homo* by a unique combination of features rather than unique features *per se* (Trinkaus 1988; Wolpoff and Caspari 1996). A similar case in the midface concerns the Neandertal nose, which combines the primitive trait of a wide nasal aperture (and presumably a

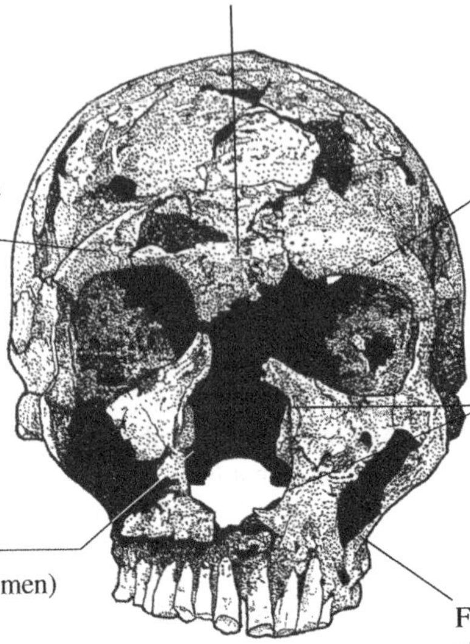

FIGURE 2.3

Anterior view of the La Ferrassie 1 male Neandertal cranium, illustrating some of the craniofacial morphology characteristic of Neandertals. Drawing © Matt Cartmill, used with permission.

voluminous nasal cavity) with a derived, strongly projecting external nose (this latter feature being shared, perhaps homoplasticly, with modern Europeans) (Franciscus and Trinkaus 1988a, 1988b; Franciscus 1995).

A number of details of Neandertal facial morphology are likely developmental or functional consequences, or both, of these two architectural changes in the face (a shift from total to midfacial prognathism and enlargement and greater projection of an already wide nose). The absence of a maxillary canine fossa (resulting in what is often described as "puffiness" of the maxilla) likely resulted from a coronal-to-parasagittal shift in the orientation of the zygomatic process of the maxilla (Figure 2.5) coincident with posterior migration of the zygomatic root, in combination with a maintenance of large maxillary sinuses (while the maxillary sinuses are absolutely large, they are not larger than those of modern humans relative to the size of the face: Rae et al. 2011). Other details of the morphology of the maxilla and zygomatic follow from this shift, including the parasagittal orientation of the infraorbital plates, anterolateral flattening of the zygomatics, absence of a zygomaticoalveolar notch, and the relatively horizontal orientation of the nasal bones (Trinkaus 1988). Distinctive mandibular features (Figure 2.6), such as a posteriorly positioned mental foramen and a retromolar space, likewise reflect the spatial or biomechanical consequences of a posterior shift in the position of the temporomandibular joint while maintaining an anterior,

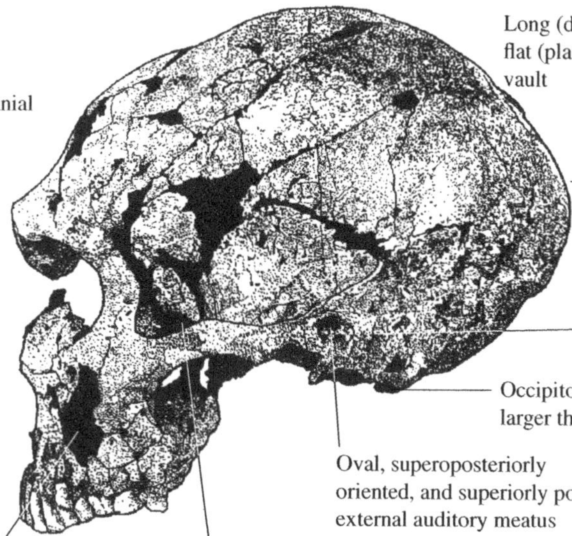

FIGURE 2.4

Lateral view of the La Ferrassie 1 Neandertal cranium. Drawing © Matt Cartmill, used with permission.

prognathic position of the dentition (Trinkaus 1988; Franciscus and Trinkaus 1995; Walker et al. 2010). Along these lines, the Neandertal tendency to have large medial pterygoid tubercles may reflect hypertrophy of some of the muscles of mastication relative to loss of mechanical leverage that accompanied their peculiar facial architecture (Antón 1990; O'Connor et al. 2005).

Recent studies of variation in Neandertal and modern human skeletal nasal morphology has called into doubt the taxonomic utility of three nasal features argued by Schwartz and Tattersall (1996a) to be Neandertal autapomorphies. The existence of two of these proposed autapomorphies – a medial projection on the internal nasal margin and the lack of an ossified roof over the lacrimal canal – have been called into question based on taphonomic considerations and patterns of variation across Neandertals (Franciscus 1999; Yokley 1999, 2006). Of the traits identified by Schwartz and Tattersall (1996a), only one – a medial swelling of the lateral wall of the posterior portion of the internal nasal cavity – may in fact occur with any regularity in a large sample of Neandertals (Yokley 1999) and thus may be autapomorphic. Since Neandertal facial sinuses scale with facial shape in the same way as do those of modern humans (Rae et al. 2011), this feature is unlikely to be a simple developmental or architectural correlate of expanded maxillary sinuses.

Before discussing the second major complex of traits that distinguish Neandertal skulls from those of other hominins, a few other facial features deserve mention. In

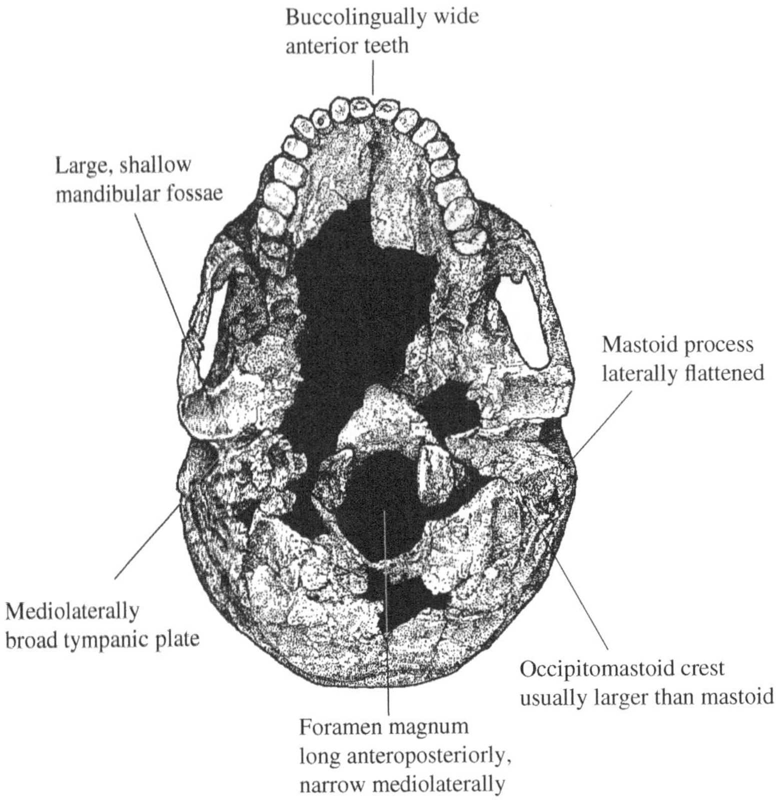

FIGURE 2.5

Inferior view of La Ferrassie 1 cranium. Drawing © Matt Cartmill, used with permission.

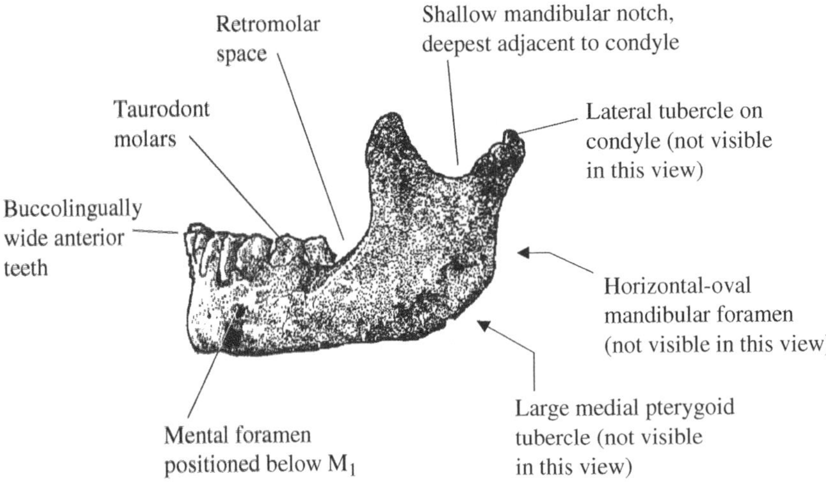

FIGURE 2.6

La Ferrassie 1 mandible in lateral view. Drawing © Matt Cartmill, used with permission.

addition to being midfacially prognathic, Neandertal faces were tall superoinferiorly yet narrow mediolaterally (relative to their height) due to the posterior displacement of the zygomatics. Neandertals also retained the primitive condition of pronounced supraorbital tori. As in *H. erectus* and other early humans, Neandertal brow ridges were continuous, both across glabella and over each orbit (among modern humans, our reduced brow ridges can be seen to be divided over each orbit into a more projecting supercillary arch and a less projecting supraorbital margin). However, Neandertals differed from earlier humans in having brow ridges that were double arched and that thinned laterally, unlike the often horizontal and relatively uniformly thick bar seen in earlier *Homo*. The overall constellation of a tall, midfacially prognathic face with a large, projecting nose, large double arched brow ridges and chinless mandible must have created a striking countenance indeed (at least by modern standards)!

The second complex concerns the overall shape of the cranial vault, and again reflects the intersection of primitive and derived features (Trinkaus 1988; Manzi et al. 2000). Like modern humans, Neandertals had large brains relative to early Middle Pleistocene *Homo*. Unlike modern humans, however, the necessary expansion of endocranial volume occurred in the context of the ancestral neurocranial bauplan of dolicocephaly and platycephaly (long and low cranial vaults, respectively), resulting in vaults that were particularly long and broad (Manzi et al. 2000) (Figure 2.4). In contrast, endocranial expansion in modern humans occurred in conjunction with a shortening and heightening of the vault (brachycephaly), resulting in accentuated vault globularity (Lieberman et al. 2002). These differences in cranial vault architecture suggest that endocranial expansion emerged homoplastically (independently) in the two taxa. The low frontal angle of Neandertals (reflecting the low vault overall) and their retention of large supraorbital tori (possibly the biomechanical consequence of hafting a prognathic face onto a low neurocranium: Russell 1985) lend a primitive appearance to an otherwise derived cranium.

The unique Neandertal constellation of primitive and derived neurocranial characters also results in the rounded, "en bombe" shape of the cranial vault as seen in rear view (Figure 2.7). This shape represents a departure from the pentagonal ("en tente") and straight-walled, high-domed ("en maison") form seen in earlier and modern *Homo*, respectively. Neandertals share (synapomorphically) with modern humans a degree of endocranial expansion and reduction of the cranial base and vault superstructures, yet share (symplesiomorphically) with earlier *Homo* platycephaly and an absence of parietal bosses (Trinkaus 1988). To this mix Neandertals add the uniquely derived feature of medially sloping mastoid processes (Trinkaus 1988), which accentuates the rounded outline of the neurocranium in rear view (Figure 2.7). Thus the "en bombe" morphology, like midfacial prognathism, results from a mosaic of ancestral, shared-derived and uniquely derived traits – again a unique combination of traits rather than a combination of unique traits (Wolpoff and Caspari 1996).

While inferring endocranial volume from the neurocranium is relatively straightforward (estimates of Neandertal brain size are provided in Chapter 4.2), making inferences about neuroanatomy from cranial vault morphology is more difficult (Holloway 1976, 1978). Studies of endocranial morphology in Neandertals reveal brains that, while primitive in overall shape (Gunz et al. 2012), were like those of modern humans in terms of asymmetries, extent of cortical folding (gyrification), and widening of the frontal lobes (Holloway 1981; Bruner and Holloway 2010), and that some of

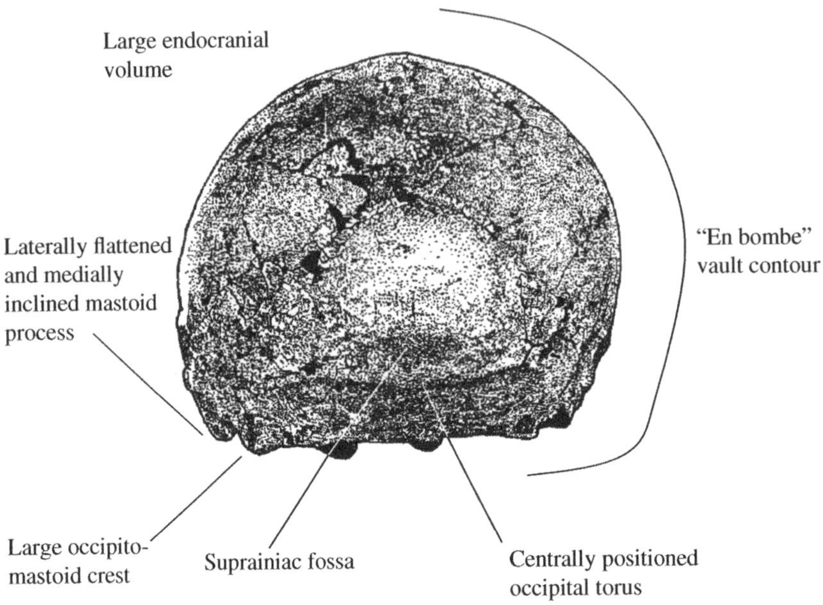

FIGURE 2.7

La Ferrassie 1 cranium in posterior view. Drawing © Matt Cartmill, used with permission.

the apparent morphological differences with modern humans are the consequence of allometry and do not necessarily imply cognitive differences (Holloway 1985). Neandertals brains do appear to differ from those of modern humans in terms of having smaller and less-projecting temporal poles, relatively narrower orbitofrontal cortices, smaller olfactory bulbs, laterally widened and relatively flat parietal lobes, and relatively larger occipital lobes (Bastir et al. 2011; Bruner 2010; Balzeau et al. 2012). It has also been suggested, on the basis of differences in orbital volume between Neandertals and modern humans, that Neandertals had relatively more visual cortex than modern humans (Pearce et al. 2013), which is consistent with their having had relatively large occipital lobes (Balzeau et al. 2012). Proportionally greater investment in visual cortex, combined with relatively great lean body mass (Chapter 4.2) and thus more neural tissue devoted to somatic systems, has been argued to have resulted in relatively less neural tissue for other functions, such as social cognition (Pearce et al. 2013).

The list of distinguishing cranial features is rounded out by a number of traits, seemingly unique to the Neandertals (and their near ancestors), of the posterior occipital and occipitomastoid regions (Table 2.1). The extent to which these features are developmentally or functionally correlated with overall cranial vault architecture has never been addressed, nor has the impact of dental attrition and masticatory biomechanics on morphologically labile regions of the cranium (such as the area of the temporomandibular joint: Trinkaus 1988). Thus, while it may be tempting to seek taxonomic or phylogenetic significance in the long list of Neandertal neurocranial and mandibular autapomorphies (e.g., Tattersall 1992), we must bear in mind that these features may reflect only a small number of genetic differences between Neandertals and other members of our genus.

Neandertal teeth also possess some characteristic features, which proves useful in the diagnosis of isolated dental remains or partial jaws (e.g., Lebel et al. 2001; Bailey and Hublin 2006a). As with other aspects of their morphology, the ways in which Neandertal teeth differ from those of modern humans or other hominins involve some uniquely derived features (autapomorphies), some unique combinations of primitive and derived features, and some traits that are shared with other groups but which differ in their frequency of expression in Neandertals. Neandertal anterior teeth tend to be large, and their maxillary incisors uniquely combine the primitive condition of well-developed marginal ridges (producing marked shoveling) with the derived conditions of marked labial convexity and prominent lingual tubercles (Crummett 1995; Bailey et al. 2008). Maxillary canine teeth tend to mirror the incisors in having well-developed marginal ridges, prominent lingual tubercles, and labial convexity (Bailey et al. 2008), and have mesial ridge frequencies above the range of variation seen in a global sample of recent humans (Bailey 2002). Neandertal lower second premolars (P_4) possess a unique and complex occlusal topography that results from the combination of a well-developed metaconid (symplesiomorphically shared with *H. erectus*), the presence of a transverse crest, and an asymmetrical lingual crown outline (the latter two traits being Neandertal autapomorphies) (Bailey 2002). Neandertal molars tend to exhibit high frequencies of taurodontism (enlarged pulp chambers) (Trinkaus 1978), and their M_1s and M_2s tend to possess mid-trigonid ridges (a Neandertal autapomorphy: Bailey 2002), large anterior fovea and relatively high frequencies of cusp 6 (Bailey and Hublin 2006b). Their M_2s are characterized by high frequencies of Y-grooves, well above the range of variation of this feature in modern humans (Bailey 2002). Neandertal maxillary first molars generally have a hypocone that is larger than the metacone, have skewed occlusal outlines produced by a metacone that is shifted lingually and mesially relative to its position in modern humans, and cusp apices which are internally positioned (resulting in a small occlusal polygon connecting the apices) (Bailey 2004). Added to this, their M^1s have higher frequencies of Carabelli's cusp and cusp 5 than do modern humans, and a zero frequency of lack of a hypocone, falling outside of the range of variation in the frequencies of these traits in a global sample of modern humans (Bailey 2002).

Postcranially, Neandertal skeletons also exhibit a mosaic of primitive (discussed above), shared-derived and uniquely derived traits (Figure 2.8; Table 2.1). Neandertals share with modern humans from cold regions an overall body form that includes: short stature, wide trunks (reflected in mediolaterally wide rib cages and pelves), "barrel-shaped" chests (anteroposteriorly and mediolaterally expanded thoraces), foreshortened distal limb segments (low brachial and crural indices) and short limbs relative to the length of the trunk (Trinkaus 1981; Ruff 1994; Holliday 1997b; Franciscus and Churchill 2002). While enlarged thoraces are found in cold adapted recent humans (So 1980), in Neandertals this appears to involve the retention of primitive morphology seen also in tropical archaic humans (Gómez-Olivencia et al. 2009), and thus may have more to do with high ventilatory rates related to a high metabolic rates, active lifestyles, or both (see Chapter 10.6) rather than cold adaptation. The short

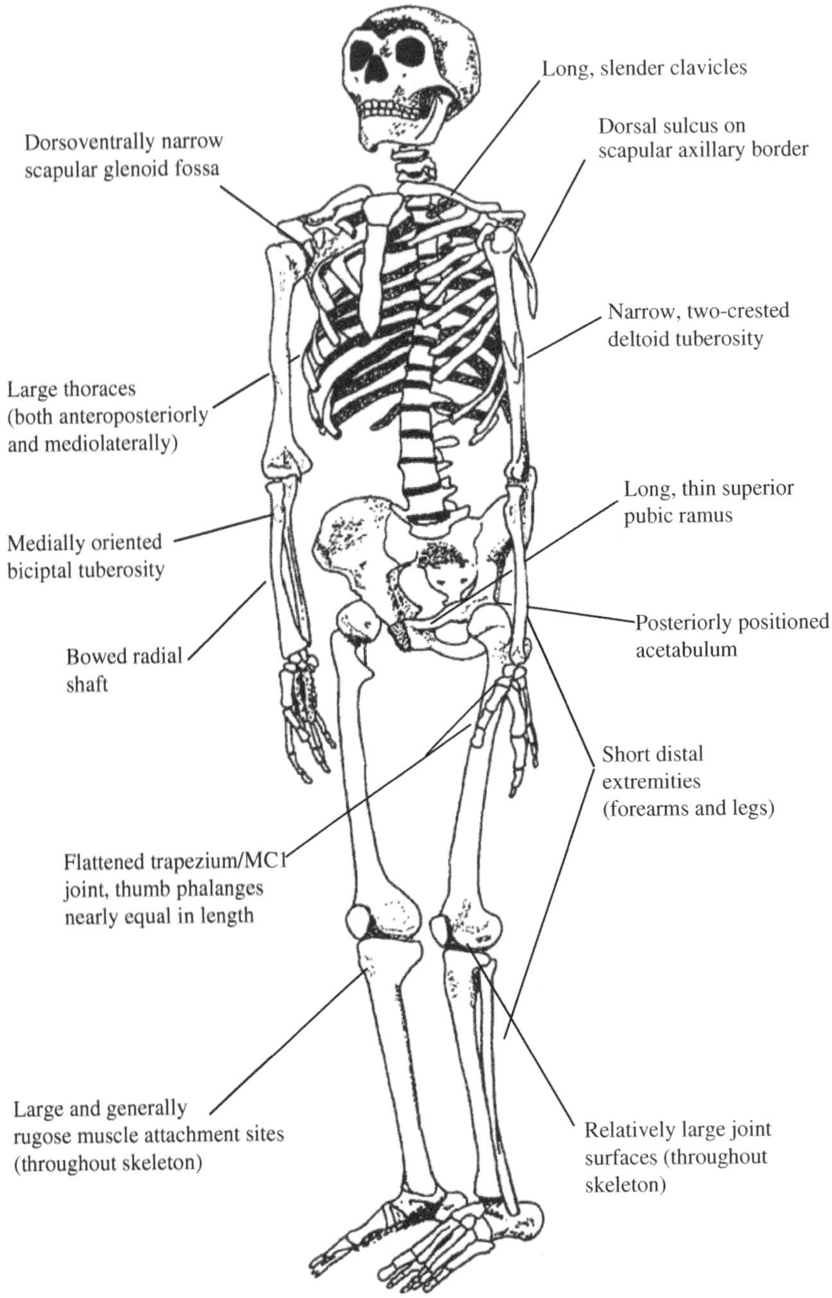

FIGURE 2.8

Postcranial characteristics of Neandertals. Drawn after photograph in Andrews and Stringer 1989.

and stocky postcranial architecture of Neandertals was accentuated by the retention of (albeit in slightly reduced form) the primitive traits of robust appendicular bones and heightened muscularity (Trinkaus 1983a, 1986; Ruff et al. 1993). Fewer autapomorphic features have been identified in the postcranial skeleton: most of the postcranial features that characterize Neandertals are either found in earlier members of the genus (primitive) or are shared with modern humans (Trinkaus 2006b). The few postcranial features that do appear to be uniquely derived in Neandertals include robust spinous processes of the lower cervical vertebrae, high frequency of scapular axillary border dorsal sulci, marked lateral curvature of the radius, presence of an *M. opponens pollicis* flange (not a crest) on the first metacarpal (Trinkaus 2006b), and reduced lumbar lordosis (Been et al. 2011).

2.3 NEANDERTAL DNA

Studies of Neandertal ancient DNA (aDNA) began with the publication, in 1997, of a 379 base sequence of mitochondrial DNA (mtDNA) from the Feldhofer 1 Neandertal (Krings et al. 1997). Subsequent improvements in aDNA methodology have increasingly overcome the problems inherent in studying ancient genetic material (which include degradation of the DNA molecule into tiny fragments, often about 100 base pairs long; deamination of bases causing base substitutions; and contamination with DNA from modern humans, soil microbes, and other sources), such that by 2010 a draft sequence (at approximately 1.3-fold coverage) of the entire Neandertal genome had been produced (Green et al. 2010). Although not yet published at the time of writing, Svante Pääbo (Max Planck Institute for Evolutionary Anthropology) has recently announced the availability of a high-coverage (50-fold) complete genome sequence of a Neandertal from Denisova Cave, Russia. This illustrates how rapidly aDNA genomic research on Neandertals (and other hominins) is developing, which means that any review of this area will quickly become dated. Still, it is worth covering the major, if still somewhat preliminary, findings that have come from this important area.

The 379 bases sampled from the Feldhofer Neandertal represented a small segment of the hypervariable region I of the mitochondrial control region (D-loop). Since then, portions of the D-loop have been sampled in an additional 26 individuals (Condemi et al. 2013), and complete mtDNA sequences have been determined for four of these previously sequenced individuals plus two new individuals (Green et al. 2008; Briggs et al. 2009). The Neandertals sampled range in age from about 100 Ka BP (Scladina, Belgium: Orlando et al. 2006) to about 38 Ka BP (Vindija, Croatia: Green et al. 2008) and span the Eurasian landmass from the northern Iberian peninsula (El Sidrón, Spain: Lalueza-Fox et al. 2005) to the Altai Mountains of southern Siberia (Okladnikov, Russia: Krause et al. 2007b). A number of conclusions about the evolutionary history and demography of the Neandertals can be drawn from the mtDNA data thus far. First, all of the Neandertal sequences fall outside of the range of variation of living humans (Green et al. 2008) and cluster into a single clade. Based on the average number of nucleotide differences between groups in the complete mtDNA sequences, Neandertal/modern human population divergence can be estimated to have occurred around

440–410 Ka BP (Endicott et al. 2010). Second, sequence diversity is low relative to that of modern humans (Briggs et al. 2009), who in turn have low genetic diversity relative to apes (Garrigan and Hammer 2006). Low genetic diversity suggests to some that Neandertals may have experienced one or more population bottlenecks since their divergence from modern humans (Krings et al. 2000), although inferred rates of protein evolution suggest instead that long-term effective population sizes were small (Green et al. 2008; Briggs et al. 2009). Third, geographic and temporal variation in mtDNA sequences provides some insights into Neandertal population structure and demographic history. Bayesian coalescent modeling of geographic variation in mtDNA sequences suggests the presence of three Neandertal subpopulations: one in eastern Europe/western Asia, one in northwestern Europe, and one in southwestern Europe (Fabre et al. 2009). It is important to note that mtDNA has not yet been sequenced from Near Eastern Neandertals, and it is possible that they might comprise a fourth subpopulation. When temporal variation is added the picture is somewhat more complex: eastern Neandertals and older specimens (>48 Ka BP) cluster together, with younger western European specimens forming their own clade (Dalén et al. 2012). The younger western European individuals (from Feldhofer, Germany; Vindija, Croatia; and El Sidrón, Spain) are much less variable (six times so!) than the older and eastern specimens, and are even less variable than just the older western European sample (from Scladina, Belgium; Monti Lessini, Italy; and Valdegoba, Spain). Bayesian demographic modeling suggests that Neandertal populations in western Europe went locally extinct sometime during MIS 3, and that the region was recolonized from some refugia shortly before the first arrival of modern humans in Europe (Dalén et al. 2012: see also Hawks 2012). Finally, the absence of Neandertal-specific mtDNA sequences in either European early modern human fossils or living humans (Serre et al. 2004), and the finding that living Europeans do not have mtDNA sequences that are more similar to those of the Neandertals than do living Africans or Asians (Krings et al. 2000), suggests that Neandertals did not contribute to the mtDNA gene pool of living peoples.

Despite a consistent lack of mtDNA evidence for interbreeding between Neandertals and modern humans (Krings et al. 1997; Ovchinnikov et al. 2000; Serre et al. 2004; Orlando et al. 2006; Briggs et al. 2009), potential evidence for introgression did come with the first nuclear DNA genome sequence (Green et al. 2010). Green and colleagues (2010) found, from a draft sequence (ca. 1.3-fold coverage) derived from three individuals from Vindija Cave (Croatia), that living Eurasians were genetically more similar to Neandertals than are living Africans. These results, and the degree of similarity, are consistent with a small degree of introgression occurring on the order of 80–50 Ka BP, before the divergence of European and Asian modern human populations (and thus likely occurring in the Near East [Green et al. 2010] or the Caucasus region between the Black and Caspian Seas [Pennisi 2013]). Surprisingly, modern Europeans were not found to be genetically closer to Neandertals than are modern Asians, suggesting little intermixing after modern humans reached Europe. The small contribution (about 1–4%) of Neandertal genes to the modern Eurasian gene pool may be the product of a very small number of matings (less than 2%: Currat and Excoffier 2011), since any Neandertals alleles introduced into the colonizing modern human gene pool could "surf" to higher frequencies as the population expanded (Green et al. 2010). This low level of introgression suggests a degree of avoidance of interbreeding, and would be consistent with the observed lack of mtDNA introgression (Currat and Excoffier 2011). It is also important to note that these results need not imply any interbreeding whatsoever, as they are also consistent with ancient genetic substructure within African early modern humans (the same results might obtain if north African early modern humans were genetically more similar to Neandertals, and they gave rise to non-African modern human populations: Green et al. 2010).

Ancient DNA analysis of hominin material from Denisova Cave, in the Altai Mountains of southern Siberia, has revealed additional complexity to population relationships in the Eurasian Paleolithic. A manual phalanx from a 48–30 Ka BP level at Denisova produced an mtDNA sequence distinct from that of all known Neandertal sequences as well as those of modern humans (Krause et al. 2010), and this specimen also provided a complete nuclear genome (Reich et al. 2010; Meyer et al. 2012). The cave has also produced two molars of uncertain (but seemingly non-Neandertal) affinity, both of which are genetically aligned with the manual phalanx (Reich et al. 2010; Pennisi 2013), as well as a pedal phalanx that has produced mitochondrial and nuclear sequences that align it with Neandertals (Pennisi 2013). The Denisovans, as they have come to be called, appear to represent a population which diverged from the Neandertals on the order of about 640 Ka BP and that represent a sister group to Neandertals (Reich et al. 2010). Based on the Neandertal nuclear sequence from the Denisovan pedal phalanx, it is estimated that as much as 17% of the Denisovan genome was derived from local Neandertal populations, suggesting substantial admixture between these groups.

While the analysis of Neandertal aDNA will no doubt continue to produce profound insights into the biology and demography of these hominins, a few early findings merit highlighting. Sequencing of the melanocortin 1 receptor (MC1R) in Neandertals from El Sidrón (Spain) and Monti Lessini (Italy) revealed a mutation (in both individuals) that reduces the activity of the receptor (Lalueza-Fox et al. 2007). Reduced activity of MC1R results in lighter pigmentation and hair color, and individuals homozygous for this Neandertal allele (estimated to be about 1% of Neandertals) would have had red hair and fair complexions (note, however, that it is a different mutation to the *mc1r* gene that produces red hair in living humans). Two of the individuals from El Sidrón also revealed that Neandertals share with modern humans the derived version of the *FOXP2* gene (Krause et al. 2007a), which controls for a host of neurological developmental events, some of which underlie language ability in modern humans (Lai et al. 2001). The site of El Sidrón in northern Spain has produced the remains of 12 individuals who are thought to represent a single social group (Rosas et al. 2006b; Lalueza-Fox et al. 2011). Three maternal mtDNA lineages are represented among the individuals, and all three of the adult males at the site shared the same mtDNA lineage (Lalueza-Fox et al. 2011). The three adult females, on the other hand, each carried different mtDNA. This suggests that the Neandertals of El Sidrón practiced a patrilocal residence system (the extent to which this can be generalized to other Neandertal groups is unclear – modern human groups vary in residence systems, and there's no reason to suspect that Neandertals didn't as well). Finally, average nuclear DNA divergence between Neandertals and modern humans allows an estimate of the time of population divergence on the order of 370 Ka BP (Noonan et al. 2006).

2.4 NEANDERTAL TAXONOMY

One of the thorniest issues in Neandertal studies concerns their classification: were they a regionally and temporally distinct subspecies of *Homo sapiens* (thus *H.s. neanderthalensis*) or were they a separate, sister species to *H. sapiens* (thus *H. neanderthalensis*)? Obviously one's position on this issue is dependent upon choice of species concept (see for example Kimbel 1991; Tattersall 1992; Kimbel and Martin 1993; Jolly 2001), opinions concerning appropriate methods of classification and phylogenetic reconstruction, and opinions about the interpretation of paleogenetic and molecular studies (e.g., Krings et al. 1997; 1999; Jorde et al. 1998; Nordburg 1998; Relethford 2001; Hawks and Wolpoff 2001a). Arguments for both of these positions (subspecies vs. separate species) are too numerous to review in detail here. There is no question that Neandertals were morphologically distinct from both their contemporaries outside of Europe and from the modern humans that succeeded them there. Thus, researchers that equate morphological distance with biological (reproductive) distance are prone to infer genetic distance and relative reproductive isolation (Currat and Excoffier 2004, 2011) between Neandertals and early modern humans (see, for example, Tyrrell and Chamberlain 1998; Schillaci and Froehlich 2001; Tattersall and Schwartz 2006). Tattersall (1992: 347) raised a compelling point when, referring to the list of derived features that characterize Neandertals, he noted: "If a list of autapomorphies ... this long is regarded as insufficient to demonstrate distinct specific status, it is hard to imagine what it would take to do so."

There is little doubt that the Neandertals fit a phylogenetic or morphological species concept (see Tattersall 1992). However, despite their morphological distinctiveness, there is mounting evidence, both morphological (Churchill and Smith 2000a; Trinkaus 2007) and genetic (Green et al. 2010) that Neandertals were not reproductively isolated from early modern humans. Thus, from a strict biological species concept (see Tattersall 1992) Neandertals may best be considered as a temporally and regionally proscribed subspecies of *H. sapiens*. Taking these two seeming contradictions together, I agree with Jolly (2001) that Neandertals and early modern humans fit the concept of *allotaxa*[6] – populations that are distributed as a "patchwork" of nonoverlapping but often parapatric forms and that are morphologically diagnosable yet not reproductively isolated. As noted by Jolly, allotaxa would be considered as distinct species when using the phylogenetic species concept but as subspecies when using the biological species concept, a situation that fits the Neandertals well. On the basis of morphological distinctiveness and evolutionary history I feel it is most appropriate to consider Neandertals a distinct species (*H. neanderthalensis*), but with the proviso that this makes no assumptions about their potential or actual interbreeding with early modern humans (*H. sapiens*).

2.5 REGIONAL AND TEMPORAL VARIATION IN NEANDERTAL MORPHOLOGY

The features that define the Neandertal clade are best expressed in the western European early Würm (MIS 4) "classic" Neandertal sample (Howell 1957; Trinkaus 1988), exemplified by western European and predominantly later specimens from La Ferrassie, La Chapelle-aux-Saints, Spy, Feldhofer, and Monte Circeo. Neandertal features are more variably expressed in earlier specimens as well as in specimens from central Europe and the Near East. The relatively early (MIS 5e) central European Neandertals from Krapina, for example, lack anterior mastoid tubercles, are variable in the formation of the entoglenoid (involving in some cases a contribution from the sphenoid), have a number of specimens with the modern human configuration of the auditory canal (laterally oriented and inferiorly positioned external auditory meatus) (Vinyard and Smith 2002), and have an elevated frequency of bisulcate scapular axillary borders relative to later Neandertals (Trinkaus 2006a).[7] Likewise, the Near Eastern specimens are variable in certain defining characteristics. For example, tapering mastoid processes are present on Shanidar 1 and Tabun C1 but are absent in Shanidar 2 and 5 and Amud 1 (Trinkaus 1988).

This variable expression of features in the Levantine sample has led to the suggestion that the later Pleistocene human populations of the Near East, although closely related, were not Neandertals, the latter being confined to the European subcontinent (Arensburg and Belfer-Cohen 1998). While it is undeniable that all the regional clade features (autapomorphies) that define the Neandertals are not expressed in combination in a single Near Eastern specimen (as they are in the classic Neandertals), the very presence of these features, along with the overall facial, neurocranial and postcranial similarities with the Würmian European fossils, suggests that we are simply seeing geographic variation along the southern extent of the Neandertal range. This possibility is best expressed in the observation by Trinkaus (1988: 20): "...given the geographic distance between them and western European populations, some morphological differences between these Near Eastern Neandertals and western European ones would be expected. In fact, the degree of similarity between these Near Eastern Neandertals and their western European counterparts is rather surprising, given the contrasts between them and their not very distant north Africa contemporaries."

The discovery of regional and temporal structure in the mtDNA variation in Neandertals (see Section 2.3) provides an opportunity to examine the degree to which genetic variation maps onto morphological variation across space and time. The claim that the Neandertal metapopulation was divided into three regional subpopulations (Fabre et al. 2009) finds some broad support in the morphological differences between the classic Neandertals in western Europe and the less derived Neandertals to the east. Fabre and colleagues (2009) also take as support of their model claims that the Neandertals of Mediterranean Europe may represent a morphological group distinct from Neandertals to the north and east (e.g., de Lumley 1973; Condemi 1992; Rosas et al. 2006b), although this possibility is not well established. When the temporal dimension is also considered (Dalén et al. 2012), the link between genetic and morphological variation breaks down. Patterns of mtDNA variation show that most (but not all) of the later western European Neandertals form a clade distinct from the earlier and eastern specimens, which may support the idea of a morphologically and genetically distinct group of classic Neandertals in MIS 3 western Europe. However, this clade also includes the less derived Neandertals from Vindija Cave, yet does not include the ca. 50 Ka BP remains from Monti Lessini in northern Italy (Hawks 2012). While further work on autosomal aDNA may help to resolve the seeming disparity between morphological and genetic subgroups within the Neandertals, at present the evidence suggests that morphology-based subgroupings of Neandertals specimens may have little relevance to understanding their population biology (Hawks 2012). Complexity in the aDNA patterns also appears to be a common finding among European Pleistocene mammals (Hofreiter et al. 2004), as the result of population migrations driven by climatic cycles. Hofreiter and colleagues (2004) have suggested that the lack of clear phylogeographic signals in Neandertal mtDNA may be the result of repeated cycles of retreat to southerly refugia during glacial episodes and subsequent re-expansion of Neandertal populations following climatic amelioration (see Hublin and Roebroeks 2009). An alternative possibility is that populations were more stable in the more productive regions to the south, while those to the north and east were more prone to experience population declines and even local extinctions (Dennell et al. 2011), such that persistant migration from areas of stable core populations to areas of population instability disrupts phylogeographic patterns in the DNA.

2.6 THE EVOLUTIONARY HISTORY OF THE NEANDERTALS

It is not entirely clear when the ancestors of the Neandertals first ventured into the European subcontinent. Initial hominin colonization events may have begun as early as 1.77 million years (Ma) BP (Gabunia et al. 2000), but the geographic extent and duration of these early occupations remain unknown. Soon after the emergence of the genus *Homo*, one or more of its species had expanded out of Africa and into the subtropical and warm-temperate zones of Asia (Swisher et al. 1994; Gabunia et al. 2000). Remains of *H. erectus* (or *H. erectus ergaster*: Asfaw et al. 2002) have been found in 1.77 Ma BP deposits at Dmanisi, in the Lesser Caucasus Mountains of southwestern Asia. If the Early Pleistocene occurance of Oldowan-style pebble tools and manuports at various western and central European sites (Ventana Micena, Cueva Victoria and Duero Valley in Spain, Chillac in France, Monte Poggiolo in Italy, Stránska Skála and Beroun in the Czech Republic, and Korolevo in Russia: see Bosinski 1992; Carbonell et al. 1996b) can be confirmed, it would place hominins in Europe at roughly the same time. However, the archeological evidence for this early human presence is not universally accepted, both because of doubts about the artifactual status of lithics from some of these sites and because of uncertainties about the dates of others (see for example Roebroeks and Kolfschoten 1994a, 1994b). Even if this handful of Lower Pleistocene sites does mark the first human incursion into Europe and southwestern Asia, these early forays were of either short duration or low population density, or both, because with the exception of these few remarkable occurrences the earliest human presence left little archeological and no paleontological trace. The site of Barranco León at Orce in southern Spain has produced a single deciduous molar that is thought to be ca. 1.4 Ma old (on the basis of ESR, paleomagnetic and biostratigraphic dating: Toro-Moyano et al. 2013). Deposits at the Sima del Elefante at Atapuerca (in central Spain) have produced a mandible, dated by a variety of methods to 1.2 – 1.1 Ma BP, with morphological affinities to the Dmanisi specimens as well as African Plio-Pleistocene specimens of early *Homo* (Carbonell et al. 2008). If the Barranco León and Sima del Elefante fossils represent the first wave of colonization of Europe, it would indicate that the Mode 1 (Oldowan-like) assemblages discussed above date to this time as well. Alternatively, and perhaps more likely, the earlier human presence in Europe during the Lower Pleistocene may have been so sporadic and of such low density that few archeological and paleontological traces were left behind.

The European hominin fossil record remains sparse throughout the Early Pleistocene. At Ceprano, an Italian site between Rome and Naples, a single calvarium, initially thought to date to 900–800 Ka BP, suggested that *H. erectus* had penetrated at least into the Mediterranean fringe of southern Europe by MIS 22–19 times (Ascenzi et al. 1996, 2000; Clarke 2000). However, subsequent dating work at the site indicates that the specimen dates to about 430–385 Ka (Muttoni et al. 2009; Manzi et al. 2010), raising questions about its taxonomic attribution and its relationship to other European Middle Pleistocene specimens. Morphometric analysis of the specimen suggests that, despite being a relatively morphologically primitive member of the group, the Ceprano cranium clusters with other Middle Pleistocene specimens from Europe, such as Petralona, Steinheim and the SH5 cranium from Sima de los Huesos (see below), as well as African and Asian representatives of *H. heidelbergensis* (Kabwe, Dali, and Jinniushan) (Mounier et al. 2011).

A small collection of human fossils from the Gran Dolina of Atapuerca (Spain), dated to the terminal Early Pleistocene at 780 Ka BP, may represent the beginning of long-term, continuous human occupation of Europe (Figure 2.9). Although artifactual

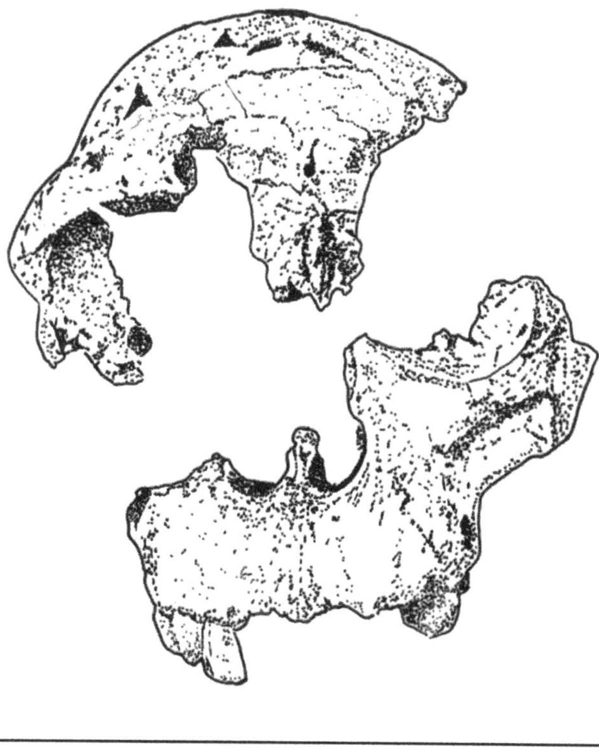

FIGURE 2.9

Anterior cranial vault and facial skeleton of TD6, from ca. 780 Ka BP deposits at Gran Dolina, Spain. This material (representing a juvenile) and other fossils from the site have been argued to represent a separate species – *Homo antecessor* – that was the last common ancestor of Neandertals and modern humans (Bermúdez de Castro et al. 1997). Most workers consider it a representative of *H. heidelbergensis*.

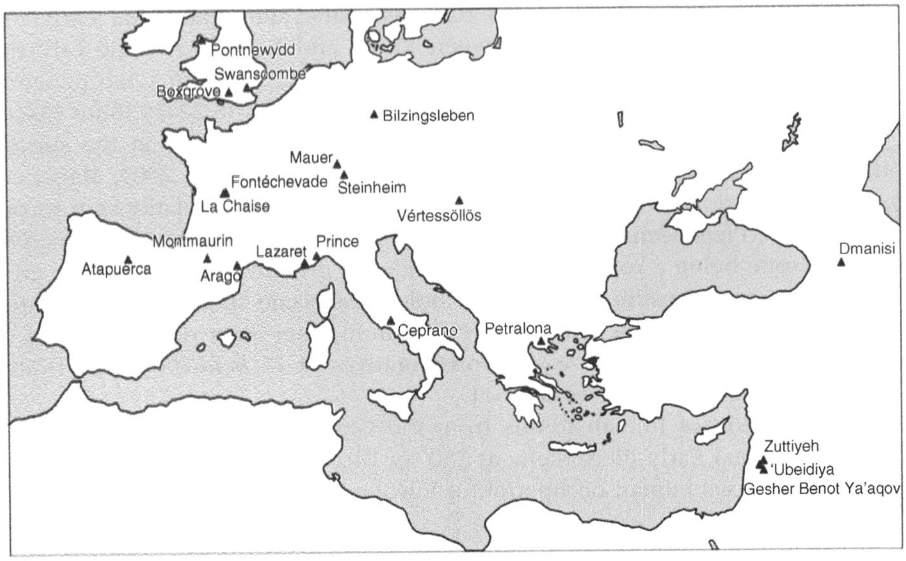

FIGURE 2.10

Important Middle Pleistocene fossil hominin sites of Europe and western Asia.

evidence can be mustered to argue that the Gran Dolina hominins represent only another temporary wave of human migration into Europe, a couple of synapomorphic traits in the face link these fossils with later European *H. heidelbergensis* (Bermúdez de Castro et al. 1999a: 698), suggesting permanent establishment of humans on the subcontinent at this time.

The interval between ca. 500–250 Ka BP (MIS 13–8) preserves a rich human fossil record for Europe (Figure 2.10). These specimens (from the sites of Boxgrove and Swanscombe [England], Bilzingsleben, Mauer, Reilingen, and Steinheim [Germany], Arago and Montmaurin [France], Sima de los Huesos at Atapuerca [Spain], Ceprano [Italy], Apidima and Petralona [Greece], and Vértesszöllös [Hungary]) share phenetic similarities with Middle Pleistocene fossils from Africa (e.g., Bodo [Ethiopia], Kabwe [Zambia] and Elandsfontein [South Africa]) and are considered by some to represent the European clade of a single, wide-ranging species, *H. heidelbergensis* (see Rightmire 1998; Stringer 2012) (Figure 2.11). Roughly contemporaneous fossils from Asia (e.g., Zuttiyeh [Israel], Dali, Maba, Yunxian and Jinniushan [China]) may also belong to this taxon. All of these specimens share primitive aspects of cranial vault and brow ridge morphology with *H. erectus*, yet are more similar to modern humans in endocranial volume, cranial base morphology and in some aspects of the face (Rightmire 1998). If one considers both the European and African fossils to be representatives of a single species, this species in all likelihood represents the last common ancestor of Neandertals and modern humans (Stringer 2012: but see Arsuaga et al. 1999b).

From the European branch of these Middle Pleistocene humans arose the Neandertals, in a morphological transition (or perhaps more appropriately, speciation event: see above) that was "... gradual in tempo and mosaic in nature" (Trinkaus, 1988:12).

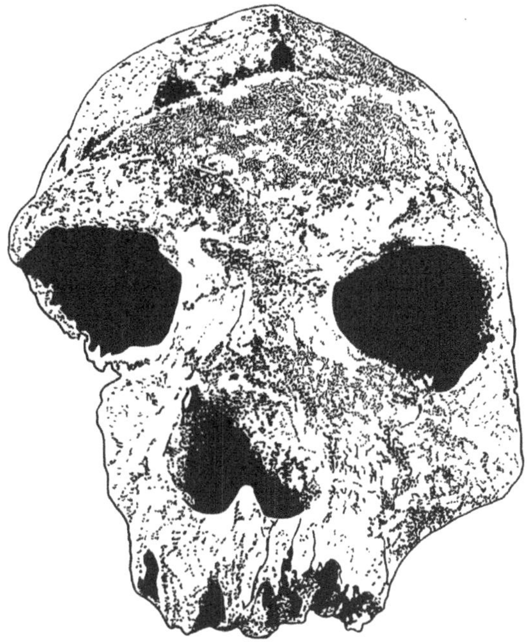

FIGURE 2.11

Anterior view of the ca. 600 Ka BP cranium from Bodo, Ethiopia, representing an early member of the species *Homo heidelbergensis* (thought by many to be the last common ancestor of Neandertals and modern humans).

This makes difficult, if not impossible, the task of pinpointing the emergence of Neandertals in time. In a model that has come to be called the "accretion" hypothesis (Hublin 1988a, 1998; Stringer 1995; Dean et al. 1998), Neandertal features are argued to have coalesced over time, beginning with aspects of facial morphology, followed in time by derived features of the occipital, and culminating with derived temporal morphology (this model is not universally accepted: for alternative perspectives, see Hawks and Wolpoff 2001b; Rosas et al. 2006a; Voisin 2006). Variable combinations of primitive and derived features during this period of coalescence suggest that the accretion of Neandertal characters involved changing frequencies of derived traits rather than the abrupt emergence of new morphologies (Stringer and Hublin 1999). This accretion is argued to have been the end result of genetic drift and local selection acting on a small later Middle Pleistocene population cut off from extra-European gene flow by glacial conditions on the subcontinent (Hublin 1996; Weaver et al. 2007; *cf* Howell 1952, 1957; but see Hawks and Wolpoff 2001b).

The first stage of "Neandertalization" of European Middle Pleistocene hominins involved incipient midfacial prognathism (evident in flattening or convexity of the infraorbital plates) and widening of the occipital torus, as can be seen in the MIS 13–12 specimens from Petralona (Dean et al. 1998) and Arago (Hublin 1996; Arsuaga et al. 1997a; although some would argue that these specimens may belong to a separate, non-Neandertal lineage: Tattersall and Schwartz 2006). If the large sample of human fossils from the Sima de los Huesos (more than 5,500 specimens, including one complete skull and several partial crania, representing at least 32 individuals) date to 600–530 Ka BP as has been claimed (Bischoff et al. 2007), this would demonstrate the establishment of a suite of Neandertal derived facial and dental features as early as MIS 15–14 (contra the accretion model: Martinón-Torres et al. 2012). Some (e.g., Dean et al. 1998), however, see the Sima de los Huesos sample, along with the fossils from Bilzingsleben, Steinheim, Reilingen, Swanscombe and Vértesszöllös, as representing the second stage of accretion during MIS 11–9. Naturally, the issue of the age of the fossils from the Sima de los Huesos is critical to understanding the evolutionary emergence of Neandertal morphology and, not surprisingly, the very early age claimed for the sample has been challenged on a variety of grounds, including taphonomy (Andrews and Fernández Jalvo 1997; Fernández Jalvo and Andrews 2003), morphology (Stringer 2012) and paleogenetics (Endicott et al. 2010). Morphologically, the fossils from the Sima de los Huesos cluster with other MIS 11–9 fossils in having greater midfacial prognathism, incipient suprainiac fossae, incipient "en bombe" cranial vault shape, and other derived Neandertal features (Dean et al. 1998), although the fossils in this group are still primitive in some facial features (Hublin 1988b; Arsuaga et al. 1993; Rosas 2001; Bischoff et al. 2003). By the late Middle Pleistocene (MIS 7 and 6) there appear a number of specimens that display a sufficient number of Neandertal autapomorphies that they can be considered likely candidates for the earliest true Neandertals, including the fossils from Biache-St-Vaast (France), Ehringsdorf (Germany) and La Chaise Suard and La Chaise Bourgeois-Delaunay (France) (Hublin 1980; Smith 1984; Dean et al. 1998; Condemi 2001; Guipert et al. 2011). These and other MIS 7–6 fossils, along with MIS 5 specimens from sites like Saccopastore (Italy), Krapina (Croatia) and Shanidar (Iraq), represent the third, "early Neandertal," stage of accretion (Dean et al. 1998; Bruner and Manzi 2006). The fourth and final stage ("classic Neandertals") is represented by MIS 4–3 specimens that show the full suite of derived Neandertal features, including Feldhofer (Germany), La Chapelle-aux-Saints, La Quina, Le Moustier, and La Ferrassie (France), Spy (Belgium), Monte Circeo (Italy) and others (Dean et al. 1998).

Additional insights into the evolutionary origins of the Neandertals have recently come from paleogenetic studies. Coalescence dates derived from variation in mtDNA suggest that the Neandertals distributed across Eurasia in the Late Pleistocene (MIS 5–3) last shared a common ancestor between 250–200 Ka BP (Lalueza-Fox et al. 2005; Caramelli et al. 2006; Hawks 2012), roughly coincident with the first appearance of fossils possessing a more complete suite of diagnostic Neandertal features (stage 3 of the accretion model) at the sites of Biache-St-Vaast (250–170 Ka BP: Guipert et al. 2011) and Ehringsdorf (MIS 7, > 200 Ka BP: Street et al. 2006). Some researchers (Lalueza-Fox et al. 2005) see these late coalescence dates as inconsistent with a model of gradual accretion of Neandertal morphology in Middle Pleistocene *H. heidelbergensis* populations, but rather supportive of a clearly defined speciation event (during which Neandertal diagnostic characteristics were rapidly fixed) around 300–250 Ka BP (Rightmire 2001). Such a model, however, does not account for the clear expression of Neandertal features in specimens prior to 300 Ka BP (such as the MIS 11–9 fossils that constitute stage 2 of the accretion model). While the 600–530 Ka BP age of the Sima de los Huesos sample has been questioned, the fossils are still likely to be older than 300 Ka BP (Arsuaga et al. 1997a; Bischoff et al. 2003), and possess a constellation of derived features sufficient to cause some to view them as early Neandertals (Stringer

2012). Estimates of the timing of Neandertal/modern human population divergence from mitochondrial and autosomal aDNA suggest that the Neandertal clade was most likely established during MIS 11 or 10 (Noonan et al. 2006; Endicott et al. 2010), and no earlier than MIS 12. This divergence date is consistent with the accretion model (since the real establishment of Neandertal features – stages 2–4 – occurred from MIS 11 onwards), although it would suggest that a small number of incipient Neandertal features (stage 1 of the accretion model) characterized the European Middle Pleistocene hominin population before full separation of the Neandertal and modern human clades. These MIS 12–10 divergence estimates also cast strong doubt on the claimed MIS 15–14 age of the Sima de los Huesos hominins (Endicott et al. 2010; Stringer 2012).

While the evolutionary origin of the European Neandertals is not yet fully resolved, the nature of the emergence of Neandertals in the Levant and Middle East is even less clear. This issue hinges to a large extent on dating of key sites and specimens, most notably that of the female Neandertal skeleton from Tabun Cave (Israel). This specimen (Tabun C1) has been variously attributed to levels C and B at the site (see Bar-Yosef and Callander 1999), with age estimates ranging anywhere from 104 +33/−18 Ka BP cal (Grün and Stringer 2000) to 171 ± 17 Ka BP cal (Mercier et al. 1995). Attempts at direct dating of the Tabun C1 skeleton have produced wildly different estimates: ≤ 40 Ka BP cal based on gamma spectrometric U-series on the complete skeleton (Schwarcz et al. 1998) to between 112 (± 29) and 143 (± 37) Ka BP cal based on ESR of dental enamel (Grün and Stringer 2000), the latter range being the best supported on archeological and dating methodological grounds (see Grün and Stringer 2000). While this issue is not completely resolved, the weight of the evidence suggests an early date (MIS 6 or 5e) for the initial Neandertal occupation of southwestern Asia, with a subsequent alternating occupation or co-occupation of this region by Neandertals and early modern humans (represented by the fossils of Skhul and Qafzeh Caves in Israel [McCown and Keith 1939; Vandermeersch 1981] and the Tamsara Hill specimen from Egypt [Vermeersch et al. 1998]).

An early date for the Tabun Neandertal would be consistent with the "local emergence" hypothesis (see Trinkaus 1988), in which the Middle Pleistocene emergence of Neandertals from *H. heidelbergensis* occured in western Asia as well as in Europe. Proponents of this model see a gradual shift in facial morphology from Near Eastern Middle Pleistocene humans (such as represented by Zuttiyeh, Israel) to somewhat primitive Neandertals (represented by Shanidar 2 and 4 in Iraq and the Tabun specimen in Israel) to fully Neandertal morphology in MIS 4 (represented by Amud 1 [Israel] and Shanidar 1 and 5 [Iraq]). By this model, the occurrence of anatomically modern humans in the Near East between 100–80 Ka BP (Valladas et al. 1988; Mercier et al. 1993) may represent a temporary incursion of north African populations (with perhaps a northward retreat of Neandertals) during the relatively mild climates of MIS 5d-a. Importantly, this model sees the Levantine sequence not as an autonomous lineage, but rather one in clinal genetic contact with Neandertals to the north and west (Trinkaus, 1988; see also Voisin 2006). Thus, the more variable expression of Neandertal features in the Levantine sample is a function of geographic distance, not separate evolutionary history.

Alternatively, a late (MIS 4 or 3: after 75 Ka BP) date for the Tabun C1 skeleton would be concordant with the "in-migration" hypothesis (Vandermeersch 1981; Bar-Yosef 1989). By this model, Neandertals were late-comers to the Near East, having evolved in Europe prior to the last Interglacial (MIS 5e) and having expanded southward as cold conditions gripped Europe in the early Würm (MIS 4). The extension of European biomes into the Near East during MIS 4, along with the influx of Neandertal populations, may have displaced the indigenous modern humans southward to north Africa. Proponents of this model see regional differences in Neandertals as attributable to the derivation of Near Eastern Neandertals from the more generalized (non-classic) Neandertals of the last Interglacial (as seen at sites like Krapina and Saccopastore). Thus, the classic Neandertals of western Europe and the less-derived Neandertals of the Levant represent two branches of the larger Neandertal clade.

The fossil and cultural record of the Neandertals is richest from last Interglacial (MIS 5e) through early mid-Würm (to MIS 3) deposits, with this later interval representing the "heyday" of the classic Neandertals of glacial Europe. Our knowledge of morphological and behavioral evolution during this interval is obscured somewhat by difficulties (historically) with assigning accurate dates to Late Pleistocene deposits, many from sites that were excavated early in the 20th century. The advent of radiometric methods suitable to this time range (TL, OSL, ESR, and U-series: see Aitken et al. 1992) and refinement of archeological and sedimentological methods and analyses are ameliorating this situation. Accordingly, important behavioral shifts (in the realm of subsistence), once thought to have occurred at the Middle/Upper Paleolithic transition, are now recognized to have occurred during Middle Paleolithic times (e.g., Stiner 1994; but see Mussi 2001a).

The fate of the Neandertals remains an issue of some debate. They leave the fossil record during the later part of MIS 3 (but at somewhat different times in different parts of Europe: see above), apparently several millennia after early modern humans (the "Cro-Magnon")[8] had arrived in Europe (Churchill and Smith 2000a; Mellars 2005). Redating of fossils that were once thought to represent the earliest modern Europeans (from Vogelherd, Hahnöfersand, Cro-Magnon, Zlatý Kůň) has now removed them from consideration, but in the meantime a small but growing sample of well-dated early modern humans has emerged to take their place (from Peştera cu Oase [Trinkaus et al. 2003; Rougier et al. 2007], Peştera Muierii [Soficaru et al. 2006], and Cioclovina [Olariu et al. 2003]). The earliest of these fossils (from Peştera cu Oase, Romania) have associated AMS radiocarbon dates of ca. 34.9 Ka BP (corresponding to a calendar age of about 40.5 Ka BP: Trinkaus et al. 2003; Rougier et al. 2007), while the late Neandertals from level G_1 at Vindija (Croatia) have been dated to 32.4 Ka BP (Higham et al. 2006; Janković et al. 2011). This corresponds with a calendar age (using CalPal: http://www.calpal-online.de/index.html) of ca. 37.3 Ka BP cal, suggesting at least a couple of millennia of co-occupation of Europe. However, a modern human maxillary fragment from Kent's Cavern (England) has recently been indirectly redated to 44.2–41.5 Ka BP cal (Higham et al. 2011: but see White and Pettitt 2012, who argue this date should be rejected), and two deciduous teeth from Grotta del Cavallo in southern Italy, initially thought to be those of a Neandertal (Palma di Cesnola and Messeri 1967; Churchill and Smith 2000a), have been reclassified as modern human and dated to 45–43 Ka BP cal (Benazzi et al. 2011). Combined with evidence that the Neandertals may have persisted in Iberia until ca. 35–32.5 Ka BP cal (see above), the period of overlap may have been considerably longer, perhaps on the order of 10,000 years or more. In western Asia, the period of overlap may have been longer still (see above, and Chapter 3.1), although discerning between true co-occupation of the Near East versus shifting of geographic boundaries (ecogeographic vicarism: Shea 2003a) is not possible with the data currently available.

While the extent to which Neandertal and modern human contemporaries were exchanging cultural ideas remains an open question (see Chapter 3.1), the fossil evidence suggests that these groups were certainly exchanging genes (although the aDNA evidence, while supporting the inference of Neandertal/modern human interbreeding, does not at present support hybridization in Europe proper: Green et al. 2010). Early modern European specimens, while clearly representing *H. sapiens*, exhibit distinctive Neandertal traits or primitive features that had been lost in African early modern humans but were retained in Neandertals (Trinkaus 2007, 2011a). The persistence of Neandertal genes in the early modern European gene pool has also been argued to account for the Neandertal-like morphological features in a 24.5 Ka BP Gravettian child from Lagar Velho (Portugal) (Duarte et al. 1999). In addition, late Neandertals (from Vindija G_1 and G_3, Kůlna, Barakaevskaïa, Grotte du Renne, St. Césaire, Monti Lessini, and Zafarraya) are unequivocally identifiable as Neandertals, yet have morphological details that approach the modern human condition (Smith 1984, 1994; Smith and Trinkaus 1991; Wolpoff 1999; di Vincenzo et al. 2012; Condemi et al. 2013). Although these details may reflect parallelisms, the most parsimonious explanation is low level gene flow from modern populations that were beginning to colonize Europe during MIS 3 times (or from populations on the peripheries of Europe prior to the period of active colonization) (Churchill and Smith 2000a; di Vincenzo et al. 2012).

By about 32 Ka BP cal the Neandertals were gone. The question of why the Neandertals disappeared, after hundreds of thousands of years of successful adaptation to Pleistocene Europe, while the Cro-Magnon – the adaptive new-comers to European Ice Age conditions – persisted is far from resolved. As we saw in Chapter 1, there is no deficit of hypotheses as to what did in the Neandertals. Personally, I suspect that the combination of persistently low population density, rapid climatic fluctuation that marked the Interpleniglacial (MIS 3), and new competitive pressures exerted by modern humans was more than the Neandertals could handle (cf. Müller et al. 2011). As I argue in the chapters to follow, their low populations density was itself a function of tight energetic budgets (which limited the intrinsic rate of population growth) and their ecological relationships with resource competitors in the carnivore guild (which also served to keep population sizes in check), and that by the time modern humans reached Europe the Neandertals were already vulnerable. Regardless of why they disappeared, it is clear that their extinction was part and parcel of the dynamic processes that gave rise to the Upper Paleolithic, and that they left their genetic mark (albeit a small one in the evolutionary long run) on the modern human populations that came to have sole possession of Eurasia.

NOTES

1. This position is not universally accepted – see below.
2. Archeolgical remains dated to 40 Ka BP at the site of Mamontovaya Kurya (Pavlov et al. 2001) and 34–31 Ka BP cal at Byzovaya (Slimak et al. 2011), both in the polar portion of the Ural Mountains in Russia, *might* indicate Neandertal penetration into areas above the Arctic Circle.

3. Because of uncertainties surrounding radiocarbon calibration curves beyond 20 Ka (Pettitt and Pike 2001), reported dates can be assumed to be uncalibrated unless otherwise indicated (as "Ka BP cal").
4. Trinkaus (2006b: Tables 1–6) identifies 17 traits that appear to be uniquely derived in Neandertals, compared to 33 such traits in modern humans.
5. For a fuller consideration of evolutionary developmental integration in the Neandertal craniofacial skeleton, see Rosas et al. 2006a.
6. Or, to use the terminology of Holliday (2003, 2006), Neandertals and modern humans can be considered members of a *syngameon*.
7. However, the Saccopastore 1 specimen from MIS 5 Italy appears to display the hallmark characteristics of the "classic" Neandertals (Bruner and Manzi 2006).
8. Throughout the book I use the term "Cro-magnon" as a synonym for "European early modern humans." Ironically, recent dating of artifacts associated with the human fossils from the Cro-magnon rock shelter (France) suggests that they date to the Gravettian (after 28,000 years ago), and thus the eponymous fossils do not actually represent the *earliest* modern Europeans that we call "Cro-magnons" (Henry-Gambier and White 2003).

CHAPTER THREE

Neandertal Material Culture

The Neandertals are associated with a rich and well-studied archeological record, comprised of items of material culture (mainly in the form of stone tools), dietary residues (primarily in the form of animal bones), and domestic features (such as hearths and, in a few cases, remnants of structures). The material culture component of this record allows for inferences about the level of technological sophistication and cultural complexity of the Neandertals (Hayden 1993; Ambrose 2001; Langley et al. 2008), their mobility patterns, settlement systems, and strategies for raw material procurement (Féblot-Augustins 1993, 1997, 1999; Slimak and Giraud 2007; Fernandes et al. 2008; Daujeard and Moncel 2010; Spinapolice 2012), aspects of their domestic and subsistence behavior (Kuhn 1993, 2011; Shea 1998; Vaquero et al. 2001b; Daujeard and Moncel 2010), and, possibly, cognitive processes related to tool manufacture and use (Simek 1992; Wynn and Coolidge 2004; Burke 2012). Likewise, domestic features and intra-site spatial patterning tell us something about thermal buffering, group size, and social organization (Farizy 1994; Pettitt 1997; Vallverdú et al. 2005, 2010; Speth 2006).

Thin on the Ground: Neandertal Biology, Archeology, and Ecology, First Edition. Steven Emilio Churchill.

For living humans, material culture is bound up in all aspects of our behavior, serving everything from social and symbolic to adaptive needs. From an adaptive perspective, technology serves as the interface between humans and the environment. While this chapter provides a brief overview of Neandertal material culture generally, its focus is on the adaptively-significant aspects of Neandertal technology.

3.1 NEANDERTAL-ASSOCIATED LITHIC INDUSTRIES

Neandertal remains are generally, but not exclusively, associated with Middle Paleolithic or "Mode 3" (Clark 1968)[1] artifact assemblages across their entire range. While the time of origin of the Middle Paleolithic in Eurasia is as nebulous as that of the Neandertals themselves, this technological stage is roughly coincident with the reign of the Neandertals as a biological group (lest one be tempted to correlate the emergence of the Middle Paleolithic with that of the Neandertals, it should be noted that Mode 3 assemblages appear in sub-Saharan Africa, in the form of the Middle Stone Age, at roughly the same time). In Europe, Middle Paleolithic technology may emerge somewhat gradually out of the preceding Lower Paleolithic/Mode 2, in a process parallel to the accretion of Neandertal morphological features (see comments in Moncel et al. 2011), making it difficult to draw a clear boundary between these cultural periods. The emergence of Levallois core preparation (see Section 3.2, below) has often been taken as a chronological marker for the beginning of the Middle Paleolithic (Richter 2011), and by this view the origins of Mode 3 in Europe would date to about 300 Ka BP (Tuffreau 1982, 1992; Delagnes et al. 2007; Moncel et al. 2011). The Middle Paleolithic of Europe would thus appear to have emerged slightly earlier than the Neandertals themselves, likely at the hand of Neandertal-like late *H. heidelbergensis* at the beginning of MIS 8. If, on the other hand, one uses the predominant establishment of Levallois and other forms of core preparation, rather than their first occurrence, then the Middle Paleolithic proper appears to have arisen later, during MIS 7 or 6 (Richter 2011), making the first emergence of Mode 3 technology roughly coeval with that of the first true Neandertals (see Chapter 2). This later date would be consistent with the picture in the Levant, where the preponderance of dated sites also places the beginning of the Middle Paleolithic between the end of MIS 8 and the middle of MIS 7 (Porat et al. 2002; Shea 2003a).

The Middle Paleolithic was variable across space and time (Mellars 1996), with assemblages varying both typologically and technologically between regions and temporally within regions (often within single sites). Most of the assemblages associated with Neandertal fossil remains represent Mousterian traditions, which tend to be flake-based, with the principal retouched tool forms being sidescrapers (Figures 3.1 and 3.2) points (Figure 3.3) and denticulates (Figure 3.4) (Wymer 1982; Mellars 1996) (Table 3.1). Mousterian tools were commonly produced with Levallois or disc core preparation techniques (Debénath and Dibble 1994; Mellars 1996). Frequencies of Levallois core preparation are highly variable across Mousterian assemblages (ranging from being the dominant method of core reduction to being entirely absent: Dibble 1991; Mellars 1996), as are the relative abundances of scrapers, points, backed knives, blades and denticulates. This type of variation (and in particular the relative abundance of sidescrapers) in assemblages between layers within single sites and between sites

FIGURE 3.1

Representative Mousterian tool forms: scrapers (*racloirs*). Numbers in parentheses refer to tool type number in Bordes' typology (Bordes 1972) (Table 2). A. single-edged, straight scraper (9); B. single-edged, convex scraper (10); C. single-edged, concave scraper (11); D. double-edged, straight/concave scraper (14); E. double-edged, biconvex scraper (15); F. double-edged, convex/concave scraper (17); G. double-edged, straight/convex scraper (13). Scale bar in centimeters.

in southwest France led Bordes (1961a) to identify four distinct variants, or facies, of the Mousterian. These include:[2] (1) Mousterian of Acheulean Tradition (MAT or MTA), characterized by the presence of triangular or heart-shaped (cordiform) hand axes, and subdivided into two variants; an earlier phase (MAT type A) with abundant hand axes and numerous sidescrapers (*racloirs*) and denticulate tools, as well as some backed knives;[3] and a later phase (MAT type B) with considerably fewer hand axes (all of which are cordiform) and scrapers, but with more knives (some made on blades) and denticulates. (2) Typical Mousterian, which is similar to the MAT except hand axes and knives are generally lacking (with sidescrapers, denticulates, notched

FIGURE 3.2

Representative Mousterian tool forms: scrapers (*racloirs*). Numbers in parentheses refer to tool type number in Bordes' typology (Bordes 1972) (Table 2). A. convergent scraper, convex (19); B. convergent scraper, straight (18); C and D. déjeté scraper (21); E. transverse scraper, straight (22); F. scraper with alternate retouch (29). Scale bar in centimeters.

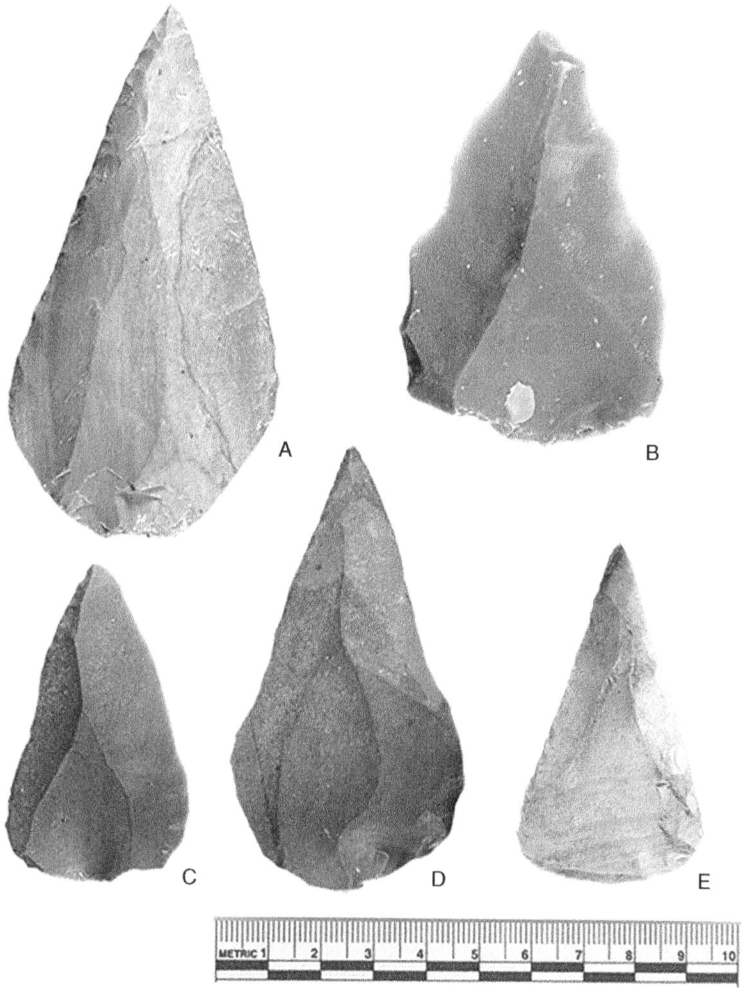

FIGURE 3.3

Representative Mousterian tool forms: points. Numbers in parentheses refer to tool type number in Bordes' typology (Bordes 1972) (Table 2). A. Mousterian point (6); B. Levallois point (3); C and D. retouched Levallois point (4); E. pseudo-Levallois point (5). Scale bar in centimeters.

pieces and points being correspondingly more abundant). (3) Denticulate Mousterian, mainly comprised of notched and denticulate pieces (up to 80% in some assemblages) and with relatively low frequencies of *racloirs* and other tool forms. (4) Quina-Type Mousterian (now more commonly known as Charentian or Quina-Ferrassie), with a high frequency (up to 75% or more) of sidescrapers, some of which are of the Quina type (that is, showing a distinctive steep, scalar ["Quina"] retouch), along with some denticulates and notched tools, burins, borers, carinate or nosed end scrapers, and

FIGURE 3.4

Representative Mousterian tool forms: various. Numbers in parentheses refer to tool type number in Bordes' typology (Bordes 1972) (Table 2). A. denticulate (43); B. typical burin (32); C. limace (slug) (8); D. natural backed knife (38); E. typical backed knife (36). Scale bar in centimeters.

limaces. Within these typological categories the underlying technology may vary, perhaps best recognized in the varying use of Levallois reduction across these facies (for example, within the Ferrassie variant of the Charentian, Levallois flakes may be fairly common, whereas they are rare in the Quina variant). The various facies and their subvariants are often found to interstratify across levels with single sites (the classic example being the deeply stratified site of Combe Grenal in southwest France: Bordes 1972), and with the exception of variation within the MAT (MAT A vs. MAT B), Bordes failed to recognize temporal or climatic patterning to the observed typological

TABLE 3.1 ■ Typological classification of Lower and Middle Paleolithic tool forms according to Bordes[a]

1. Levallois flake – typical	32. Burin – typical
2. Levallois flake – atypical	33. Burin – atypical
3. Levallois point	34. Drill – typical
4. Retouched Levallois point	35. Drill – atypical
5. Pseudo-Levallois point	36. Backed knife – typical
6. Mousterian point	37. Backed knife – atypical
7. Elongated Mousterian point	38. Natural backed knife
8. Limace	39. Raclette
9. Racloir – single-edged, straight	40. Truncated flake
10. Racloir – single-edged, convex	41. Mousterian tranchet
11. Racloir – single-edged, concave	42. Notch
12. Racloir – double-edged, straight	43. Denticulate
13. Racloir – double-edged, straight/convex	44. Alternate beaked burin
14. Racloir – double-edged, straight/concave	45. Retouch on bulbar surface
15. Racloir – double-edged, biconvex	46-7. Abrupt, thick, alternate retouch
16. Racloir – double-edged, biconcave	48–49. Abrupt, thin, alternate retouch
17. Racloir – double-edged, convex/concave	50. Bifacial retouch
18. Convergent racloir, straight	51. Tayac point
19. Convergent racloir, convex	52. Notched triangle
20. Convergent racloir, concave	53. Pseudo-microburin
21. Déjeté racloir	54. End notch
22. Transverse racloir, straight	55. Mincer (hachoir)
23. Transverse racloir, convex	56. Plane
24. Transverse racloir, concave	57. Tanged point
25. Racloir on bulbar face	58. Tanged tool
26. Racloir with abrupt retouch	59. Chopper
27. Racloir with thinned back	60. Chopper – inverse
28. Racloir with bifacial retouch	61. Chopping tool
29. Racloir with alternate retouch	62. Diverse
30. End-scraper – typical	63. Bifacial leaf point
31. End-scraper – atypical	

[a] Numbers follow Bordes 1972; types are defined in Bordes 1961b. Bordes' system of classification was based on the formal attributes of tools (that is, it is a typological rather than technological system), including the shape of the tool in plan view, the number of retouched edges, the position, location, steepness, and invasiveness of retouch, and the shape of retouched edges in plan view (Shea 2013).

and technological variation, attributing it instead to enduring cultural traditions among the Neandertals (Bordes 1961a). Subsequent work has revealed a strong chronological pattern to some facies (namely, Ferrassie-type Charentian assemblages predate those of Quina-type, which in turn are earlier than the MAT) but not others (Denticulate and Typical Mousterian assemblages tend to crop up periodically throughout the Mousterian sequence) (Mellars 1965, 1986, 1996b). The potential adaptive significance of temporal (and regional) variation in Mousterian facies is explored below (see Section 3.2).

Intra- and inter-site variation in Mousterian assemblages of southwestern France is just part of larger spatial and temporal variation in the Eurasian Middle Paleolithic.

The Mousterian itself, broadly defined, is regionally variable. In western Italy there is a diminutive variant of the Mousterian, fashioned out of small flint pebbles, called the Pontinian (Blanc 1937; Taschini 1979). The Pontinian is effectively a non-Levallois facies (although Levallois flakes and points are not unknown: Kuhn 1995) that is rich in sidescrapers, leading some to see it as basically a Quina-type Mousterian as expressed on small nodules of raw material (Taschini 1979). The Pontinian is perhaps the best known of several similar sets of assemblages from coastal Italy (from both the Tyrrhenian and Adriatic coasts), including some Levallois-dominant complexes, that tend to be made on small beach pebbles and that have been given various regional names (Kuhn 1995; Mussi 2001a).

The situation is equally complex in western Asia, where at least three variants of the Middle Paleolithic have been identified. Early efforts to discriminate between assemblage complexes in the Near East identified three major typological groups (Skinner 1965) which today are recognized as the temporally-early Yabrudian, the Levantine Mousterian, and the Zagros Mousterian (Dibble 1984b). The Yabrudian, generally considered a facies of the Acheulo-Yabrudian or "Mugharan Tradition" (Jelinek 1981, 1982a), is an early, Lower/Middle Paleolithic transitional industry that first emerged during MIS 11 or 10 (Barkai et al. 2003; Rink et al. 2004) and which appears largely confined to the Levant. The Acheulo-Yabrudian combines elements of the Lower and Middle Paleolithic: the presence of the Lower Paleolithic forms (namely bifaces), in assemblages that tend to be dominated by Quina scrapers (with abundant transverse and déjeté scrapers [sidescrapers with asymmetrical convergent edges]), has drawn typological comparisons to the European Mousterian of Acheulean Tradition (Perrot 1968). As a distinct facies, the Yabrudian is apparently distinguishable only by the lack of a bifacial component, which has led to the suggestion that it is not in fact distinct from the broader Acheulo-Yabrudian (Jelinek 1981, 1982a). Another notable component of the Acheulo-Yabrudian is the typologically-distinct, blade-dominated Amudian industry (Barkai et al. 2009). The Amudian (sometimes also called the Pre-Aurignacian), while preserving variable frequencies of scrapers and bifaces and a significant component of flake debitage, is remarkable in having a major component comprised of shaped blades (Barkai et al. 2009), including burins, backed knives and endscrapers (tool forms which, typologically, are more characteristic of the Upper Paleolithic). While assemblages rich in blades[4] do occur with some regularity in the Lower and Middle Paleolithic (Bar-Yosef and Kuhn 1999), such laminar or "leptolithic" assemblages are more characteristic of the Upper Paleolithic, and in fact are the defining technological feature of Mode 4 (Clark 1968).

As for the Levantine Mousterian, the key site that documents the sequence of lithic complexes is the deeply stratified site of Tabun Cave on the western slopes of Mount Carmel (Israel). Tabun preserves eight archeological horizons that span the later Acheulean to the Bronze Age (Garrod and Bate 1937; Jelinek 1982b), and three of the Lower and Middle Paleolithic levels (E, C and B) have produced hominin remains (McCown and Keith 1939; Coppa et al. 2005). The sequence at Tabun has served as the basis for classifying Middle Paleolithic assemblages along the eastern Mediterranean, and thus the Levantine Mousterian is conventionally divided into three temporally-arrayed facies known (from oldest to youngest) as "Tabun D," "Tabun C" and "Tabun B" (Bar-Yosef 1998: also known as Phase 1, 2 and 3, respectively: Copeland 1975). Unlike the situation in Europe, where there is only partial temporal ordering to the appearance of Mousterian facies, the three facies of the Levantine Mousterian are strongly temporally-arrayed (Bar-Yosef 1998; Hovers 2006). At Tabun the Amudian (in level E) is overlain by a Mousterian level (D) which has a greater proportion of Levallois products than found in the Acheulo-Yabrudian (Jelinek 1982b), but which continues the laminar quality of the Amudian with a high proportion of blades and elongated points (essentially pointed blades) (Bar-Yosef 1998). Thermoluminescence dates on burnt flint place the transition from the Acheulo-Yabrudian to the Tabun D Levantine Mousterian at the end of MIS 8 or beginning of MIS 7 (Mercier and Valladas 2003). At roughly the beginning of MIS 6 Tabun D assemblages give way to those of Tabun C-type, which tends to be made on large Levallois flakes and which includes broad triangular points (albeit in low percentages) as well as sidescrapers, notches and denticulates (Hovers 2009). Tabun B, likely emerging at the beginning of MIS 4 (Shea 2003a; see also Mercier and Valladas 2003) is typologically similar to Tabun C (the main typological distinction being a higher percentage of triangular points in Tabun B), the major difference between the assemblages being a technological shift from predominantly bidirectional and centripetal (radial) flake removal from Levallois cores to flakes removed from unipolar, convergent Levallois cores (Bar-Yosef 1998). While most see a clear distinction between Tabun C and B (Meignen and Bar-Yosef 1988, 1991), the overall typological and technological similarities between the two facies has led some to suggest that they are not in fact distinct, and that they should be considered as a single complex variously called the "Later Levantine Mousterian" (with Tabun D representing the "Early Levantine Mousterian": Jelinek 1982a; Marks 1992) or "Tabun C/B" (Culley et al. 2013).

In addition to the question of the distinctiveness of Tabun C versus B, it is not currently clear which of the lithic assemblages of the Levantine Mousterian were created by Neandertals and which were produced by early modern humans. Levantine Neandertal fossils have been recovered from Tabun Cave associated with both Tabun C and Tabun B type tools (McCown and Keith 1939; Coppa et al. 2005: but see below), from the cave sites of Kebara (with a Tabun B assemblage: Bar-Yosef and Vandermeersch 1991) and Amud (with Tabun B: Suzuki and Takai 1970) in Israel and from Dederiyeh Cave in Syria (with Tabun B: Akazawa et al. 1995). Early modern human remains have been recovered from Skhul and Qafzeh Caves in Israel, in association with Tabun C-type assemblages at both sites (McCown and Keith 1939; Vandermeersch 1981). Taxonomically less certain but possibly modern human remains have also been recovered from Geula Cave (Israel), associated with Levantine Mousterian tools and possibly dating to ca. 45 Ka BP (Arensburg 2002), which temporally at least would correspond with Tabun B-type assemblages. Undiagnostic human remains have also been recovered associated with Tabun C-type lithics in Level E of Hayonim Cave (Israel) (Arensburg and Nathan 1980; Arensberg et al. 1990). On the surface, the dual records (fossil and lithic) suggest that both assemblages (Tabun C and B) were produced by both types of hominin, which in turn implies a troubling degree of cultural and adaptive identity between two anatomically distinct populations (Shea 1989, 2003a; Kaufman 1999) (although an argument can be made that parity in lithic technology and typology does not necessarily mean parity in culture: Shea 2005). Since the facies of the Levantine Mousterian are also temporally arrayed, the nature of the hominin-lithic assemblage relationships tells us something about when the Neandertals first moved into this region (see Chapter 2.6), as well as about the nature of the evolutionary relationship between Neandertal and modern human populations in the Levant. It is not currently clear if they were sympatric populations that were either direct competitors or were engaging in some form of long-term niche partitioning, or whether they were parapatric populations practicing ecogeographic vicarism (Shea 2003a). The hominin-assemblage associations, again if taken at face value, would be most consistent with long-term co-occupation of the Levant, or with alternating occupations (vicarism) of relatively short duration. But questions exist about the perceived pattern of fossil and lithic associations, and again the site of Tabun appears to be central to resolving these questions.

As noted above (Chapter 2.6), the age of the adult female Neandertal skeleton from Tabun is uncertain. The specimen was attributed to level C, but lay close to the contact with level B, and may have been an intrusive burial from that level. Uncertainty about the stratigraphic provenance of the skeleton was noted by its excavator, Dorothy Garrod (Garrod and Bate 1937), and an argument can be made that the skeleton more likely derives from layer B (Bar-Yosef and Callander 1999). However, Trinkaus (see comment in Lieberman 1993) notes that Layer C produced a distal right radius and a right hamate that are virtually perfect mirror images of the same bones in the left arm of the Tabun C1 skeleton, suggesting that the skeleton was properly attributed to level C. Furthermore, there is a femoral diaphysis from level C that morphologically falls with Neandertal specimens and not with modern humans (Trinkaus in Lieberman 1993). Particularly germane to this debate are the morphological affinities of an isolated mandible securely attributed to level C, Tabun C2. The specimen combines both Neandertal features (retromolar space, mental foramen below the M_1, HO mandibular foramen) and modern human features (prominent mental eminence, deep mandibular notch, medially-placed condyle), and it has been seen variously as representing a Neandertal (Trinkaus 1983b, 1995b; Smith 1985; Smith et al. 1989; Stefan and Trinkaus 1997) or an early modern human (Vandermeersch 1981; Bar-Yosef and Pilbeam 1993; Rak 1998). Despite its archaic characteristics, the Tabun C2 mandible clusters with early modern humans in multivariate space, supporting the suggestion that it represents a member of the Skhul-Qafzeh population of early modern humans, or possibly a Neandertal–early modern human hybrid (Quam and Smith 1998). If the Tabun C1 skeleton does derive from level C, if the Tabun C2 mandible and other isolated remains from level C represent Neandertals, and if the relatively late remains recovered from Geula represent modern humans, then the inescapable conclusion is that both of the later facies of the Levantine Mousterian (Tabun C and B) were produced by both types of hominin. This would also suggest that (1) Neandertals arrived early in the Levant, before or at roughly the same time as the first early modern humans, and (2) the two groups either coexisted for an extended period of time, or alternated their occupation of the region over relatively short intervals. Alternatively, if the Tabun C1 skeleton derives from layer B, the C2 mandible and other fossils from level C represent modern humans, and the fossil material from Geula Cave derives from Neandertals, then there appears to be a clear assemblage-hominin correlation: early modern humans as represented at Skhul and Qafzeh made Tabun C assemblages, while Neandertals were the authors of Tabun B (Lieberman 1993). Such an association would allow us to make inferences about early modern human and Neandertal culture, technology and adaptation on the basis of levels with Tabun C- and Tabun B-type artifacts (respectively) at sites that do not preserve associated hominin fossils (that is, the vast majority of the Levantine Middle Paleolithic sites). Furthermore, it would imply that (1) Neandertals were late-comers to the Levant, arriving at the earliest at the beginning of MIS 4, and (2) Neandertal and modern human occupation of the Levant represents a case of ecogeographic vicarism played out over long time periods (modern humans having dispersed northward during the warm conditions of MIS 5, and Neandertals moving into the region during the cold conditions of MIS 4 and 3). If this is the case, co-occupation of the Levant may have been brief or nonexistent (although the genetic evidence suggests that the two groups did encounter each other there at least briefly during late MIS 5 or MIS 4: see Chapter 2.3), and both groups likely occupied the same ecological niche when present in the region (but perhaps with a period of intense competition during the period of overlap: Shea 2003a, 2003b, 2005).

The Zagros Mousterian, known primarily from Zagros Mountain sites in northern Iraq and western Iran, is typologically distinct from the late Lower and Middle Paleolithic assembles from the Levant (the Acheulo-Yabrudian and the Levantine Mousterian). Skinner (1965) distinguished the Zagros Mousterian from its counterparts to the west by its generally non-Levallois character and its high percentage of points and sidescrapers and low percentage of denticulates. In terms of the proportion of sidescrapers (especially in having a high frequency of double and convergent *racloirs*), plus a high frequency of retouched tools, the Zagros Mousterian is typologically similar to the Ferrassie-variant of the Charentian facies of the European Mousterian (Dibble 1991). While the Zagros Mousterian also shares some technological similarities (high occurrences of platform faceting and blades) with Ferrassie assemblages, it differs in having a much lower representation of Levallois flakes (Dibble 1991). The use of Levallois reduction in the Zagros Mousterian, however, appears to have been variable: while generally low, at least one site (Bisitun, Iran) has produced assemblages with frequencies of Levallois flakes that approach those of assemblages of the Levantine Mousterian (Dibble 1984b; Dibble and Holdaway 1993). Two Zagros Mousterian sites have produced hominin remains: 10 partial Neandertal skeletons were recovered from Shanidar Cave (Iraq) (Solecki 1971; Trinkaus 1983b; Cowgill et al. 2007), while Bisitun Cave (Iran) has yielded a radius fragment that likely derives from a Neandertal (Trinkaus and Biglari 2006).

While the Mousterian was certainly geographically widespread, extending from western Europe to the Middle East, it was not the only Middle Paleolithic technocomplex produced by the Neandertals. In central and eastern Europe, Neandertal remains are often associated with biface-rich assemblages of the Micoquian (Bosinski 1967; Svoboda et al. 1996) or Keilmessergruppe. Although originally identified by the presence of distinctive "Micoquian" hand axes (or "picks," sharply pointed bifaces, often with rounded bases: Bordes 1961b) in Layer N at the French site of La Micoque (Hauser 1916), the Micoquian as now understood has little to nothing to do with these characteristic bifaces or the eponymous French site (Jöris 2006). The Micoquian is characterized by a variety of bifacal tools with flat retouch, perhaps the most typical being bifacially backed knives (*Keilmesser*), but also including bifacial scrapers (*Faustkeilschaber*), foliate (leaf) points, and small hand axes (*Fäustel*) (Ruebens 2006, 2009). Larger, asymmetrical hand axes (*Micoquekeile*), as well as some unifacial hand axes (*Halbkeile*) are also occasionaly recovered in Micoquian assemblages. Micoquian hand axes as found in earlier deposits in western Europe, such as at La Micoque, now appear to be terminal Lower Paleolithic tools forms (Rolland 1986), and it thus seems that the central/eastern European Micoquian should be dissociated from the western European assemblages (Conard and Fischer 2000; Jöris 2006). Accordingly, some workers prefer the term Keilmessergruppe rather than Micoquian (Reubens 2009). It should also be noted that bifacial pieces characteristic of the Micoquian occur in Mousterian assemblages in western Europe, sometimes with high enough frequencies to invoke designations of "Mousterian of Micoquian tradition" (Freund 1968) or "mixed assemblages" (Kind 1992), and leading to a certain degree of difficulty in taxonomically diagnosing some assemblages (Ruebens 2006, 2009). It is also the case that Mousterian and Micoquian assemblages interstratify at some sites, such as the Bavarian site of Sesselfelsgrotte (Rots 2009).

The Eurasian Middle Paleolithic ended at different times in different places, without a clear geographic pattern to its disappearance. The first Initial Upper Paleolithic (IUP) assemblages appear to emerge in southwest Asia on the order of 47–45 Ka BP (Bar-Yosef 2002), but the Mousterian seems to have persisted there, at some sites, to as late as 33.3 Ka BP (Richter et al. 2001). In eastern Europe, the youngest Mousterian and Micoqian dates tend to fall around 37–36 Ka BP (Soffer 1989a; Pinhasi et al. 2011, 2012), although dates as young as 35 Ka BP have been reported for sites near the Black Sea (Lioubine 1993) and Mousterian tools have been found in 34–31 Ka BP cal contexts at Byzovaya (Russia) in the Ural Mountains above the Arctic Circle (Slimak et al. 2011). In Central Europe, the dates for terminal Mousterian levels tend to fall between 41–36 Ka BP, with most of them clustering around 40–39 Ka BP (see Table 1 in Churchill and Smith 2000a). However, dates as recent as 35 – 33 Ka BP have been reported from two Hungarian sites (Vogel and Waterbolk 1972), and a late Middle Paleolithic layer at a Romanian site has produced an age of ca. 30 Ka BP (Cârciumaru 1989). In western Europe the Mousterian seems to disappear on the order of 35 – 34 Ka BP (see Table 1 in Churchill and Smith 2000a), with the Iberian Peninsula possibly serving as a Mousterian refugium until late into MIS 3, persisting perhaps as late as 24 Ka BP at Gorham's Cave (Gibraltar) (Finlayson et al. 2006, 2008).

On the western (Iberian) and eastern (Russian) fringes of their range, the last Neandertals seem to have held to a Mousterian way of life to the bitter end, but in the areas in-between it is possible (although debated) that they were players in the development of Upper Paleolithic technologies and culture. Neandertal remains have been found in association with an Initial Upper Paleolithic (IUP) industry, the Châtelperronian, in southern France (Lévêque and Vandermeersch 1980; Hublin et al. 1996; Bailey and Hublin 2006a, 2006b), and on limited evidence the Neandertals have also been suggested to have been the authors of similar industries in Italy (the Ulluzian), Croatia (the Olschewian) and central/eastern Europe (the Szeletian) (see Churchill and Smith 2000a). These assemblages and other IUP technocomplexes (such as the Altmühlian, Streletskian and Jerzmanowician of central and eastern Europe) share typological attributes with later (and possibly contemporaneous) Early Upper Paleolithic industries, but generally have both typological and technological ties with earlier, regional Middle Paleolithic industries (for example, the use of Levallois reduction methods in the Châtelperronian [Lucas et al. 2007] or the persistence of bifacial leaf points from the

Micoquian to the Szeletian, Altmühlian, and Jerzmanowician [Kozlowski 2003; Richter 2009]).[5] These Initial Upper Paleolithic assemblages, appearing roughly during the Hengelo temperate oscillation of MIS 3 (also known as the Würm II/III interstadial, beginning ca. 38 Ka BP) and thus at the end of the Neandertal's reign, may have emerged in response to the influx of anatomically modern humans in Europe and the climatic instability of the Interpleniglacial (see Churchill and Smith 2000a; Mellars 2005), or perhaps as an independent invention prior to contact with modern humans (Zilhão and d'Errico 1999). In addition, revision of the dates of key early modern human specimens (Henry-Gambier and White 2003; Conard and Bolus 2003; Street et al. 2006), along with a very poor fossil record of Aurignacian-associated humans, makes it impossible to rule out the idea that Neandertals contributed to the development of the Aurignacian (d'Errico et al. 1998). However, the association of Neandertals with the Châtelperronian has been challenged (Bar-Yosef 2006; Bar-Yosef and Bordes 2010; Higham et al. 2010; but see Hublin et al. 2012), as has their association with the Ulluzian (Benazzi et al. 2011). Resolving this issue is central to understanding the nature of Neandertal/modern human interaction, and the role that the Neandertals might have played in the development of the Upper Paleolithic (d'Errico 2003).

3.2 VARIATION IN THE EURASIAN MIDDLE PALEOLITHIC: TECHNOLOGY AS ADAPTIVE INTERFACE

The Eurasian Middle Paleolithic was variable across time and space. Temporally, there was both dynamic variation (as when interstratification at a single site demonstrates a pattern of alternation between two or more facies) and directional change over time. Geographic variation was manifested as broad regional "flavors" in Middle Paleolithic technocomplexes, from a predilection for bifacial tool forms in the Micoquian of central Europe to a heavy emphasis on the production of Levallois flakes in the Levantine Mousterian. Although not as pronounced as the stylistic variation in lithic tools seen among Holocene foragers (Wiessner 1983; Barton 1997), or perhaps even among the contemporary makers of the African MSA (Clark 1992; McBrearty and Brooks 2000), the regional variation seen in the tools made by Neandertals can fairly be said to reflect regional styles, at least in an incipient way (Ruebens 2013). Regional differences also exist in temporal patterns of variation: in some cases (such as the Mousterian of southwest France or the Levant), temporal sequences show a distinct evolutionary trajectory; in other cases (such as the Micoquian of the Crimean peninsula: Marks and Chabai 2006) there seems to be no clear trajectory. Some of the observed variation is no doubt idiosyncratic, resulting from historical contingencies and transient cultural norms, as occurs in any culture. But some of the variation, and perhaps much of it, was likely to have an adaptive basis. Extant and historically-known foragers are also highly variable in their material culture, and of course much (but not all) of this variation relates to the ecological and environmental circumstances faced by various groups (Oswalt 1976; Kelly 1995; Binford 2001). Thus, variation in the technology employed by Neandertals gives us a valuable window through which to explore their adaptive dynamics.

If modern foragers are used as the frame with which to examine Neandertal behavior, one aspect of their adaptive systems stands out as unusual (Kuhn 1995, 2011): subsistence tools (or "extractive technology") always comprise a very small component of any assemblage. While lithics that could have or did function as weapon armatures and butchery tools are present in virtually every assemblage, they (1) do not show the characteristics of specialized extractive tools that one commonly sees among modern foragers, (2) generally present ambiguous evidence that they were actually used in hunting, carcass processing, or other subsistence activities, and (3) are relatively invariant across space and time, and thus don't show the type of ecogeographic patterns seen in the toolkits of modern hunter-gatherers. With respect to this last point, it is also the case that, although the Middle Paleolithic was geographically and temporally variable, the level of technological complexity does not vary predictably with latitude or environment as it does in modern human foragers (Bocquet-Appel and Tuffreau 2009: Kuhn 2011). The nature of the Neandertal subsistence toolkit will be explored more fully in Section 3.4, but this overall pattern suggests a Neandertal/modern human difference in their adaptive relationship to technology. Among modern human foragers, technology is modulated dynamically in response to changes in environmental conditions or to adapt to the ecological realities encountered in different parts of the world (for example, toolkits tend to be much more diverse in areas where extractive foraging is more difficult, such as the arctic). Among the Neandertals, subsistence technology is temporally static and geographically invariant, which means that they *must have* responded to ecological and environmental variation in other ways, perhaps through social mechanisms (by modulating the size and composition of labor parties: Stiner 1994; Kuhn and Stiner 2006) or foraging intensity (for example, by modulating their foraging ranges: Kuhn 1995).

Much of the technological variation seen in the Eurasian Middle Paleolithic revolves around core preparation. In some places and at some times there was a heavy emphasis on Levallois reduction methods; in other contexts it was rare or absent. Even among knappers employing Levallois techniques, there was variation in basic reduction strategies (lineal versus recurrent reduction) and in the directions from which flakes were driven off of the core (unidirectional, bidirectional, or radial). Typological variation relates in part to technological variation, including the extent to which laminar technology was employed, or to varying emphases on *façonnage* (in which the core or flake is reduced into a particular tool form, in a process often likened to sculpting) versus *débitage* (where the core is divided up into a number of flakes which are used, either modified or unmodified, as tools) in tool manufacture. Typological variation also derives in part from reduction intensity (that is, the extent to which tools were re-sharpened prior to discard). Much of this technological and typological variability might be a manifestation of the adaptive dynamics in play across the Neandertal temporal and geographic range, and in particular the mobility strategies they used to strategically position themselves relative to the distributions of resources across varied environments. Before we explore the connection between technology and mobility, it might be useful to summarize the various sources of variation in lithic assemblages as mentioned above.

As we have seen, an important technological characteristic of Middle Paleolithic assemblages concerns the extent to which the Levallois method was employed. What, exactly, constitutes Levallois reduction? Boëda (1988, 1993) described the "Levallois concept," which consists of two main components (Figure 3.5). The first is the preparation of a continuous striking platform around the perimeter of the core. This is accomplished by successive removals around the edge of the core, in a direction

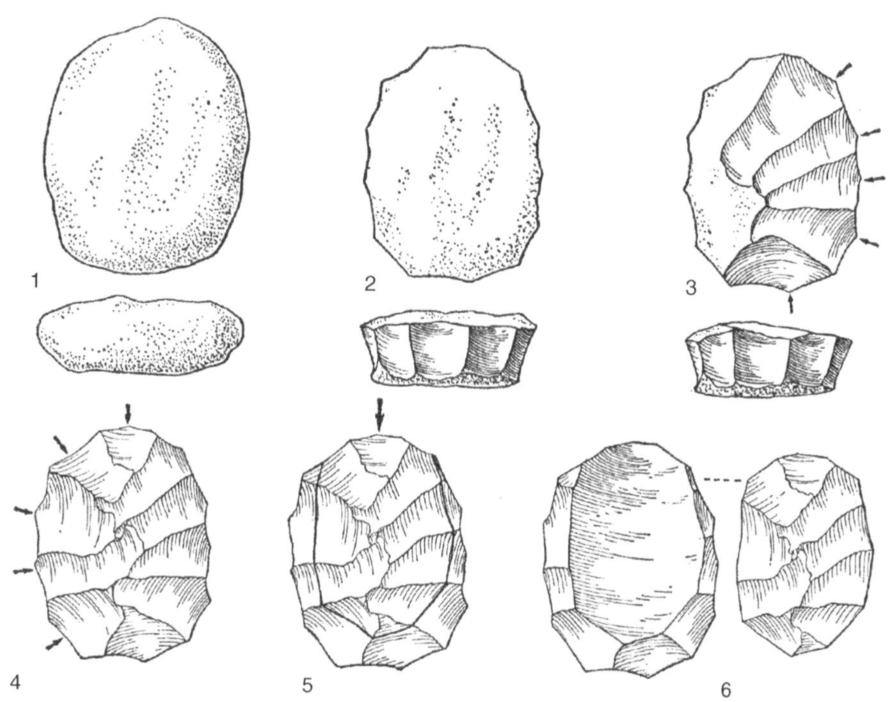

FIGURE 3.5

Stages in the production of a Levallois flake (figure from Bordes 1961a. Reprinted with permission from AAAS): 1. unmodified core; 2. removal of flakes around the periphery of the core; 3. preparation of the surface, using the previous removals as striking platforms; 4. continuation of surface preparation; 5. removal of the central or 'privileged' flake; 6. the Levallois core and flake.

perpendicular to the axis along which the flakes will be struck. This first step represents, in its entirety, the process of preparing a disc-core: that is, if the knapper were to begin driving flakes off of the core at this stage, it would represent a non-Levallois form of core preparation. The second component of the Levallois concept involves shaping the upper surface of the core, so as to further define the shape of the flakes that will be removed. Thus, Levallois reduction requires two distinct stages of core preparation, and it can be seen that a continuum exists between disc-core and full Levallois methods, and different workers might reasonably draw the line between the techniques in different places (see Mellars 1996). The basic Levallois concept could also be applied to cores of different shape, or varied in the way the striking platforms or surface removals were detached, such that flakes of a variety of predetermined forms could be produced. Accordingly, different types of Levallois cores (ovoid, round, rectangular, point, and blade cores: Mellars 1996) can be found in Middle Paleolithic assemblages. Given that both Levallois and non-Levallois methods of core preparation exist, what value did knappers gain by the use of the former? Levallois methods appear to produce more usable edge per nodule than do non-Levallois forms of reduction (Brantingham and

Kuhn 2001) and thus offer an advantage in terms of raw material economy. If this were the only benefit to Levallois reduction, we might expect its use to reflect raw material availability and quality, as well as group and individual mobility patterns as they relate to raw material provisioning. There is no doubt that the Neandertals' raw material provisioning strategies and economics were dynamic and flexible, and that levels of tool curation, resharpening and reworking were conditioned by raw material availability (see references in Kuhn 2011). However, the choice of Levallois reduction methods was likely determined by more than just raw material economics. Levallois techniques also produce flakes that are larger on average, that have a high number of usable edges, and that are relatively thin (Sandgathe 2005). Large Levallois flakes and points have the highest utility/weight ratios (that is, greatest length of edge per unit weight) of typical Middle Paleolithic products (Kuhn 2011). This characteristic was no doubt very important with respect to mobility, since Levallois flakes would be expected to provide a mobile forager with a variety of tool shapes and cutting edges, suitable for many different purposes, in a toolkit that was otherwise relatively light (Pettitt 2003). In this regard, Pettitt (2003) notes that, while most Mousterian assemblages were made on locally available material (usually within about 5 km of the site), when material is curated over larger distances (up to 100 km in western Europe, possibly 300 km in central Europe) it is generally in the form of Levallois products (Geneste 1985; Féblot-Augustins 1993; see also Kuhn 2011).

Within the "Levallois concept," knappers had substantial latitude in reduction techniques (Boëda 1993), allowing them to modulate functionally important morphological attributes of blanks (size, length and angle of cutting edge, ventral curvature, and overall robusticity: Sandgathe 2005) to accommodate a variety of needs. This latitude included a choice of how many flakes were removed between episodes of core reshaping (with lineal techniques, the core was reshaped after the removal of each "privileged" flake; with recurrent techniques, a repeated succession of flakes were removed without remodification of the core), a choice of the directions in which flakes were removed (unipolar, bipolar, or centripetal reduction), as well as a choice of overall core size, shape, and surface preparation (see Mellars 1996 for an excellent review of Levallois technology). Non-Levallois methods could also be employed to produce end products with specific desired attributes, and these methods were sometimes applied alongside Levallois methods or sometimes instead of them. The Quina variant of the Charentian Mousterian is notable for its near absence of Levallois reduction, and for an abundance of tools with thick, triangular cross-sections. These thick blanks, which could withstand abundant resharpening while maintaining a thick rear edge suitable for grasping or hafting (Mellars 1996), were produced with reduction techniques that were particularly well suited to the elongated nodules of flint found in southwestern France. This "Quina technology," involving either the removal of slices perpendicular to the long axis of the nodule ("citrus slices" or "salami slices") or parallel to it (elongated flake removal), produced wedge-shaped flakes whose maximum thickness was located directly opposite the retouched edge of the tool (Turq 1989, 1992). Finally, Middle Paleolithic knappers could also control the laminarity of flakes (along a continuum from flakes to flake-blades to blades) through the use of various types of Levallois blade cores ("classic Levallois" versus "specialized Levallois" blade technology: Mellars 1996) and non-Levallois blade cores (Mellars 1996; Bar-Yosef and Kuhn 1999).

This brief review of Middle Paleolithic technology shows that the Neandertals had multiple technological avenues available to them to produce desired tool forms in response to varied extractive and domestic tasks (although there is some limited evidence to suggest that, in some cases, the goal of flint knapping was to produce a suitable working edge which was positioned in a way to maximize ease and comfort of use, rather than to produce tools with specific formal attributes: Bisson 2001). Thus much of the typological variation seen in the Middle Paleolithic relates to underlying technological variation. Before exploring the adaptive inferences that have been generated from consideration of this technological variation, it is also important to note that some of the typological variation seen across assemblages might be related to another technological variable: resharpening intensity. In his classic typology of the Mousterian, Bordes (1961b) identified 17 separate types of *simple* sidescrapers, in addition to *déjeté* sidescrapers and three forms of convergent sidescrapers (Table 3.1). In a series of papers, Rolland and Dibble (1990; Rolland 1977, 1981; Dibble 1984a, 1987; Dibble and Rolland 1992) have argued that the various types of *racloirs* identified by Bordes simply represent stages in the use-life and resharpening history of flakes. Tools may have entered the systems as unretouched flakes, but with progressive use and resharpening their shapes changed, resulting in the classic sidescraper types identified by Bordes. By this argument, much of the typological variation observed between assemblages is a function of resharpening intensity prior to discard: scraper-poor assemblages (such as the denticulate Mousterian) were lightly retouched, while scraper-rich assemblages (such as the Charentian Mousterian) were more heavily retouched. This suggests that much of the interassemblage typological variation in the Middle Paleolithic can be explained with respect to raw material availability and curation (where quality flint was a commodity, or in heavily curated assemblages, retouching intensity would have been high, and thus sidescrapers are expected to be abundant). There is good evidence (reviewed in Mellars 1996) to support the claim that reduction intensity was an important component of Mousterian typological variation, but there are also good reasons (see Mellars 1996: 101–104) to believe that resharpening only accounts for a portion of that variation. Still, this suggests again that some of the variation seen in the Middle Paleolithic is attributable to variation in mobility, site settlement systems, and raw material provisioning strategies, since all of these influence retouch intensity (with high mobility or short-term site occupation, there was a greater emphasis on curated toolkits, and thus resharpening is expected to be more frequent; tools were also more heavily resharpened when raw material was scarce: Kuhn 2011).

The possible relationships between mobility patterns and technological and typological aspects of Middle Paleolithic assemblages suggest that some – and perhaps most – of the observed interassemblage variation can be understood in adaptive terms. Pettit (2003) has articulated such a model for western Europe, arguing that the interplay of climate, prey ecology, and Neandertal mobility produced the temporal trends in the Mousterian from MIS 5 to 3. Faunal biomass was lower and thus prey was more dispersed during the relatively closed conditions of MIS 5 (see Chapter 6), necessitating relatively high mobility among the Neandertals. This mobility is reflected in raw material movement during this interval, in which some lithic pieces were moving distances up to 100 km. Under these conditions of closed cover and high mobility, Neandertals relied heavily on Levallois technology, as it provided them with large flakes and blades that were adaptable to a range of foraging tasks (high flexibility), but that were also thin (and thus light relative to their utility) (see also Delagnes and Rendu 2011). Accordingly, the MIS 5 deposits of western Europe tend to be dominated by Levallois-rich Ferrassie-type Charentian Mousterian assemblages (Pettitt 2003). Climatic downturn at the end of MIS 5 and beginning of MIS 4 reduced forest cover and simultaneously improved prey biomass while reducing prey diversity (see Chapter 6), promoting conditions that may have constrained Neandertal mobility. Raw material procurement became more local, and lithic reduction strategies shifted from an emphasis on Levallois to methods that were geared towards most effectively reducing the locally-available resources (without regard to utility/weight ratios). This corresponds with the initial MIS 4 "Ferrassie-Quina" transition (Mellars 1969), involving a reduction in Levallois reduction and increased "Quinarisation" (use of Quina technology: Turq 1989, 1992) and the dominant establishment of Quina-type Charetian Mousterian assemblages (Pettitt 2003). The heavily "recycled" nature of Quina-type assemblages, in which flakes were heavily retouched and blanks were frequently recycled for the production of smaller blanks (Delagnes and Rendu 2011), might be consistent with reduced mobility (with lower frequencies of raw material collection, and thus greater reduction intensity of cores and greater use/retouch intensity of tools). The ameliorated conditions of MIS 3 (Chapter 6) resulted in a richer and more diverse faunal community, which would support Neandertal social groups within a relatively smaller home range. As in MIS 4, MIS 3 assemblages are composed of local raw materials. Here the extractive technology did not need to be highly flexible (because the Neandertals were foraging within well-known territories where resources were predictable), and there is a technological re-emergence of *façonnage* and the production of a dominant single, reliable tool form (the biface), as seen in the Mousterian of Acheulean Tradition (Pettitt 2003). However, Monnier (2006) examined the typological and technological characteristics of 89 assemblages from 26 western European sites, and found that the Levallois Index (the percentage of all [retouched and unretouched] Levallois flakes relative to the entire flake assemblage) was on average higher during cold intervals, whereas the production of bifaces tended to be higher during warm periods. Median values of the Levallois Index increase from MIS 5 (8.32, n = 7 assemblages) to MIS 4 (9.71, n = 12), contra the idea of a "Ferrassie-Quina" transition, and median biface frequencies are lower in MIS 3 (0.00, n = 14) and MIS 4 (0.00, n = 12) than in MIS 5 assemblages (0.18, n = 7). Monnier's (2006) sample was constructed to examine temporal changes across the Lower to Middle Paleolithic, not to specifically examine the MIS 5-3 interval of Pettitt's model, and it is not clear how much the picture would change if a larger number of Late Pleistocene assemblages were examined. Furthermore, associations between lithic technology and faunal remains in western France suggested to Delagnes and Rendu (2011) a scenario opposite to that proposed by Pettitt, namely that Levallois reduction was practiced during periods of low mobility (but was associated with hunting of diverse prey, as also suggested by Pettitt), while Quina reduction was related to high mobility and a hunting focus on migratory game. Thus, while the details are currently unclear, the initial picture seems to be one in which technology was in some way dynamically responding to changing environmental conditions and mobility demands, suggesting a high level of behavioral flexibility to Neandertal adaptive systems (Kuhn 2011). This also implies that similar adaptive scenarios may underlie some of the other regional and temporal variation seen in the Eurasian Middle Paleolithic.

3.3 Composite Technology, and the Archeologically Less-Visible Component of Technology

The emergence of composite technology, in which lithics were bound to handles of wood or bone, is generally considered to be an important landmark in the evolution of technology. Composite tools appear, at a minimum, to reflect some understanding of principles of leverage – that when a tool is set in a handle, greater force can be exerted on the target object (hide being scraped, wood being worked, etc.). In this respect, the first use of composite tools was perhaps the first major step in the evolutionary trajectory of substituting technology and leverage for somatic effort in our interactions with the environment. Hafting also dramatically improved the effectiveness of hunting weapons, since the addition of a sharp lithic armature to a spear would have increased tissue damage and hemorrhaging in prey (see Salem and Churchill 2014). Finally, the advent of composite technology also represents a major axis along which technological complexity evolved. Hafting involves at a minimum three "technounits" (Oswalt 1976) – a lithic, a handle, and binding material. Often binding materials of two types were used in conjunction, such as when a lithic was glued into a handle with plant resins or bitumen and then bound about with plant fibers or animal tissues (sinew or leather). Thus, even the simplest of hafted tools involve a level of complexity greater than that involved in the direct prehension of stone tools. But beyond that, "conjunctive technologies" (Ambrose 2001) also allow for unique combinations of components and thus greatly facilitate the invention of novel tools.

There is now considerable evidence to suggest that hafting of lithics onto wooden or bone handles was a regular occurrence during the Middle Paleolithic. Direct evidence of the production of composite tools comes in the form of hafting glues or their residues on stone tools. The first such evidence was found at Königsaue (Germany), where two pieces of hardened birch-bark resin were recovered from Micoquian levels (Mania and Toepfer 1973). The larger of the two pieces (directly dated to 43.8 ± 2.1 Ka BP and 48.4 ± 3.7 Ka BP [Hedges et al. 1998], although possibly older [Koller et al. 2001]) preserves the imprint of a backed bifacial tool on one side, and what appears to be the imprint of a wooden haft and a fingerprint on the other (Mania and Toepfer 1973). The direct recovery of hafting glues is rare, but a similar case was discovered at Campitello Quarry in central Italy (Mazza et al. 2006). In this case, hardened birch-bark pitch was found surrounding the butt end of one large flake, and patches of pitch were found adhering to a second flake. The artifacts from Campitello are associated with a fauna that suggests a terminal Middle Pleistocene (MIS 6) age for the deposits, which makes this the earliest known example of hafting in the archeological record. While the recovery of pieces of hafting resins are rare, traces of their residues on lithics have now been detected in a number of Middle Paleolithic contexts. Plant resins and naturally-occurring tar (bitumen) have been reported on Middle Paleolithic tools from Israel (Friedman et al. 1995), Syria (Boëda et al. 1996, 1998, 2008), the Crimea (Hardy et al. 2001), Romania (Cârciumaru et al. 2012), Germany (Pawlik and Thissen 2011) and Poland (Cyrek and Cyrek 2009). Tools are not the only place these glues appear: chemical signatures consistent with a small degree of ingestion of bitumen (as might happen when using one's mouth as a vice to secure a haft or tool while working on the binding) have been detected in the dental calculus of a young adult male Neandertal from the MIS 3 site of El Sidrón (Hardy et al. 2012).

There is also indirect evidence for hafting. When stone tools are set in handles and used, the forces transmitted through the tool cause the stone to move against the handle, producing distinctive wear traces (from both the handle and the binding agents) on the non-active parts of the tool (that is, away from the working edge: Moss 1987; Rots 2008, 2010). Microwear referable to hafting has now been identified on Middle Paleolithic tools from several sites (Beyries 1987a, 1990; Anderson-Gerfaud and Helmer 1987; Anderson-Gerfaud 1990; Rots 2009, 2013; Pawlik and Thissen 2011). These traces occur on a variety of tool types that exhibit varied types of edge-wear (Rots 2013), indicating that hafted tools were used in the performance of a variety of subsistence and domestic tasks. Impact damage has also been seen on the business end of Middle Paleolithic points, consistent with their use as weapon armatures (Shea 1988, 1993, 1998; Shea et al. 2001; Rots 2009), which implies that they functioned as the tips of multicomponent weapon systems (most likely thrusting spears: see Section 3.4). Impact damage, often in the form of spalls removed from the tip, generally occurs when a point makes forceful contact with bone. Similar evidence comes from the Mousterian site of Umm el Tlel (Syria), where a fragment of a Levallois point was found embedded in the neck vertebra of a wild ass (Boëda et al. 1999). It is difficult to imagine Neandertals generating characteristic impact damage, or driving a point into bone, with hand-held Levallois or Mousterian points, and again the inference is that these points were hafted onto spear shafts to form composite hunting weapons. The preponderance of evidence thus suggests that hafting was a common component of Mode 3, Middle Paleolithic technology, and that it had its origins in the latter part of Mode 2 (Lower Paleolithic).

Hafting requires handles, which necessarily need to be made of a stiff material such as wood, bone, or antler. Bone and antler working was apparently rare in both Mousterian and Micoquian assemblages, although not entirely unknown (Montet-White 1996; Gaudzinski 1999; Villa and d'Errico 2001; Soressi et al. 2013). Most of these examples involve bone points or *lissoirs* (hide-working tools), plus some bones from several sites that show modifications thought to reflect their use as retouchers for sharpening flint tools (Chase 1990; Mallye et al. 2012). There is also a seemingly isolated case from Cueva Morín (Spain) of Neandertals flaking and retouching bone (Freeman 1978), although the evidence for this has been challenged (see Freeman 1983). Woodworking, on the other hand, appears to have been common during Middle Paleolithic times, judging from recovered wooden implements (Tode 1954; Oakley et al. 1977; Carbonell and Castro-Curel 1992), natural casts of wooden items (Bordes 1972; Castro-Curel and Carbonell 1995; Vallverdú et al. 2010), and micro-wear and residue analyses of stone tools (Beyries 1987b, 1988; Anderson-Gerfaud 1990; Pawlik and Thissen 2011). Lithic micro-wear and residues indicative of hide preparation (Beyries 1987b; Anderson-Gerfaud 1990; Loy and Hardy 1992; Lemorini 2000; Hardy et al. 2001; Soressi and Hays 2003; Hardy 2004; Pawlik and Thissen 2011) also indicate that animal skins were regularly used, most likely for clothing, containers or satchels, shelters and windbreaks, and bedding material, and also possibly for cooking (Speth 2012). In combination, the evidence suggests that Neandertals had a rich technological component made on perishable materials, of which wood and hides were the predominant materials.

While it is not currently possible to explore the question of how variation in the lithic component of Middle Paleolithic technology related to that of the Neandertals' total material culture, and how perishable components of their technology were being used to solve adaptive problems, we are at least beginning to assemble an inventory of some of the wooden and bone tools in the Neandertal toolkit. In some cases, wooden artifacts themselves have been recovered, in other cases it is natural casts of them (or "pseudomorphs": Castro-Curel and Carbonell 1995) left behind in the sediments. From this evidence it appears that Neandertals were at least occasionally producing (but with enough frequency that some evidence is being found despite the taphonomic factors militating against the preservation of such evidence: Castro-Curel and Carbonell 1995) a variety of wooden artifacts, although because of taphonomic damage or partial burning the original morphology and function of many of the pieces is uncertain (e.g., two cylindrical wooden objects from the site of Kärlich, Germany: Kulemeyer 1988). A possible fire starting drill, made of beech and burnt on one end, was recovered from the Croatian site of Krapina (Oakley 1956). Shallow wooden bowls or trays, made of pine and juniper and possibly used as vessels for carrying food, or as shovels, have been recovered from the Abric Romaní, Spain (Carbonell and Castro-Curel 1992). Abundant pseudomorphs have also been uncovered at the Abric Romaní, and while most of these appear to represent firewood, a set of three cylindrical and straight sticks in association with a hearth may represent a tripod used for some sort of food processing (Castro-Curel and Carbonell 1995), such as curing meat over the fire. Another 510 cm-long, 3–6 cm-diameter imprint may represent a pine post used in the construction of a dwelling structure within the rock shelter (Vallverdú et al. 2010). This interpretation is consistent with the discovery of archeological features interpreted as postholes from Mousterian levels of both open air (La Folie, France: Bourguignon et al. 2002) and cave/abri sites (Combe Grenal, France: Bordes 1961a).

Another important class of wooden artifacts includes partial or complete sharpened wooden staves (spears), which have been recovered from both Lower and Middle Paleolithic sites. These artifacts, along with bone points, are discussed in Section 3.4. Bone tools that may have served a hide working purpose, along with the use of hides generally, are discussed in more detail in Section 3.5.

3.4 SUBSISTENCE TECHNOLOGY

All modern foragers (those extant today as well as those known historically) use technology in foraging. Its use is especially prevalent in hunting – as a means of acquiring prey that is hard to access, subduing and killing fast, large, or otherwise difficult-to-capture prey, reducing the risk of failure and of personal injury, and reducing the search and handling costs associated with hunting. The toolkits of most modern foragers are relatively complex compared with Mousterian and Micoquian technology (as inferred from its more durable components), and specialized and often complex tools for hunting frequently comprise a substantial segment of these toolkits, especially in environments in which prey capture is critical to survival (Oswalt 1976; Bleed 1986). Mousterian technology, although simpler and not varying in complexity in response to foraging complexity in the way that modern human toolkits do (see Section 3.2), also had elements that played a role in prey capture.

A handful of Middle and early Late Pleistocene sites in Europe have yielded simple sharpened wooden staves – generally interpreted as spears – as well as a single wooden artifact that was possibly used as a throwing stick (Table 3.2). The most remarkable

TABLE 3.2 ■ Spears and throwing sticks from European Lower and Middle Paleolithic sites

Site	Item	Age	Reference
Clacton, England	Distal 0.39 m of yew spear	early MIS 11 (possibly > 400 Ka BP)	Warren 1911; Oakley et al. 1977; see also White 2000
Schöningen 13, Germany	Seven 1.8–2.5 m-long spruce spears	400 Ka BP	Thieme 1997, 1999
Schöningen 13, Germany	Possible throwing stick fragment, 0.78 m long, of spruce	400 Ka BP	Thieme 1999
Lehringen, Germany	2.4 m-long yew spear	ca. 125 Ka BP	Movius 1950; Tode 1954

of these sites is the MIS 11 site of Schöningen 13, near Helmstedt, Germany, which has produced seven largely intact wooden spears that are roughly 400 Ka old (Thieme 1997, 1999). These spears are relatively long (1.8–2.5 m) and thick (maximum diameters of 29–47 mm) (Thieme 1999), and share with modern track and field javelins a couple of important design features (distally-positioned center of mass and tapering of the proximal end). The morphological similarities between the Schöningen spears and javelins led Thieme (1997, 1999) to argue that the Schöningen spears were purposefully designed for aerodynamic stability, and thus for use as projectile weapons. At least 15 horses – some of which may bear skeletal lesions from spear point trauma – are represented in the same layer (level 4), suggesting the spears were used in the procurement of these open country grazers. Similar artifacts have been recovered from Hoxnian Interglacial-aged (MIS 11) deposits at Clacton-on-Sea, England (the distal end of a sharpened yew stave: Warren 1911; Oakley et al. 1977) and from Eemian Interglacial (MIS 5) deposits at Lehringen, Germany (a broken spear of yew, found between the ribs of a straight-tusked elephant: Movius 1950; Tode 1954).[6]

While some of these fossil spears exhibit design elements that may reflect a concern for aerodynamics, their overall size and shape is most similar to those of ethnographically-known thrusting spears (Table 3.3: see also Oakley et al. 1977; Schmitt et al. 2003; Shea 2006). The Schöningen and Lehringen spears are relatively thick compared to recent throwing spears, and the preserved portion of the Clacton spear point (which has a maximum diameter of 39 mm: Oakley et al. 1977) suggests that it was as well. The average maximum diameter of the fossil spears for which dimensions have been published is almost 3.5 standard deviations above the mean for a sample of recent thrusting spears, and is more than 7 standard deviations above the mean for modern throwing spears (Table 3.3). If one creates a thickness index by taking the ratio of maximum diameter to spear length, the fossil spears fall between ethnographically-known thrusting spears and stout digging sticks. While the mass of the fossil specimens has not been published, the great length and diameters of the fossil spears suggests that they would have been heavy relative even to modern thrusting spears, and easily 2–4 times the weight of modern throwing spears (Table 3.3, Figure 3.6). A 2 m-long spear with an average diameter of 30 mm would have a

TABLE 3.3 ■ Dimensions of fossil and modern spears

	Length (mm)	Mass (g)	Maximum diameter (mm)	Thickness index[a]	Source
Lehringen	2400	—	34	1.417	Oakley et al. 1977
Schöningen I	2250	—	47	2.089	Thieme 1999
Schöningen II	2300	—	37	1.609	Thieme 1999
Schöningen III	1820	—	29	1.593	Thieme 1999
Mean of fossil spears (n=4)	2193 ± 256	—	39·3 ± 6·8	1.677 ± 0.3	
Modern thrusting spears (n=8)	2170 ± 308	772 ± 386	26·8 ± 3·6	1.251 ± 0.227	Oakley et al. 1977
Modern throwing spears (n=28)	2161 ± 284	258 ± 110	14·6 ± 3·5	0.685 ± 0.181	Oakley et al. 1977
Modern digging sticks (n=16)	1189 ± 385	559 ± 387	28·4 ± 8·6	2.574 ± 1.055	Oakley et al. 1977

[a] 100*maximum diameter/length.

volume (modeling the spear as a cylinder) of about 0.0014 m^3 which, given a density of spruce of about 450 kg m^{-3}, would have a mass of about 640 g. Clearly, if these were designed to be projectile weapons, they were larger and heavier than any comparable projectile used by modern foragers.[7]

Replicative experiments based on the Schöningen spears reveal that they perform effectively as short- to middle-range (≤ 15 m) weapons (Rieder 2001, 2003). My own efforts to test replicas of some of these spears (the Lehringen spear and spears 2 and 3 from Schöningen)[8] have met with less success. Hundreds of trials by both inexperienced throwers and experienced javelin throwers (all of them given a one hour practice session with the spears before the trials began) produced only a 13.6% hit rate at 15 m (a throw was scored as a hit if the tip of the spear penetrated the target – a hapless bale of hay). More often than not, the spears tumbled in flight, ending up striking the target sideways or butt-first, or missing the target altogether. Admittedly this experiment probably says more about the ineptitude of college students with primitive technology than it does about the way the Schöningen spears were used: still, my impression from this experience was that these spears lacked the kind of flight stability required for accuracy when thrown from anything more than a few meters (cf. Guthrie 1983; Odell and Cowan 1986). Based on the size and shape of the Pleistocene spears, and based on my own experiences in trying to throw them, I believe that they most likely represent close-range weapons (5–10 m) which were primarily hand-held and thrust at prey from close quarters.

This is not to say the Schöningen spears weren't sometimes thrown. When it comes to hand-delivered spears, the distinction between thrusting and throwing spears is largely artificial – among recent foraging groups, hand-held spears are sometimes thrust at prey and sometimes thrown (Churchill 1993, 2002; Hitchcock and Bleed 1997; Kortlandt 2002). Human use of hunting technology is highly variable from situation to situation (and hence highly adaptable to circumstance), making it naive to talk about the function of any given subsistence tool in any sort of absolute terms (see, for

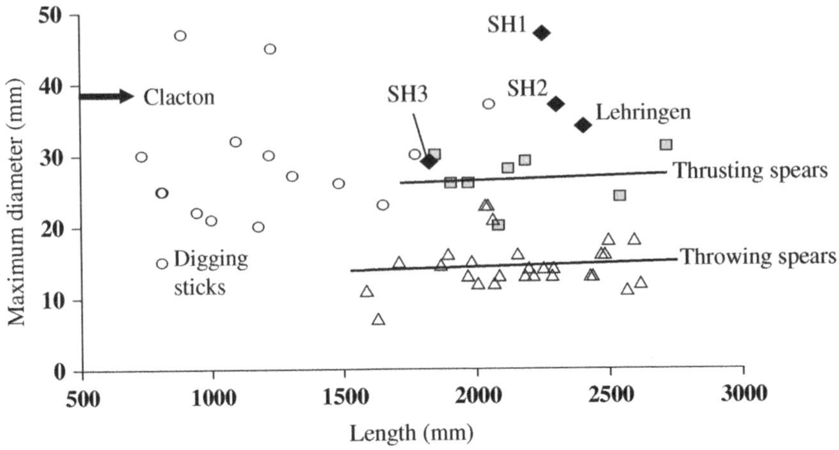

FIGURE 3.6

Relative thickness (maximum diameter on length) of fossil and recent spears and digging sticks. Fossil spears are indicated by ◆ or ➡. Data from ethnographically-known digging sticks (○), throwing spears (△) and thrusting spears (■), as well as for the fossil spears from Clacton and Lehringen, from Oakley et al. 1977. Data for Schöningen spears 1–3 from Thieme 1999. The length of the Clacton spear is unknown, but the preserved portion has a maximum diameter of 39 mm (Oakley et al. 1977). Ordinary least squares regression lines are provided for the recent throwing and thrusting spears.

example, Hill and Hawkes 1983; Bleed 1986; Greaves 1997). Still, whether thrust or thrown, hand-deployed spears *tend* to be used at close range. When effective weapon distances (the actual distances over which weapon systems are deployed in hunting, as opposed to the more oft-reported maximum range of the weapon) are considered it is apparent that even hand-thrown spears are close range weapons (Table 3.4). Note in Table 3.4 that, of the 14 groups in the ethnohistoric literature for which the effective distance of hand-thrown spears was reported, one is an extreme outlier. Tasmanians

TABLE 3.4 ■ Mean effective distances[a] of various weapon systems

Weapon system	Effective distance (m)[b]
Thrusting spear	Contact
Hand-thrown spear	7.8 (SE = 2.2, n = 14)
Hand-thrown spear, "corrected"[c]	5.7 (SE = 0.9, n = 13) [range = 2–10]
Spearthrower and dart	39.6 (SE = 5.5, n = 9)
Bow and arrow	25.8 (SE = 2.5, n = 25)

[a] Effective distance = distance from which weapon was deploy in hunting, as reported in the ethnographic record (as opposed to maximum range of the weapon). Data from Churchill 1993.

[b] Mean and standard error of the mean; n = number of groups from which data derives.

[c] See text for details.

are reported to have thrown very light spears very long distances (40 m) at very small prey (Roth 1890) – a use of hand-delivered spears which is fairly atypical. When the data are "corrected" by removing this outlier, the average effective distance of hand-thrown spears drops to 5.7 ± 0.9 m. Again, regardless of how they are actually deployed (thrown or thrust), hand-delivered spears are close-range weapons.[9]

As reviewed in Section 3.1, lithic points of various kinds (Mousterian points, Levallois points, and elongated Levallois points) are common components of Middle Paleolithic assemblages and, as noted in Section 3.3, there are several lines of evidence indicating that the Neandertals were sometimes hafting these points into handles to make composite tools. Middle Paleolithic stone points were undoubtedly multifunctional – likely variously used as weapon armatures, knives, scrapers and wood-working tools among other things (multifunctionality may have been characteristic of both the class of tool and of the use life of a given point: see for example the discussion of the multipurpose use of Australian "men's knives" in Shea 1997) – and this multifunctionality no doubt contributed to the wide range of variation in size seen even within narrowly defined point types (e.g., Levallois or Mousterian points: see Shea 1988, 1997; Shea et al. 2001).[10] Edge and tip damage, hafting trace evidence, and point tips embedded in the bones of prey indicate that at least some of these points were hafted and used in hunting (Shea 1988, 1993, 1998; Boëda et al. 1999; Shea et al. 2001; Rots 2009, 2013; Villa et al. 2009a; Pawlik and Thissen 2011). In their earliest use in this context, points were most likely hafted onto relatively large spears like those discussed above and thrust or thrown at prey from close range (Shea 1997), and thus represent improvements to an existing weapon system (hand-delivered, close-range sharpened wooden spears). This simple improvement increased effectiveness – by increasing organ damage and hemorrhaging (see Shea et al. 2001; Salem and Churchill 2014) – and most likely improved hunting return rates and, marginally at least, safety.

Based on the available archeological record, the hunting toolkit of Neandertals appears to have been relatively limited. Large, heavy and sometimes stone-tipped spears seem to be the mainstay of the arsenal, but we can reasonably assume that they also had other simple tools at their disposal. Given evidence for woodworking (Section 3.3), it is probably safe to assume that Neandertals also had at their disposal various simple wooden tools like bludgeons, digging sticks (which can be used in hunting to dig small animals out of burrows), and throwing sticks (such as the 0.78 m-long stick recovered from Schöningen: Thieme 1999). Stones no doubt were also used – perhaps thrown directly at game or perhaps flung to flush game or control the movement of animals during a hunt. Slings, which could be easily constructed from hide and sinew, may have also been used to launch stones with greater velocity (no evidence for the use of slings has been found, and given the perishable nature of the material we would not expect to ever find evidence of this weapon even if it was regularly employed). It should also be noted that bone points have been found in a limited number of Middle Paleolithic contexts (Montet-White 1996; Gaudzinski 1999; Villa and d'Errico 2001). Ethnohistoric and experimental research suggests that the choice of bone or antler for weapon armatures reflects a concern for point durability among recent foragers (Guthrie 1983; Knecht 1997). Where osseous points are tough, those of stone are brittle and thus break more readily than organic points (see references in Knecht 1997). However, the brittleness of flint and other stone allows for sharp edges that compound tissue damage and hemorrhaging, and thus the durability advantage provided by osseous points must outweigh the loss of wounding effectiveness. Among recent foragers, osseous points are generally associated with reliable weapon systems (Bleed 1986) employed in mass harvesting of game. In this regard, it is interesting that one of the few Middle Paleolithic sites to produce bone points, Salzgitter Lebenstedt, Germany (Gaudzinski 1999), appears to be an autumn reindeer hunting camp with the remains of at least 86 reindeer (Gaudzinski and Roebroeks 2000). Unfortunately, the period over which these remains accumulated is unknown, and thus it is not possible to determine if Salzgitter Lebenstedt represents a "one off" mass kill site or a palimpsest of multiple hunting seasons. Regardless, the rareness of bone points in Middle Paleolithic assemblages indicates that the conditions favoring the choice of osseous weapon armatures occurred infrequently, and discerning the contexts in which bone points were used presents an interesting challenge.

What appears to be lacking from the Mousterian toolbox, based largely but not entirely on negative evidence, was true, long-range projectile weaponry (Shea 2006; Churchill and Rhodes 2009; Rhodes and Churchill 2009) and specialized equipment for taking small prey (lagomorphs, larger rodents, birds, and fish) (Straus 1993; Holliday 1998). The issue of projectile weaponry in the Mousterian is an important one (and one that has stimulated a good amount of attention: Shea 1988, 1990, 2006; Holdaway 1989, 1990; Solecki 1992; Villa and Lenoir 2009; Villa 2010) because the ability to "kill at a distance" (or at least wound prey from a distance) has significant implications for the types of hunting techniques that can be employed and the behavioral characteristics of the prey that can be pursued (see below). The earliest direct evidence for true, long-range projectile weaponry comes in the form of partial spearthrowers from the European Upper Paleolithic (Breuil 1912; Cattelain 1989), contemporaneous with lithic weapon armatures whose morphology betrays a projectile function (Straus 1990; see also Churchill 1993). Comparative analyses of Middle Stone Age and Middle Paleolithic points with projectile armatures made by Holocene forager (from the ethnographic and recent archeological records) suggest that long-range projectile weapons may have first come into use in Africa or the Near East by late MIS 5, but that they did not enter Europe until Upper Paleolithic times, presumably in the hands of modern humans (Brooks et al. 2005; Shea 2006). The preponderance of evidence from spear and lithic point morphology suggests a lack of long-range projectiles in the Mousterian, a conclusion supported by consideration of Neandertal skeletal morphology (reviewed in Chapter 8).

The question of specialized equipment for small game is important because such technology, among modern foragers, represents a response to the high handling costs associated with small, fast, or otherwise difficult-to-access game. Despite being abundant and generally having high rates of productivity, this class of prey is usually of low rank in terms of forager prey selection, because the high energetic and time costs of catching and processing them make their net return rate low. Technological solutions to this problem reduce prey handling time, and thus improve the net return rate and make prey that was once eschewed consequently worth capturing. Not surprisingly, a florescence of subsistence technology focused on this type of prey – including nets, traps, fish leisters, hooks, and weirs (Straus 1993; Holliday 1998) – accompanies the broadening of human diets in the Upper Paleolithic (Stiner et al. 1999, 2000; Stiner 2001; Richards et al. 2001). Again based largely on the absence of evidence (but also on zooarcheological considerations), this aspect of subsistence technology appears to have been lacking in the Mousterian.

3.5 DOMESTIC TECHNOLOGY

As noted in Section 3.4, abundant use wear and residue evidence on Mousterian tools attests to some degree of hide working among the Neandertals (Beyries 1987b; Anderson-Gerfaud 1990; Loy and Hardy 1992; Lemorini 2000; Hardy et al. 2001; Soressi and Hays 2003; Hardy 2004; Pawlik and Thissen 2011), and frequently-observed heavy wear to the anterior dentition (Trinkaus 1983b) may signal that they also regularly engaged in hide-chewing or some other form of oral processing of skins. Wear traces on bone tools from Pech-de-l'Azé I, Abri Peyrony, and possibly La Quina (all in France) suggest that they were used as *lissoirs* (hide-smoothers) to make skins more supple and impermeable (Soressi et al. 2013). Edge damage on scrapers and other tools suggests that hide working predominantly involved fresh skins (see Kuhn and Stiner 2006), and probably entailed removal of fat and connective tissue from the inner surface (probably in conjunction with the use of preservative agents such as brain matter, urine, or possibly ground ochre) rather than removal of the hair from the outer surface (see White 2006), thus retaining more of the insulating value of the fur. While hides can be used for a variety of purposes, most of their uses (clothing, shelter walls, and bedding material) serve a thermal buffering purpose (explored more fully in Chapter 5.5), and thus we might expect hide working to have been more intensive and important during cold intervals (especially since the need for hides may have at times been great: see Speth 2012). Scraping may have also been an important way of maintaining the pliability of hides that had become wet and frozen (see Moran 1981). Along these lines, it is interesting that frequencies of scrapers tend to be greater in assemblages from cold intervals (Moncel et al. 2005; Monnier 2006). Monnier (2006) attributes this to reduced mobility related to ice and snow cover during cold intervals, resulting in reduced access to raw materials and thus greater reduction intensity (producing higher proportions of *racloirs*: see Section 3.2). Others note that the scraper-rich Quina variant of the Charentian tends to be broadly associated with faunal species, such as reindeer, with high quality hides (Chase 1986b; White 2006), perhaps belying a greater focus on hide working during cold episodes (White 2006; Gilligan 2007, 2010b; see also Mellars 1996). It is important to bear in mind that scrapers were certainly multifunctional tools, employed as often in woodworking as in processing animal resources (Rots 2013), and thus we should be cautious about making inferences about the activities performed at a site based on frequencies of scrapers or other tools. Also, if a high frequency of scrapers (i.e., a high *Racloir* Index) is a technological by-product of cold climate reductions in access to raw materials (Monnier 2006), it would not be surprising to find scraper-rich assemblages associated with cold climate faunas (with abundant reindeer) in the absence of a functional correlation between the two. Also, if the *Racloir* Index reflects hide working intensity, we might expect to find hide finishing tools (such as *lissoirs*) more commonly in assemblages with high *Racloir* Indices. Interestingly, the recently-identified possible *lissoirs* from south-west France (Soressi et al. 2013) derive from Mousterian of Acheulean Tradition assemblages, which tend to be relatively scraper-poor (Mellars 1996).

Another aspect of the domestic technology of Neandertals was fire. Fire provides a host of well-known benefits, from warmth and light to its uses in cooking, defense from predators, and as a material-transformative agent in the production and maintenance of technology. In this latter role, fire can be used to heat treat lithic nodules to improve the flaking qualities of the flint (Domanski and Webb 1992), to make wood easier to shape into artifacts (Cosner 1956), and to cook pitches or bitumen to produce mastics for hafting (Binford 1984a). Among some foragers, fire may also be used to manage ecosystems, that is, to enhance the primary productivity of vegetation patches by periodically burning them off (Deacon 1993). There is no evidence to suggest that Neandertals (or early modern humans, for that matter) were burning the landscape to tweak productivity (Daniau et al. 2010), and fire in the Middle Paleolithic appears to have been restricted to domestic use. Hearths and associated indicators of domestic fires show that fire was regularly used by Eurasian hominins from about 400–300 Ka BP onwards, and that "Middle Paleolithic Neandertals…had the ability to make, conserve, and transport fires during successive occupations or at different sites" (Roebroeks and Villa 2011: 5211). This conclusion, however, is not universally accepted, as some workers see the record as reflecting irregular use of fire, perhaps suggesting that Neandertals could not make it upon demand (and had to wait for access to naturally-occurring sources: Sandgathe et al. 2011). In addition to wood (Castro-Curel and Carbonell 1995), Neandertals also occasionally used bone (Callow et al. 1986) or lignite (a soft form of coal: Théry-Parisot and Meignen 2000) as fuel sources. The accumulating evidence for hafting (Section 3.3) indicates that Neandertals comprehended the utility of fire as an "engineering tool" (Roebroeks and Villa 2011), as the production of pitch resins requires some fairly sophisticated pyrotechnology (Koller et al. 2001; Pawlik and Thissen 2011). Their pyrotechnical know-how, however, appears to only marginally extend to hearth structures themselves. Stone-lined or stone delimited hearths have been found at a handful of Middle Paleolithic sites (Roebroeks and Villa 2011), but by-and-large Mousterian fireplaces tend to be informal in construction (consisting of fires made on flat surfaces or shallow pits: Vaquero et al. 2001b; Vallverdú et al. 2012). In some cases, these shallow pits (*cuvettes*) were enhanced with an excavated tail area that presumably enhanced aeration of the fire (Perlès 1976).

Of course, one of the important benefits of fire is the ability to cook foodstuffs, rendering them more palatable and greatly enhancing the availability of their nutrients and calories (Wrangham et al. 1999). Burnt bones are ubiquitous at Mousterian and Micoquian sites (see Stiner et al. 1995), and no doubt much of this is attributable to cooking of meat. The pseudomorph of what appears to have been a tripod erected over a hearth at l'Abric Romaní (see Section 3.3) is also suggestive of fire curing of meat, perhaps by smoking. It is also likely that Neandertals were using fire as an aid to rendering grease from cancellous bone in the vertebrae and long bone epiphyses of prey animals (Costamagno 2013). While the evidence (reviewed in Speth 2012) for stone boiling as a means of rendering fat is scarce to non-existent in the Middle Paleolithic, it is quite possible that Neandertals were using direct-heating of skin or wooden containers to boil liquids and render nutritionally important fats (Speth 2012). Straches from grass seeds extracted from the dental calculus of the Shanidar 3 Neandertal (from Iraq) showed modifications characteristic of cooking, and appear to provide some direct evidence that Neandertals were, at least at times, cooking vegetal foodstuffs (Henry et al. 2011).

Although the archeological residues left behind by Neandertals are not easy to interpret, as currently understood they depict a material culture that (1) was technologically flexible and capable of responding – via adjustments in raw material acquisition strategies, technological modes of production, and typological variation in tool forms – to changing environmental conditions and mobility demands, (2) manifested the rudiments of real technological complexity, in the elaboration of *debitage* systems, in the development of conjunctive technology (composite tools) with unlimited potential for novel tool forms, and in the elaboration of the use of wood and bone as raw materials, (3) provided, at least minimally, solutions to the greater extractive foraging challenges posed by cold temperate and glacial environments, especially with respect to technologically-aided hunting, (4) buffered the Neandertals from the cold (more on this in Chapter 5), and (5) enhanced their ability to assimilate calories and nutrients from their foods.

NOTES

1. The phrase "Middle Paleolithic" denotes a cultural period in a typological classification system; that is, a system based on the presence and frequency of certain tool types, which in turn are defined by their formal attributes. Typological attributes of the Middle Paleolithic include an emphasis on flake (rather than core) tools, an abundance of points and scrapers, and the possible presence of triangular bifaces (Dibble and Rolland 1992). "Mode 3," on the other hand, refers to a technological classification system (Clark 1968) that emphasizes aspects of tool manufacture rather than formal attributes. Mode 3 assemblages are characterized by an emphasis on prepared core methods of reduction, and have variably elaborated debitage systems (some flakes are retouched into various formal tool types, others are used without modification). The two classification systems are complementary, and the two phrases are used interchangeably here.
2. The names and defining characteristics of the facies presented here follows Bordes (1961a). In earlier work (Bordes 1953) Bordes presented a slightly more complicated scheme involving three main assemblage groups, each comprised of two subgroups (see Dibble and Rolland 1992).
3. For details about the tool forms recognized by Bordes, which serves as the basis for typological classification of Mousterian assemblages, see Bordes (1961b).
4. Generally defined as a flake that is more than twice as long as it is wide, but technically defined as *débitage* with parallel or slightly convergent edges from elongated blanks (Bar-Yosef and Kuhn 1999).
5. It is also clear from the regular occurrence of blades in Middle Paleolithic assemblages (Bar-Yosef and Kuhn 1999) and from the occasional occurrence of microlithic assemblages (such as the MIS 5 Taubacian from central Europe: Moncel 2001) that the technological basis of the Upper Paleolithic (that is, Mode 4 [laminar] and Mode 5 [microlithic] technology) was not beyond the reach of Neandertals.
6. The Lehringen spear, found as it was in association with the remains of an elephant, has often been cited as *prima facie* evidence of Neandertal predation on large game. While I have no objection with the conclusion that Neandertals hunted large game, a more likely explanation for the association of the Lehringen spear and the bones of an elephant is that the carcass was found frozen (perhaps by Neandertals who were out looking for frozen carcasses, and possibly using their spears as snow probes) and the spear was broken in

the act of trying to leverage part of the carcass or prize off frozen chunks of flesh (Gamble 1987). The alternative claim that the spear was broken during the killing of the elephant is not consistent with evidence that the carcass was minimally processed (the skeleton was in articulation and had only a few cut-marks around one shoulder).

7. An argument could be made that the larger average size of Pleistocene prey animals necessitated larger, heavier spears (to maximize momentum and thus improve penetration ability of a thrown projectile), although it should be noted that this is not how modern human foragers deal with large prey (see Churchill 1993). More work needs to be done on these Middle Pleistocene spears to better define their capabilities and probable function – at present we can only note that, if indeed they were used as javelins, they were larger and heavier than anything modern humans are known to have used in subsistence hunting – whether thrown or thrust.
8. The replicas were created from saplings using published data on length, tip length, maximum diameter and the position of maximum diameter from the distal end, midpoint diameter, and diameters at 1/3 and 2/3s of length. No effort was made to match the wood used in the construction of the fossil spears, and thus the replicas may differ in important ways in density and mass from the originals. Naturally the replicas have their own idiosyncratic kinks and bends, but do match the originals in terms of the dimensions listed above.
9. As noted by Villa (2010; Villa and Lenoir 2009), Roman soldiers are known to have thrown heavy spears 15 m or more in warfare. While I admit that this lends credence to the notion of heavy spears used in hunting, I also wish to note that the economics of warfare and subsistence hunting are quite different, and the nature of target acquisition, target size and weapon accuracy may vary in important ways (see, for example, Guthrie 1983; and discussion of shock vs. surgical weapons in Churchill 1993). In general, I agree that there is much to learn from military examples of weapon use (see Kortlandt 2002 vs. Churchill 2002), but I would also argue that the specifics of direct comparisons need to be carefully considered before conclusions are reached.
10. This multifunctional aspect of lithic tools makes it difficult to make inferences about hunting intensity based on the presence or abundance of pointed lithics (see Kuhn 1989; Shea 1998 and comments therein; see also Rolland and Dibble 1990).

CHAPTER FOUR

The Body Neandertal

Body size is the single most important determinant of an organism's biology, impacting everything from physiological processes (McNab 1983, 1990; Schmidt-Nielsen 1984) to ecological relationships (Eisenberg 1990; Calder 1996). Physiologically, body size and shape are key determinants of the absolute and relative metabolic rates of organisms, and thus dictate caloric needs and dietary quality. Size governs energy through-flow and storage (and thus productivity) in heterotrophs, and accordingly size co-varies with important life history variables, such as growth rate, fecundity, and life span. Size is also central to thermoregulation, with larger being better when it comes to maintaining thermal constancy in cold environments (McNab 1974, 1983) – which forms the physical basis for Bergmann's (1847) well-known rule (see below). Ecologically, body size is related to trophic strategies, substrate choice, and day- and home-range sizes (Eisenberg 1990). Thanks to the laws of physics and the realities of scaling, size also dictates the types of food items that can be accessed, ingested, and assimilated by heterotrophs, and is thus one of the major determinants of niche. Given a diet that

Thin on the Ground: Neandertal Biology, Archeology, and Ecology, First Edition. Steven Emilio Churchill.

was rich in animal tissues (Chapter 7), Neandertal body size would have been central to predatory success, as well as to the outcome of competitive interactions with carnivore competitors (that is, their body size determined in part their social dominance rank in the Eurasian large-bodied carnivore guild, and their attendant susceptibility to the depredations of other carnivores: Chapter 9). Neandertal size would have also factored into the economics of prey size choice, social hunting, and carcass transport.

The focus of this chapter is on the energetic consequences of Neandertal body size (in terms of the caloric demands of basal metabolism), on the nutritional requirements of large-brained, large-bodied primates living in cold-temperate conditions, and on the ecological context of Neandertal body size. In subsequent chapters, we will explore Neandertal body size and shape as it relates to heat loss in a hypothermic environment (Chapter 5), locomotor kinematics and bioenergetics (Chapter 10), and predatory behavior (Chapter 8).

4.1 NEANDERTAL BODY SIZE: SHORT BUT MASSIVE

How big were the Neandertals? There exist multiple methods for both mass and stature estimation in fossil populations, and little consensus on the most appropriate methods to be applied to Neandertals (see, for example, Kappelman 1996; Ruff et al. 1997, 2005; Ruff 2000b; Pearson et al. 2001 on mass; Feldesman et al. 1990; Feldesman and Fountain 1996; Holliday and Ruff 1997 on stature). Estimates of stature from long bone lengths are highly dependent upon the choice of reference sample, which should match the fossil sample under consideration as closely as possible in body proportions (Holliday and Ruff 1997). Since Neandertals had body proportions similar to those of modern, cold-adapted peoples (Holliday and Trinkaus 1991), stature estimation techniques based on cold or temperate climate peoples are most appropriate. The regressions of Trotter and Gleser (1952) for European-Americans provide the largest reference samples for temperate adapted samples. However, Feldesman et al. (1990) note that femur/stature ratios based on geographically heterogeneous samples do a better job of predicting stature in people of European descent than do Trotter and Gleser's regression equations (see also Feldesman and Fountain 1996; but see Holliday and Ruff 1997). Therefore the stature estimates presented here are, in most cases, averages of estimates derived from femoral bicondylar length regressions derived from European Americans (Trotter and Gleser 1952) and estimates based on the femur/stature ratio derived from a global sample of modern humans (Feldesman et al. 1990). For males, three stature estimates were averaged (estimates from the femur/stature ratio, and from the regression equations for military personnel and Terry Collection males developed by Trotter and Gleser), while for females two estimates were averaged (femur/stature ratio and Terry Collection female regression equation). In some cases, stature was estimated from long bones other than the femur, using the appropriate equation in Trotter and Gleser (1952: as reported by Ruff et al. 1997, supplementary data). The resultant stature estimates (Table 4.1) produce a mean Neandertal male stature of 166.1 ± 4.9 cm, and a mean female stature of 156.6 ± 5.6 cm.

A number of methods also exist for estimating body mass from skeletal remains. Methods based on postcranial elements, whether "mechanical" (e.g, involving regressions based on articular dimensions) or "morphometric" (e.g., geometric modeling

TABLE 4.1 ■ Estimated body mass and stature in Neandertals

	Mass[a] (kg)	Stature[b] (cm)	Mass/Stature
Males			
Amud 1	75.3	179	0.42
La Chapelle-aux-Saints 1	77.3[d]	163	0.47
La Ferrassie 1	85.0	172	0.49
Fond-de-Foret 1	83.9	162	0.52
Kebara 2	75.6[d]	166[e]	0.46
Kiik-Koba 1	78.1	164[e]	0.48
Krapina 213	80.6	–	–
Lezetxiki 1[c]	73.9	166[e]	0.45
Neandertal 1	78.9	167	0.47
Regourdou 1[c]	72.1	164[e]	0.44
Saint-Cesaire 1	78.9	165[e]	0.48
Shanidar 1	80.5	172	0.47
Shanidar 2	75.2	160[e]	0.47
Shanidar 3	79.9	166[e]	0.48
Shanidar 4	72.0	161	0.45
Shanidar 5	68.5	169	0.41
Spy 2	83.6	162	0.52
Male Mean	77.9	166.1	0.47
(SD, n)	(4.7, 17)	(4.9, 16)	(0.03, 16)
Females			
Grotte du Prince	74.8	–	–
La Ferrassie 2	67.0	155	0.43
Krapina 208	68.4	–	–
Krapina 209	63.7	–	–
Krapina 214	62.2	–	–
Palomas 92[f]	62.6[g]	152	0.41
Palomas 96[f]	61.7[h]	154[i]	0.40
La Quina 5	71.2	163[e]	0.44
La Quina 18	–	164[e]	–
Shanidar 6	59.4	148	0.40
Spy 1	67.5	161[e]	0.42
Tabun C1	63.2	156[e]	0.41
Female Mean	65.6	156.6	0.42
(SD, n)	(4.6, 11)	(5.6, 8)	(0.01, 7)

[a] All mass estimates, unless otherwise noted, are from Ruff et al. (1997: supplemental data).

[b] Stature estimates, unless otherwise noted, are the average of regression predictions from femoral length in European-Americans (Trotter and Gleser 1952) and estimates based on femur/stature ratios (Feldesman et al. 1990).

[c] Sex indeterminate, considered here as male based on relatively high mass estimate.

[d] Mass estimate from Ruff et al. 2005.

[e] Stature estimate from Ruff et al. (1997: supplemental data).

[f] Sex indeterminate, considered here as female based on relatively low mass estimate.

[g] Mass estimate calculated following Ruff et al. (1997) using femoral head diameter from Walker et al. (2011a).

[h] Mass estimate from Walker et al. (2011b).

[i] Stature estimated using European American regression (Trotter and Gleser 1952) from femoral bicondylar length reported in Walker et al. (2011b).

using stature and pelvic breadth), tend to give results that are broadly consistent with one another (Auerbach and Ruff 2004). Methods based on cranial dimensions (such as orbital dimensions), on the other hand, generally produce higher estimates (see Kappelman 1996). Table 4.1 provides mass estimates for individual Neandertals based on modern human-derived regressions of femoral head diameter on mass, or from modeling the body as a cylinder of height equal to estimated stature and diameter equal to bi-iliac breadth (necessitating the reasonable assumption that the human body has an average mass density of 1 gm cm^{-3}), or averaging the two (Ruff et al., 1997: supplemental data). This combination of postcranial-based approaches produces an average mass of 77.9 ± 4.7 kg for male Neandertals, and 65.6 ± 4.6 kg for females (Table 4.1). Mass estimation based on orbital area (Kappelman, 1996), in contrast, indicates a mean mass of 85.5 ± 12.7 kg for males and a mass of 93.4 kg for the only female Neandertal (Forbes Quarry 1) used in the sample (but if the somewhat earlier, and presumably female, specimen from Steinheim is also used, the average drops to 77 kg) – mean estimates some 10–18% higher than obtained from postcranial-based methods. This method is based on a regression of orbital area on mean body mass in a mixed-species sample of hominoids, and thus Neandertal deviation from other apes and humans in brow ridge and orbital morphology (Pearce et al. 2013) may introduce bias into the estimation; for this reason, only the postcranial-based estimates are used here.

It is clear from these estimates that Neandertals were relatively massive humans. Mean mass measured in a large sample of males of mixed ancestry (n = 1,121; ca. 23% African American, 63% European American and 14% Asian American/Hispanic) by Armstrong et al. (1990) was 76.8 kg ± 11.3 kg, very close to the Neandertal male mean. However, Neandertal males, if similar to modern arctic people, probably averaged about 13% body fat (Shephard et al. 1973), whereas the sample in the study by Armstrong and colleagues averaged 20.3 ± 6.7% body fat. The recent humans were also nine centimeters taller on average (175.1 ± 6.9 cm) than the Neandertals. By way of comparison, males from Inuit[1] and other arctic populations of Asiatic descent tend to be similar in mean stature to Neandertal males (160–179 cm), yet about 10–20 kg lighter on average (Table 4.2; Figure 4.1). It should also be noted that the inclusion of two specimens of indeterminate sex – Lezetxiki 1 and Regourdou 1 – into the male sample may cause an underestimation of average male body mass. Elsewhere (Churchill, 1994a), I considered Lezetxiki 1 to be female based on the small size and relatively slight rugosity of the muscle scars on this isolated humerus, and it is quite possible that it represents a large female. Excluding these two specimens produces an average Neandertal male mass of 78.6 ± 4.6 kg (n = 15).

The study by Armstrong et al. (1990) also reported an average mass for 305 females (again using a mixed-ancestry sample: ca. 43% African American, 57% European American) of 59.9 ± 8.0 kg, almost 9% less massive than the estimates for the female Neandertal sample (Table 4.1). As before, the percentage body fat was fairly high in this sample (mean 27.3 ± 5.6%: compared to an average of about 22% in Inuit females [Shephard et al. 1973], and presumably Neandertal females), and the average stature (162.4 ± 6.4 cm) was a good five centimeters taller than that of the Neandertals. Compared to arctic groups, Neandertal females were of similar mass and stature to females in some Inuit groups (namely the sample from Wainwright: Table 4.2), but were generally taller and heavier than the females of most arctic populations (of Asiatic descent)

TABLE 4.2 ■ Mean (SD, n) stature and mass in Neandertals and circumpolar modern human groups

	Stature (cm)	Mass (kg)	Mass/St[a]	Reference
Males				
Neandertals	**166.1 ± 4.9 (16)**	**77.9 ± 4.7 (17)**	**0.47**	
Wainwright Inuit	166.3	67.2	0.40	Eveleth and Tanner 1976
Igloolik Inuit	165.8	69.3	0.42	Eveleth and Tanner 1976
Inupiat	167.4 ± 0.8 (27)	68.6 ± 1.5 (27)	0.41	Ruff et al. 2005
Aleut	164.2	67.7	0.41	Laughlin 1951
Yakut	170.1 ± 6.1 (50)	66.6 ± 12.9 (50)	0.39	Snodgrass et al. 2005
Evenki herders	160 ± 5 (26)	59.9 ± 7.6 (26)	0.37	Leonard et al. 1996
Evenki herders	159.9 ± 7.2 (19)	55.7 ± 4.0 (19)	0.35	Galloway et al. 2000
Finn	182.1 ± 1.2 (27)	82.9 ± 2.3 (27)	0.46	Ruff et al. 2005
Females				
Neandertals	**156.6 ± 5.6 (8)**	**65.6 ± 4.6 (11)**	**0.42**	
Wainwright Inuit	155.8	66.3	0.43	Eveleth and Tanner 1976
Igloolik Inuit	157.5	57.8	0.37	Eveleth and Tanner 1976
Inupiat	155.9 ± 0.7 (40)	59.6 ± 1.5 (40)	0.38	Ruff et al. 2005
Aleut	152.8	53.4[b]	0.35	Laughlin 1951
Yakut	156.7 ± 5.5 (75)	60.4 ± 13.8 (75)	0.39	Snodgrass et al. 2005
Evenki herders	151 ± 4 (10)	52.2 ± 4.0 (10)	0.35	Leonard et al. 1996
Evenki herders	148.9 ± 4.9 (39)	50.2 ± 6.3 (39)	0.34	Galloway et al. 2000
Finn	164.4 ± 1.1 (27)	58.8 ± 1.6 (27)	0.36	Ruff et al. 2005

[a] Mass/stature ratio as determined from mean mass and stature values.

[b] The original publication erroneously listed a value of 80.5 kg rather than the correct value presented here (Laughlin, personal communication to C. Ruff).

(Figure 4.1). Again, if either of the specimens from Lezetxiki or Regourdou belong with the female sample, the average mass of the Neandertal females would be higher still.

In terms of stature, Neandertals had levels of sexual dimorphism comparable to, and in fact slightly lower than, that seen in the comparative modern human arctic samples in Table 4.2 (female Neandertals average 94% of the stature of males, compared to a range of 90–94% in the modern groups). Dimorphism in mass is more variable in the modern sample (average female mass ranging from 71–99% that of the males: Table 4.2), likely reflecting sex differences in muscularity and fatness between groups. Neandertal females appear to have been about 84% the mass of their male counterparts, on average. The Neandertal sample encompasses a considerable temporal and geographic range, and thus ecogeographic patterns in body size (see below), combined with unequal representation of the sexes across temporal and geographical categories, might be producing a biased picture of sexual dimorphism in the Neandertals. Confining the consideration to European Neandertals from cold–temperate conditions (the comparison that produces the greatest sample sizes: Table 4.3) results in the unexpected finding of virtually no dimorphism in stature (females 97% of male stature: Table 4.4), but a 12.5 kilogram difference in average mass (females 84% the mass of males). Despite being the best represented period in the Neandertal fossil record, the

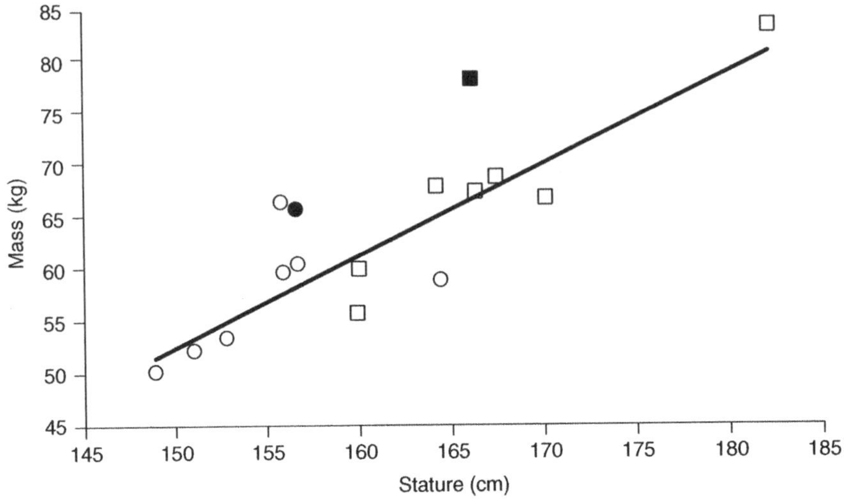

FIGURE 4.1

Mean body mass on stature in Neandertals and circumpolar modern humans. Data from Table 4.2. Ordinary least squares regression line ($y = 0.8716x - 78.316$; $r = 0.8964$) based on modern human sample means only. ○ = modern females; □ = modern males; ● = Neandertal females; ■ = Neandertal males.

cold–temperate sample is still too small to rule out sampling effects: thus we have no reason to reject, based on the current sample, the claim that levels of sexual dimorphism in Neandertals were no different than those of modern humans (Trinkaus 1980: see also Arsuaga et al. 1997b).

We might expect geographic or temporal variation in Neandertal size if body mass is indeed related to ecological and climatic circumstances (as larger bodies are generally adaptive in colder climates: see Section 4.4). Unfortunately, most of the specimens for whom body mass can be reasonably estimated derive from the cold–temperate conditions of MIS 5d-a or 3, making comparison across climatic conditions difficult (Table 4.3). Furthermore, dividing this small sample up by geographic region and climatic conditions makes for sample sizes that are too small to allow confident inferences about ecogeographic patterning. Even so, such an exercise produces some intriguing, if non-conclusive, results (Table 4.4, Figure 4.2). Among the males, the most massive individuals from Europe and the Near East (La Ferrassie 1 and Kebara 2, respectively) are both from glacial intervals. Beyond that, expectations break down: the male samples actually show a reverse of expectation, wherein glacial period European Neandertals are on average less massive (77.0 ± 7.0 kg, n = 3) and less stocky than are those from the cold–temperate periods (80.1 ± 2.9 kg, n = 6), who in turn have an average mass slightly below that of the single interglacial period male (Krapina 213, at an estimated 80.6 kg). Among the females, European Neandertals from glacial and cold–temperate episodes tend to be more massive than individuals from the interglacial MIS

TABLE 4.3 ■ Neandertal individuals with estimated body masses by geographic region and climatic conditions[a]

	Europe	Near East
	Males	
Glacial (MIS 6 & 4)	La Ferrassie 1[b]	Kebara 2
	Lezetxiki 1[c]	
	Regourdou 1[d]	
Cold temperate (MIS 5d-a & 3)	La Chapelle-aux-Saints 1	Amud 1
	Fond-de-Foret 1	Shanidar 1
	Kiik-Koba 1	Shanidar 2
	Neandertal 1	Shanidar 3
	Saint Cesaire 1	Shanidar 4
	Spy 2	Shanidar 5
Interglacial (MIS 5e)	Krapina 213	–
	Females	
Glacial (MIS 6 & 4)	La Ferrassie 2[b]	Tabun C1
Cold temperate (MIS 5d-a & 3)	Grotte du Prince	Shanidar 6
	Palomas 92	
	Palomas 96	
	La Quina 5	
	Spy 1	
Interglacial (MIS 5e)	Krapina 208	–
	Krapina 209	
	Krapina 214	

[a] Using dates from Ruff et al. (1997: supplemental data) and Holliday (1997b).
[b] Stratigraphic comparison with dated levels at Le Moustier (Mellars 1986) places the specimens from La Ferrassie at the MIS 5a/4 boundary. Even if these individuals lived during the final millennia of stage 5a, climatic conditions were likely more like those of the succeeding glacial episode than the conditions at the peak of stage 5a.
[c] Following Baldeon (1993).
[d] Following Gómez-Olivencia et al. (2012).

5e (represented entirely by specimens from Krapina) (Figure 4.2), but with the cold–temperate females being on average only 2.8 kg heavier than their interglacial counterparts. The single female from a glacial episode (La Ferrassie 2: see footnote b in Table 4.3) has a mass that falls between the mean values of the cold–temperate and interglacial samples. Here the Near Eastern samples are just too small to say anything at all (Table 4.3).

The limited mass data also suggest that there is a geographic effect in play. Among males, the Europeans are more massive than the individuals from the Near East for each climatic condition for which a comparison can be made (Figure 4.2). Most interesting (because it's the only comparison for which sample sizes are large enough to provide even a modicum of confidence) is the cold–temperate interval Europeans, who average almost five kilograms heavier than their Near Eastern contemporaries. Although

TABLE 4.4 ■ Mean stature and mass/stature ratios (M/S) in male and female Neandertals, by geographic region and climatic episode (mean, SD, n)

	Males				Females			
	Europe		Near East		Europe		Near East	
	Stature	M/S	Stature	M/S	Stature	M/S	Stature	M/S
Glacial	167.3	0.46	166	0.46	155	0.43	156	0.41
	4.2	0.03	–	–	–	–	–	–
	3	3	1	1	1	1	1	1
Cold–temperate	163.8	0.49	167.8	0.45	158.8	0.42	148	0.40
	1.9	0.02	7.1	0.03	5.4	0.02	–	–
	6	6	6	6	5	4	1	1

Sample compositions as in Table 4.3, with the addition of La Quina 18 (for whom stature but not mass can be estimated) to the female European cold–temperate group.

the sample sizes are very small, a similar pattern seems to obtain among the females (Figure 4.2). The slight tendency for European Neandertals to be heavier for a given stature is also apparent in Figure 4.3, as is their generally greater mass/stature ratio relative to Near Eastern Neandertals (Table 4.4). Again, without making too much of small samples, the data at hand do suggest ecogeographic patterning in body size and shape, presumably reflecting colder climates in Europe.

Thanks in large part to discoveries made at the Middle Pleistocene site of Sima de los Huesos (Arsuaga et al. 1999a; Bischoff et al. 2007), the body size of the ancestors of the Neandertals is becoming increasingly known. From some admittedly limited data, it appears that Neandertal body size may represent a mass reduction relative to their

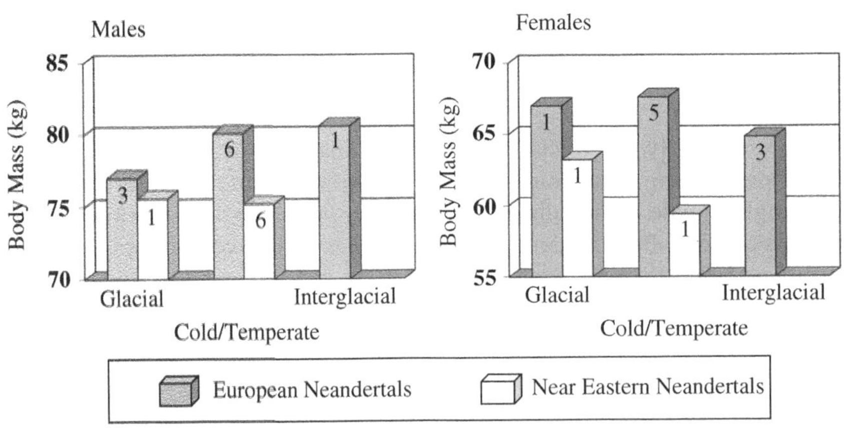

FIGURE 4.2

Mean body mass (kg) in male and female Neandertals by geographic region and climatic conditions. Sample sizes are given inside each bar.

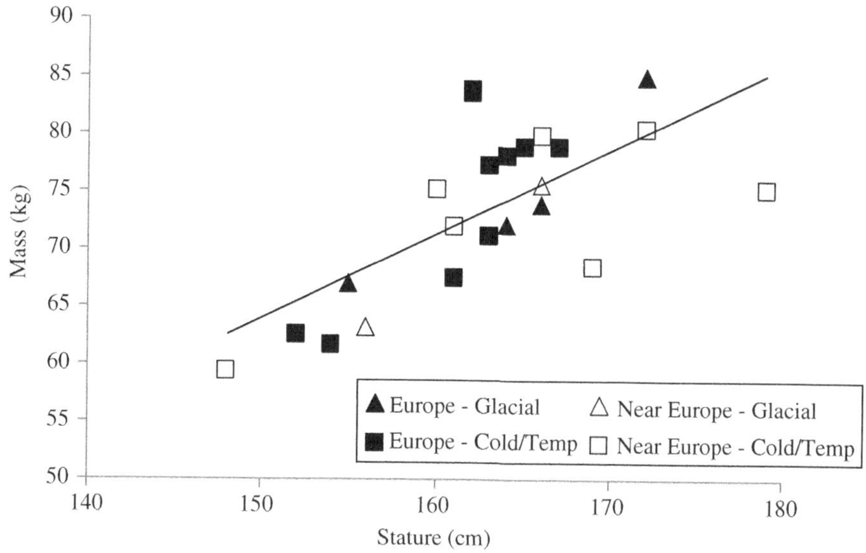

FIGURE 4.3

Mass (kg) on stature (cm) for 23 Neandertals (sexes mixed: data from Table 4.1). European specimens are represented by solid symbols, Near Eastern specimens by open symbols; triangles represent glacial condition specimens, squares represent Neandertals from cold–temperate conditions. Ordinary least squares regression line ($y = 0.7334x - 45.95$; $r = 0.6863$) based on the total sample. Samples composed as in Table 4.3.

Middle Pleistocene ancestors. A nearly complete male pelvis with associated partial femora from Atapuerca produces a mass estimate between 93.1 and 95.4 kg, making this the largest known individual in the human fossil record (Arsuaga et al. 1999a). At an estimated stature of 174.5 cm (Ruff et al. 2005), the Atapuerca hominin is also perhaps the stockiest human in the fossil record (mass/stature ratio between 0.53 and 0.55). The inference of large body size in this population is supported by estimates of mass from the presumably male, 500 Ka BP tibial shaft from Boxgrove and the 400 Ka BP male partial os coxa from Arago, estimated at 86.7 and 79.2 kg, respectively (Ruff et al. 1997: supplemental data). These three males produce a mean mass estimate of 86.7 ± 7.6 kg, well above the mean value for European Neandertal males.

4.2 BODY COMPOSITION: "SCALED UP" INUIT?

It's clear from the mass/stature ratios in Table 4.2 that Neandertals, like modern humans indigenous to the arctic, were massive relative to their stature. Short but with very wide trunks (Ruff 1991, 1993, 1994), Neandertals packed more mass into each centimeter of stature than do most Inuit, leading Holliday (1997b) to refer to their body form as "hyperpolar." But what, if anything, can we infer about the nature of that mass? Were Neandertals like "scaled up" Inuit, with similar body compositions?

Or did Neandertals have higher proportions of insulating, energetically inexpensive body fat?[2] Perhaps they packed on more muscle, which also has insulating as well as heat-generating advantages, but which is more metabolically costly. Unfortunately, we have no direct evidence of Neandertal soft tissue anatomy, and must make what inferences we can from their skeletal morphology, and from reasonable consideration of the physiology of cold-adapted modern humans.

While the insulative properties of fat are well appreciated (see references in Steegmann et al. 2002), arctic-adapted modern humans appear to take minimal advantage of this potential adaptive solution to cold exposure. Inuit living traditional lifestyles (that is, not acculturated to a high fat, high calorie diet and sedentary way of life) tend to have relatively small deposits of subcutaneous fat (Shephard et al. 1973 and references therein), which has been explained by some as a function of their high activity levels, high energetic demands, and consequent difficulty in acquiring surplus calories sufficient to accumulate body fat (Rode and Shephard 1971). However, when methods are used that assess body composition without relying on skinfold thickness (which measures subcutaneous fat), Inuit are found to have greater than expected stores of visceral fat (Shephard et al. 1973).[3] One such method (Deuterium oxide dilution) revealed average body fat of 11.8–15.3% in unacculturated Inuit males aged 15–39, and 19.4–26.0% in comparably aged females (grand means = 13.4% males, 21.7% females: Shephard et al. 1973). Slightly higher mean values (12.9–19.6%) were reported in "moderately acculturated" males, based on direct hydrostatic estimates of body composition (Rode and Shephard 1994). Body composition no doubt varied seasonally in Neandertals as it does in modern foragers (e.g., Wilmsen 1978; Hurtado and Hill 1990), but as a working supposition we can assume that adult Neandertals in good condition had body fat stores similar to those reported for Inuit: ca. 13.5% for males and ca. 22% for females.

The size of muscle attachment sites in the Neandertal postcranial skeleton indicates a greater degree of muscularity in them than seen among most modern human groups (Churchill 1994a; Churchill and Rhodes 2006). This suggests in turn that Neandertals may have devoted a greater proportion of their body mass to muscle tissue than do modern humans. Deriving a reasonable idea of muscularity is important to understanding Neandertal biology for three reasons. First, since muscle is a metabolically active tissue (even at rest), muscularity has implications for our estimates of metabolic rates and energetics in Neandertals. Second, muscle has important insulating properties, and thus muscularity is an important variable in cold adaptation. Third, the amount and distribution of muscle tissue is fundamental to the behavioral ecology of mammals (Grand 1997), and Neandertal muscularity was no doubt central to their predatory tactics, mobility patterns, and other aspects of their adaptation to the cold–temperate and glacial environments of Eurasia (see below).

Evidence of heightened muscularity can be found throughout the Neandertal postcranial skeleton. Their vertebra exhibit robust spinous and transverse processes, attesting to hypertrophy of the erector spinae muscles, splenius muscles, and other dorsal and nuchal muscles (Trinkaus 1983a). Likewise, large and rugose muscle markings on their ribs attest to large and strong scapulothoracic musculature (Franciscus and Churchill 2002). Hypertrophy of upper limb muscles is evident in the size and proportions of the scapula (reflecting enlarged rotator cuff muscles), and in the size and rugosity of the insertion scars of the scapulothoracic, brachial, and antebrachial muscles (*Mm. pectoralis major, latissimus dorsi, teres major, biceps brachii*, and *pronator quadratus*) as well as the extrinsic and intrinsic muscles of the wrist and hand (Stoner and Trinkaus 1981; Trinkaus 1983a, 1983b, 1989; Senut 1985; Vandermeersch 1991; Villemeur 1992; Hambücken 1993b; Churchill 1994a; Vandermeersch and Trinkaus 1995). In the lower limb, muscular hypertrophy is indicated by enlargement of the area of origin of the femoral adductors, enlarged femoral gluteal tuberosities, wide patellae (suggesting a large *M. quadriceps femoris* tendon), large navicular tuberosities (for the insertion of *M. tibialis posterior*), and fifth metatarsals with large tuberosities (for the insertion of *M. peroneus brevis*), and rugose muscle attachment areas at midshaft (for *M. peroneus tertius*) (Trinkaus 1982, 1986, 2000; Trinkaus and Hilton 1996; Black 1999).

So it seems clear that Neandertals were more muscular than are most living humans today, but just how much more muscular? We should be able to arrive at some idea by comparing the size of muscle scars across groups, since the size of muscle tendons (at least for some muscles) are highly correlated with the physiological cross-sectional area of the muscle ($r = 0.91$: An et al. 1991; but see Zumwalt 2006). Unfortunately, there have been few systematic attempts to quantify the size of muscle scars in fossil and recent humans. Consideration of two upper limb muscle attachment sites (humeral epicondylar breadth standardized to humeral length as a measure of the size of the humeral head of *M. pronator teres* and the common flexor and extensor tendons; and radial bicipital tuberosity area as a measure of the development of the tendon of *M. biceps brachii*) suggests that Neandertals males were about 16–26%, and females about 19–38%, larger than their recent European–American counterparts in the size of muscles crossing the elbow (Churchill 2006; Churchill and Rhodes 2006). A similar consideration, this time incorporating the area of the body of the scapula (a measure of the size of the rotator cuff muscles), shows Neandertal males to have muscle attachment sites about 13% larger than found in a sample of fairly muscular male Aleutian Islanders (Churchill 2006). Neandertal females have attachment areas for the muscles about the elbow that are 16–26% larger than those of Aleut females (Churchill 2006; Churchill and Rhodes 2006). While noting that not all muscle scars follow this pattern (the humeral deltoid tuberosity, which is the insertion site for *M. deltoideus*, tends to be smaller in Neandertals than in modern humans: Endo 1971; Churchill and Rhodes 2006), the values above provide a rough idea of the overall muscularity of Neandertals relative to modern humans.

But the question remains, does this reflect a greater proportion of muscle mass as a percentage of body mass than seen among modern humans? Neandertal males were about 15% heavier on average than Aleut males, while the females were about 23% more massive (Table 4.2). Assuming geometric similarity (that is, that Neandertals *were* like "scaled up" Inuit or, in this case, Aleuts. We know this isn't strictly correct because the two groups were of similar stature but different masses; still, it may hold largely true for aspects of body composition), muscle cross-sectional areas should scale to the 0.67 power of mass (mass $\propto$ volume). Thus the proportionally greater masses in Neandertals should equate with ca. 10% and ca. 16% increases in muscle cross-sectional areas (and thus muscle scar sizes) in male and females, respectively. This is roughly what we find, and accordingly we can't rule out the proposition that Neandertals were similar in their overall body composition to modern humans indigenous to circumpolar regions. Make no mistake: the Neandertals were strong. Their muscles had larger physiological cross-sectional areas and were thus capable of generating more force than the muscles of the average modern human, and these muscles were coupled with more mechanically efficient bony lever systems, such that they could exert absolutely more force on the environment than can most modern humans (Churchill 2006; Churchill and Rhodes 2006). The present, and very limited, data suggest that the strength of Neandertals relative to average modern humans was a function of their more massive bodies which, while retaining body compositions similar to modern arctic peoples, had absolutely greater masses of muscle tissue. Muscle scars in the appendicular skeleton indicate that the greater mass/stature ratios of the Neandertals weren't entirely a function of their having had wider trunks: more muscle was also distributed in the limbs, likely contributing further to a squat, lateral body build. Combined with a relatively short and biomechanically efficient skeleton, the greater muscle mass resulted in foragers that were strong, and most likely powerful – undoubtedly key physical attributes for dealing with the Ice Age world.

A final aspect of body size composition meriting consideration concerns Neandertal brain size. It has become almost axiomatic in paleoanthropology that Neandertals have the largest average brain size of any hominin, modern humans included. The average estimates of endocranial volume in Table 4.5, when compared to the grand means for global samples of modern humans (1,427 cm^3 for males, 1,272 cm^3 for females: Beals et al. 1984), bear this out. However, relative brain size (brain/body size) is smaller in Neandertals than in modern humans (Ruff et al. 1997), suggesting that the absolutely large brains of Neandertals are a function of their large body size. This may be so either because thermoregulatory stresses selected for larger bodies and larger brains in parallel, as each aid in maintaining thermal constancy in cold environments (Beals et al. 1984), or because the greater lean muscle mass of Neandertals demanded a greater amount of neural tissue to control it (the "meat head" hypothesis: Holloway 1981). Brain size in modern humans follows ecogeographic patterns similar to those seen for body size (Beals et al. 1984), and some of the largest reported sample means for endocranial volume come from circumpolar populations (1,527–1,566 cm^3 for various Inuit: see Holloway 1981). This does suggests a thermoregulatory role for brain size in Neandertals (which we will explore in Chapter 5), but also points to one difference in body composition – brain mass relative to body mass – between Neandertals and Inuit. Table 4.5 provides endocranial volume (ECV) estimates for the more complete Neandertal crania. Brain mass can be estimated from ECV using a regression derived from primates (McHenry 1992):

$$\text{Brain mass} = 1.147\text{ECV}^{0.976}$$

This method was used with the data in Table 4.5 to derive brain mass averages (Table 4.6) as a preliminary to exploring the metabolic and nutritional demands faced by Neandertals.

4.3 THE COST OF SIZE: FEEDING A LARGE BODY AND LARGE BRAIN

Given an idea of the size and composition of the Neandertal body, we can ask about dietary need, both in terms of daily caloric requirements and daily need for

TABLE 4.5 ■ Endocranial volume (ECV) in Neandertals

Specimen	ECV (cm^3)	Reference
Males		
Amud 1	1740–1750	Suzuki and Takai 1970; Kappelman 1996
La Chapelle 1	1625	Kappelman 1996
La Ferrassie 1	1641–1689	Caspari and Radovčić 2006; Kappelman 1996
Guattari	1350	Recheis et al. 1999
Krapina 5	1535	Caspari and Radovčić 2006
Le Moustier 1	1590–1650	Ponce de León and Zollikofer 1999
Neandertal 1	1452–1525	Beals et al. 1984; Hawks and Wolpoff 2001b
Saccopastore 2	1280–1300	Holloway 1985; Manzi et al. 2001
Shanidar 1	1600	Beals et al. 1984
Spy 2	1553	Holloway 1981
Male mean[a]	**1547** ± **140 (10)**	
Females		
Ehringsdorf H	1450	Beals et al. 1984
Gibraltar 1	1200–1270	Kappelman 1996; Caspari and Radovčić 2006
Krapina 3	1200	Smith 1976
La Quina H5	1345	Beals et al. 1984
Saccopastore 1	1174	Bruner and Manzi 2006
Spy 1	1305	Holloway 1981
Tabun C1	1271	McCown and Keith 1939
Female mean[a]	**1283** ± **94 (7)**	
Sex indeterminate		
Biache	1200	Hawks and Wolpoff 2001b
La Chaise (Suard)	1065	Hawks and Wolpoff 2001b
Gánovce	1320	Beals et al. 1984
Reilingen	1430	Hawks and Wolpoff 2001b
Teshik-Tash[b]	1565	Beals et al. 1984
Pooled sex mean[a]	**1411** ± **186 (22)**	

[a] For specimens in which a range of ECV values are given, midpoint of range was used.
[b] Juvenile.

physiologically necessary and nonsynthesizable vitamins, minerals, trace elements and essential fatty acids. Of course, Neandertals may have differed from extant humans in some aspects of physiology, and thus an exact accounting of their nutritional demands may not be attainable: still, we can begin to get a general picture of energetic and nutritional costs of their massive bodies and big brains.

4.3.1 Nutrition and Somatic Maintenance: Neandertal Bodies were Energetically Costly

Daily caloric requirements are a function of basal metabolic rates and caloric expenditures for activity (including some activity devoted to staying warm, like shivering). Basal metabolic rate (BMR) also varies with climate in important ways (Leonard et al. 2002; Froehle 2008), and thus some physiological aspects of cold adaptation (such as

TABLE 4.6 ■ Predicted organ mass and metabolic rate in an adult male Neandertal of average body mass (M = 77.9 kg)

Organ	Organ mass prediction method[a]	Predicted mass (g)	Mass-specific metabolic rate[a] ($W \cdot kg^{-1}$)	Total organ metabolic rate (W)
Heart	$5.2M^{0.987}$	383	32.3	12.4
Lungs	9.0M	701	6.7	4.7
Gut	17.0M	1324	12.2	16.2
Liver	$32.2M^{0.94}$	1932	12.2	23.6
Kidney	$6.3M^{0.87}$	279	23.3	6.5
Brain[b]		1487.5	11.2	16.7
Skeletal muscle	415M	32329	0.5	16.2
Skin	77M	5998	0.3	1.8
Total				**98.1**

[a] Organ mass prediction methods and mass-specific metabolic rates from Table 1 and Figure 3 in Aiello and Wheeler 1995.

[b] Average male Neandertal brain volume calculated from values in Table 4.5: see text for details.

non-shivering thermogenesis) must be factored in to estimates of Neandertal caloric need. In this chapter, we'll consider only BMR under benign conditions – in subsequent chapters, we'll take into account increases to BMR brought about by cold adaptation, and the activity costs of Neandertal foraging and other behaviors, to develop estimates of total energy expenditure in these hominins.

A number of methods exist for estimating BMR from mass (elsewhere [Churchill 2006], I have argued that surface area is a better predictor of BMR. We'll consider how Neandertal body shape affects BMR in Chapter 5.3). Following Kleiber (1961) for placental mammals generally, BMR (in Watts)[4] scales with the $\frac{3}{4}$ power of mass (M, in kg) as:

$$BMR = 3.39M^{0.75}$$

Thus, an adult male Neandertal of average mass (77.9 kg) would be expected to have a BMR of about 89 W, burning about 1836 kcal d^{-1}, while a female of average mass (65.6 kg) would be expected to have a BMR of about 78 W, corresponding to about 1611 kcal d^{-1}. Human specific equations (Schofield 1985; Sorensen and Leonard 2001) produce comparable results for males (1868–1870 kcal d^{-1}) but lower rates for females (1457–1460 kcal d^{-1}). An alternative approach would be to estimate the masses of metabolically active tissues and their contributions to total BMR (using mass-specific organ metabolic rates), and to sum them (Table 4.6). Assuming these tissues in combination contribute about 89% of total BMR (see Table 1 in Aiello and Wheeler 1995), the BMR of an average-sized Neandertal male can be estimated at about 110 W, or 2270 kcal d^{-1}. For an average-sized female (Table 4.7) this method produces an estimate of about 93 W, or ca.1921 kcal d^{-1}.

Neither of these BMR estimation methods (regressions based on empirically-determined BMR-mass relationships versus using organ-specific metabolic rates)

TABLE 4.7 ■ Predicted organ mass and metabolic rate in an adult female Neandertal of average body mass (M = 65.6 kg)

Organ	Predicted mass (g)[a]	Total organ metabolic rate (W)
Heart	323	10.4
Lungs	590	4.0
Gut	1115	13.6
Liver	1643	20.0
Kidney	240	5.6
Brain[b]	1239	13.9
Skeletal muscle	27224	13.6
Skin	5051	1.5
Total		**82.6**

[a] Organ mass prediction methods and mass-specific metabolic rates as in Table 4.6.
[b] Average female Neandertal brain volume calculated from values in Table 4.5: see text for details.

account for the effects of climate on metabolic rate in humans (Froehle 2008). While this will be explored in greater detail in Chapter 5.4, it bears upon our efforts to determine the average male and female BMR ranges that will enter into subsequent considerations of Neandertal thermoregulation and energetics. Among modern humans, BMR varies inversely with mean annual temperature (TMEAN, in °C), such that equations incorporating TMEAN might be expected to do a better job of predicting BMR than equations based on mass (M, in kg) alone. Using human data from 103 different locations around the world, Froehle (2008) developed BMR (in kcal d^{-1}) prediction equations for both males (BMR = 14.7M–5.6TMEAN + 735) and females (BMR = 9.2M–3.8TMEAN + 852). These equations can be applied to a sample of Neandertals if TMEAN is estimated using geological age-appropriate climatic reconstructions (Froehle and Churchill 2009). Surprisingly, this approach produces BMR estimates that, on average, are below (females) or at the low end (males) of the mass-only based approaches. These climate-adjusted equations (using a slightly different sample than employed by Froehle and Churchill 2009) suggest an average male Neandertal BMR of 1824 ± 78 kcal d^{-1} (n = 17), and a female average BMR of 1417 ± 47 kcal d^{-1} (n = 11). In combination, these methods suggest BMRs – for average-sized adult Neandertals – on the order of about 1800–2300 kcal d^{-1} for males and about 1400–1950 kcal d^{-1} for females.

As for other nutritional demands, it is reasonable to assume that Neandertal dietary needs were not unlike those of modern humans living in arctic conditions, bearing in mind, of course, that living human populations vary in their specific nutritional demands, that minimum daily requirements of many vitamins and minerals are not well established, and that human populations are capable of adapting physiologically to chronic shortfalls in some nutrients (see Keene 1985; Kormondy and Brown 1998). Estimates of daily nutritional requirements of arctic people (from Keene 1985) are provided in Table 4.8 which, along with the following short review, is intended to identify the major limiting nutrients in Neandertal diet, and to provide a consideration of how limiting nutrients may have factored into prey selection and scheduling of subsistence activities.

TABLE 4.8 ■ Daily nutritional requirements for arctic conditions

Age	kcal		Protein (g)		Calcium (mg)	Fat (g)	Vit. A (I.U.)		Vit. B_1 (mg)		Vit. B_2 (mg)		Vit. C (mg)	Iron (mg)	
	M	F	M	F			M	F	M	F	M	F		M	F
0–2	1200		28		550	25	2500		0.7		0.8		30	15	
2–10	2200		48		700	25	3000		1.2		1.2		32	10	
10–18	2800	2100	70	60	500	25	6000	5000	1.5	1.2	1.8	1.4	36	18	18
18–50	2800	2100	70	60	500	25	6000	5000	1.5	1.2	1.6	1.2	36	10	18
50+	2700	2000	70	60	500	25	6000	5000	1.4	1.1	1.5	1.1	36	10	

Average minimum daily requirements:

2322 kcal	58.6 g protein	552 mg calcium
25g fat	4720 I.U. vitamin A	1.27 mg thiamine
1.35 mg riboflavin	34.6 mg ascorbic acid	13.2 mg iron

Data from Keene 1985.

In order to maintain a balanced nutritional state, humans require: (1) sufficient calories to fuel basal (resting) and active (muscular) metabolism; (2) sufficient protein or amino acids to replenish bodily tissues (somatic maintenance) or develop new ones (growth), and to produce enzymes, antibodies, and certain hormones (peptide and protein hormones) without developing a negative nitrogen balance; (3) sufficient essential fatty acids for growth and maintenance of neural tissue; (4) sufficient water and minerals to compensate for losses or incorporation into new tissues; and (5) sufficient essential vitamins (those that can't be synthesized) to maintain physiologically important metabolic pathways (Eckert 1988). Ingested calories can come in the form of protein (4 kcal g^{-1}), carbohydrates (4 kcal g^{-1}) or fat (9 kcal g^{-1}). Carbohydrates, which are used by the body for chemical storage of energy (glycogen) or as immediate sources of energy (glucose 6-phosphate), derive from plant foods, and are notoriously lacking in arctic diets and potentially in Neandertal diets as well, at least seasonally (Cachel 1997). In the absence of carbohydrates, metabolic energy must be derived from protein and fat, and thus only surplus protein above that needed to meet energy requirements is available for anabolism (synthesis of new tissues) (Chaney and Ross 1972). However, as has been discussed in detail by Speth (1987, 1989; Speth and Speilmann 1983), protein alone is insufficient to meet energy demands, in part because protein has a very high (20–30%: Tappy 1996) specific dynamic action (SDA: the rise in metabolism that accompanies digestion of food). Fats and carbohydrates, in contrast, have much lower SDAs (0–3% for fats, 5–10% for carbohydrates: Tappy 1996). Thus, if protein from lean meat is the only source of energy, up to 3.5 kg per day may have to be consumed to meet metabolic demands plus the increased SDA (meat consumption of this quantity is documented in multiple historic cases of fat starvation: Speth 1987) Furthermore, excessive protein intake (more than about 300–400 g d^{-1}) overwhelms the liver's capacity to catabolize amino acids, and outpaces the body's ability to synthesize and excrete urea, with a host of negative sequelae: hypertrophy and failure of the liver, elevated (even toxic) levels of circulating ammonia, dehydration and electrolyte imbalance, kidney failure, calcium loss, micronutrient deficiencies, and lean tissue loss (references in Speth 1987). Arctic explorers and ethnographers of Inuit people have documented the ill-effects of inadequate fat in the arctic diet, resulting in the short term in insatiable hunger, abdominal cramps and diarrhea (references in Cachel 1997; see also Speth 1987). The "fat sickness" that follows from prolonged ingestion of excessively lean meat (and no carbohydrates) may be ameliorated with fat intake possibly as low as one gram per day (Davidson et al. 1986).

Protein quality is unlikely to have been an issue in a diet that was fairly heavily meat dependent (see Chapter 7). Humans lack the ability to synthesize eight of the twenty amino acids, and thus must get these eight directly from dietary sources. Meats and other animal products contain complete proteins, i.e., proteins that provide various amino acids in roughly the proportions in which the body needs them, and thus protein quality is never an issue for foragers eating substantial amounts of meat.

If protein quality is not a limiting factor, the one major limiting resource in carbohydrate-poor arctic diets (and by inference Neandertal diets at times) is adequate fat. In addition to providing the calories necessary for basal metabolism and the catabolism (breakdown) of dietary protein in a diet devoid of carbohydrates, fats improve the palatability of food, produce feelings of satiety (hence the insatiable hunger associated with fat sickness), and are critical to the absorption and transport of fat-soluble vitamins (A, D, E and K) (Keene, 1985). Furthermore, certain essential fatty acids are critical for the growth and maintenance of neural tissue. The long-chain polyunsaturatued fatty acids (LC-PUFA) docosahexaenoic acid (DHA: an *n*-3 fatty acid) and arachidonic acid (AA: an *n*-6 fatty acid) are considered the predominant limiting nutrients for brain growth (Broadhurst et al. 2002). The precursor PUFAs linoleic acid (LA) and α-linolenic acid (LNA) can be derived from plant tissue, but in pre-agricultural plant sources the quantities are generally insufficient for human demands. Pre-formed LC-PUFAs (both DHA and AA) as well as the precursors (LA and LNA) can be obtained directly from animal tissues, although their abundance is much more limited in terrestrial than marine ecosystems (values for shellfish being between 2.5 and 100 times higher than found in land animals: Broadhurst et al. 2002). Egg yolk and animal muscle and organs are good terrestrial sources of AA, while small mammals (which have relatively high brain-to-body mass ratios) and the brain tissue of larger mammals would have been the primary terrestrial sources of DHA (Broadhurst et al. 2002).

Minimum daily requirements of essential fatty acids are not well established, although some general guidelines have been put forth for people eating a modern western diet (INFORM 1998). These guidelines recommend an adult diet in which 0.5% of total dietary energy (en%) is derived from *n*-3 fatty acids (of which half should be preformed DHA or its immediate precursor, eicosapentaenoic acid or EPA) and 3.0 en% derives from *n*-6 fatty acids (AA and LA). For pregnant or lactating females and growing children, these values should perhaps be doubled, and with high activity-level foragers like the Neandertals, in which more fats are oxidized for energy and thus unavailable for building or repairing neural tissue, the correct values could be substantially higher (see Broadhurst et al., 2002). We can get an idea of what a fatty acid-sufficient diet might look like for a Neandertal with a total daily energy expenditure (TEE) of 4000 kcal (about middle of the road for a female Neandertal: see Chapter 10.6), assuming the diet was calorically sufficient (this was clearly not always the case [see below], but the assumption allows us to establish some minimum requirements). If adequate calories for basal and active metabolism were being provided by monounsaturated and saturated fats and protein (and perhaps some carbohydrates), then probably only about 50% of ingested LA and LNA would be oxidized for energy (Leyton et al. 1987) and only 15% of ingested preformed AA and DHA would be oxidized (Sinclair 1975). For such a diet, about 120 kcal a day (3.0 en%) should derive from *n*-6 PUFAs, that is, ca. 13.3 g of AA + LA d^{-1}. If we assume half of this requirement comes from preformed AA (with a 15% oxidative loss) and half comes from the precursor LA (at a 50% oxidative loss), the diet must provide 7.8 g AA and 13.3 g LA d^{-1} to be sufficient. Likewise, to attain 0.5 en% from *n*-3 PUFA, 20 kcal of a 4000 kcal d^{-1} diet would need to come from these fatty acids. Assuming again half of the *n*-3 PUFAs come from preformed DHA or its immediate precursor EPA (at an oxidative loss of 15%) and half from LNA (50% oxidative loss), tissues providing 1.3 g of DHA or EPA and 2.2 g of LNA would need to be consumed each day to meet demands.

Table 4.9 provides estimates of the absolute quantities of these essential fatty acids available to the consumers of an 80 kg reindeer. These estimates are based on values of these PUFAs in the meat, viscera and adipose fat of African ungulates (from data in Table 2 of Broadhurst et al. 2002), and on the amount of edible tissue of these types on a reindeer carcass (Foote 1965). Although the amounts of the essential fatty acids in different tissues vary considerably between ungulate species (largely varying

TABLE 4.9 ■ Estimated amounts of essential fatty acids provided by an 80 kg reindeer (in good condition)

	Edible mass[a] (kg)	AA[b] (g)	DHA[c] (g)	LA[d] (g)	LNA[e] (g)
Meat	28.1	36.8	2.4	112.4	–
Viscera	16.1	63.4	4.3	–	–
Adipose fat	7.8	0.0	0.0	260.5	195.0
Total[f]	52.0	100.2	6.7	372.9	195.0

[a] Based on percentages of edible caribou tissue from Foote 1965.
[b] Based on average values of 1.31 g/kg in meat, 3.94 g/kg in viscera, and little/no AA in adipose tissue in African ungulates (Table 2 in Broadhurst et al. 2002).
[c] Based on average values of 0.085 g/kg in meat, 0.264 g/kg in viscera, and little/no DHA in adipose tissue in African ungulates (Table 2 in Broadhurst et al. 2002).
[d] Based on LA composition (0.4%) in meat of North American antelope (Cordain et al. 2002) and average LA composition (3.34%) in adipose fat of African ruminants (Table 2 in Broadhurst et al. 2002).
[e] Based on average LNA composition (2.5%) in adipose fat of African ruminants (Table 2 in Broadhurst et al. 2002).
[f] Excluding contribution of brain tissue and marrow fat.

with diet), the averages used here should provide a rough estimate of the potential values of these nutrients in European Pleistocene ungulates. From the values in Table 4.9 it's clear that the preformed LC-PUFAs are the most limiting aspect of fatty acid nutrition. Four-hundred grams of adipose tissue would provide the daily requirement of LA and about four times the required amount of LNA. However, about two kilograms of viscera (the tissue highest in both AA and DHA) would be needed to meet demands for dietary AA, while it would take almost five kilos to satisfy DHA requirements! However, the lack of DHA in muscle and viscera is easily remedied by the consumption of brain tissue which, for North American cervids at least (deer and elk: Cordain et al. 2002), is comprised of about 9.3% DHA (Table 4.10 provides representative values from African ungulates). Accordingly, only about 14 g of brain per day would be needed to meet DHA requirements, which could also probably be fairly easily met with capture and consumption of small prey (which have naturally higher DHA compositions per unit body mass: Broadhurst et al. 2002). No such concentrated sources of AA exist in large ungulate tissues. Consumption of marrow fat, which is composed of 75% monounsaturated fats, would have provided an important source of calories and fat (the major long bones of a 165 kg wildebeest, for example, contain about 2700 kcal stored as fat: see Table 4.10. See also Brink 1997 for bison, Outram and Rowley-Conwy 1998 for horse, and Madrigal and Holt 2002 for deer), but marrow fat is poor in AA (or its precursor LA) relative to meat and viscera (compare Tables 4.9 and 4.10). Thus, despite prodigious amounts of bone cracking and marrow extraction (Marean and Kim 1998), and possibly even fat rendering (Speth 2012), Neandertal diets may have been limited in AA. This deficiency may have been rectified at times by consumption of fish (which provide in the range of 84–3135 mg AA per 100 g meat: Table 1 in Broadhurst et al. 2002). Shellfish, while being rich sources of DHA, are no richer in AA than terrestrial ungulate meat (20–145 mg AA per 100 g shellfish meat vs. 74–190 mg AA per 100 g meat in African savanna ungulates: Tables 1 and 2 in Broadhurst et al. 2002). Clearly, Neandertal diet provided, *on the average*, sufficient DHA, AA and other fatty

TABLE 4.10 ■ Mean live weight (kg), wet marrow weight (g) and brain weight (g) of select adult African bovids, and amounts of essential fatty acids provided from consumption of marrow and brain tissue

	Live weight (kg)[a]	Total wet marrow weight (g)[a]	CLA provided by marrow (g)[b]	Brain weight (g)[c]	DHA provided by brain (g)[d]
Warthog (n = 1)[e]	61.4	90.8	0.908	204	18.972
Thompson's gazelle (n = 1)[e]	16.3	57.9	0.579	87	8.091
Grant's gazelle (n = 3)	57.3	202.4	2.024	195	18.135
Impala (n = 2)	55.4	173.2	1.732	191	17.763
Topi (n = 1)	113.6	353.0	3.530	303	28.179
Wildebeest (n = 3)	165.1	299.1	2.991	385	35.805
Zebra (n = 2)	272.7	85.4	0.854	533	49.569

[a] Data from Blumenschine and Madrigal (1993). Live weights were determined directly for Thomson's and Grant's gazelles, impala and wildebeest; live weights for topi, warthog and zebra represent midpoints of weight ranges from Dorst and Dandelot (1970). Wet marrow weights are total weights from 12 bones.
[b] CLA = conjugated linoleic acid, based on contribution of CLA to marrow fat in North American cervids (deer and elk: Cordain et al. 2002) of 1.0%.
[c] Brain weight estimated from average body masses provided in column one, using the equation of Economos (1980) (following arguments in Calder 1996): $\underline{m}_{brain} = 14.3\ M_{body}^{0.645}$ (where $\underline{m}$ is mass in grams, M is mass in kilograms).
[d] DHA = docosahexaenoic acid, based on average percentage (9.3%) amount of DHA in brain tissue of North American cervids (deer and elk: Cordain et al. 2002).
[e] Data from a single female.

acids to allow for the development and maintenance of large and metabolically costly brains, and did so consistently over the course of several hundred thousand years. However, the metabolic expense of large brains *may have constrained the degree to which Neandertals could have relied on plants,* regardless of the abundance of edible plant foods in the environment. As noted by Broadhurst and coworkers (2002: 663), in the absence of a diet with a significant component of marine foods, "... a prodigious amount of muscle tissue of terrestrial game needs to be consumed, on the order of 100 en%" to meet essential fatty acid requirements.

There is also a host of micronutrients that must be obtained dietarily to ensure effective physiological functioning (Hockett and Haws 2003). Most of these fall into the category of vitamins – coenzymes that play important roles in a variety of metabolic pathways. Certain minerals are also important in human biochemistry. Many of these substances (both vitamins and minerals) are recycled repeatedly in the body, and thus need only be ingested in minute quantities. Vitamins critical to human physiological functioning include retinol (vitamin A), thiamine (B_1), riboflavin (B_2), niacin (B_3), pyridoxine (B_6), cyanocobalamin (B_{12}), ascorbic acid (C), calciferol (D_3), tocopherol (E), biotin, folic acid, and naphthoquinone (K). Some of these vitamins are obtainable from non-dietary sources, such as those that are synthesized by intestinal flora and absorbed in the gut (biotin, vitamins B_{12} and K) or synthesized directly by body tissues (vitamin D_3). Other vitamins or their precursors must be consumed in food, and dietary insufficiency of any given vitamin results in specific pathological conditions (see, for

TABLE 4.11 ■ Nutritional composition of arctic foods (per 100 g)

	kcal	Protein (g)	Fat (g)	Calcium (mg)	Vit. A (I.U.)	Vit. B_1 (mg)	Vit. B_2 (mg)	Vit. C (mg)	Iron (mg)
Reindeer									
meat	119	26.7	1.2	28.0	0.1	0.17	0.5	2.0	1.9
fat	900	–	100.0	–	–	–	–	–	–
guts	125	15.0	5.0	98.0	2000	0.1	0.2	15.0	8.0
Musk ox									
total	365	10.8	34.2	9.0	200	0.08	0.19	0.5	1.9
Polar bear									
meat	135	25.6	3.1	17.0	1400	0.23	0.57	2.0	1.0
fat	810	2.8	88.7	3.3	0	0	0	1.0	0.0
Hare									
total	144	24.2	8.0	20.7	0.2	0.08	0.21	5.0	2.2
Small game									
total	104	18	3.0	14.0	0.2	0.05	0.19	1.0	2.2
Bird									
average	112	22.4	2.1	17.5	150	0.3	0.43	9.8	2.1
Trout									
meat	133	22.3	4.5	35.8	0.1	0.08	0.21	5.0	1.0
entrails	94	22.2	0.3	16.0	1500	0.02	0.08	3.4	0.0
roe	126	22.1	4.0	23.0	0	0.02	0.08	44.3	1.6
Oysters	75	9.4	2.2	57.9–63.2	–	98–170	98–269[a]	–	6.2

Data from Keene (1985), except values for oysters, which are from Tressler (1951).
[a] For shellfish generally (Tressler 1951).

example, Harrison et al. 1988). Arctic diets, however, seem to be largely adequate with respect to vitamins (see for example Table 22.5 in Harrison et al. 1988), and in fact nutritional status is argued to have declined in contemporary Inuit as reliance on imported foods has increased (Draper 1977). Inuit diets tend to be poor in carotenoids – the precursors of vitamin A – but this is hardly problematic given the ready availability of vitamin A in the organs of fish, seals, polar bear, and reindeer (indeed, some arctic explorers have suffered from excesses of vitamin A: Harrison et al. 1988). All of the dietarily essential vitamins can be obtained in adequate amounts through the consumption of meat (Table 4.11), provided that at least some fresh meat is regularly eaten raw (since cooking and storage will rapidly break down vitamin C: Draper 1977; Keene 1985). Mineral deficiencies are also generally not problematic with arctic diets, with the exception of calcium sufficiency (Keene 1985; E. Smith 1991), which may be as low as 50% of daily requirements (Harrison et al. 1988). This is especially problematic with a meat-based diet, since meat is high in phosphorus and because high levels of protein metabolism may lead to calciuria (excretion of calcium in the urine), both contributing to hypocalcemia (Draper 1977: but see Keene 1985). Since calcium is critical to nerve conduction, muscle contraction, cell membrane permeability, and blood clotting (Eckert 1988), hypocalcemia usually results in mining of calcium phosphate from the skeleton and hence increased rates of bone resporption. Calcium was likely especially important to Neandertals to maintain their robust skeletons, and they may well have needed to supplement meat, fat and viscera intake with blood (which has about 10 mg of calcium per liter) and bone[5] – in other words, there may have been little of an animal that they didn't eat!

4.3.2 Nutrition and Reproduction: Were Neandertal Mothers Like Polar Bears?

The energetic and nutritive needs outlined above have been for adult Neandertals: having a large adult body increases the energetic and nutritive costs of development as well. These costs come in the form of both maternal contributions to producing and suckling larger babies and in the form of greater dietary needs during childhood. The World Health Organization estimates that it takes a woman an additional 80,000 kcal of dietary intake to produce a term infant (corresponding to 285 kcal d^{-1}), and 500 additional kilocalories per day for successful lactation (FAO/WHO/UNU 1985). Are these values representative of the energetic cost of Neandertal reproduction, and what can be inferred about the additional non-caloric nutritional requirements of Neandertal mothers?

Modern cold-adapted people give birth to large babies, and it is likely Neandertals did as well (Rosenberg 1988; Anderson 1989). Birth weight among arctic and sub-arctic populations tends to be about 9–10% heavier on average than that of neighboring, lower latitude populations (see, for example, Munroe et al. 1984).[6] Mean maternal pre-gravid and infant birth weight data from various modern human samples (Figure 4.4) allows us to predict an average neonatal size for Neandertal babies of 3445 g based on mean estimated adult female body mass. The regression equation in Figure 4.4, however, underestimates the mean birth weight of the cold-adapted samples (Cree Indians, Yup'ik Eskimos and Nunavik Inuit) by about 9%, which suggests that a more reasonable estimate of mean Neandertal neonatal mass would be around 3755 g. Assuming that 25% of Neandertal neonatal mass was composed of muscle and 14.5% of fat (the norm for modern human babies in good condition), the energy cost of producing a new Neandertal would have been about 11,265–15,020 kcal (at 3–4 kcal g^{-1}: Shephard 1991). Added to this would have been energy demands for the development of the placenta, deposition of maternal fat (in anticipation of lactation), increased maternal basal metabolic rate, and increased uterus and breast size, blood volume, amniotic fluid, and extracellular fluid, totaling between about 11,500 and 59,000 kcal (Table 5.6 in Ulijaszek 1995). The great range in this last value reflects a number of physiological and behavioral strategies a pregnant woman can employ when calories are in short supply, including limiting pregnancy-induced increases in BMR, limiting storage of maternal and fetal body fat, and limiting activity levels (Lawrence and Whitehead 1984; Durnin et al. 1985; Ulijaszek 1995). For a well-fed Neandertal not faced with substantial food shortages, the total cost of pregnancy was likely around 70,000–74,000 kcal (250–264 additional kcal d^{-1}) – close to but lower than the FAO/WHO estimates. During lean times, Neandertal females may have been able to produce smaller and leaner babies, and limit their own activity levels, fat storage and BMR increases in order to bring their infants to term. Nutritionally-stressed Gambian women, for example, are able to complete pregnancy at about 27–30% the total energetic cost of someone from a developed nation (Ulijaszek 1995): this implies a lower limit on Neandertal energetic costs of pregnancy of about 19,000–20,000 kcal (amounting to dietary supplements on the order of only 68–71 kcal d^{-1}).[7] Thus Neandertal mothers may have been able to

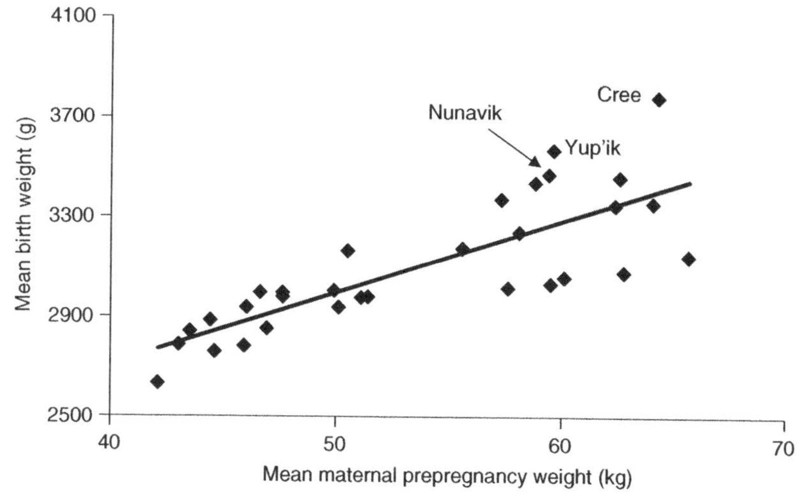

FIGURE 4.4

Mean birth weight (g) on mean maternal pregravid weight (kg) in a global sample from Asia, Europe and North and South America. Data from Mohanty et al. 2005; Dufour and Sauther 2002; Dewailly et al. 2000; Partington and Roberts 1969; Murphy et al. 1993. Birth weight mean for central Alaskan Yup'ik (excluding mothers with gestational diabetes mellitus: Murphy et al. 1993) was paired with mean female body mass from western Alaskan Inupiat (Ruff et al. 2005); mean birth weight for Cree (Partington and Roberts 1969) was paired with mean female body mass for Blackfoot Indians (Eveleth and Tanner 1976). Ordinary least squares regression $y = 28.734x + 1560.0$; $r = 0.8034$.

bring infants to term even during times of nutritional stress (which may have been a yearly occurrence: see below).

However, nutritional shortfalls during pregnancy can result in low birth weight infants at higher risk of postnatal mortality, and can compromise the mother's ability to develop fat stores sufficient for nursing. While maternal fat storage (as much as 3.4 kg during good times: Ulijaszek 1995) is continuous across the pregnancy, the third trimester appears to be the critical time for growth completion and fat storage in the infant (Chiswick 1985). Neandertal babies born in the late winter or early spring may have faced low survival odds, as they were likely underweight (having completed their intrauterine tenure during a time of potentially severe nutritional stress), had mothers who likely had insufficient fat stores to produce adequate milk, and may have had weeks or months of continued cold stress (in which a small, lean baby would have been especially disadvantaged) and starvation to survive before things got better. Under these conditions, there may have been strong selective pressures favoring behavioral or physiological mechanisms that ensured late spring or early summer deliveries.

Alternatively, Neandertal females may have adopted the "polar bear strategy" of accumulating maternal fat stores in the late summer and fall sufficient for overwintering while gestating and nursing an infant. Polar bear mothers may enter a denning period with a body composition close to 50% fat, and during the ensuing eight month fast (during which they gestate and deliver one or two cubs, which they then suckle for two to three months) lose on average 43% of their body mass (Atkinson and Ramsay 1995). However, to accomplish this feat a Neandertal mother would have had to add substantial amounts of body fat during the first two trimesters of pregnancy, while also meeting the increased energy demands of the pregnancy (polar bears manage by delaying implantation until the start of the denning period, after the mother has fattened up). To meet her own basal metabolic needs, complete gestation, and produce milk over a four-month lean period (January–April), a Neandertal would need to have about 26–30 extra kilograms of body fat, which to develop would demand dietary surpluses on the order of 1770–2300 kcal d^{-1} over the first six months of the pregnancy.[8] Assuming that any decreases in activity related to the pregnancy were offset by the increased cost of transport of her ever-increasing mass, she would have to consume these calories on top of the ca. 4000 kcal she needs for her daily energy expenses. To put this into terms of prey consumption (see Table 5 in Churchill 2006), a Neandertal mother-to-be would have needed to eat about 2.5 kg of animal tissues per day for 190 days – a consumption rate within the range of female lions (Packer et al. 1990)! These values make it unlikely that expectant Neandertals could have stored sufficient fat to make such a strategy work (see also Hockett 2012).

Assuming that Neandertal neonates, like those of modern humans, were born with brains that were between 22 and 28% of adult size (see Hublin and Coqueugniot 2006), average brain mass at birth would have been on the order of 300–380 g (using brain mass predicted from the pooled endocranial volume estimates in Table 4.5). Virtual reconstruction of Neandertal neonates from Mezmaiskaya (Russia: Ponce de León et al. 2008) and Le Moustier (France: Gunz et al. 2011) produced estimated ECVs (which are slightly larger than actual brain size) of 382–416 cm^3 (Mezmaiskaya, estimated ECV at birth) and 408–428 cm^3 (Le Moustier, estimated ECV at time of death). Both of these reconstructions produce neonatal ECV values somewhat above the high end of the 300–380 range, suggesting that the estimates above are fairly conservative. Over the course of the ensuing five years, the average Neandertal child would have added another 850 or so grams of grey matter (growing to about 85% of adult brain size: see Coqueugniot et al. 2004). During this same interval, the child would have added another 18–22 kg of body mass (based on growth rates in Evenki herders, in which 5-year-olds are ca. 33% of the average mass of individuals of the same sex at age 20: see Tables 2 and 3 in Leonard et al. 1994), at a total caloric cost of about 145,800–178,200 kcal (at 8.1 kcal g^{-1} energy cost of postnatal growth: Shephard 1991). Bear in mind, however, that these are the costs for new tissue production only: the basal metabolic processes and the activity of growing children need to be fueled as well (see below). Neandertal mothers appear to have been bearing much of this energetic cost directly: based on various lines of evidence (see Chapter 11.2), supplementation with solid foods may have started very early (before one year of age), but complete weaning *may* not have occurred until children were 3.5–4.5 years of age (which is relatively late for foragers in general [Marlowe 2005], although comparable weaning ages have been reported in Inuit [Kennedy 2005]). Reported total daily caloric demands (total energy expenditure plus energy deposition [growth]) of children in the arctic are quite high (Table 4.8), and consideration of likely Neandertal growth patterns (Figure 4.5) suggests that Neandertal infants and children also had very high caloric demands.[9] The mass estimates in Figure 4.5, which provide the basis for estimating daily caloric needs,

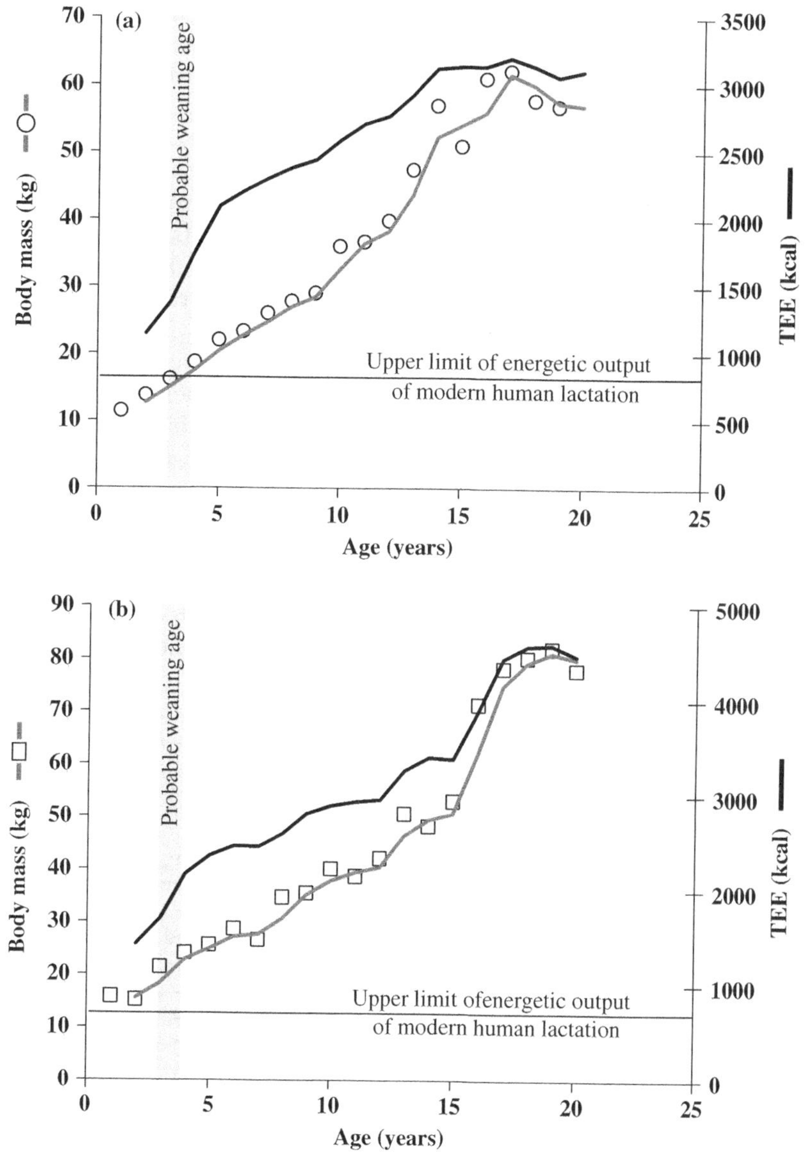

FIGURE 4.5

Hypothetical distance curves for growth in body mass (kg) for Neandertal females (a) and males (b) and corresponding daily total energy expenditures (TEE: in kcal). Grey lines represent 2-year moving averages of body mass points, black lines represent moving averages of TEE estimates. Mass data were generated using proportional distance data by 1-year intervals based on data for cold-adapted Evenki herders (from Tables 2 and 3 in Leonard et al. 1994) applied to mean adult male and female Neandertal body mass. Growth in stature (not shown) was calculated in the same way for purposes of estimating TEE. TEE determined from age- and sex-specific equations from Institute of Medicine (2002).

are based on sex-specific proportional growth in mass (that is, the percentage of adult mass attained each year) in cold-adapted Evenki herders (Leonard et al. 1994) applied to mean adult Neandertal male and female mass estimates. This approach assumes that Neandertal growth patterns were like those of modern humans (Guatelli-Steinberg et al. 2005; Macchiarelli et al. 2006; vs. Ramírez Rozzi and Bermúdez de Castro 2004; Dean et al. 1986), and that growth in modern humans from cold environments (in which less energy may be allocated to growth because of higher thermoregulatory demands: Leonard et al. 1994) are most similar to those of the Neandertals.[10] The results suggest that Neandertal children would have attained considerable weight at an early age (e.g., a three-year-old male at about 21 kg), and would have had correspondingly high daily energy expenditures.

By these estimates, to meet the energetic demands of a three-year-old male child, a Neandertal mother would have to produce almost three liters of milk a day (at a caloric value of 700 kcal l^{-1}: Binns 1998) – about four times what the average modern human mother produces (Kennedy 2005)! A more conservative estimate can be had by beginning with estimated Neandertal birth weight and using empirically-determined infant and child growth rates in modern humans (Dewey et al. 1992; Cole 1995) to estimate early childhood milk demands (Table 4.12). This approach is conservative both in terms of growth rate (resulting in estimated three-year-old body mass well below that estimated by the use of Evenki proportional growth data), and in ignoring metabolic rate increases related to thermogenesis in cold conditions. Based on the estimates in Table 4.12, it appears that a lactating Neandertal could meet her child's energetic and protein demands with milk production within or only slightly above the maximum outputs seen among modern humans (1.2 l d^{-1}: Kennedy 2005). However,

TABLE 4.12 ■ Nutritional demands of Neandertal infants and young children

Age (m)	Mass (kg) at end of interval	Energy need (kcal d^{-1})[a]	Milk needed to meet demand (l)[b]	Protein demand (g d^{-1})[a]	Milk needed (l)[c]	DHA demand (g d^{-1})[d]	Milk needed (l)[e]
0–3[f]	6.07	615	0.879	5.3	0.408	0.4	1.429
4–6[g]	7.64	636	0.909	6.6	0.508	0.4	1.429
7–12[h]	9.45	763	1.090	8.2	0.632	0.5	1.786
13–36[i]	13.43	1115.27	1.593	11.7	0.900	0.6	2.143

[a] Based on age-specific requirements for modern children (Institute of Medicine 2002).
[b] Based on energetic value of 70 kcal/100 ml for mature milk (Binns 1998).
[c] Based on protein content of 1.3 g/100 ml for mature milk (Binns 1998).
[d] Calculated as 0.5 en%, equivalent to 200% of adult daily requirements (Broadhurst et al. 2002).
[e] Based on 28 g milk fat l^{-1} breast milk, and DHA ca. 1% of milk fat (Innis and Kuhnlein 1988).
[f] At an average growth rate of 867.6 g $month^{-1}$ and birth weight of 3.469 kg. Growth rate represents the average over the first three months of all babies in the DARLING study (Dewey et al. 1992).
[g] At an average growth rate of 523.7 g $month^{-1}$. Growth rate represents the average over months 4–6 of all babies in the DARLING study (Dewey et al. 1992).
[h] At an average growth rate of 301.5 g $month^{-1}$. Growth rate represents the average over months 4–6 of all babies in the DARLING study (Dewey et al. 1992).
[i] At an average growth rate of 166 g $month^{-1}$, based on recent British boys (Cole 1995).

adequate amounts of critical fatty acids – namely DHA – might have been more difficult for a Neandertal mother to supply (since DHA is the most limited fatty acid in terrestrial food webs [see above], and since it is critical to brain and eye development, it is the only fatty acid considered here). Levels of DHA in breast milk, though, are a function of the mother's diet, and with increased consumption of ungulate brain and small mammals (the best terrestrial sources of DHA) a Neandertal may have been able to increase the proportion of DHA in her breast milk above the 0.28 g l^{-1} average for modern women (Innis and Kuhnlein 1988). Assuming that breast milk production was a function of the infants caloric (rather than DHA) needs, a lactating Neandertal may have needed an additional 770–1395 kcal d^{-1} (assuming a conversion efficiency of 80% in milk production: Dufour and Sauther 2002) over the course of the first three years of breast feeding (cf. the estimated energetic cost of lactation of 1.39*TEE for *H. erectus* by Aiello and Key 2002). Although Neandertal females, like their modern human counterparts, had behavioral and physiological ways of reducing the energetic cost of lactation (reducing activity levels, reducing BMR, and reducing the thermogenic [specific dynamic] effect of food: Ulijaszek 1995), lactation was no doubt much more energetically demanding than pregnancy (Dufour and Sauther 2002). Together, these two attempts to estimate the caloric demands of small Neandertal children suggest that either (1) milk production by Neandertal females was many times that of modern women (sustaining rapid gains in infant mass, as suggested by the Evenki proportional growth observations), which may have necessitated increased food intakes on the order of 2500 kcal d^{-1}, or (2) supplementation with high calorie, protein- and DHA-rich solid foods began much earlier than the dental wear data (Skinner 1997) would suggest (see Chapter 11.2), or (3) Neandertal babies, although starting out large, added mass at rates comparable to, or even slower than (Martín-González et al. 2012), that seen in modern human children. The latter two possibilities seem the most likely from an energetic point of view, but even if one of these two scenarios obtained, Neandertal mothers would have faced substantial energetic needs while lactating (the reproductive and demographic consequences of this are explored in Chapter 11).

4.3.3 Dietary Shortfalls: Hard and Lean Winters

Although Neandertal dietary adaptations were sufficient to sustain the lineage for hundreds of thousands of years, it's clear that, like modern human foragers in seasonal environments, they occasionally experienced severe shortfalls in calories. Among modern hunter-gatherers in temperate zones, the late winter and early spring is a time of nutritional stress, because plant foods are unavailable, the nutritional quality of prey is low (ungulates may lose upwards of 77% of their fat stores: Gerhart et al. 1996) and the associated search costs are often high (mobility costs can be further compounded by snow cover), and because winter stores of food have run low. Seasonal weight loss is common in foraging groups (see references in Speth 1987), and starvation is an ever-present risk during the lean season for many groups. The analysis of skeletal and dental non-specific indicators of stress (so called because their presence in hard tissues indicates only the number of systemic stress episodes severe enough to retard growth experienced by an individual during childhood, and not the nature of the stress [starvation vs. malnutrition vs. infection vs. trauma]: see Larsen 1997) show the Neandertals to have suffered seasonal deprivations as well. Harris lines, which are radio-opaque transverse lines in the ends of long bones and represent periods of halted growth at the epiphyseal plate, are abundant in the limbs of Neandertals from France (Brennan 1991). Dental enamel defects (hypoplasias and hypocalcifications) also indicate regular periods of systemic stress throughout childhood (presumably reflecting seasonal starvation that was affecting every member of the social group, but which is only recorded for the childhood years because these are when dental enamel is forming). In the largest study of this kind, Ogilvie et al. (1989) examined 669 Neandertal teeth and found that 75% of the individuals represented had at least one hypoplasia. The age distribution of defects (the age of insult can be determined for a hypoplasia from the distance of the defect to the cervoenamel junction of the crown, combined with knowledge of the tooth-specific developmental rate) showed two peaks, one centered on age 4 (perhaps reflecting weaning: see Chapter 11.2) and the other centered on age 11 (perhaps reflecting onset of adolescence and reduced nutritional dependency on adults in the social group). However, the pattern observed may also be an artifact of the timing of crown formation of the permanent dentition, since little amelogenesis occurs between 5 and 11 years of age or after age 12 (Ogilvie et al. 1989), and thus perhaps not all yearly starvation episodes would be recorded in the adult dentition. A similar study focused on the last interglacial-aged (MIS 5e) dental sample from Krapina (Croatia), but examined hypocalcifications as well as hypoplasias, reporting similar results (Hutchinson et al. 1997). Here, 86% of the individuals represented had at least one form of enamel defect, whereas this value is a little lower if hypoplasias alone are used (80%: Ogilvie et al. 1989). Hypoplasia incidence has been found to be roughly comparable between Neandertals and Inuit, indicating parity between the two groups in the frequency of physiological stress episodes (Guatelli-Steinberg et al. 2004), with perhaps a longer average duration of stress episodes in the Inuit. While controlled studies of defect frequency by climate and environment have not been undertaken, it is interesting to note that Brennan (1991) found only 39% of Neandertals from southwest France had hypoplasias. While Brennan's sample was not temporally restricted, it was heavily biased towards MIS 4 and 3 Neandertals, and thus the difference in frequency may relate to differences in the degree of seasonal nutritional stress in colder stadial/interstadial conditions versus warmer interglacial environments. At the risk of making too much of some admittedly limited data, on the surface it suggests that seasonal starvation may have been more problematic for Neandertals living in forested environments during warmer times, a not altogether unexpected finding (see Chapter 6).

From the foregoing analysis, it appears that calories and certain types of fats (LC-PUFAs) may have been the limiting dietary factors in Neandertal productivity (production and reproduction). Noting that variation in nutritional status among and between populations of living humans is a major determinant of variation in body proportions and stature, Bogin and Rios (2002) have suggested that nutritional shortfalls may account in large part for the somatic morphology of Neandertals (namely, short stature combined with femora and tibiae that are short relative to stature). It has also been suggested that the monotonic nature of Neandertal diets (see Chapter 7) led to chronic nutritional insufficiencies that may have reduced their fitness (via increased rates of fetal and infant mortality) *vis-à-vis* early modern Europeans with more varied diets (Hockett and Haws 2005). Neandertals in the southerly stretches of their geographic range may have had more varied diets (see Chapter 7), which may have reduced their fetal/infant mortality and decreased their susceptibility to extinction when faced with competition from modern humans (Hockett and Haws 2005). Since cold climate and caloric and nutrient shortfalls may tend to co-occur, it becomes especially challenging to disentangle genetically-determined morphological adaptations to climate from epigenetic morphological consequences of the nutritional environment – again the problem of equifinality (see Preface) that plagues studies of adaptation in fossil and recent humans rears its head.

4.4 THE BENEFITS OF SIZE: NEANDERTAL BODY SIZE IN ECOLOGICAL CONTEXT

Having examined the energetic costs of large body size (specifically, great mass) in Neandertals, we can ask what benefits accrued to larger hominins in glacial Europe – that is, what benefits were sufficient to outweigh those costs? Putting this question into the proper context requires acknowledging that Pleistocene faunas in general – in Europe as on all the other continents – were characterized by larger average sizes of most of the mammals. This was in part because of the existence of now-extinct megaherbivores (>1000 kg), which in Europe included straight-tusked elephant (*Palaeoloxodon antiquus*), steppe and wooly mammoth (*Mammuthus trogontherii* and *M. primigenius*), Merck's, steppe, and wooly rhinos (*Dicerorhinus kirchbergensis, Stephanorhinus hemitoechus* and *Coelodonta antiquitatis*), giant deer (Irish elk: *Megaceros giganteus*), and aurochs (*Bos primigenius*) (Kurtén 1968; Sutcliffe 1985; Owen-Smith 1987). Size reduction in mammalian faunas was also a function of a reduction in the size of species that survived the end Ice Age "blitzkrieg" (the terminal Pleistocene extinction of roughly half of all the mammalian species larger than 5 kg: Owen-Smith 1987; Brook and Bowman 2004) and that are still extant (e.g., Davis 1981; Guthrie 2003). While the degree of size reduction in European forms is uncertain, size reduction in African ungulates across this same time period suggests that Pleistocene forms may have been 25–50% larger than their modern counterparts (Brink 1993; Peters et al. 1994; Churchill et al. 2000). Large size was not the universal rule in the Pleistocene – under certain conditions it pays for smaller (<5 kg) mammals to become smaller still (see Klein and Cruz-Uribe 1996; Damuth 1993) – but it was certainly the general rule followed by larger mammals. Given that larger forms have absolutely greater energy demands (Calder 1996), and that they have slower reproductive rates and lower population densities than smaller animals, and are thus at greater risk of extinction (Johnson 2002), there must have been considerable countervailing benefits to gigantism during Pleistocene times.

In general, larger animals tend to be better at interspecific competition (they are the ecologically-dominant members of their communities and feeding guilds) and, accordingly, their populations tend to control more of the energy flowing through an ecosystem – by as much as two orders of magnitude – than do those of smaller forms (Brown and Maurer 1986; Silva and Downing 1995; but see Damuth 1981, 1993). But in addition to this general advantage, larger-sized mammals enjoy specific benefits – in terms of thermoregulation, predator avoidance, and dietary ecology – when climates become colder and drier, as characterized much of the Pleistocene (Chapter 5.1). A general model relating climate change and associated selective factors favoring larger size in herbivores and carnivores is presented in Figure 4.6. Climatic and nutritional factors

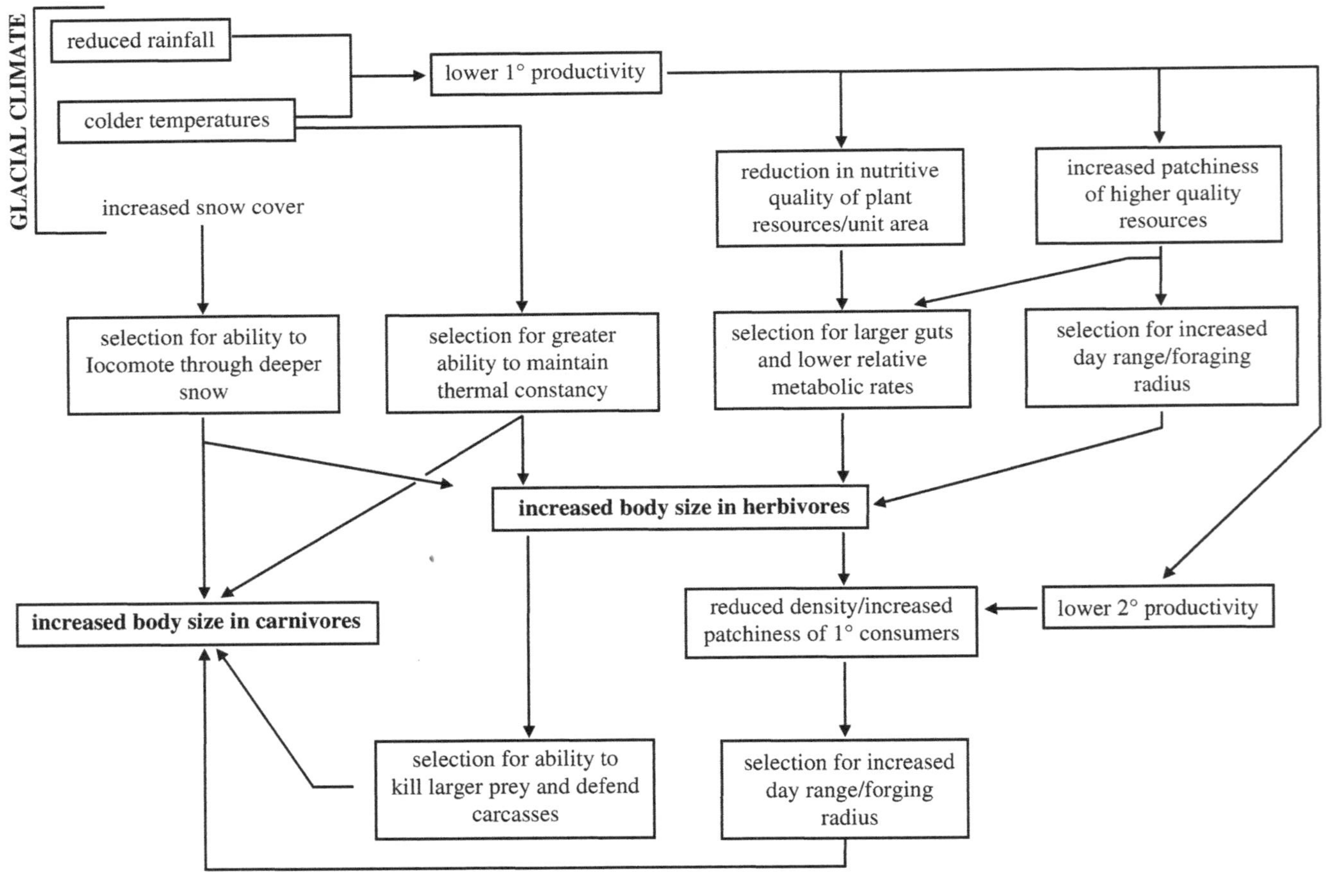

FIGURE 4.6

Model relating glacial climatic conditions with environmental factors that favor large body size in both herbivore and carnivore communities. See text for details.

favor larger body size in herbivores, which in turn places a premium on larger size at the next trophic level (carnivores and Neandertals).

Reduced rainfall and lower temperatures during glacial periods brought reductions in net primary productivity (Chapter 6) and changes in the density, distribution, seasonality, nutritive quality, and palatability of edible resources for herbivores. In grassland ecosystems, colder and drier climates generally see an increase in herbaceous plants using the C_3 photosynthetic pathway relative to those using the C_4 pathway, with an associated reduction in overall grassland productivity (Epstein et al. 1997). Changes in the overall productivity of these ecosystems, and of the composition of the herbaceous vegetation (proportions of C_3 vs. C_4 grasses, and forbes vs. graminoids: see Briggs and Knapp 2004) likely affected both bulk grazers (like bison) and more selective grazers (like horses) in equal measure by reducing the net nutritive quality of edible resources per unit area. Colder and drier climates also result in an overall reduction in nonherbaceous, closed-cover vegetation relative to open country herbaceous plants, reducing the abundance and increasing the patchiness of forage available to browsers (see Guthrie 1990). Larger individuals have larger day- and home-ranges (see Burness et al. 2001), and thus are able to more effectively exploit larger areas as net productivity goes down. While larger herbivores require absolutely more food, their mass-specific metabolic rates are lower (McNab 1983), and their guts are longer (permitting longer retention of ingesta and higher digestive efficiency: Illius and Gordon 1987, 1992; Demment and Greenwood 1988), allowing them to subsist on lower-quality forage than can smaller individuals. With colder climates comes increased temporal patchiness in the form of greater seasonal variation in plant productivity and nutritional quality. These longer periods of plant dormancy – long, lean winters – also place a selective premium on larger body size in herbivores. Within a species or within taxa of ecologically-similar mammals, larger individuals may be better able to survive periods of resource shortage because they are better able to store fat during good times and, given their lower mass-specific metabolic rate, are better able to conserve energy stores during times of starvation (Lindstedt and Boyce 1985; Reynolds 2007: however, see Dunbrack and Ramsay 1993 for a discussion of how the "fasting endurance hypothesis" may not apply across ecologically- and behaviorally-diverse taxa).

In addition to the selective advantages large body size might confer in situations of reduced primary productivity, cold climates also create thermoregulatory challenges that favor larger animals. Assuming geometric similarity in body shape across different sizes, larger animals will have more volume relative to surface area than do smaller animals. In endotherms, this means that larger forms will have higher proportions of metabolically-active heat-producing tissue relative to heat-dissipating surface area than their smaller counterparts, and thus will be better able to maintain thermal constancy in low ambient temperatures (but conversely will have a more difficult time dissipating heat at higher ambient temperatures). This relationship between surface area/volume (SA/V) ratios and thermoregulatory competence in cold environments forms the explanative basis for the empirical observation that body size tends to increase with latitude within species or among closely-related species (congeners). This observation was originally proposed as a general rule to explain body size variation across endothermic taxa by Bergmann (1847), and while the resulting ecogeographic rule has since been reformulated to explain body size variation within species (Rensch 1938), the rule still bears Bergmann's name. Critiques of both the empirical and theoretical basis of Bergmann's rule have since arisen (Rosenzweig 1968; McNab 1971; Geist 1987), as has a growing realization that large numbers of exceptions to the rule can be found (see Meiri and Dayan 2003) and that the rule may not hold in certain taxa (carnivores seem least to obey the rule: Dayan et al. 1991; Meiri et al. 2004). While the validity of the rule is still open to question, global body size patterns in endothermic animals support the general proposition that larger animals are found in colder environments (Blackburn et al. 1999; Ashton et al. 2000; Meiri and Dayan 2003; Blackburn and Hawkins 2004; Rodríguez et al. 2006), although the claim that thermoregulatory competence is the mechanism behind the relationship remains controversial. Nonetheless, it is likely that thermoregulatory benefits accrue to larger mammals living in cold environments, since larger animals can support thicker and heavier layers of fur and subcutaneous fat, and thus have lower conductance values in addition to having more favorable SA/V ratios (see references in Blackburn et al. 1999).

Another potentially important benefit to large body size in herbivores involves their ability to deal with snow cover. While total annual precipitation was reduced during glacial and cold–temperate intervals (relative to today), winters likely saw more days with snow cover and increases in the average depth of snow compared to European winters today (Barron et al. 2003). Snow accumulation can influence the amount of forage available, for both consumers whose food is arrayed horizontally (grazers) and those with vertical distributions of forage (browsers). Grazers often deal with snow by digging (cratering) with hooves or head movements to uncover buried forage. Among caribou, who may spend up to two hours per winter day cratering, the net energy cost of pawing the snow is inversely related to body weight (see Table 2 in Fancy and White 1985). For browsers, larger size equates with a greater ability to reach browse above the snow. Accumulations on the order of 60 cm, for example, may reduce the availability of browse for deer by up to 40%, while having little to no impact on that available to elk or moose (Moen 1973). Furthermore, the ability of an animal to locomote through snow (assuming it is too large to walk on top of crusted snow), and the energetic cost of doing so, is a function of its leg length. When snow reaches the chest, the animal can no longer lift its legs above the surface, which greatly increases the energetic cost of walking (Guthrie 1990) or effectively impedes forward progress. Smaller animals are much less able to flee predators as snow depth increases, both because they are in poorer condition and because they are impeded by the snow, and predation rates upon them may increase 5–7-fold (Huggard 1993; DelGiudice 1998).

Colder temperatures and increased snow cover may select for larger body size in carnivores in the same way that it does in herbivores (although it has proved harder to demonstrate Bergmann's rule among carnivores: Dayan et al. 1991; Meiri et al. 2004). Likewise, reduced secondary productivity (consequent to reduced primary productivity) may decrease prey density (or increase patchiness of prey), favoring larger-bodied carnivores with greater day- and home-ranges. We would also expect overall increases in body size in herbivore communities to have a direct effect on carnivore body size. Body size and muscular strength are major determinants of prey body size selection, and of hunting success rates with larger prey items in the carnivore's diet (Gittleman

1985; Vézina 1985; Pierce et al. 2000; Funston et al. 2001). Smaller carnivores are limited to smaller classes of prey, while larger, stronger carnivores – by being able to take both small prey and those of larger size – can feed on a greater diversity of prey species and thus enjoy a competitive advantage over smaller predators (Gittleman 1985; Wilson 1975). Gittleman (1985) notes that the ability of a carnivore to pursue, pull down, and consume prey is a function of a number of size-dependent characteristics, one of these being powerful forelimbs. The absolute strength of a carnivore is proportional to its mass (Gambaryan 1974; Alexander 1985a), which provides larger, stronger carnivores with the ability "…for persistent tracking of prey and, once caught, continual attacking" (Gittleman 1985: 549). Social hunters (which would include the Neandertals) can, by working cooperatively, overcome somewhat prey size constraints imposed by their own body size (see Stiner 1994). Nonetheless, even among social carnivores, size (and strength) matters: male lions, despite hunting in smaller groups or solitarily, attempt more attacks on buffalo and have a higher success rate with this large prey than do the more gregarious females (Funston et al. 2001). Likewise mountain lions, despite being solitary hunters, prey on mule deer much more frequently than do pack hunting but smaller coyotes, and the larger male mountain lions prey on larger male deer more often than do the smaller females (Pierce et al. 2000). Finally, as prey size increases, so does the value of defending carcasses against kleptoparasitism by other members of the carnivore guild (see Chapter 9). Because the energetic cost and injury risk associated with killing prey have already been paid, dead animals are more valuable to carnivores than live ones. As prey density decreases, and the energetic cost of acquiring prey increases (as a function of lowered secondary productivity), interference competition between sympatric carnivore species may increase. The increased need to defend carcasses against rivals may also favor larger body size among members of the carnivore guild (Linnell and Strand 2000).

As illustrated in Figure 4.6, climatic conditions work in parallel on primary and secondary consumers, such that coordinated change in average body size occurred across whole mammalian communities as glacial episodes waxed and waned. While earlier interglacials did not see the dramatic reduction of average body size that characterized the onset of the present interglacial (a function of the mass extinction of megaherbivores, which was at least partly anthropic in nature: Brook and Bowman 2004), some degree of body size reduction may have characterized these warmer intervals (see Guthrie 1990 on MIS 3 bison). Overall, Pleistocene faunas were larger than those of today, and the relative massiveness of Neandertals must be seen in this context. Great body mass was no doubt part of the adaptive package that allowed Neandertals to survive glacial cold (as we'll explore further in the next chapter), and overall size and muscular strength – especially upper body strength – may have been critical to Neandertal predatory strategies and success rates (Churchill and Rhodes 2006). In speaking of African bovids, Grand (1997:375) noted that muscle mass was central to habitat choice, dietary specialization, social structure, and male agnostic behavior, and was thus "part of the fabric of behavioral ecology" in these antelopes. While muscle mass, and for that matter body size itself, may no longer be part of the behavioral ecology of humans living in industrialized, mechanized agricultural economies, they most certainly were for humans living in the more technologically-limited foraging economies of the later Pleistocene.

NOTES

1. Throughout the book I endeavor to use the term "Inuit" to refer to Inuit-speakers (Inupiaq, Inuktitut, Kalaallisut) of northern Alaska, Canada and Greenland, and "Eskimo" to refer to Yu'pik-speakers (Yupiat, Alutiiq) of central, western and southern Alaska.
2. Body fat may require a surplus of calories to acquire but, as tissues go, fat has a low metabolic rate, and thus per unit mass contributes less to basal metabolic rate than more costly tissues like muscle. Of course, activity costs (such as locomotion) increase as a function of body mass, and so even metabolically inactive tissues like fat have an energetic cost.
3. Inuit apparently don't use body fat as a significant adaptive solution to heat loss. Shepard et al. (1973) offer two explanations: in situations involving high work output, it is better to have insulating layers that can be added and subtracted as needed (i.e., clothing) rather than an ever-present layer of subcutaneous fat; and in subsistence contexts in which fat is a critical source of calories and meals may be sporadic, it pays to have more labile, internal fat stores (i.e., visceral fat stored in the omentum rather than subcutaneous fat).
4. Watts (Joules s^{-1}) are a measure of power (rate of energy expenditure) and are used in this book when discussing heat production and work output. The measure of food energy used here, and the basic currency with which we will explore Neandertal energetics, is the kilocalorie (kcal = 4186.8 J). A watt is thus equivalent to 0.2388×10^{-4} kcal s^{-1}.
5. There's nothing to suggest that Neandertals were gnawing bones from the large ungulates on which they regularly preyed. However, it is possible that periodic consumption of smaller prey involved ingesting some bone.
6. With increasing acculturation to western diets circumpolar populations have seen increased incidences of obesity, gestational diabetes mellitus, and macrosomia (excessive birth weight). A recent study of Cree from James Bay (Trevors 2001) reported an obesity rate of 54.2% among pregnant females, who had an average pregravid weight of 80.7 ± 17.3 kg (obese + non-obese women, n = 969) and whose infants had an average birth weight of 3844 ± 510 g (n = 510), some 16.5% larger than babies born to the general Canadian population (Munroe et al. 1984). Macrosomia secondary to gestational diabetes confounds the relationship between maternal pregravid weight and infant birth weight as it likely obtained in Pleistocene foragers, and thus caution must be exercised in using and interpreting data from recent circumpolar populations.
7. In considering energetic costs of reproduction in *Homo erectus*, Aiello and Key (2002) modeled gestational needs at 25% over normal daily energy expenditure. Assuming midrange female Neandertal daily energy expenditure at 4000 kcal, such a cost would require an additional 1000 kcal d^{-1} (and a total additional energetic expense, assuming a 280 day gestation length, of 280,000 kcal) – a value that seems high based on empirical data in modern humans.
8. These estimates are based on the female completing the last trimester of pregnancy in the "den," and lactating for one month off her stored energy reserves before food becomes available. Thus, she would need fat stores to cover 120 days of her own basal metabolism (at 1400–1950 kcal d^{-1}), the third trimester of pregnancy (23,100–24,420 kcal), and 30 days of lactation (at 500 kcal d^{-1}). At 9 kcal g^{-1} of fat, 22.9–30.4 kg of body fat would be required to meet these needs. To accumulate this fat, a Neandertal mother would have to consume 288,540–382,788 extra calories (thermodynamic inefficiencies in the conversion of food to stored fat result in a storage cost of 3.6 kcal g^{-1}, thus it takes 12.6 kcal to store 9 kcal as fat: see Shephard 1991) over the first 190 days of her pregnancy (assuming a 280-day gestation). Thus, she would need 1520–2014 kcal extra per day towards the coming winter, while also consuming an extra 250–264 kcal d^{-1} to meet the immediate demands of gestation.
9. To meet the higher metabolic demands of thermoregulation in the arctic, polar mammals produce breast milk that is higher in fat and protein than that of temperate analogues. Four

terrestrial arctic mammals (polar bear, reindeer, musk ox, and arctic fox) average 18 ± 10% fat and 11 ± 1% protein in their milk, compared with 9 ± 4% fat and 7 ± 3% protein in the milk of four temperate species (wolf, dog, sheep, and cow) (Blix and Steen 1979). It is possible that Neandertal breast milk was similarly enriched to help meet the high caloric and protein needs of growth plus thermoregulation.

10. This exercise is also undertaken with full cognizance that growth rates are highly variable among and between populations, and that a host of factors that are unknowable for Pleistocene humans (such as daily caloric expenditures for activity) dictate total energy expenditures. As with similar exercises throughout the book, the estimates provided are considered first-order approximations that provide a rough idea of the energetic realities of Neandertal biology.

CHAPTER FIVE

Surviving the Cold

As we saw in previous chapters, Neandertals had an unusual body build relative to most people alive today. Short of stature yet massive and muscular, with deep chests, wide bodies, and short limbs, Neandertals took the body form seen in modern arctic peoples one step further, producing a build that has been called "hyperpolar" or "hyperarctic" (Holliday 1997b; Trinkaus et al. 1998b). The similarities in body shape and climatic conditions between Neandertals and modern circumpolar groups like the Inuit, along with the presumption that intense cold created a large selection coefficient for body form following well-known ecogeographic rules (Holliday 1997b), has led to the widespread perception that Neandertal morphology reflects an adaptation to the cold of glacial Europe (Howell 1952; Coon 1962; Trinkaus 1981; Ruff 1994; Holliday 1997b, 1997a; Churchill 1998; Weaver 2003, 2009). Cold adaptation has also been invoked to explain characteristic aspects of Neandertal craniofacial anatomy (Coon 1962; Wolpoff 1968, 1999; Steegmann 1970; Beals et al. 1983; Trinkaus 1987a; Franciscus and Trinkaus 1988b). It would not be a stretch to say that one of the central paradigmatic tenets of Late Pleistocene paleoanthropology is the idea that the morphological and behavioral characteristics of the Neandertals were born of the adaptive challenges presented by glacial climates. Recent studies, however, have increasingly questioned the thermoregulatory adaptive benefit of various Neandertal traits, or the role of climate in producing those traits (Czarnetzki 1995; Pearson 2000; Blumenfeld 2002; Bogin and Rios 2002; Chang and Meyer 2007; Weaver et al. 2007; Weaver 2009; Rae et al. 2011). Additionally, the ability of Neandertals to endure episodes of severe cold has been questioned on morphological (Finlayson 2004), energetic (Aiello and Wheeler 2003; see also Churchill 2006), and biogeographic (Stewart et al. 2003; Stewart 2004) grounds. Nowhere have these doubts about Neandertal cold adaptation been more strongly expressed than by Finlayson, who describes the Neandertals as "people of mild climates" (2004: ix) (cf. Stewart et al. 2003; Serangeli and Bolus 2008), and whose review of Neandertal-climate associations through time concludes that cold adaptation is an inadequate explanation for their morphology (Finlayson 2004: 87). Resolving this issue is critical to both our understanding of the biology of Mousterian humans, and to our efforts to model the energetic costs of thermoregulation in Neandertals.

In this chapter, we will first examine climatic reconstructions for the glacial cycles between 240–30 Ka BP, and then review the evidence for cold adaptation in Neandertals – including the inferences we can make about morphological adaptation from their skeletal morphology and about behavioral adaptations from the archeological record, as well as reasonable inferences we might be able to make about physiological adaptations based on considerations of extant circumpolar people. This review will be undertaken with an eye towards (1) resolving the controversy concerning the degree to which Neandertal morphology (and possibly behavior) reflects adaptation to cold, (2) evaluating the efficacy of proposed cold-adapted features (that is, how well would various cold-adapted traits work in keeping a Neandertal warm?), and (3) estimating the energetic costs of those features.

5.1 HOW COLD WAS IT?

The Neandertals inhabited a climatically-dynamic Eurasia, which cycled between warm interglacial and cold glacial periods, generally with protracted temperate periods contained within the glacial episodes. These second order oscillations can be seen in the marine oxygen isotope record (Figure 5.1): the last interglacial (Eemian) is represented by stage 5e, while stages 5d-a and 3 represent temperate events within the Weichselian glacial (also known as the Würm in the now dated Alpine terminology), which had its severest downturns during stages 4 and 2. The second order climatic events of the Weichselian were themselves punctuated by third order oscillations – short warm intervals on the order of 100–1000 years' duration – known as Dansgaard/Oeschger (D/O) oscillations (Dansgaard et al. 1993; GRIP 1993). Twenty-five distinct D/O events within the Weichselian have been recognized in Greenland ice cores (Grootes et al. 1993). In addition, six distinct short-lived periods (generally less than 1000 years) of extreme cold, known as Heinrich events, were interspersed between D/O events during the last glacial episode (Bond and Lotti 1995). In combination, these larger and smaller duration oscillations created variation in the growth and size of the Fennoscandian ice sheet

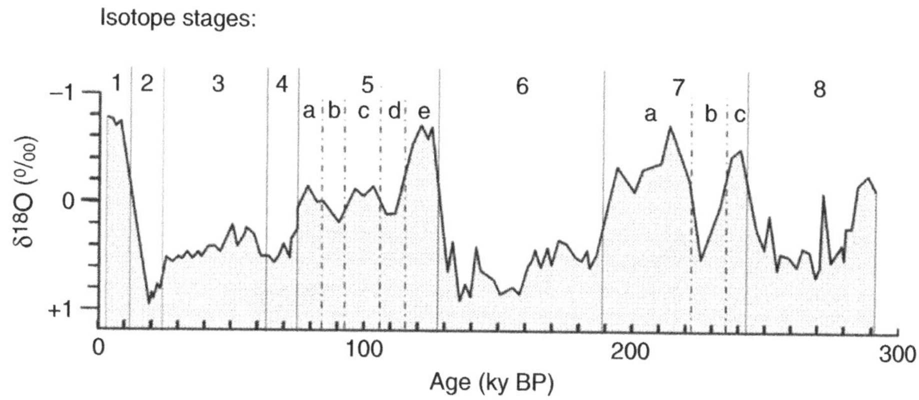

FIGURE 5.1

Later Quaternary fluctuations in oxygen isotope ratios ($^{18}O/^{16}O$), based on an average of five separate ocean cores (redrawn from Figure 2.1 in Mellars 1996). Greater amounts of ^{18}O (low points on the curve) reflect cold periods and accumulation of ice in continental glaciers; reduced ^{18}O (high points on the curve) indicate warmer intervals and reduction of ice sheets.

and in temperature and rainfall clines across time (van Andel 2003). Climate-driven changes in ocean temperatures and circulation patterns created additional variation in climatic conditions across Europe, further complicating efforts to model Pleistocene climate change across the continent (Barron et al. 2003). This temporal and spatial variation, in conjunction with imprecision in radiocarbon dates and uncertainty about ^{14}C calibration curves beyond 20 Ka BP (Pettitt and Pike 2001; Pettitt et al. 2003), makes difficult the characterization of the climates to which Neandertals were exposed. Still, we can say something in general about the climatic characteristics of Eurasia (although most work has centered on western Europe) during interglacial, cold–temperate and cold glacial periods. Thanks to the climatic modeling done by members of the Stage 3 Project (van Andel and Davies 2003), we can also connect Mousterian sites with climatic variables for at least MIS 4 and 3, giving us a better idea of the actual climatic tolerances of the Neandertals. Because the temperature extremes of the colder intervals were the likely source of selective pressures for Neandertal body form (if indeed they were cold adapted), only the climatic data for those intervals (MIS 6, 5d-a, 4 and 3) are reviewed here (climatic reconstructions for the interglacial periods MIS 7 and 5e are reviewed in Chapter 6).

5.1.1 Cold–Temperate Conditions (MIS 5d-a and 3)

Both the latter part of MIS 5 (substages 5d-a) and MIS 3 were marked by fluctuations between cold stadials and warmer interstadials as climates deteriorated towards the glacial maxima of MIS 4 and 2, respectively.[1] For MIS 5d-a (ca. 118–75 Ka BP: dates from Bassinot et al. 1994) climatic oscillations waxed and waned with a frequency of about ten thousand years. The Würm Interpleniglacial, or MIS 3 (ca. 57-24 Ka BP: Bassinot et al. 1994), was both colder and less stable than the latter part of MIS 5.

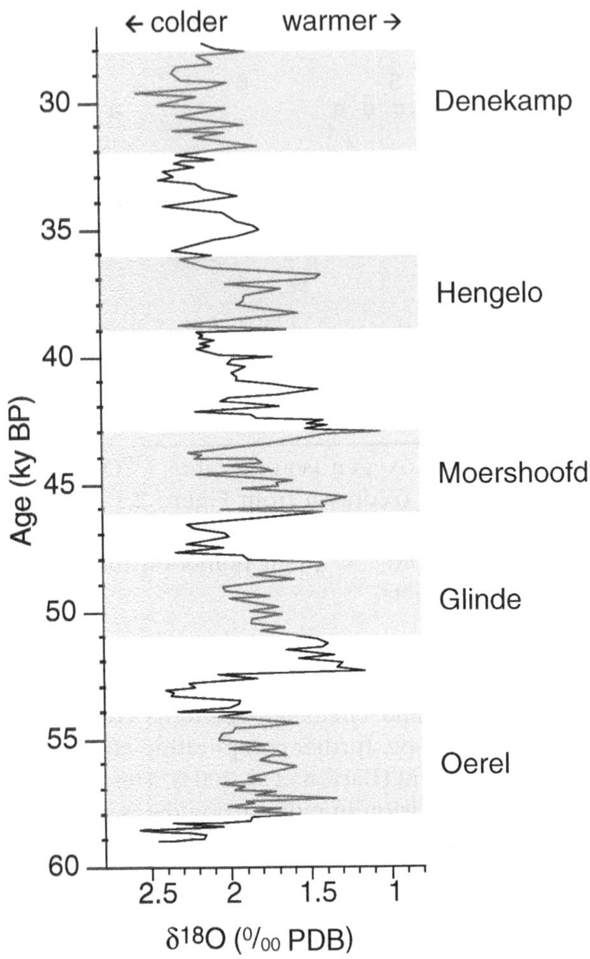

FIGURE 5.2

Correlation of named D/O events as inferred from terrestrial records with climatic change as recorded in the Greenland ice cores. Dates of D/O events following Behre and van der Plicht 1992, isotopic curve redrawn from Kennett et al. 2000, based on GRIP data (GRIP 1993). PDB = Peedee belemnite.

While most of the Interpleniglacial D/O events were on the order of 100–1000 years' duration (Bond et al. 1993), a number of longer D/O interstadials occurred during MIS 3, including (from oldest to youngest, following named intervals in NW Europe) the Oerel, Glinde, Moershoofd, Hengelo and Denekamp episodes (Figure 5.2). Although these mild episodes account for roughly half of the MIS 3 interval (van Andel and Tzedakis 1996a), colder conditions also prevailed, and even the warmest interstadials of the Interpleniglacial were likely colder than the coldest points of stage 5 (Figure 5.2: note, however, that evidence from beetles suggests that the warm interval centered on 45 Ka BP may have been as warm as today: Coope 2002). Computer modeling suggests that during these warmer interstadials mean winter (December, January and February) temperatures tended to be some 4–6 °C colder than today, at least across the more continental parts of Europe (Barron et al. 2003). The short-lived (ca. 2–4 Ka) cold episode separating the Hengelo and Denekamp interstadials probably saw conditions every bit as cold and dry as full glacial intervals: pollen spectra from Les Echets and La Grande Pile in France record mean annual temperatures 12–8 °C, respectively, below modern values and rainfall about 500 mm less than occurs today (Guiot et al. 1989). However, oxygen isotopic estimates derived directly from phosphates in animal teeth suggest that, for north-central Europe at least, conditions were likely warmer (by 2–4 °C) during MIS 5 and 3 than conventionally thought (Skrzypek et al. 2011).

5.1.2 Cold Glacial Conditions (MIS 6 and 4)

Although the cold climates of the penultimate glacial of MIS 6 (the Saale, also known as the Riss, Wolstonian, or Illinoian glacial) and the first major ice advance of the last glacial during MIS 4 differed in their severities and durations, both of these intervals saw the build-up of large continental ice sheets and reductions in sea level of 75 m or more (Shackleton 1987). MIS 6 (186–127 Ka BP: Bassinot et al. 1994) was both more uniformly cold and longer than MIS 4 (71–57 Ka BP: Bassinot et al. 1994) and, accordingly, the extent of the Fennoscandian ice sheet and the distribution of vegetation zones differed somewhat at the climaxes of these two cold spells. Although not reflected in the oxygen isotope record of deep-sea cores, the positions and extents of periglacial geological features suggests that MIS 6 may have experienced a greater accumulation of ice, and a larger drop in sea level (perhaps −140 to −150 m relative to current sea level) than that of either MIS 4 or 2, such that MIS 6 may be considered a phase of *intense* as opposed to *restricted* (i.e., MIS 4 and 2) continental glaciation (Frenzel 1973; van Andel and Tzedakis 1996b). Throughout much of this interval, permanent ice sheets extended as far south as northern Belgium and Germany, while in the British Isles (at that time joined to the continent by lowered sea levels) the glaciers extended almost to London (van Andel and Tzedakis 1996b).

While periglacial features and models of ice sheet formation suggest that MIS 6 was climatically more severe than the last glacial maximum (MIS 2) (Frenzel 1973), the oxygen isotope and pollen records are somewhat equivocal in this regard. Oxygen isotope analyses (references in Frenzel 1973) indicate no significant difference in mean annual temperature between the penultimate and last glacial maxima. The temperature and moisture tolerances of plant taxa represented in the core from Les Echets suggest that the latter part of MIS 6 (the core samples sediments dating to about 140 Ka BP at its lower limit), which was likely the coldest portion of the penultimate glacial (Figure 5.1), saw mean annual temperatures as much as 13 °C below the current local mean, only slightly colder than the maximum recorded for MIS 2 (12 °C below current: Guiot et al. 1989). The pollen from La Grande Pile actually suggests that MIS 6 was the warmer of the two glacial periods (mean annual temperature 8 °C below current compared to 11°C below for MIS 2: Guiot et al. 1989). The distribution of chenopods on the eastern Russian plain points to colder conditions in the penultimate glacial, but the totality of pollen and oxygen isotope evidence indicates no important difference in European mean annual temperatures during the last two glacial episodes (Frenzel 1973). The pollen spectra from La Grande Pile and Les Eschets (Guiot et al. 1989) also

TABLE 5.1 ■ Reductions in mean July temperatures of the last glacial maximum (MIS 2) compared to contemporary conditions, based on snow line elevations and polar limits of tree growth

	Elevation of climatic snow line	Polar limit of tree growth in plains
Southern France	−10 °C	−10 to −11 °C
Northeastern/central Europe	−7 to −8 °C	?
Vienna Gate	−9 °C	−10 °C
Southern Carpathians	−8 to −9 °C	−8 to −10 °C
Southern Ukraine	−8 to −9 °C	−10 °C
Southern Urals	−5 °C	?
Middle Yenessi and west Siberian lowlands	−6 °C	−7 °C
Baikalia	−7 to −8 °C	−6 to −10 °C
Northern Yakutia	−5 °C	?
Northern Sakhalin	−8 °C	?

Data from Table 16 in Frenzel 1973.

suggest similar reductions in rainfall during the two glacials relative to the present (ca. 200–300 mm per annum, about 600–800 mm below current values).

These uncertainties in climatic reconstruction make estimation of seasonal variation in temperature difficult. Given a paucity of temperature reconstructions for continental landmasses during the penultimate glaciation, estimates from the last glacial maximum are used here in their place (working on the assumption, as outlined above, that the two glacial episodes were roughly similar in climate). Mean summer temperatures are the primary determinant of the elevation of the climatic snow line in modern Europe (e.g., the 1000 m snow line follows the 10° July isotherm, the 1500 m line follows the 14° July isotherm, etc: Büdel 1949), such that the position of the snow line in antiquity can be used to estimate mean summer temperatures. Likewise, the position of the polar tree line on the plains corresponds to mean summer temperatures, again providing a basis for climatic reconstruction of past glacials. Table 5.1 provides estimates of the reductions in mean July temperature (relative to present-day means) at the height of the last glacial maximum for various areas of Europe and Asia. Taking the average of both estimates for the areas west of the Urals suggests a mean difference of about −9 °C in mid-summer temperatures relative to today, suggesting July averages of 6–9 °C across most of mid-latitude Europe. Beetle data from an MIS 4 site in Great Britain suggests a mean July temperature of about 13 °C (Boismier 2006), reflecting both the somewhat warmer nature of MIS 4 and the more oceanic conditions that obtained in the British Isles. Mean winter temperatures are somewhat more difficult to estimate, but computer models suggest that the peak of MIS 2 saw reductions (relative to modern temperatures) of 10 °C in average January temperatures across Europe (Kutzbach and Wright 1985: cf Frenzel 1973 for a similar estimate based on pollen), or possibly by as much as 12–16 °C across the continental portion of mid-latitude Europe (Barron et al. 2003). This would suggest mean January temperatures of about −8 to −10 °C for the western part of mid-latitude Europe (see also Boismier 2006), with possibly colder conditions further inland.

These reconstructions provide a general idea of the climatic conditions that prevailed during the colder portions of the last two glacial cycles – but what can we infer about the climatic extremes endured by the Neandertals? If Finlayson (2004) is correct, the Neandertals may simply have followed the warmer isotherms as they drifted south while the glacials waxed (or simply gone locally extinct in colder regions: Hublin and Roebroeks 2009), ending up in southern refugia that were not much colder than most parts of Europe today. Consideration of the geographic distribution of Mousterian sites by temporal intervals shows that Neandertal distributions did contract southward as the ice sheet advanced: sites are far less numerous during the MIS 4 glacial period than during warmer episodes, and are generally constrained to the south of the continent below 45° N latitude (van Andel et al. 2003). Climatic modeling, used in conjunction with site distribution data, suggests that Neandertals were able to tolerate *average* winter temperatures on the order of −20 °C (Davies and Gollop 2003). However, while most sites during MIS 4 fall below 45° N latitude, a handful of more northerly sites from this period have been found, extending as far north as Norfolk in England and Westphalia in Germany (Uthmeier et al. 2011). The recent identification of possible Neandertal remains at Okladnikov Cave in the Altai Mountains of southern Siberia (Krause et al. 2007b) supports the inference that at times they persisted in areas of bitter cold. Uncalibrated radiocarbon dates on a juvenile humerus there indicate an age between ca. 30–38 Ka BP (calibration would push these dates back in time), corresponding to either the early cold phase or a transitional phase of MIS 3 (van Andel et al. 2003). Even if the Neandertal occupation of Siberia occurred during a warm D/O event (the Hengelo or Denekamp), average winter temperatures – without taking into account wind chill – were likely on the order of −25 °C (Shichi et al. 2009).

As noted by Aiello and Wheeler (2003), wind chill temperature (WC: also known as wind chill equivalent temperature, T_{eq}) is a better measure of human cold exposure since it takes into account both ambient temperature and convective heat losses from airflow over the body. Mousterian sites are associated with median winter WC temperatures on the order of −13.5 to −18.4 °C during MIS 4 and the colder portions of MIS 3 (between 37–34 Ka BP: Aiello and Wheeler 2003). The overall pattern of Mousterian sites across space and time suggests that Neandertals had wind chill tolerances down to about −34 °C (Davies and Gollop 2003), although they generally preferred to occupy warmer areas (for example, they seem not to have recolonized north-central Europe after cold episodes until mean annual temperatures there had climbed past 6 °C: Skrzypek et al. 2011). By way of comparison, a survey of weather data for 25 sites above the Arctic Circle on a mid-January day in 2008 (www.athropolis.com/map2.htm) produced a median WC of −34.9 °C. While some of these arctic sites had wind chill temperatures well below the Neandertal tolerance (the coldest spot on the day the survey was conducted was Cambridge Bay, Nunavut, at −67.6 °C), the data suggest that Neandertals were regularly experiencing winter wind chill temperatures comparable to those encountered by modern humans living above the Arctic Circle.

Human exposure to low temperatures and wind chill is certainly attenuated by cultural buffering (shelter and clothing) and behavior patterns (reducing outside activity on particularly cold days), and probable differences in these parameters between modern circumpolar peoples and Neandertals makes a direct comparison of relative cold exposure impossible. Still, two things seem certain given the preceding review. First, Neandertals were *at times* exposed to extremely low temperatures – every bit as cold as cold-adapted circumpolar humans have experienced in recent times – which no doubt created selective filters as or more extreme than those encountered by modern arctic peoples (who are presumably better buffered against the cold). Second, while it is clear that Neandertals migrated to control as much as they could their climatic and ecological surroundings, and that to an extent they were able to capitalize on the more mild climates of southern refugia during glacial maxima, they also found themselves in places of bitter cold at times. This conclusion is supported by the association of cold climate fauna and Mousterian assemblages at some sites (Gaudzinski and Roebroeks 2000; Slimak et al. 2011). While large-bodied mammals tend to be relatively eurythermic, the modern distributions of holarctic fauna reflect ranges of temperature and biotic tolerances (see, for example, Ion and Kershaw 1989) that likely characterized their Pleistocene ancestors. In this regard, the widespread occurrence of reindeer (see Chapter 7) and the occasional presence of musk oxen (Stewart 2004) in Mousterian and Micoquian assemblages support the conclusion that Neandertals sometimes inhabited relatively cold climates. Furthermore, faunal evidence shows that even areas below 45° N latitude were not uniformly mild in climate, as revealed by the cold-resistant fauna from the MIS 6 site of La Parte in Asturias, Spain (Álvarez-Lao and García-García 2006).

While accepting that Neandertals preferred milder climates (Davies and Gollop 2003; Stewart et al. 2003) and that, for energetic reasons, their tolerance of the cold may not have been as high as that of early modern Europeans (Aiello and Wheeler 2003; Churchill 2006; but see Table 8.3 in Davies and Gollop 2003), we must admit that they did at times endure arctic-like temperatures – temperatures that are assumed to have brought about cold-adapted morphology and physiology among circumpolar peoples. Pleistocene Europe was at times warm and at times cold, but the persistence of the Neandertals in relatively cold places through the colder intervals of MIS 4 and 3 (and presumably the earlier stadials as well) makes untenable the claim that Neandertals were "people of mild climates" (Finlayson 2004: ix).

5.2 HUMAN ADAPTATION TO THE COLD

Ecogeographic patterning in human body size and shape and in thermally-relevant physiological processes constitutes perhaps the best-studied aspect of human adaptation (see reviews in Frisancho 1993; Ruff 1994; Leonard et al. 2005). Selection operating on body size and shape has, following the oft-cited Bergmann's (1847) and Allen's (1877) rules,[2] produced human morphologies that minimize the amount of heat-losing skin surface relative to heat-generating mass in cold environments, and maximize it in hot environments. Selection has also acted on physiology in people living in cold environments, both to enhance their heat generating capabilities and to reduce the flow of heat from their body core to skin surface. Before reviewing Neandertal features that have been claimed to be adaptations to cold, a few words about thermoregulation in general are in order.

For an endothermic animal, heat exchange with the environment is primarily a function of four factors; evaporation (E), conduction (K), convection (C), and radiation (R). (Additional modes of transferring heat – such as ablative cooling and gas expansion cooling – are available to organisms [see Vogel 2005a] but do not figure prominently in heat conservation strategies in large mammals.) Evaporation, conduction and radiation are all surface area-dependent processes, and thus the greater the exposed surface of the animal the more rapidly heat is transferred between organism and environment. Heat loss through evaporation into respired air may represent a special case in which evaporation is more mass than surface area dependent (since lung volume and gas exchange tend to scale isometrically with mass in mammals: Schmidt-Nielsen 1984); however, surface area-dependent anatomical mechanisms in the upper respiratory tract (discussed below) serve to recapture most, if not all, of these losses. Heat loss through convection, which for an endotherm in a cold environment generally involves heat flow from warmer body core to cooler peripheral tissues via circulation or heat lost in expired air, is only indirectly related to surface area (see below). (Of course, convection in the air surrounding the body – especially forced convection in the form of wind – lowers the temperature at the air/skin interface, but this only serves to increase *conductive* heat loss from the organism. For this reason only respiratory and circulatory convection are considered here.) Because internal convection is driven by the circulatory system, the organism can exert a degree of control over it: heat loss from the core can be reduced by selective vasoconstriction in peripheral tissues. Cold-induced vasoconstriction has its limits, however, for two reasons. First, overly cool extremities – especially those with high SA/V ratios like fingers and toes – are prone to freezing, and so must at a minimum be periodically rewarmed to prevent frostbite.[3] Second, during bouts of activity muscles must be supplied with oxygen, which necessitates increased blood flow to peripheral tissues. While muscle itself is an effective insulating tissue, its insulation value decreases with activity (Steegmann et al. 2002). Still, even with high rates of convective heat transfer to peripheral tissues, all this serves to do is increase the thermal gradient between subcutaneous tissues/skin/environment, which in turn increases the rate of *conductive* heat loss. As for respiratory heat losses through convection, these technically arise from conductive transfers across the surfaces of the lung alveoli (although gas exchange rates may scale with mass), and are attenuated by mechanisms that enhance recapture of heat along the surface of the respiratory mucosa (see below). Thus, ultimately, all of the relevant heat exchange processes are surface area dependent, and for our purposes the external surfaces (skin) probably matter more than the internal surfaces (respiratory passageways).

While heat exchange rates with the environment are surface area dependent, heat production is volume dependent, as all of the metabolic pathways available for heat generation are dependent upon the mass of various organs (and since human tissues in general have a specific density of 1 gm cm^{-3}, mass and volume can be used interchangeably: Katch et al. 1967). Figure 5.3 provides a model of energy flow through an endothermic organism, and reveals the sources of heat that arise from the thermodynamic inefficiencies which attend the transfer of energy from one form to another. These inefficiencies, or heat increments, provide the avenues by which an endotherm can generate additional heat when losses to the environment exceed normal metabolic heat production (at what is called the lower critical temperature: see Aiello and Wheeler 2003). Consideration of Figure 5.3 shows that anabolic processes (synthesis of complex organic molecules from simpler precursors) have relatively low heat increments, regardless of whether they involve growth, maintenance of bodily tissues, fat storage, or reproduction (growth of fetal tissues or milk production). Anabolism is "uphill" or endergonic (requiring input of energy for the work of production), thus most of the free energy entering the process is stored as free energy in the new tissues, and heat

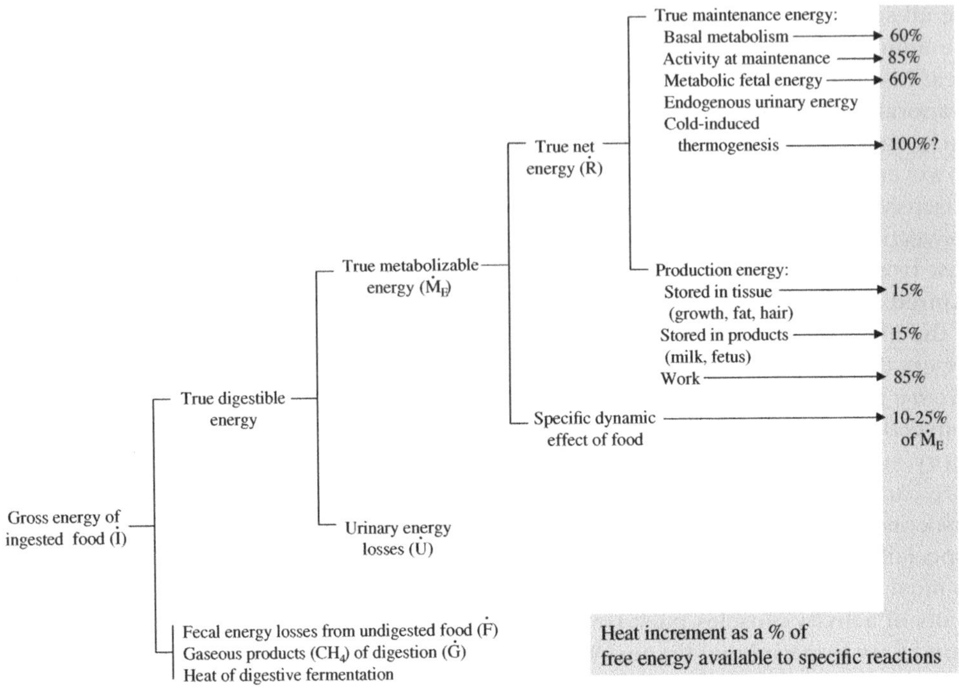

FIGURE 5.3

Biological partitioning of food energy in an endothermic animal. The right-hand column (shaded) reflects the heat increment (the relative inefficiency of the process: that is, the amount of heat produced as by-product relative to the amount of free energy entering the reaction) of various metabolic processes. Some processes (e.g. digestive fermentation) obviously apply more to some types of animals than others. The heat increment of endogenous urinary losses of energy (from secretion of nitrogen-containing substances such as urea, uric acid, creatine and creatinine) is unknown but is likely to be negligible. Adapted from Moen 1973.

production is negligible (Blaxter 1989). Catabolic processes (breakdown of complex organic molecules into simpler molecules), which involve the disassembly of energy-yielding substrates, are notably better at heat production. For example, the synthesis of adenosine triphosphate (ATP) – needed to drive most cellular processes – requires the exergonic (energy-yielding) oxidation of glucose, glycogen, or other substrates. The conversion of energy from one chemical form to another occurs at about 40% efficiency, thus more than half of the free energy entering the reaction is liberated as heat (Martini 1998). Additional heat is liberated as ATP stores are used to perform cellular work. At temperatures between 29–31 °C (the zone of thermal neutrality for humans: Hardy and DuBois 1938) heat added to the body by basal metabolism equals heat losses from the skin and respiratory system, and net change in heat is zero. Above these temperatures, vasodilation and evaporative cooling (sweating) must be called upon to eliminate excess heat, and even then there may be a net gain in body temperature (the body might store heat). Below these temperatures, metabolic heat production must be increased or a net loss of temperature will ensue.

In a hypothermic environment two metabolic pathways – both with high heat increments (Figure 5.3) – can be called upon to ramp up heat production. The first is to disregard work efficiency and simply burn fuel to produce heat in what is called cold-induced or non-shivering thermogenesis. Under the combined effect of the hormones thyroxine and norepinephrine, cellular respiration can be revved as much as 40% above basal levels (see references in Sorensen and Leonard 2001: work on rats suggests that values 80% above basal metabolism are attainable by some mammals; see references in Frisancho 1993). This response can occur within days of exposure to hypothermic environments, and apparently is accomplished by increasing the rate of active transport of cellular substances.[4] The second pathway is muscular activity, which is performed at a fuel efficiency of only 15–25% (Heglund and Cavagna 1985, 1987). Thus, muscular activity can be drawn upon – voluntarily through work or involuntarily through shivering – to provide a rapid source of heat.

Change in stored heat (S) is a function of the additive effects of surface area-dependent heat transfers and volume-dependent heat production. Depending on ambient temperatures, heat can be gained or lost by radiation (R), conduction (K) and convection (C), whereas under most circumstances evaporation (E) only serves to cool the body (objects gain heat through condensation, thus under conditions of ambient air temperatures above skin temperature and very high relative humidity – as may occur in tropical jungles – humans may gain heat as atmospheric water condenses on their skin). Because heat and mass transfers can occur across both skin surfaces and those of the respiratory tract, any model of thermoregulation should consider these pathways separately. The effect of active metabolism (M_a: here including both voluntary and involuntary contributions) and basal metabolism (M_b: here including both basal metabolism proper and cold-induced thermogenesis) is always additive:

$$\Delta S = \pm R_{skin} \pm K_{skin} \pm C_{skin} - E_{skin} \pm C_{resp} - E_{resp} + M_b + M_a$$

Thermoregulation is a matter of balancing the variables in this equation.

5.3 COLD ADAPTATION AND NEANDERTAL MORPHOLOGY

Circumpolar peoples are characterized by a suite of morphological features that have consistently been interpreted as adaptations to cold stress. These include mediolaterally wide trunks and anteroposteriorly deep chests; extremities that are short relative to the height of the trunk and that have foreshortened distal portions (forearms and hands; legs and feet); and large, brachycephalic crania (Hrdlicka 1930; Newman 1953; Coon 1962; Beals 1974; Roberts 1978; Auger et al. 1980; Ruff 1991, 1993, 1994; Lazenby and Smashnuk 1999; but see Holliday and Hilton 2010 with respect to deviations from expected ecogeographic patterns among some circumpolar groups). The great breadth and depth dimensions of the trunk, combined with shortened limbs, result in a relatively short yet massive body. These features characterized the Neandertals, but to an even greater degree (Holliday 1997b; Churchill 1998). The observed convergences in body form between Neandertals and circumpolar modern humans, coupled with the evidence that Neandertals inhabited cold–temperate environments that *at times* were severely cold, forms the inferential basis for cold adaptation in these hominins.

In Chapter 4, we asked if Neandertals were like "scaled up" Inuit, at least in terms of body composition, and the best answer we could reach was that there's no evidence to negate the working hypothesis that the two groups were similar in composition. With respect to overall body form, we already know that Neandertals were more massive per unit stature than are circumpolar peoples, and thus in important ways the groups differ in shape as well as size. Comparison of a suite of skeletal measures between Neandertals, Inuit (Alaska) and Sami (Norway) revealed generally greater skeletal robusticity in the lower limbs, but not necessarily the upper limbs, of Neandertals (Pearson 2000), which likely reflects between-group differences in habitual activity patterns and upper and lower limb biomechanical loading histories. Pearson (2000) also reported that, relative to the two arctic-adapted modern human samples, Neandertals had remarkably large hands, long claviculae and wide scapulae.[5] These differences may reflect adaptive differences (having to do either with thermoregulation or activity patterns) between groups, or may be consequences of size-and-shape scaling (allometry). Scapular breadth, as discussed in Chapter 4.2, may reflect proportional increase in muscle size with increased mass – given similar stature – in Neandertals relative to modern Inuit. These skeletal differences demand further study to decipher their significance. Nonetheless, in terms of overall body shape – involving limb proportions and mass-to-stature relationships – Neandertals, Inuit and Sami all converged (apparently independently[6]) on similar post-cranial morphology. So while Neandertals may not be like modern circumpolar peoples in every way, the striking similarities between groups in body shape strongly suggest convergent evolution under selective pressures stemming largely from cold stress.

Features thought to reflect cold adaptation in Neandertals have been well-documented (Trinkaus 1981; Beals et al. 1983, 1984; Ruff 1994; Holliday 1995, 1997b, 1997a; Porter 1999), and only a brief review will be provided here. That Neandertals were massive relative to stature – even in comparison with arctic peoples – was demonstrated in the previous chapter. We have only two Neandertals with sufficiently preserved ossa coxae to even reasonably estimate bi-iliac breadth (BIB: as a direct measure of mediolateral trunk width). Both specimens are male – La Chapelle aux Saints 1 and Kebara 2 – and both have BIBs that are equal to or above the mean value for similar-stature Inupiat males (Inupiat male mean: 29.1 ± 0.3 cm, n = 27, Ruff et al. 2005; La Chapelle: 29.2 cm, Trinkaus 2011c; Kebara: 31.8 cm, Ruff 1994). The recent virtual reconstruction of the very fragmentary pelvic remains of the Tabun C1 female (Weaver and Hublin 2009) additionally allows for a rough estimate of her BIB (based on their Figure 2) of 27.0 cm, which in this case is below the mean value (but within the range) for Inupiat females (28.6 ± 0.2 cm, n = 40, range 26.6 – 31.4; Ruff et al. 2005). Based on the relationship of BIB to mass and stature in high latitude samples (Ruff et al. 2005), the average estimated mass and stature of Neandertals would correspond with trunk widths on the order of 32.5–32.9 cm, roughly 9–15% greater than the mean values for Inuit and Aleut reported by Ruff (1994) and Ruff et al. (2005). In addition to having had mediolaterally wide bodies, Neandertals appear to have had anteroposteriorly deep thoraces as well. On the basis of the relative lengths of the medial and lateral portions of the clavicle, I have argued that the elongation evident in Neandertal claviculae is in large part a function of their having had anteroposteriorly deeper superior

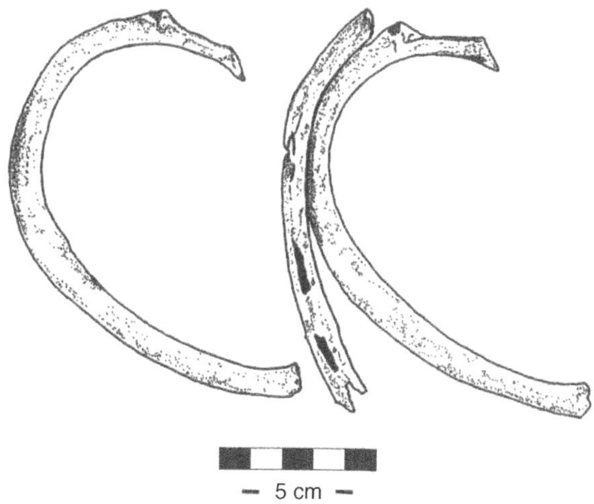

FIGURE 5.4

The right second rib of the Shanidar 3 Neandertal (center) compared to those of two recent European–American males from the collections of the Maxwell Museum of Anthropology, University of New Mexico. In this superior view, the ribs have been aligned with one another using anatomical landmarks. The two complete comparative ribs are from individuals of similar stature to Shanidar 3 (based on humeral length) yet clearly enclose a thoracic volume smaller than that of the Neandertal (see Franciscus and Churchill 2002).

rib cages (Churchill 1994a, 1994b). The deeper superior rib cage is also reflected in the size and relative lack of curvature in the superior ribs (Churchill 1994a), as is evident in Figure 5.4. This pattern holds for the lower ribs as well in some Neandertals (Shanidar 3: Franciscus and Churchill 2002), suggesting a greater anteroposterior depth along the entire thorax in at least some individuals. Other individuals may evince a greater mediolateral expansion of the thorax (Tabun C1: Weinstein 2008), or a combination of expansion in the both the mediolateral and anteroposterior dimensions of the lower rib cage (Kebara 2: Gómez-Olivencia et al. 2009), although taphonomic distortion to these specimens makes this possibility difficult to evaluate. In addition, partial mesosterna of three Neandertals (Regourdou [France] and Tabun C1 and Kebara 2 [Israel]) are long for their body size relative to modern humans (Gómez-Olivencia et al. 2012), supporting the inference of large thoraces in the Neandertals. Known thoracic remains are equivocal with respect to the possibility of ecogeographic patterns in relative chest size (where the expectation might be that European Neandertals would have relatively larger thoraces than their counterparts from the warmer climates of the Near East), in that rib remains appear not to show a difference between regions (Gómez-Olivencia et al. 2009; see also Weinstein 2008), whereas the mesosternal fossils support somewhat the idea of larger chests in the European Neandertals (Gómez-Olivencia et al. 2012). Regardless, the evidence suggests that Neandertals in both regions had wide and deep thoraces for their stature, representing a short and stocky physique that had a SA/V ratio advantageous for cold climates. This physique is reflected in their high mass-to-stature ratios, and their similarities to modern circumpolar peoples in this regard (Chapter 4).

In terms of limb proportions, Neandertals again appear "hyperpolar" (Table 5.2). Allen's (1877) rule predicts that humans living in cold environments will have shorter extremities relative to mass than their warm-adapted counterparts, and among modern humans this expectation is fairly well met (variation in overall body size related to climatic and nutritional factors complicates the picture somewhat: see Katzmarzyk and Leonard 1998). Consideration of average limb bone lengths in Neandertals shows them to fall among the mean values reported for people native to the arctic (Figure 5.5). For the European Neandertals, their average humeral, radial, femoral, and tibial lengths fall among the mean values for same-sexed circumpolar samples, and below the values for modern Europeans (including European–Americans). The average femoral length of male Neandertals from the Near East is similar to the means of the European samples, but male averages for humeral, radial and tibial length and all of the female long bone length averages fall below the European means. When limb bone length is examined relative to trunk height (in this case, skeletal trunk height, which is the summed dorsal body heights of the thoracic and lumbar vertebrae plus the ventral length of the sacrum: Holliday 1995), the European Neandertals are seen to have proximal elements (humeri and femora) that are relatively long compared to Eskimos (Koniag), and not dissimilar to Europeans (Table 5.2), while the distal elements (radii and tibiae) are intermediate between the Eskimo and Europeans. This is, however, based on very small fossil samples, and the situation is no better for the Near Eastern Neandertals, who appear to have relatively long upper limb elements (both humeri and radii relative to skeletal trunk height) compared to their European counterparts, Eskimos and even Europeans (but not sub-Saharan Africans: Table 5.2).

Allen's rule is also played out in intralimb proportions, with people from colder climates generally having lower brachial and crural indices (100 * radial maximum length/humeral maximum length and 100 * tibial maximum length/femoral bicondylar length, respectively) than warm-adapted peoples (Holliday 1997b, 1997a, 1999; but see Holliday and Hilton 2010). Among modern humans, both indices show strong negative correlations with latitude (Holliday 1997b), and the mean indices in a sample of western European Neandertals would, based on morphology–climate relationships in modern humans, equate with mean annual temperatures on the order of −1.7 to 2.4 °C (Holliday 1995) (by way of comparison, the mean annual temperature of Paris today is 11.7 °C). The mean brachial and crural indices of European Neandertals (Table 5.2) compare favorably to those of Sami herders of Norway (Trinkaus 1981), Koniag Eskimo (Trinkaus 1981; Holliday 1997b), and a very small sample (n = 2) of Inupiaq Inuit (Hawkes 1916). The Koniag, however, are a subarctic sample (from Kodiak Island, about 1000 km south of the Arctic Circle), although in some respects their morphology may be more "cold adapted" than that of groups farther north (Holliday and Hilton 2010), and the Inupiaq sample is too small to provide a reliable indication of intralimb proportions in true arctic peoples. Unfortunately, published osteometric data on Inuit and Eskimos are scarce. A better sense of Neandertal body proportions relative to those of arctic peoples can be had by referring to Figure 5.5, which uses mean long bone lengths of Inuit and Siberian (data from Jørgensen 1953: from west Greenland, north-east Greenland, Labrador, Repulse Bay, Point Barrow, Point Hope, and Siberia), Sami (Lapp) and Eskimo (data from Trinkaus 1981[7] and Holliday and Hilton 2010), and Inuit

TABLE 5.2 ■ Mean limb proportion indices in European and Near Eastern Neandertals and recent modern human comparative samples (Mean, SD, n)

	Crural Index[a]	Brachial Index[b]	FL/STH[c]	TL/STH[d]	HL/STH[e]	RL/STH[f]
European Neandertals	78.5 ± 1.5 (5)[g]	73.7 ± 2.5 (6)[g]	89.1 ± 0.0 (2)	71.2 ± 1.0 (2)	64.0 ± 1.5 (3)	47.0 ± 0.2 (3)
Near Eastern Neandertals	78.6 ± 1.1 (5)[g]	75.5 ± 3.2 (6)[g]	–	–	65.2 ± 0.1 (2)[h]	51.9 - (1)[h]
Koniag (Alaska)	80.5 ± 2.3 (20)	75.3 ± 2.6 (20)	84.3 ± 3.9 (20)	67.9 ± 3.5 (21)	62.0 ± 3.5 (21)	46.6 ± 2.7 (20)
Inupiaq (Alaska)	84.4 ± 0.4 (2)[i]	73.4 ± 2.3 (2)[i]	–	–	–	–
Sami (Norway)	78.9 (269)[j]	73.7 (215)[j]	–	–	–	–
Europeans	82.7 ± 2.4 (436)	75.0 ± 2.5 (391)	88.6 ± 2.4 (123)	73.6 ± 4.3 (124)	63.6 ± 3.4 (124)	47.9 ± 2.8 (123)
Sub-Saharan Africans	86.1 ± 2.2 (66)	79.6 ± 2.5 (67)	99.3 ± 6.7 (43)	85.5 ± 5.9 (43)	70.2 ± 4.0 (43)	55.8 ± 3.7 (43)

All values are from Holliday (1997b) unless otherwise noted.

[a] 100 * tibial length/femoral length.

[b] 100 * radial length/humeral length.

[c] 100 * femoral length/skeletal trunk height.

[d] 100 * tibial length/skeletal trunk height.

[e] 100 * humeral length/skeletal trunk height.

[f] 100 * radial length/skeletal trunk height.

[g] data from Holliday (1995), with the addition of Las Palomas 96 (Walker et al. 2011b) and humeral length estimates (Churchill, personal observation) for Amud 1 and Shanidar 1.

[h] based on STH data from Holliday (2000) and upper limb data from Churchill (1994a).

[i] data from Hawkes (1916).

[j] data from Trinkaus (1981).

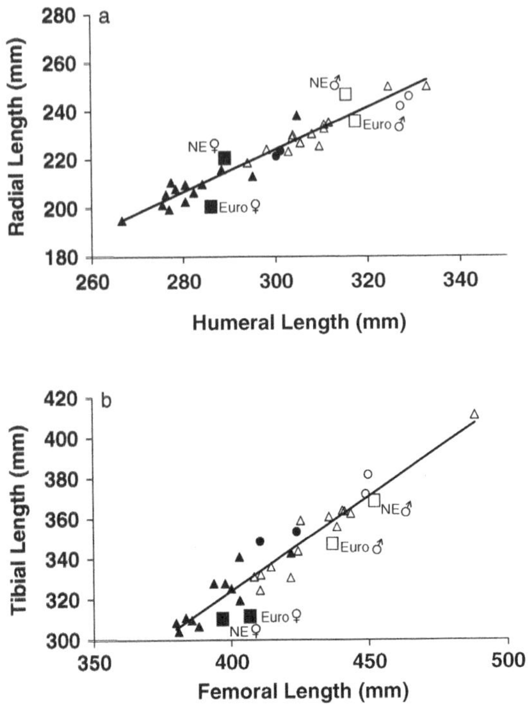

FIGURE 5.5

Mean intralimb proportions in Neandertals and circumpolar peoples. Symbols represent mean values for bone length for represented groups: triangles, arctic and subarctic populations; circles, European/European descent; squares, Neandertals: open symbols, males; filled symbols, females. Ordinary least squares regression lines are based on arctic and subarctic samples only. (a) radial length on humeral length: $y = 0.87x - 36.876$; $r = 0.9680$; (b) tibial length on femoral length: $y = 0.9378x - 51.016$; $r = 0.9670$. See text for details of sample composition.

(data from Hawkes 1916 and Holliday and Hilton 2010). The data from Hawkes (1916) and Trinkaus (1981) represent the means of indices derived from individuals – that is, from individuals preserving both the proximal and distal elements sufficient for measurement of length. The data from Jørgensen (1953) and Holliday and Hilton (2010), on the other hand, represent mean long bone lengths derived from samples of differing sizes. In some cases, the sample sizes for the two elements are similar (e.g., the Point Barrow female means were based on 25 femora and 24 tibiae), and one can imagine that the index derived from the means is very close to the mean index that would have been derived from the individuals (and thus the sampling error introduced by measuring femoral length in one sample of individuals and tibial length in different individuals would be minimal). In other cases, the sample sizes are quite disparate (for example, among the Point Hope females, the humeral length mean was based on 55 individuals, while the radial length mean was derived from 8). In these cases, it is not clear what effect the representation of different individuals in the two sample has on the position of the sample in the plots in Figure 5.5.[8] Despite these difficulties with the data, the arctic and subarctic samples produce a relatively coherent picture, with radial lengths about 73–76% of humeral length and tibial lengths about 78–84% of femoral length. In terms of radial length relative to humeral length (brachial index), the European Neandertals fall below the regression line derived from the circumpolar modern human samples, while the Near Eastern Neandertals fall above the line but still within the general cluster of the arctic samples. Interestingly, the modern Europeans also fall below, but close to, the regression line. As for tibial length relative to femoral length (crural index), all of the Neandertal samples fall below the modern human circumpolar line, with the Near Eastern Neandertals falling generally within the spread of arctic modern human samples and the European Neandertals appearing "hyperpolar." Here, the modern European samples all fall above the regression line. In general, distal limb segments appear to be more variable than proximal segments among and between groups of modern humans, with the length of the tibia being perhaps more variable than that of the radius (Holliday and Ruff 2001), and thus the crural index may reflect climatic adaptation more than does the brachial index. Regardless, Neandertals overall appear to be characterized by a body form like that of modern circumpolar peoples, with wide and massive bodies and short limbs, particularly the distal segments (Holliday 1997b).

Not to belabor the point, but foreshortened distal elements are also apparent in Neandertal hand bones. As demonstrated by Lazenby and Smashnuk (1999), Inuit have shorter second metacarpals relative to body size (using femoral head diameter as a proxy for body mass) than do Europeans, which they interpret in terms of adherence to Allen's rule. That hand proportions would be the target of selection in cold environments is not surprising, given the potentially debilitating effects of loss of fingers to frostbite (Gilligan 2010b). Neandertals are fairly variable in relative second metacarpal length (Figure 5.6), and one of them (the Tabun C1 female from MIS 4 [glacial] Israel)

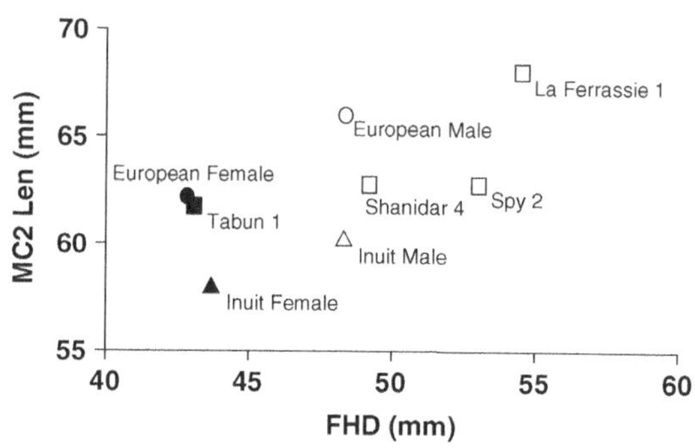

FIGURE 5.6

Second metacarpal length on femoral head diameter in individual Neandertals (□) and samples of Inuit (Southampton Island, △) and Europeans (Canadian settlers, ○). Open symbols, males; filled symbols, females. Inuit and European data points represent weighted means of right and left hand/leg data (data from Lazenby and Smashnuk 1999). Neandertal MC2 length data from Niewoehner et al. 1997, FHD data from Trinkaus 1984.

has a morphology that falls among the European females. Two of the others (the males La Ferrassie 1 from MIS 4 France and Shanidar 4 from MIS 3 [cold–temperate] Iraq) have values intermediate between the European and Inuit samples, while one (Spy 2, a male from MIS 3 Belgium) has morphology like that of the Inuit. Thus three of the four Neandertals sampled have short relative MC2 lengths compared to Europeans.

Experimental work with living humans has shown that there is an energetic (and hence thermoregulatory) benefit to shorter extremities in hypothermic environments. In situations of whole body cooling in air, people with longer extremities consume more oxygen than do those with shorter extremities – even after variation in lean body mass and fat composition have been controlled for (Tilkens et al. 2007), beginning only a few minutes after exposure to cool temperatures. An increased rate of oxygen consumption equates with increased metabolic rate, which indicates a thermoregulatory effort on the part of the subject's body to replace heat lost to the environment. This and other experiments (involving whole body cooling in water: see references in Tilkens et al. 2007) demonstrate the link between body form and thermoregulation that is assumed in Bergmann's and Allen's rule. But the question remains: just how good was the Neandertal body form at resisting cold stress?

5.3.1 The Survival Value of Neandertal Body Form

The combined effect of wide and deep trunks and short legs was a body that was massive for its stature. When the ratio of mass/stature is plotted against latitude (Figure 5.7), a clear tendency can be seen for extant humans to become less "linear" and more "lateral" with increasing latitude. At the equator, people average between about 310–330 grams of mass per centimeter of stature. Both males and females add slightly more than a gram per centimeter, on average, with every degree of latitude distance from the equator. As latitude increases and mean annual temperatures drop, local populations respond by increasing their laterality or squatness – effectively reducing SA/V by more closely approximating the shape of a sphere. Neandertals appear to have carried this strategy to an extreme. Male Neandertals from both Europe and the Near East and females from Europe had mean mass/stature ratios above values reported for modern humans and most similar to circumpolar peoples (on modern weight-for-height charts most Neandertals would fall in the "overweight" category, and some would even be classified as "obese," despite having body fat percentages that were probably low by western standards [see Chapter 4.2]).

All of this invites the question of how much good such a body shape would do for a hapless Neandertal trying to survive a long, harsh winter. It's important to be clear that Neandertals likely had a number of behavioral and physiological means of staying warm, as well as other morphological means, in their thermoregulatory bag of tricks – here, we are simply concerned with the thermoregulatory benefits (and ultimately, energetic costs) of their overall body form. SA/V ratios appear to be critical to maintaining thermal constancy in cold conditions, and high average Neandertal mass/stature ratios suggest that they had low SA/V ratios. Determining what impact this body shape had on thermoregulatory competence, however, requires some estimation of the actual surface areas of these hominins. Regression equations for estimating surface area from stature and mass are commonly employed by anesthesiologists (DuBois and DuBois 1916; Boyd et al. 1927; Gehan and George 1970; Bailey and Briars

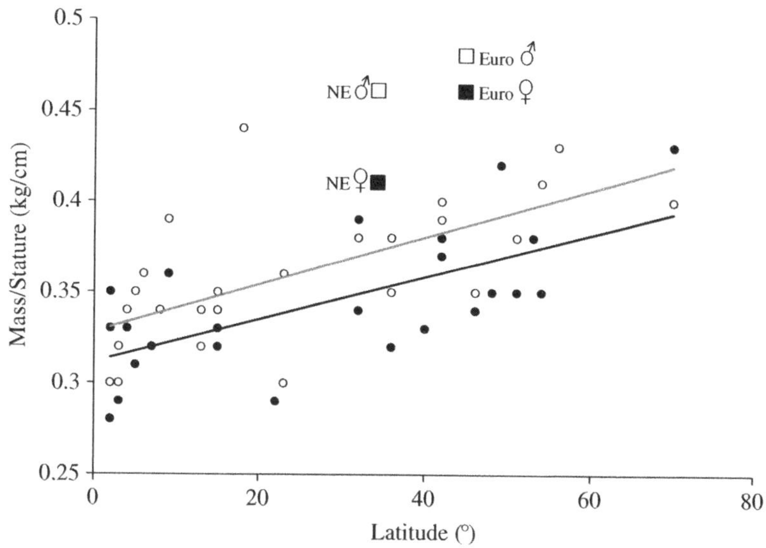

FIGURE 5.7

Mean mass/stature ratios relative to latitude in modern humans (○) and European (Euro) and Near Eastern (NE) Neandertals (□). Ordinary least squares regression lines are based on modern human sample means only: females (—, ●) y = 0.0012x + 0.3117, r = 0.6766; males (, ○) y = 0.0013x + 0.3282, r = 0.6666. Modern human data from Ruff 1994, Neandertal data from Table 5.6.

1996), because the medicinal dosages needed for anesthetics and chemotherapy are dependent upon organ surface areas and renal (glomerular) filtration rates, which are themselves proportional to skin surface area. Because these formulae are derived from modern human samples, one might wonder if they can reasonably be used to estimate surface area in humans that lie at one extreme of the range of variation in body shape. Construction of a scale model of a male Neandertal suggests the answer is "yes." Scale models have been used to estimate body size and shape parameters in other areas of paleontology (Colbert 1962; Alexander 1985b; Paul 1988; Farlow et al. 1995), and elsewhere (Churchill 2006) I report the results of efforts to use an anatomically accurate scale model to determine the applicability of clinical equations for estimating Neandertal surface area. This endeavor involved constructing a proportionately correct half-stature model of the La Ferrassie 1 male Neandertal (a fairly complete individual, whose stature, at 171 cm, is somewhat above but close to the male Neandertal mean), with the understanding that a one-half linear scale model would have one-fourth the surface area and one-eighth the volume of its full-size equivalent. Surface area and volume were measured directly on the model and scaled to life size. The volume of the model corresponds with a full-size adult of 84.4 kg, very close to the estimate of 85 kg for La Ferrassie 1 obtained by Ruff et al. (1997) based on a combination of femoral head and bi-iliac breadth + stature methods. The model study showed that equations in clinical use for estimating surface area work reasonably well with Neandertals, and one in particular works quite well (that of Gehan and George 1970) – likely

TABLE 5.3 ■ Estimated surface area (m^2) in Neandertals[a]

Females	
La Ferrassie 2	1.722
Las Palomas 92	1.649
Las Palomas 96	1.646
La Quina 5	1.815
Shanidar 6	1.588
Spy 1	1.757
Tabun C1	1.676
Female mean	**1.693**
(SD, n)	(0.077, 7)
Males	
Amud 1	1.944
La Chapelle 1	1.894
La Ferrassie 1	2.034
Fond de Foret 1	1.970
Kebara 2	1.887
Kiik Koba 1	1.909
Lezetziki 1	1.865
Neandertal 1	1.933
Regourdou 1	1.832
Saint Cesaire 1	1.924
Shanidar 1	1.978
Shanidar 2	1.852
Shanidar 3	1.941
Shanidar 4	1.816
Shanidar 5	1.807
Spy 2	1.966
Male mean	**1.910**
(SD, n)	(0.064, 16)

[a] Surface area estimated following Gehan and George (1970) using mass and stature from Table 4.1.

because it incorporated infants and children into the estimation sample and may thus be applicable to a wider range of human body shapes (Churchill 2006).

The surface area equation of Gehan and George (1970) was thus used with mass and stature estimates to predict surface area for seven female and 16 male Neandertals (Table 5.3). Results reveal that Neandertals had absolutely greater amounts of surface area compared to modern humans – even arctic-adapted modern humans – of similar stature (Table 5.4 and Figure 5.8). Inupiats are close to Neandertals in mean stature (for males: 167.4 vs. 166.1 cm, respectively; for females: 155.9 vs. 156.6 cm, respectively: Inupiat data from Ruff et al. 2005), yet Neandertals average about 0.1 m^2 – or about 5–6% – more skin SA regardless of sex (when sex-specific average Inupiat statures and masses are used to estimate SA). However, Neanderthals also tend to be more massive for a given stature – by 11–13% – than the Inupiat, and thus tend to have lower ratios of SA to mass (Figure 5.8). Of the modern human samples represented in Figure 5.8, only the two Inuit samples (represented by the two triangles closest to the regression line) fall close to the regression line that describes the relationship between SA and mass in

TABLE 5.4 ■ Body size and estimated surface area (SA) in nilotic and arctic groups (sexes combined)[a]

Population	Stature (cm)	Mass (kg)	SA (m^2)	SA/M (m^2/kg)
Nilotics				
Turkana	177	55.0	1.645	0.0299
Shiluk	179	57.6	1.693	0.0294
Ruweng Dinka	181	57.5	1.699	0.0296
Ageir Dinka	183	58.5	1.723	0.0294
Arctic				
Wainwright Inuit	161	66.8	1.747	0.0262
Inupiat	161	63.2	1.698	0.0269
Igloolik	162	63.6	1.708	0.0269
Aleut	159	60.6	1.653	0.0273
Sami	171	66.1	1.782	0.0270
Finn	173	70.9	1.857	0.0262
Evenki	158	57.8	1.609	0.0278
Evenki	153	52.0	1.503	0.0289
Yakut	162	62.9	1.699	0.0270
Neandertals[b]	**163**	**73.1**	**1.847**	**0.0251**

[a] Surface area was estimated following Gehan and George (1970) from mean stature and mass of the combined sexes. Average stature and mass values are from Roberts and Bainbridge 1963 for nilotics and Laughlin 1951; Eveleth and Tanner 1976; Leonard et al. 1996; Snodgrass et al. 2005; Galloway et al. 2000; and Ruff et al. 2005 for arctic groups.

[b] Based on mean stature and mass of the combined male and female samples.

Neandertals. One sample of Evenki reindeer herders of Siberia (Galloway et al. 2000) – represented by the triangle in the lower left corner of the plot – also appears to be similar in body shape to Neandertals. Sami, ethnic Finns, Yakuts and a second sample of Evenki are all relatively cold-adapted but fall farther away from the line (as seen in the remaining triangles). The earliest modern Europeans (Early Upper Paleolithic sample) are more similar to North Africans (North African Epipaleolithic sample) in SA/M relationships, while by the later Upper Paleolithic (Late Upper Paleolithic sample) modern Europeans had converged somewhat on the SA/M relationship characteristic of Neandertals and Inuit. Nilotic East Africans, not surprisingly, can be seen to have high SA/M ratios.

With estimates of mass and surface area, we can explore the thermoregulatory benefits of Neandertal body form. This exercise has been performed before, first by Aiello and Wheeler (2003), and then by me (Churchill 2006), with similar results. Elsewhere (Churchill 2006), I have argued that SA is better for predicting BMR than is mass, since the rate of metabolic breakdown of pharmaceuticals and chemotherapeutic agents (as a reflection of metabolic rates) is proportional to SA in humans (Gehan and George 1970). I thus used empirically determined BMR/SA relationships in humans (Winslow and Herrington 1949) to estimate Neandertal BMR from their estimated SA. It has since been pointed out to me that there is some circularity in basing estimates of both heat production and heat loss on skin surface area (oops!). To avoid this pitfall, I here follow Aiello and Wheeler (2003) in estimating BMR (and hence heat production at rest)

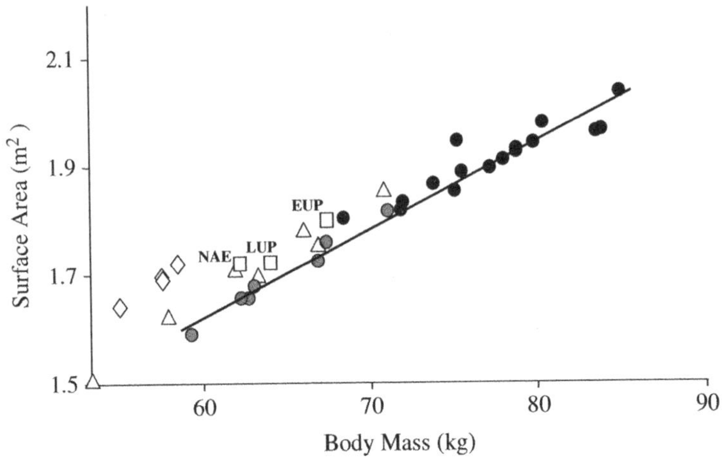

FIGURE 5.8

Skin surface area (m^2) on body mass (kg) for individual Neandertals and means of combined-sex samples of recent and fossil modern humans. Individual Neandertal male (●) and female (●) data from Tables 4.1 and 4.3. Surface areas for modern humans calculated from published mean stature and mass: diamonds, warm-adapted Nilotics (Roberts and Bainbridge 1963); triangles, cold adapted Inuit, Evenki, Yakut, Sami and ethnic Finns (Eveleth and Tanner 1976; Leonard et al. 1996; Galloway et al. 2000; Ruff et al. 2005; Snodgrass et al. 2005); squares, European Early [EUP] and Late Upper Paleolithic [LUP] and North Africa Epipaleolithic [NAE] fossil modern humans (Ruff et al. 1997 supplemental data). Ordinary least squares regression line based on Neandertals only, $y = 0.0159x + 0.6719$, $r = 0.9798$.

from body mass (from Chapter 4.3) alone and total conductance (as a measure of heat loss) from SA. Because my mass and surface area estimates differ slightly from theirs, and because they considered only males, the results here differ somewhat from those of Aiello and Wheeler (2003).

Following Aiello and Wheeler (2003), two measures of cold tolerance can be calculated from the Neandertal data: lower critical temperature (T_{lc}) and minimum ambient temperature (T_{min}). The lower critical temperature is the temperature below which natural insulation (for a naked person, so only the insulation provided by hair, skin, fat, and muscle) is no longer sufficient to maintain a constant temperature at the BMR. Below T_{lc} a person must begin to expend metabolites – through increased metabolic rate, shivering, or voluntary muscular activity – simply to stay warm. Here, the amount and insulative value of bodily tissues becomes important: skin and muscle are similar in conductivity (0.50 and 0.46 $W\ m^{-1}\ K^{-1}$, respectively) but fat is considerably better at insulation (0.21 $W\ m^{-1}\ K^{-1}$) (Vogel 2005a). Empirical data suggest that humans have an average conductance[9] of 5 $W\ m^{-2}\ K^{-1}$ (Mount 1979), but this value may overestimate heat loss in females given their greater stores of subcutaneous fat. The lower critical temperature can be calculated from BMR (in Watts), SA and conductance (C: using the value of 5 $W\ m^{-2}\ K^{-1}$) as:

$$T_{lc} = 37\,°C - (BMR/C \cdot SA)$$

Using the high end of the mean male and female BMR estimates from Chapter 4, and using SA estimates from Table 5.3 produces T_{lc} estimates of 25.8–27.2 °C for Neandertal males and females, respectively (if the low end of the BMR estimates is used the corresponding values are 27.8–29.0 °C, respectively). By comparison, Inupiat Inuit have T_{lc}s on the order of 28.0–28.1 °C, and modern French come in at about 28.2–28.3 °C (using stature and mass data from Ruff et al. 2005 for the Inuit and Eveleth and Tanner 1976 for the French, and using the equation of Kleiber 1961 to estimate BMR from mass).

The second measure of temperature tolerance, T_{min}, is the lowest temperature that a naked human could endure for a sustained period without becoming critically hypothermic. Estimates of the minimum ambient temperature are based on the finding that the maximum metabolic rate (and thus maximum heat output) that humans can sustain for long periods of time is roughly three times BMR (Burton and Edholm 1955), and are calculated as:

$$T_{min} = 37\,^{\circ}\mathrm{C} - (3 \cdot \mathrm{BMR}/\mathrm{C} \cdot \mathrm{SA})$$

Using the same values as above, T_{min} estimates for average-sized Neandertal males and females are 2.0° and 3.5 °C, respectively (9.6° and 13.0 °C if the lowest BMR estimates are used), while the Inuit have values of 10.1–10.2 °C and the French 10.6–10.9 °C.

What does this mean in terms of cold adaptation? The short and stocky bodies of Neandertals appear to have bought them a little additional buffering from the cold, but only a little. The T_{lc} of Neandertals is not that much better than that of the modern human comparators, and not what one would expect of endotherms adapted to arctic-like temperatures. In stark contrast, arctic foxes have T_{lc} below −30 °C, and large endotherms like muskoxen may have T_{lc} of −40 °C or lower (Scholander 1955). The Neandertals, despite their short and stocky build, suffered the legacy of a tropical ancestry, just as do modern Inuit and Eskimo. It is important to note, however, that the low T_{lc}s of arctic animals are not won by body size and shape alone – they are in larger part a function of effective insulation (and hence reduced conductance: a thick layer of fur may have a conductivity of 0.024 W m^{-1} K^{-1}, compared to 0.50 W m^{-1} K^{-1} for naked skin: Vogel 2005a). Human use of insulation has an enormous effect on temperature tolerances (to my knowledge, no one has yet tried the experiment of shaving a muskox and then determining its T_{lc}, but this would be the only fair comparison to the naked Neandertals considered above), and when clothing is added to the picture the thermoregulatory advantages of Neandertal body form become a bit more apparent (see below).

5.3.2 Craniofacial Morphology and Cold Adaptation

While people indigenous to cold climates share aspects of body size and shape that have been interpreted as adaptations to cold stress, documenting similar convergences in craniofacial morphology has proven a bit more difficult. Features of the face that have been argued to provide an adaptive advantage in cold climates in Asiatic people, such as reduced brow ridges and well-developed, laterally-projecting zygomatic arches (Ingman and Gyllensten 2007), have failed to confer any demonstrable advantage in experimental test cases (Steegmann 1965, 1970). Indeed, it even remains uncertain the extent to which frostbite is or ever was a selective agent in shaping human facial morphology (Steegmann 1967), and comparative studies suggest that – at least for Eskimos and Inuit – masticatory and paramasticatory biomechanics better explain facial morphology in arctic peoples (Hylander 1972, 1977; Shea 1977). Only three craniofacial features have actually been shown to exhibit either ecogeographic patterning in modern humans or correlations with climate among circumpolar peoples: head shape, sinus size, and nasal shape. Each of these features will be reviewed briefly below.

As with the flat faces of Eskimos/Inuit, the projecting faces of Neandertals have been argued to have been the product of cold stress. Coon (1962) contended that midfacial prognathism increased the thermal buffer between cold ambient air and the temperature-sensitive brain of Neandertals. While Coon's thermal buffer idea is no longer widely accepted, cold adaptation remains one of the main explanatory models in studies of Neandertal facial morphology (Steegmann 1970; Trinkaus 1987a; Franciscus and Trinkaus 1988a; FH Smith 1991; Wolpoff 1999). (The other major model is paramasticatory use of the anterior teeth: Smith 1983; Rak 1986; Demes 1987; Trinkaus 1987b; Smith and Paquette 1989; Spencer and Demes 1993: but see Antón 1994, 1996; O'Connor et al. 2005). With respect to the three features mentioned above – head shape, sinus size, and nasal shape – Neandertals evince some morphologies that are inconsistent with expectations for cold-adapted people based on modern human ecogeographic patterns. Consideration of these features is increasingly leading to claims that, at least when it comes to craniofacial form, Neandertal morphology cannot be explained by reference to cold-climate adaptation (Hylander 1977; Blumenfeld 2002; Franciscus 2003; Márquez et al. 2005; Rae et al. 2006, 2011; Chang and Meyer 2007: see also Weaver et al. 2007).

Ecogeographic patterning in cranial morphology, related to variation in temperature and humidity, has been demonstrated among recent humans (Guglielmino-Matessi et al. 1979), and variation in cranial shape in particular has been argued to reflect climatic stressors (Coon 1955; Beals 1974). Given the brain's high metabolic rate and heat production, as well as its sensitivity to temperature (Cabanac and Caputa 1979), selection would likely favor skull shapes conducive to heat dissipation in hot climates and heat retention in cold environments. Under cold conditions it may pay to have a more spherically-shaped (brachycephalic) cranial vault, which would reduce SA/V relative to an elongated (dolicocephalic) shape (Coon 1955). While considerable variation exists within climatic types, overall patterns in human cephalic indices (100 * cranial breadth/length) adhere to this expectation (Beals 1974). While modern humans from colder environments have higher (brachycephalic) indices, on average, than do people from hot environments, the pattern does not hold well for Neandertals (Table 5.5; Figure 5.9). As can be seen in Figure 5.9, modern human cranial vault shape follows a clear pattern with temperature and humidity: as reported by Beals et al. (1983: 425), "ethnic groups exposed to winter frost have a mean cephalic index that is 4.3 units higher than those living within the tropics." Two things become apparent when geographic and temporal variation in Neandertal morphology is examined against the modern human pattern (Table 5.5; Figure 5.9). First, Neandertals do not show the kind of morphological convergence with cold-adapted modern humans that was seen for aspects of body shape. Neandertals from glacial (MIS 6 and 4) and cold–temperate (MIS 5d-3) intervals have mean cephalic indices that are more similar to modern humans from hot and dry climates. Those from temperate interglacial Europe

TABLE 5.5 ■ Mean cephalic index by climatic condition for recent human samples and Neandertals; Mean, SD, n

	Dry Cold	Wet Cold	Temperate	Wet Hot	Dry Hot
Recent humans[a]	82.3 ± 3.5 (60)	80.5 ± 2.7 (34)	–	78.7 ± 3.4 (121)	76.0 ± 3.2 (124)
European Neandertals	75.2 ± 3.2 (12)[b]	–	80.4 ± 5.2 (4)[c]	–	–
Near Eastern Neandertals	75.9 ± 2.7 (4)[d]	–	–	–	–

All data from Beals 1974 and Beals et al. 1983.

[a] Grand means of group means from Beals 1974, number in parentheses represents the number of groups comprising the grand mean.

[b] MIS 5d-3 specimens: La Ferrassie 1, La Chapelle 1, Neandertal 1, Spy 1 and 2, La Quina 5 and 18, Ganovcé, Gibraltar 1, Le Moustier 1, Monte Circeo, Subalyuk.

[c] MIS 7a and 5e specimens: Ehringsdorf H, Krapina 3 and 5, Saccopastore 1.

[d] MIS 5d-3 specimens: Amud 1, Shanidar 1, Tabun C1, Teshik-Tash 1.

(MIS 7 and 5e) have a mean cephalic index more similar to modern people from somewhat colder (yet wet) climates (note that Beals [1974] did not provide data for modern humans living in mid-latitude temperate environments).

Overall, Neandertals tend to be dolicocephalic but, to be fair, the Neandertal and modern human data are not directly comparable. This is so in part because the Neandertal data represent sample means, while the modern human data represent the grand means of large numbers of samples (compiled from Appendix 1 in Beals 1974). The modern human grand means mask a considerable amount of variation in sample means and, while it is true that the mean cephalic index for European cold-climate Neandertals falls two standard deviations below the grand mean for modern humans from

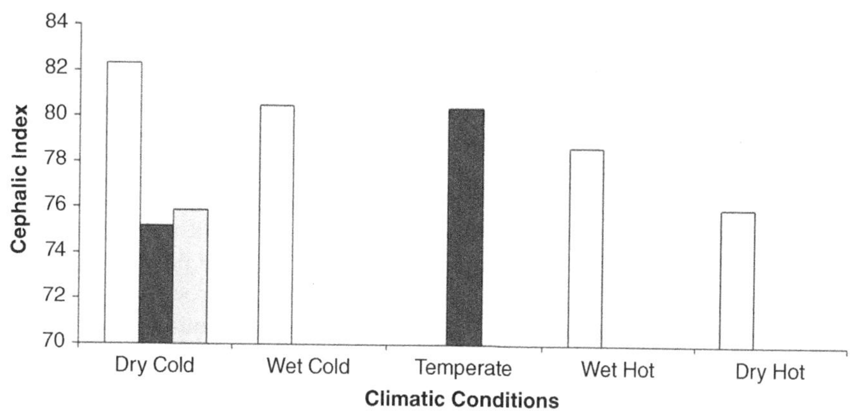

FIGURE 5.9

Mean cephalic indices (100 * cranial breadth/length) in modern humans (□), European Neandertals (■) and Near Eastern Neandertals (▒). All data from Beals 1974 and Beals et al. 1983. Sample composition as in Table 4.5.

similar environments, it is also the case that a number of modern human cold-adapted samples have mean indices quite close to the Neandertal mean (and one group – Koukpagmiut Inuit – have a mean value below the Neandertals at 74.6). Also, to be fair, the Neandertals should be compared to their contemporaries from different climatic conditions – not to modern humans – since species-level differences in cranial vault shape may be confounding any apparent climatic signal. In other words, we should ask if the Neandertals had rounder heads than their contemporaries in hot climates. According to Beals et al. (1983, 1984), they did. However, such a comparison encompasses unwelcome taxonomic diversity, and potentially conflates climatic and phylogenetic sources of variation in cranial shape. This brings us to the second thing that can be seen in Figure 5.9 and Table 5.5, which is that expected patterns of cranial vault variation with climate do not hold within the Neandertal sample. Near-eastern cold-interval Neandertals, who presumably would have experienced less severe extremes of climate than their European contemporaries, have skulls that are similar in shape to their European cousins. And Neandertals from interglacial episodes appear, in terms of cephalic indices, more cold-adapted than their glacial counterparts. Taken together, the lack of convergence between Neandertal and cold-adapted modern human morphology and the lack of clear ecogeographic patterning within the Neandertal sample argue against a cold-adaptation explanation for Neandertal head shape.

With respect to size of the facial sinuses (maxillary and frontal), Koertvelyessy (1972) documented a positive correlation between temperature (in this case, wind chill temperatures), absolute humidity, and sinus sizes among Alaskan natives. The finding that sinuses become smaller as climates become colder and drier has also been supported by other work with Eskimos and Inuit (Shea 1977), and by work with non-human primates (Rae et al. 2003). The adaptive significance of this relationship is not understood, but it has been suggested that, among highly active people in cold environments, heat dumping via the head and face is an important component of cold adaptation, and that smaller sinuses facilitate heat loss (Koertvelyessy 1972). Recent experimental work with rats, however, suggests that decreased pneumatization of the craniofacial skeleton is an epigenetic response to development in cold conditions (Rae et al. 2006). While this does not rule out the possibility that there is an adaptive component to the size of the facial sinuses, it does raise the strong possibility that ecogeographic patterns in sinus size are the passive by-product of altered facial developmental trajectories for humans growing up in cold environments – that is, that there is not a direct adaptive value to this feature. Regardless of the adaptive valence of the trait, Neandertals once again do not conform to expectations for cold-adapted humans: they tend to have absolutely large maxillary and frontal sinuses (EH Smith 1991; Hublin 1998; Harvati 2007). It has been suggested that the answer may lie in a physiological demand for increased production of nitric oxide, which acts as an aerocrine messenger that selectively dilates pulmonary arteries without producing systemic vasodilation. Nitric oxide is produced in the paranasal sinuses, and can serve to counter activity-induced pulmonary vasoconstriction. Thus, nitric oxide demand related to high activity levels may have necessitated large facial sinuses in Neandertals (with a selective advantage sufficient to override the epigenetic mechanism that results in small sinuses in cold climates) (Marquez et al. 2002). However, Neandertal facial sinus size appears to scale with overall facial size in a way similar to that seen in modern Europeans (Rae et al. 2011), suggesting that large sinuses are simply a by-product of large Neandertal faces.

While the potential adaptive value of large sinuses in Neandertals demands further research, at present all we can say is that, like cranial vault shape, Neandertal sinus size does not seem to fit expectations of cold-adapted morphology. Holton and colleagues (2011) note, however, that paranasal sinuses may not be the most appropriate feature with which to evaluate cold adaptation in the Neandertal face, and argue that the internal nasal passages of Neandertals do – when compared with their archaic human contemporaries from other regions – follow expectations for cold-adapted hominins.

Ecogeographic variation in nasal morphology stands as one of the best-documented patterns of human morphological adaptation to climate (e.g., Thomson and Buxton 1923; Davies 1932; Weiner 1954; Wolpoff 1968; Franciscus and Long 1991). The nose performs physiologically vital functions by warming and humidifying inspired air and by cleansing that air of particles and soluble gasses (pollutants) (Ingelstedt 1970; Proctor 1977; Cole 1982a; Rouadi et al. 1999). Thus, the nose (primarily the mucosa-lined, internal nasal passageways, but also including to a smaller degree the external nose) is the first and most important defense mechanism protecting the sensitive alveolar epithelium of the lungs. The nose also recaptures heat and moisture on exhalation, thereby reducing the caloric and water cost of conditioning inspired air (Davies 1932; Webb 1951). In cold or dry environments, both the need for and the challenge of conditioning inspired air are greater. Since exchange of heat and moisture, as well as filtration of particles and gases, occurs at the air/mucosa interface, improved air conditioning performance can be won by either increasing the absolute amount of mucosa (as by elongating the nasal passageway, such that inspired and expired air is in contact with the mucosa over a longer distance) or by increasing mucosal surface area relative to nasal cavity volume, or both. In the noses of people living in cold or dry environments we see a number of morphological features that serve these ends, including: (1) a more projecting external nose, which serves to increase the total length of the nasal passageway (Wolpoff 1968; Carey and Steegmann 1981; Franciscus and Trinkaus 1988a); (2) relatively narrow airways (reflected in absolute and relative width of the skeletal piriform aperture: Thomson and Buxton 1923; Davies 1932; Weiner 1954; Wolpoff 1968; Franciscus and Long 1991); and (3) larger turbinates, especially the inferior and middle nasal conchae (Shea 1977).

Other nasal features show strong ecogeographic patterning, but do not appear to appreciably alter surface area-to-volume (SA/V) ratios in the nose. These features include: (1) the orientation of the nostrils (Courtiss and Goldwyn 1983; Calhoun et al. 1990); (2) the relative size of the nasal valve (Courtiss and Goldwyn 1983; Calhoun et al. 1990; Ohki et al. 1991); (3) the height of the inferior nasal sill (Cottle 1955); and (4) the mediolateral placement of the inferior and middle turbinates (Courtiss and Goldwyn 1983). These features are thought by some to regulate airflow dynamics in the nose, specifically to either moderate the degree of turbulence in the nasal air stream or to direct the air stream into particular regions within the nose. Turbulent airflow is advantageous for air cleansing and conditioning, since it destabilizes the laminar boundary layer at the mucosal surface and disrupts the temperature and humidity gradients that exist with smooth, laminar flow (Cole 1982b; Hahn et al. 1993; Simmen et al. 1999). This keeps the average gradient between the air and mucosa high (resulting in more rapid heat and mass transfer rates), and also increases particle impaction on the mucosal surface. However, drag forces along the walls of the passageway are greatly increased with turbulent flow (Whitaker 1968), increasing the work of breathing relative to laminar nasal airflow (Cole 1982b; Simmen et al. 1999). Thus, in cold/dry environments the conditioning benefits of turbulence outweigh the energy costs of turbulent flow, and populations adapted to these environments exhibit features thought to induce turbulence: downwardly directed nostrils, narrow nasal valves, high nasal sills, and projecting turbinates (Cottle 1955; Cole 1982b; Courtiss and Goldwyn 1983). Furthermore, enlarged turbinates are thought to direct the air stream into the mid-passage where SA/V relationships are most conducive to conditioning cold, dry air (Uliyanov 1997a, 1997b). In hot and humid environments, air modification is less critical, and it is thought that in populations living there selection has favored features conducive to less costly laminar flow: more anteriorly directed nares, a wide or absent nasal valve, minimal or no development of a nasal sill, and less projection of the inferior turbinates (Cottle 1955; Courtiss and Goldwyn 1983). The smaller size of the turbinates results in airflow predominately along the lower passageway (Uliyanov 1997a), where airflow resistance is less and the energetic cost of breathing lower.

Neandertals had large noses, both in terms of large and projecting external noses and large, voluminous internal passageways. Great projection of the external nose is indicated by the skeletal morphology around the piriform aperture, which typically includes large, projecting anterior nasal spines, strong eversion of the lateral margins of the piriform aperture, and horizontal orientation of the distal nasal bones (Franciscus and Trinkaus 1988b; Márquez et al. 2005). Great size of the nose is reflected in the size of the piriform aperture, which was both mediolaterally broad and superoinferiorly tall (Coon 1962; Franciscus and Trinkaus 1988b; Franciscus 1995). Nasal height in Neandertals scales with facial height, while nasal breadth can be seen to be large relative to any measure of facial size (Franciscus 1995; see also Blumenfeld 2002), resulting in "platyrrhine" (broad-nosed) external nasal morphology (Márquez et al. 2005). The internal nasal passageways of Neandertals were long and relatively broad (and thus voluminous), tended to have bilevel nasal floors (a marked drop-off of the floor posterior to the incisor crest), likely had large and projecting inferior turbinates (based on the size and orientation of the roots), and may have been somewhat mediolaterally compressed posteriorly by a medial swelling of the lateral nasal wall (Coon 1962; Franciscus and Trinkaus 1988b; Franciscus 1995, 2003; Schwartz and Tattersall 1996a; Yokley 1999). The presence of two additional traits (a medial projection on the internal nasal margin of the piriform aperture and the lack of an ossified roof over the lacrimal groove: Schwartz and Tattersall 1996a), and inferences about Neandertal respiratory physiology and behavior that have been based on these traits (Schwartz et al. 1999), have not been supported by subsequent studies (Franciscus 1999; Yokley 1999).

Despite considerable research attention over the last several decades, the functional significance of Neandertal nasal morphology remains obscure. Neandertals appear to meet the expectations of cold-adapted (and/or dry-adapted) nasal form in having had projecting external noses and, internally, large and projecting inferior turbinates. However, in terms of the shape of the nose (as reflected by the dimensions of the piriform aperture), Neandertals exhibit the converse of what would be expected in cold climates (Figure 5.10). Nasal breadth, both absolute and relative to nasal height, tends to decrease with latitude among modern humans (see above), such that a broad, "platyrrhine" nose is not what would be expected in archaic humans adapted to cold glacial conditions (Hylander 1977; Franciscus and Trinkaus 1988b; Márquez et al. 2005). Indeed, a perceived lack of variation in nasal breadth among the Neandertals

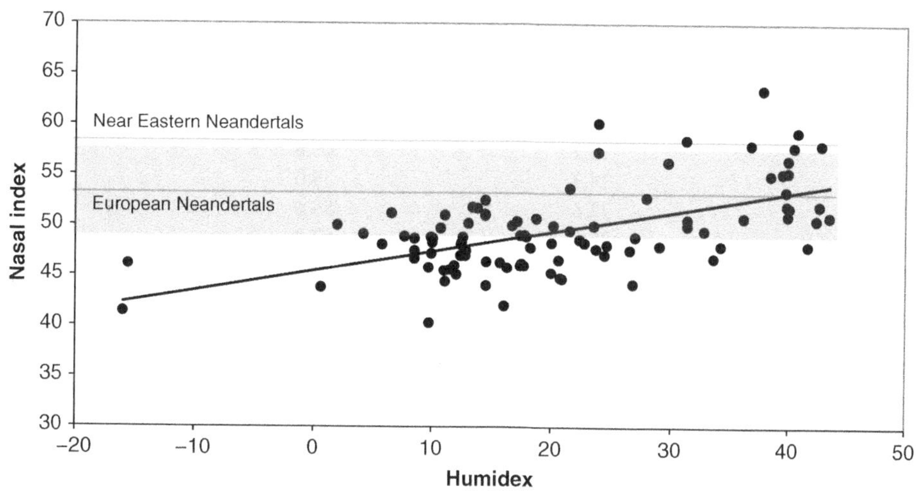

FIGURE 5.10

Mean skeletal nasal indices (100 * nasal breadth/length) on a composite measure of temperature and humidity (humidex) in recent humans. Modern human sample means (●) and climatic data from Thomson and Buxton (1923). Humidex, as a measure of thermal stress, was calculated from mean annual temperature and mean annual relative humidity following Masterton and Richardson (1979). Ordinary least squares regression line ($y = 0.1956x + 45.441$, $r = 0.5839$) based on modern human data only. Grey lines represent mean nasal indices for Neandertals. Grey shading represents one standard deviation variance around European Neandertal mean (similar shading around Near Eastern Neandertal mean omitted for clarity's sake).

(which would not be expected if nasal breadth were varying with climate across the Neandertal geographic and temporal span) has been seen as evidence that their nasofacial morphology is the result of genetic drift rather than adaptation (Maureille and Houët 1998). Hawks and colleagues (2000a), however, present nasal breadth means for temporal samples that do conform, at least minimally, to climatic-based expectations. Also, geographic patterning within Neandertals in the shape of the piriform aperture (as indicated by the nasal index: 100 * nasal breadth/height) conforms to expectations derived from ecogeographic patterning in modern humans: European Neandertals have relatively narrower piriform apertures than their Near Eastern counterparts (Table 5.6). However, the few fossils we have from interglacial Europe have relatively narrower nasal apertures than Neandertals from glacial and cold–temperate intervals (Table 5.6). In addition to relatively narrower nasal apertures in European Neandertals (that is, relative to Near Eastern Neandertals and other archaic humans from hotter environments), further narrowing of the internal fossa may also have been accomplished to a degree by a more medial positioning of the posterior lateral wall (the wall separating the nasal fossa from the maxillary sinus) (Schwartz and Tattersall 1996a; Schwartz et al. 1999; Yokley 1999). Based on dimensions of the internal nasal passageway, particularly in the upper portion of the internal fossa, Neandertals also appear to have had narrower airways than did their contemporaries from hotter

TABLE 5.6 ■ Nasal dimensions and nasal indices in Neandertals

	■ Nasal breadth	Nasal height	Nasal Index
European Cold and Cold–temperate (MIS 5d-3)			
La Chapelle 1	33.2	60.8	54.6
La Ferrassie 1	34.3	59.0	58.1
Gibraltar 1	33.6	55.6	60.4
Monte Circeo	35.0	62.2	56.3
La Quina 18	22.4	48.3	46.4
Roc de Marsal 1	21.9	40.9	53.5
		Mean	**54.9 ± 4.8 (6)**
■ European interglacial (MIS 5e)			
Saccopastore 1	31.2	62.8	49.7
Saccopastore 2	33.6	66.9	50.2
		Mean	**50.0 ± 0.4 (2)**
■ Near Eastern Cold and Cold–temperate (MIS 5d-3)			
Shanidar 1	30.2	62.2	48.6
Shanidar 5	38.5	68.3	56.4
Tabun 1	34.0	58.2	58.4
Teshik Tash	31.0	45.2	68.6
		Mean	**58.0 ± 8.2 (4)**

All data from Franciscus (1995).

climates (Franciscus 1995). Thus, there may have a degree of ecogeographic patterning among later Pleistocene hominins that parallels that seen among recent humans, albeit without the kind of marked convergence in nose shape between Neandertals and circumpolar peoples that we see in body shape.

A number of the features seen in Neandertal noses have also been suggested to either increase mucosal surface area relative to internal nasal volume, or to enhance air stream turbulence, and thus to improve heat and moisture transfer efficiency in cold climates. These features include a bilevel nasal floor (in conjunction with a pronounced nasal sill), larger and more projecting inferior turbinates, and projecting external noses with downwardly directed nares (Franciscus and Trinkaus 1988b). Experimental work (Churchill et al. 2004) on a small sample (n = 10) of modern European–American noses suggests that a downward orientation of the nostrils may enhance turbulence (cf. Proctor 1982; Cole 1982b; Hahn et al. 1993), while projection of the inferior turbinate actually serves to decrease turbulence by introducing laminar streamlines into otherwise partially turbulent flow (cf. Girardin et al. 1983; Scherer et al. 1989; Elad et al. 1993). The height of the nasal sill above the floor of the nasal fossa was unrelated to turbulence production (Churchill et al. 2004), which is consistent with comparative work which shows that, among modern humans, bilevel nasal floors attain their highest frequencies among people living in hot climates (Franciscus 2003). Thus, in sum, the great projection of the external nose (and with it the likely downward orientation of the nostrils) may have served to elongate the nasal portion of the upper respiratory tract and to enhance somewhat turbulent airflow within the passageway. Heat and moisture exchange would have also been facilitated by elongation of the internal nasal fossa, by increased mucosal surface area related to larger turbinates, and by a degree of narrowing of the mediolateral dimensions of the passageway (relative to their contemporaries and, presumably, their ancestors from hot climates). The greater size and projection of the turbinates may have also helped in directing airflow into regions of the internal fossa that are most conducive to heat and moisture transfers.

Thus there appears to be something of a signal of cold adaptation in Neandertal nasal morphology, but the great breadth of Neandertal noses still invites question. Given that nasal breadth/height relationships in modern humans are responsive to climatic factors, why didn't the Neandertals – if indeed they were cold-adapted – evolve narrower noses than they did? Arguments for constraints based on overall facial architecture (such as nasal breadth being architecturally tied to intercanine breadth: Coon 1962; Wolpoff 1968) can be easily ruled out (Franciscus and Trinkaus 1988b; Holton and Franciscus 2008). Two non-exclusive explanations, both consistent with cold adaptation, remain viable. The first of these is that Neandertal cold adaptation involved the dual problems of conserving heat under certain circumstances and dumping heat under others (as it does for all endotherms that experience variation in ambient temperature and levels of metabolic heat production). At times of high muscular activity, even potentially when ambient temperatures were low, Neandertals may have needed to vent heat to regulate both body core and brain temperature (Trinkaus 1987a; Franciscus and Trinkaus 1988b; Dean 1988; Holton and Franciscus 2008). Under conditions of cold or dry ambient air, the need for proper conditioning of inspired air and recapture of moisture upon inhalation would have favored nasal breathing rather than orotidal breathing (since the oral cavity lacks the competence for air conditioning seen in the nasal passageways). Thus, Neandertals may have needed upper respiratory tracts that could both adequately condition flowing air while also providing the volume flow necessary to sustain relatively great body mass (Yokley et al. 2009), and for high aerobic workloads and heat dumping (see Chapter 10.6). An alternative explanation for the wide noses of Neandertals is that they evolved alternative solutions to the problem of SA/V relationships within the nose. Whereas modern humans narrow the entire passageway, Neandertals may have achieved the same functional results by either greater projection of mucosa-lined bony features into the air stream (via enlarged turbinates, medial placement of the lateral wall of the chamber, or – controversially – by the use of novel structures such as the marginal projection: Schwartz et al. 1999) or by physiological management of mucosal vasodilation (see Yokley 2006). Given that engorgement of the mucosa can serve to entirely occlude the nasal passageways, it seems likely that orchestrated action of the sympathetic (vasoconstriction) and parasympathetic (vasodilation) nervous systems (Undem et al. 1999) could modulate internal nasal fossa lumen diameter relative to aerobic demands and thermoregulatory circumstances. Thus, by retaining the wide nasal apertures and internal passageways of their more tropically-adapted ancestors, Neandertals could have enjoyed nasal functioning that was physiologically adaptable to circumstances (but which raises the question as to why modern humans don't similarly rely on such a system).

5.4 PHYSIOLOGICAL SOLUTIONS TO COLD STRESS

Physiological processes don't fossilize, and so we are left having to make what reasonable assumptions we can about Neandertal physiology based on the metabolic and other physiological responses of circumpolar modern humans (and other cold-adapted mammals) to cold stress. As pointed out by Steegman et al. (2002), we are likely to be on safe ground in assuming similar physiological adaptations to cold in Neandertals and modern humans, since mammalian physiology – especially that involving homeostatic systems – is highly phylogenetically conserved. Accordingly, we can assume that Neandertals had the same physiological options for dealing with cold as we see among cold-adapted modern humans: increased metabolic heat production (non-shivering thermogenesis), and vasoconstriction in peripheral areas of the body, perhaps with periodic rewarming of extremities by cold-induced vasodilation. Seasonal increases in subcutaneous fat, thus reducing the thermal conductance of the body shell, may have also been a thermoregulatory strategy among Neandertals, although a number of considerations argue against this device (see below). Vasoconstriction, as a means of reducing internal convection and limiting heat loss by conduction, has already been discussed above, so only metabolic rate increases and fat insulation will be considered here, and these only briefly: for more thorough reviews, see So (1980), Leonard et al. (2002, 2005), and Steegman et al. (2002).

One hallmark of circumpolar peoples is elevated BMR relative to people living in warmer climates (Roberts 1978), and relative to non-indigenous people living in the arctic (Leonard et al. 2002). Increased metabolic rate after several days to a few weeks of cold exposure appears to be a basic mammalian acclimatization response to cold stress, and research with monkeys and humans suggests that developmental acclimatization and perhaps genetic adaptation may increase the responsiveness of groups indigenous to circumpolar regions. Work with macaques (reviewed in Steegmann et al. 2002) indicates that tropical monkeys raised in cold climates show stronger increases in resting metabolic rate than genetically-similar monkeys raised in warmer conditions. Indigenous arctic populations have been shown to have stronger increases in basal or resting metabolic rate than individuals of European descent living in the arctic (Rode and Shephard 1995; Leonard et al. 2002, 2005), but it has been difficult to tease apart the effects of developmental acclimatization from genetic adaptation.[10] Efforts to assess the broad sense heritability of thyroid function (perhaps the major avenue by which BMR is altered: see below), however, suggest that there is a significant genetic component to differences in metabolic response to cold (Meikle et al. 1988; Oppert et al. 1994; Tremblay et al. 1997), such that the physiological components of cold adaptation in circumpolar peoples appear to reflect gene-by-environment interactions. Regardless of the adaptive basis of cold stress responses, a number of conclusions can be drawn from studies of metabolic rate in circumpolar peoples. First, increased basal metabolic rates, on the order of 3–19% (Rode and Shephard 1995; Leonard et al. 2002), are characteristic of arctic peoples relative to individuals of European descent (even after acclimatization to cold environments, and differences in mass and body composition have been controlled for). Second, metabolic rates appear to vary seasonally, such that while arctic peoples may have stronger increases in BMR than non-indigenous people because of developmental acclimatization and/or genetic adaptation, they still experience seasonal acclimatization to cold stress (see review in Leonard et al. 2005).

Again, we can't directly study physiological responses to cold among the Neandertals, but it is reasonable to assume they were similar to extant circumpolar peoples in their degrees of seasonal acclimatization to the cold. Indirect evidence suggests that they may have had an even stronger response. Cold-induced elevation in BMR, known as "nonshivering thermogenesis," appears to be controlled by the concerted effects of the neurotransmitter norepinephrine and the thyroid hormones triiodothyronine (T_3) and thyroxine (T_4). At the risk of greatly oversimplifying a complex and poorly understood process, it appears that cold stress increases the secretion of thyroid stimulating hormone from the pituitary, which in turn induces the thyroid gland to secrete increased levels of T_3 and T_4. Thyroxine up-regulates the synthesis of "uncoupling proteins" (de Lange et al. 2001; Queiroz et al. 2004), which bind with the inner membrane of mitochondria in skeletal muscle and elsewhere to reduce oxidative phosphorylation efficiency, reducing the effectiveness of ATP production but greatly increasing the burning of calories for heat (Mishmar et al. 2002). Triiodothyronine and norepinephrine (which is released from both the central nervous system and from sympathetic nerve endings in the peripheral nervous system in response to cold) together stimulate the transcription of specific uncoupling proteins (Bianco et al. 1988) that increase mitochondrial heat production in brown adipose tissue (BAT: while BAT is a major component of thermoregulation in human neonates, the amounts of BAT in adult humans remains uncertain. Studies suggest that cold-adapted humans have larger adult stores of BAT than other groups: see Steegmann et al. 2002). Thus while norepinephrine, T_3 and T_4 all work together to regulate metabolic rate increases, T_4 (thyroxine) appears to be especially important in mediating whole-body metabolic responses to cold stress (and, indeed, BMR increases appear to be correlated with T_4 levels in arctic groups: see Leonard et al. 2002, 2005). Interestingly (and herein lies the indirect evidence for Neandertal BMR responses to cold), thyroxine is a potent morphogenic agent, having multiple effects on gene expression and physiological processes during growth and development. Crockford (2002, 2003) has suggested that some aspects of Neandertal skeletal morphology, including cranial size and shape, postcranial robusticity and foreshortened extremities, can be explained by reference to elevated levels of growth-active thyroid hormones (particularly T_4). While endocrine-shift models generally do a poor job of explaining Neandertal–modern human contrasts in morphology (Churchill 1994a, 1998, 2006), there have as yet been no systematic attempts to test a thyroid hormone-specific model, and this might prove fertile ground for future research.

Skeletal morphology hints at a pronounced thyroxine response among Neandertals, and comparative studies of circumpolar groups suggest that they likely exhibited seasonal increases in BMR. Modern arctic people have BMRs that are on average some 9–10% (Rode and Shephard 1995) or 3–7% (Leonard et al. 2002) higher than what would be estimated from their body mass (which is why it is surprising that our climate-adjusted Neandertal BMRs were lower than the mass-based estimates: Chapter 4.3). Thus, we can conservatively assume 3%, or more liberally, 10% increases in Neandertal BMR for five or six months of the year. Applying this correction to the most conservative mass-based estimates of BMR, this would raise the caloric cost of living (basal metabolism only, and averaged across the year) to between about 1900–1960 kcal d^{-1} for an average-sized male Neandertal and to between 1480–1530 kcal d^{-1} for an average-sized female. These estimates are also conservative in that nonshivering thermogenesis has been shown in some cases to increase cellular respiration by as much as 40% above basal levels (see references in Sorensen and Leonard 2001: work on rats suggests that values 80% above basal metabolism are attainable by some mammals; see references in Frisancho 1993). All together, this suggests that our climate-corrected average BMR estimates (1824 kcal d^{-1} for males, 1417 kcal d^{-1} for females) are likely to be conservative.

It is tempting to think that Neandertals may have used seasonal fat storage as a means of increasing thermal insulation during the winter months, as the white adipose tissue of subcutaneous fat has a low conductance relative to other bodily tissues (roughly half that of muscle or skin: Cohen 1977; Vogel 2005a). Strong negative correlations between percent body fat (as measured by skinfold thicknesses) and heat loss in hypothermic conditions (see Steegmann et al. 2002) demonstrate the thermal advantages of insulative fat. It has also been observed that some arctic peoples actively put on weight in the late summer and fall, presumably in anticipation of a long and cold, calorie-lean winter (e.g., Siberian Evenki herders consume 18–37% more calories than needed to meet daily energy expenditures in the late summer: Leonard et al. 2005). However, a number of considerations suggest that using body fat as insulation is of limited effectiveness. First, fat is a bulky insulator (Davenport 1992) and, as pointed out by Aiello and Wheeler (2003), Neandertals would have had to sport oppressive amounts of fat to match the insulation value of even a fairly modest amount of clothing. Second, the fat stores themselves would become progressively depleted as winter wore on (assuming that the balance of dietary intake to expenditure was negative for at least a portion of the winter) and energy demand for activity and thermogenesis outpaced foraging returns. Finally, with high activity levels a thick layer of subcutaneous fat may have been a liability, since it limits heat dumping by conduction and increases the need for evaporative cooling (sweating) – a mechanism that can quickly lead to hypothermia at cold temperatures (Noakes 2000). This is presumably one of the reasons that modern Inuit have fairly modest levels of subcutaneous fat (Shephard et al. 1973). For these reasons, it seems likely that subcutaneous fat added to a small degree to the thermoregulatory competence of Neandertals, just as it does in extant circumpolar peoples, but that it never figured prominently as a solution to cold stress (cf. Steegmann et al. 2002; Aiello and Wheeler 2003).

Heightened muscularity, on the other hand, probably was a significant component of Neandertal cold adaptation. Muscle is not particularly good as an insulator (at rest its conductivity is slightly better than skin: Vogel 2005a), but it makes up in thickness what it lacks in insulation (Veicsteinas et al. 1982), and in Neandertals – who had more muscle per unit stature (Chapter 4.2) – it was probably an effective insulator (Steegmann et al. 2002). Heightened muscularity also has the advantage of providing plenty of metabolically-active tissue that can be called upon for thermogenesis. At rest, muscle tissue does not produce much heat (but is decidedly better than fat), but when active (either during shivering or during voluntary activity) or under the action of thyroxine (see above), it can be called upon to increase metabolic heat output. Thus, greater muscularity may have done double duty as insulator and heat generator, and in this latter role it may have provided substantial adaptive flexibility by allowing Neandertals to express a potentially very strong metabolic response to cold stress. As mentioned above, however, muscle tissue loses much of its insulation value when active (because heightened blood flow to the muscles increases convective heat transfer from the body core to periphery: Veicsteinas 1987), which may at times have necessitated pacing of exertion in the cold in order to stem excessive heat loss (Steegmann et al. 2002).

5.5 COLD STRESS AND NEANDERTAL BEHAVIOR

The morphological and physiological mechanisms of thermoregulation reviewed above share a common physical constraint, and that is that thermal constancy – in any environment – can only be achieved by balancing heat production and heat loss. Clearly, the only broad strategies available to an endotherm in a hypothermic environment are either to generate more heat or to better stem its loss to the environment. The same constraints apply to potential behavioral adaptations to cold environments, and in nonhuman endotherms behavioral options are generally limited to efforts to reduce heat loss (huddling, seeking shelter in the form of wind breaks, using forests as a winter shed, etc.). Humans, however, have devised effective behavioral mechanisms for both producing and conserving heat and, accordingly, these methods constitute the primary means by which modern humans respond to hypothermic environments. While increased heat production can be effected by voluntary activity (which works better for humans than for most mammals, given our superior aerobic abilities), humans also rely on extrasomatic means of producing heat, which along with clothing and shelter serves to produce a proximate "microclimate" that keeps us in thermal neutrality[11] most of the time. This general ability to manage the local ambient temperature (whether it be of the air inside a shelter or the air inside a layer of clothing) falls under the rubric "cultural buffering," and it is generally assumed that Neandertals – being relatively technologically simple – were not as good at it as are modern arctic people or as were early modern Europeans (Holliday 1997b; Steegmann et al. 2002). As is often the case with the archeological record of past human behavior, we are confronted with the inferential dilemma of interpreting a lack of evidence (does, for example, the lack of bone needles in the Mousterian mean that Neandertals lacked tailored clothing?). Here we examine what inferences we can reasonably make about how Neandertals responded to the cold, as well as explore what behavioral capabilities we might reasonably afford them despite the absence of any direct evidence (cf. White 2006), and then endeavor to evaluate the effectiveness of the level of "cultural buffering" attained by them.

5.5.1 Activity

As mentioned above, one effective method of generating heat is to increase – voluntarily or involuntarily – one's activity, since muscular activity is performed at a fuel efficiency of only about 15–25% (Heglund and Cavagna 1985, 1987) (meaning that 75–85% of the free energy going in is liberated as heat). Involuntarily, heat outputs as much as three times basal metabolic rate can be attained after prolonged shivering alone (Frisancho 1993). For our average male Neandertal (who, with BMR elevated 3–10% from nonshivering thermogenesis, is putting out about the heat equivalent to a 100 W light bulb), this provides roughly 260–335 W of heat production – a heat generation getting into the range of desktop computers. For a female, whose nonshivering thermogenic basal metabolism is putting off the heat equivalent of a 75 W light bulb, shivering would bring her into the 200–280 W range. With voluntary activity heat output can be increased further still. A well-conditioned (championship-level athlete) adult male modern human can sustain useful power outputs of 300–370 W for a couple of hours, with a corresponding heat production of 1020–1280 W (in the range of the heat put off by a toaster oven) (Wilkie 1960; Shephard 1978). For our relatively more muscular Neandertals, even higher heat outputs are imaginable. Thus, muscular activity could have been called upon to contribute to homeostasis under hypothermic conditions, both with heat output an order of magnitude greater than possible by metabolic processes alone (bearing in mind that not all of this heat is available for thermoregulation, since exercise may increase convection and evaporative heat loss from the skin and losses from the respiratory passages: Blaxter 1989), and with useful output of mechanical work on the environment.

Two issues, however, limit the adaptive utility of voluntary activity as a thermoregulatory strategy. First, the maximum heat outputs given above are expensive – costing between 1130–1420 kcal h^{-1} above BMR – and thus periodic bouts of exercise may have been an untenable strategy for keeping warm when calories were in short supply. Second, even in well-conditioned athletes it is difficult to sustain high activity levels for any length of time, and there is an inverse relationship between the intensity and duration of effort (and thus heat output). Finally, as mentioned above, voluntary activity increases blood flow to muscle tissue, thereby increasing rates of internal convection and heat loss by conduction. These considerations together suggest that activity was probably an important component of thermoregulation for Neandertals who were away from shelter and who were active anyway as they foraged, searched for material resources, moved camp, etc. When out and about, a Neandertal could have managed their exertion levels so as best to balance heat production and heat loss, and thus keep conditions inside their clothing close to those necessary for thermal neutrality. Activity was probably also an important survival tool in emergency situations, as when caught in a storm, unexpectedly having to spend a night away from shelter, or after accidental submersion (arctic foragers are known to have run back to their villages to stay warm after becoming soaked: Moran 1981). But given the high caloric cost of active heat production, it seems likely that Neandertal adaptive strategies would have favored insulation and exogenous heat production over active metabolism.

5.5.2 Extrasomatic Heat Production and Conservation: Fire, Shelter and Clothing

From the Middle Paleolithic archeological record (Chapter 3.5) it is clear that Neandertals sought protection from the elements in naturally-occurring shelters, that they regularly used fire for warmth and other purposes, and that they processed animal hides, presumably for their value as insulative materials. What remains uncertain is the extent to which they constructed artificial shelters (either in open-air sites or for use within caves and rock shelters), and the degree of sophistication of their clothing. Again, we are left having to make what reasonable assumptions we can. The use of natural shelters by Neandertals is safely inferred, given that the bulk of the archeological record of the Middle Paleolithic derives from caves and abris across Europe and the Near East. While such shelters no doubt offered multiple advantages (such as reducing the amount of vigilance needed to defend social space against carnivores or other groups), there is no doubt that their major draw was thermoregulatory: by providing protection from wind, rain and the night sky, and by at least minimally enclosing air that could be warmed to some degree by body heat and fire. Even in relatively mild winter conditions nonhuman primates will use caves to stay warm at night, despite an increased risk of predation associated with doing so (Barrett et al. 2003). The advantages to humans with fire, especially under severe conditions, would have made it worth the risk of potentially encountering carnivores who might already be in residence, and would probably even have made it worth the trouble (and risk of injury) of driving them out (but see Chapter 9). Rock shelters and the entrance chambers of caves are, however, a mixed blessing. The limestone walls of the shelter would themselves be heat sinks that under certain conditions would have added to the cold stress of its human occupants. The temperature of the exposed walls of a shelter would match that of the ambient air, and radiative heat transfer between wall and occupant would be a function (following the Stefan-Boltzmann law) of the emissivity of each (limestone and human skin both have emissivities close to 1), the temperature difference between the two, and the effective exposed surface areas across which the transfer is occurring. Under most conditions, shelter occupants would have lost heat by radiation to the walls, notwithstanding some degree of warming of the walls by radiation from fires that might be burning inside the shelter (in this regard, small warming hearths, positioned between sleeping individuals or couples and close to the shelter walls, may have made an important contribution to keeping warm at night: Vallverdú et al. 2010; see also Hayden 2012). Cold walls could be ameliorated to a degree by selective occupation of south-facing shelters, in which solar radiation would warm the limestone during the day, potentially even producing thermal imbalances such that the bedrock actually emitted more heat than it absorbed for part of the night. Not surprisingly, a decided preference for south-facing shelters is seen in most places. For example, 61.4% of Mousterian shelters sampled in southern France face either south, south-east or south-west (Boyle 1996), a proportion markedly different from what would be expected if shelters were selected at random with respect to orientation (assuming that shelters facing in all directions were equally available for occupation). Limestone also has a fairly high thermal conductivity (ca. 1.3 W m^{-1} K^{-1}: see Vogel 2005b) – about two times worse than water. Soils (including, presumably, sediments within a cave or shelter) have much lower emissivities and tend to have lower conductivities (compacted sediments may actually be worse than limestone: see Vogel 2005a), and thus shelters with some sediment infill would have been desirable for reasons having to do with more than just the softness of the living and sleeping substrate. Phytolith analyses from Amud (Israel) and Esquilleu (Spain) caves strongly suggest that Neandertals there may have been using grasses as bedding material (Madella et al. 2002; Cabanes et al. 2010), a practice that would be consistent with edge wear analysis of Mousterian tools across Europe suggesting they were frequently used to process high silica plant material (see review in Hardy 2004). Even so, radiative heat loss to the walls and conductive heat loss to the floor and air would have made bedding materials, both over and under a sleeping body, a necessity. The marked reduction in radiant conductivity afforded by animal furs (Cena and Clark 1978) would also have made it advantageous to cover portions of the shelter walls with skins, or to construct tents or lean-tos within the shelters. The use of seaweed as bedding material and construction of a sleeping tent have been claimed for the preNeandertal Acheulian site of Lazaret in the south of France (de Lumley 1969), and a similar tent, along with warming hearths along the back wall of the shelter, has been suggested at the Mousterian site of l'Abric Romaní in Spain (Vallverdú et al. 2010).

Of course, radiation to cold bodies is no less of a concern when camping in the open, and indeed is probably worse than what occurs in a limestone rock shelter with cold walls. The effective temperature of a clear night sky may be as low as −53 °C (Nobel 1999), and radiative heat losses to it substantial (Hammel 1956). It is unlikely that Neandertals could pass even fairly mild fall or winter nights outside without substantial insulation in the form of clothing, bedding and possibly shelter. Debate continues about the use of constructed shelter by Neandertals (see Gamble 1986, 1999; Kolen 1999; White 2006). Here I will simply note that (1) there was no doubt strong motivation – in the form of pain and discomfort – to devise ways of sheltering themselves from wind, rain, and radiative heat sinks, (2) such shelters need not have been highly sophisticated to be helpful, and (3) constructing such shelters was likely within the range of abilities of hominins who were capable of working hides and making composite tools. In this I agree with White (2006: 560), "from the narrow but frankly vital perspective of survival, it almost beggars belief that Neanderthals managed to survive on the cold and open treeless steppe without some form of artificial shelter, nowhere to take refuge from the wind, rain and snow or to dry off once wet."

The possible use of animal skins for constructed shelters raises the issue of the level of sophistication of Neandertal clothing. Tools that have an obvious function in tailoring of clothing are conspicuously lacking in Mousterian assemblages (Gilligan 2010b), raising questions about the degree of complexity of clothing donned by the Neandertals (Gilligan 2007). At its simplest, Neandertal apparel may have consisted of little more than fur cloaks, possibly augmented with fur caps and moccasins (with soft soles: Trinkaus 2005) of some sort. Here an ethnographic analogy is provided by the indigenous inhabitants of Tierra del Fuego – the Selk'nam and Ya'mana – who wore little more than fur wraps to protect themselves against fairly harsh climatic conditions (see Steegmann et al. 2002). As argued by White (2006), however, the ability to use simple stitches to bind skins together with sinew or plant fibers does not seem beyond the capabilities of humans that were able to haft lithic tools to wooden handles (see Chapter 3.3). While Mousterian attire may have lacked the sophistication of cold-weather gear worn by modern circumpolar people, it seems reasonable to think that Neandertals were able to tailor simple tunics, pants, hats, moccasins and mittens to better withstand heat losses from air convection around the body. The maintenance of a thermoneutral microclimate within the clothing would have been further facilitated by drawstrings on the front of the tunic that would have allowed for increased ventilation during times of high activity (to prevent overheating). Pelts of all large prey animals were likely valuable for one insulative function or another, but those of particular species may have been especially valued for clothing. The winter coat of caribou may have been particularly prized, as it has one of the highest insulative values (5.4 clo) of any animal pelt, and it is better at retaining heat under windy conditions than are most furs (Moote 1955). (White [2006], citing Stenton [1991], gives a value of 7 clo for the autumn pelts of caribou [prior to the full development of the winter pelage]. This value seems high given the results of experimental determinations of the insulative value of caribou pelts [Hammel 1955; Moote 1955; Cena and Clark 1978], which consistently put the *winter* fur of caribou at ≤ 5.4 clo.) This raises the possibility, reviewed by White (2006), that Neandertals may have been targeting reindeer as much or more for their pelts and sinew as for food.

How effective would Neandertal hide clothing have been? There are limits to the amount of insulation that one can wear, and also limits to the value of adding additional layers of insulation (Cena and Clark 1978). These constraints place the practical upper limit on wearable insulation at somewhere around 4.5 clo. Military arctic gear provides insulation in the range of 3.5–4.3 clo (Brajkovic et al. 2001; Cena and Clark 1978), but at a substantial cost in weight (arctic combat assemblies weight between 11–16 kg: Teitlebaum and Goldman 1972; Cena and Clark 1978). The added weight of heavy clothing, combined with its distribution over the entire surface of the body (and thus mostly away from the body's center of gravity), greatly increases the energetic cost of locomotion. The predicted energy expenditure of walking at 4.5 km h^{-1}, without an external load, for a 72 kg person equals ca. 295 W (using the regression in Myo-Thein et al. 1985); at a somewhat faster speed (5.6 km h^{-1}), a similar-sized person in an 11 kg arctic suit works at an average rate of 514 W (Teitlebaum and Goldman 1972). Caribou pelts are light relative to their insulation value, and traditional Inuit clothing of caribou hide, which may attain clo values of around 4, has been reported to be as light as 3–4.5 kg (Newman 1956). It seems unlikely that Neandertal clothing attained the insulative value of modern Inuit/Eskimo winter clothing or military arctic gear, but they were probably able to attain at least a level of protection equivalent to that of a large mammal pelt over their entire body (1.94 clo: the insulation value of the fur of a large temperate-climate mammal: Cena and Clark 1978). The selective use of caribou pelts may have provided a greater level of protection, but it is doubtful that Neandertals were able to achieve insulations of greater than about 2.5 clo without very careful fitting and tailoring of their gear. Even 2 clo of protection, however, is a substantial benefit in cold weather, and when combined with the heat-generating capacity and overall shape of the Neandertal body may have produced a high degree of cold tolerance (as we'll see below).

Of course, with effective insulation and high activity levels comes the reverse problem of becoming *too* warm, potentially dampening one's clothing through perspiration, and becoming hypothermic through uncontrolled evaporative heat loss from wet clothing (potentially five times greater than from dry clothes: see references in White 2006). Sweating is a serious concern for Inuit and Eskimo, who at times must be highly active at very low ambient temperatures, and continuous adjustment of head wear and clothing vents is used to allow for adequate heat dumping to the environment (Rennie et al. 1962; Moran 1981). Heat dumping via the head, and reduced sweating in areas covered with clothing, is also facilitated among Inuit (relative to people of European descent) by a greater number of sweat glands on the face (from which they initiate sweating earlier and more profusely) and a reduced number on the body trunk and limbs (Schaefer et al. 1974). Neandertals may have dealt with this problem by loosening or tightening articles of clothing and thus varying the amount of circumductive cooling of the air around their bodies, by carefully pacing their exertion (cf. Steegmann et al. 2002), or by actively dumping heat via the respiratory tract (Trinkaus 1987a; Franciscus and Trinkaus 1988b; Holton and Franciscus 2008).

As reviewed in Chapter 3.5, simple hearths are a fairly regular component of Mousterian sites, indicating that Neandertals had regular and controlled use of fire (Perlès 1976; James 1989; Meignen et al. 1989; Olive and Taborin 1989; Mellars 1996; Rigaud et al. 1996; Théry et al. 1996; Vallverdú et al. 2005: but see Sandgathe et al. 2011). A small degree of pyrotechnic engineering may have also taken place, through the positioning of stone blocks and flagstones around hearths (as at Abric Romaní: Vaquero et al. 2001b), which may have facilitated banking of fires, helped to control ventilation and thus regulate burning intensity, and provided an additional source of radiative warming. Fire provides an extrasomatic means of generating heat, which would have warmed Neandertals directly by radiation and indirectly by warming the walls of, and air within, their living spaces (and also by warming food, and possibly fluids, that were then consumed). By using chemically-stored energy in wood and grasses (Albert et al. 2000; Madella et al. 2002) and occasionally even coal (Théry et al. 1996), Neandertals could have abated their own energy losses and reduced somewhat the need for energetically-costly metabolic heat production. While the use of fire provides a net saving of energy, it is not a free good: metabolic energy has to be invested in gathering suitable fuel. At the Abric Romaní (Spain), a pseudomorph (natural cast) of a tree trunk – most likely of pine (see Vaquero et al. 2001b) – indicates that the Neandertals there had carried or dragged in a 3.5 m long and 0.45 m wide log for fuel wood (Carbonell et al. 1996a). Using a rough value of 500 kg m^{-3} for the density of pine, this log must have weighed about 275 kg! I suspect that wood gathering usually centered around smaller pieces, but even so there is a metabolic cost to be paid. For industrialized humans, working around the house entails increases in metabolic expenditures on the order of 3–3.5 times BMR (Montoye et al. 1996), which is probably roughly comparable to the energetics of gathering firewood. Estimating the energy expenditures involved in fuel gathering is difficult, both because fuel availability is highly variable across environments and because good time allocation data on firewood collection in hunter-gatherers is unavailable. One study of woody litter production in temperate forest found a mean annual production rate of 16.5 g m^{-2} y^{-1}, of which about 35% would be small enough to be suitable for use as firewood (Table 4 in Bratton et al. 1982). At a firewood production rate of about 5775 kg km^{-2} y^{-1}, Neandertals in forested environments would have probably been able to meet their firewood needs without traveling far from camp (see Vaquero et al. 2001b). Of course, gathering time would have increased with duration of site occupancy, as proximate resources were consumed, but if residential stays were relatively short (see Chapter 10.1) this would not have presented much of a problem. In more open environments, the energetic cost of firewood gathering may have been substantial, and it is likely that Neandertals there may have relied more heavily on grasses, animal bones, and dung as fuel sources.

5.6 THERMOGENIC CAPACITY AND COLD TOLERANCE

The same stature, mass, BMR and SA estimates that went into the comparison of Neandertals, Inuit and European T_{lc} and T_{min} above can be used to model rates of change in body core temperature at different ambient temperatures, and the insulative value of clothing can be factored in. Assuming that debilitating hypothermia occurs at a body core temperature of ca. 35 °C,[12] we can estimate the duration that Neandertals could have tolerated exposure to various ambient temperatures (Figure 5.11). Heat loss ($\dot{Q}$, in W) at different ambient temperatures was estimated as:

$$\dot{Q} = \left(\frac{(t_c - t_s)}{r_s}\right) \cdot SA$$

where t_c = body core temperature (set at 37 °C),[13] t_s = surface temperature (ambient temperature), and r_s = thermal resistance of the surface (here set to 0.31 m^2 K W^{-1}

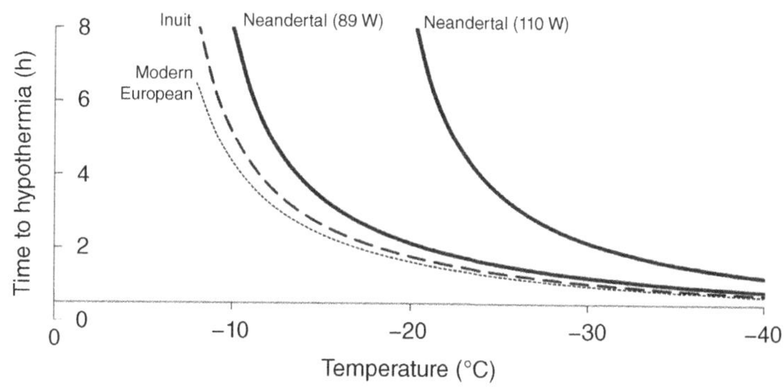

FIGURE 5.11

Approximate time to hypothermia for a clothed Neandertal, Inuit and modern European as a function of ambient temperature. Curves are based on estimated heat loss and heat generation for males of average dimensions (mass, stature, and surface area). Curves are provided for Neandertals assuming BMRs of 89W and 110W. Insulative value of clothing was modeled at ca. 2 clo. See text for details.

[1.94 clo]: see above). Heat generation was modeled at 3·BMR (the maximum metabolic rate that can be sustained indefinitely: Burton and Edholm 1955), and change in body core temperature per unit time (Δt_c) as:

$$\Delta t_c = \frac{(\dot{Q} - 3 \cdot BMR)}{M}$$

where M = mass (in grams) and with conversion of watts to calorie/seconds (and assuming that body tissues have roughly the same conductivity as water: see Vogel 2005a). This allows us to model the amount of time a clothed Neandertal or modern human could spend at various temperatures before the net heat loss dropped body core temperature the two degrees associated with the more severe stages of hypothermia (Danzl and Pozos 1994). Estimates were made for Neandertal, Inuit (Inupiat: Ruff et al. 2005) and European (French: Eveleth and Tanner 1976) males of average dimensions. Estimates for the Neandertal were run using both the low (89 W, based on the Kleiber equation applied to body mass) and high (110 W, based on estimating the mass-specific heat output of various body parts) estimates of BMR. The modern human BMRs were determined by the Kleiber equation. The resulting Neandertal curves are very divergent (Figure 5.11), and since an argument could be made that the modern human BMR estimates would also be higher if the method of using mass-specific heat output of various organs had been used, we will only consider the lower Neandertal estimate here (bearing in mind that any incremental gains in metabolic rate would have a large impact on survivability in extreme cold).

As can be seen in Figure 5.11, the benefits of body form are more pronounced at less extreme temperatures. As ambient temperatures drop below −30 °C the time lines rapidly converge, such that any benefits a more cold-adapted body might confer are measured in minutes rather than hours of exposure. At higher temperatures, however, the differences are fairly pronounced. At a clement −10 °C, our Neandertal could spend about 8 hours exposed to the elements before becoming impaired by hypothermia, whereas the Inuit could only spend about 5 hours in the cold. The European is not much worse off than the Inuit, being able to survive about 4 hours and 20 minutes. By way of comparison, a comparably-clothed human with the body shape of the Nariokotme *Homo erectus* specimen (KNM-WT 15000, as a representative of the shared tropical ancestors of Neandertals and modern humans), with a projected adult body size of 185 cm and 68 kg (Ruff 1994), would have been able to endure only about 3 hours and 40 minutes. These approximations are based on mass-derived estimates of BMR, without correction for cold-induced increases in metabolic rate, and assume the same, relatively low, degree of thermal buffering from clothing. As such, they illustrate the impact of variation on body size and shape on cold tolerance (the curves for Inuit, Europeans, and *H. erectus* provide an idea of the cold tolerance Neandertals would have had if they were of similar body size and shape). Differences on the order of hours of cold tolerance in the temperature range 0° to −20 °C could have serious consequences in terms of foraging success, and even differences on the order of 15–20 minutes can impact survival in emergency situations (for example, when trying to make it back to shelter after accidental submersion in water). Variation in body size and shape, then, may have been under strong selective pressure in hominins with a limited degree of cultural buffering to cold stress. Interestingly, if adjustments are made for cold-induced nonshivering thermogenesis (by raising the mass-based estimates of BMR by 10%) and potentially warmer clothing (2.5 clo), the risk of hypothermia for our male Neandertal does not become a real concern until temperatures drop below −30 °C. For the Inuit male, with a similarly elevated BMR but with clothing equal to 4.0 clo (Newman 1956), hypothermia does not become a significant risk until the temperature drops below −60 °C – a potent example of the value of insulation in cold weather.

5.7 THE NEANDERTALS *WERE* COLD-ADAPTED

The paleoclimatic evidence and the distribution of Neandertals sites reviewed above show that, while Neandertals may at times have enjoyed relatively mild climates in Europe, they also at times suffered relatively frosty ones. Was their exposure to cold climates sufficient to select for genetically-based adaptations to the cold? The selective coefficients imposed by cold climate remain uncertain, and hypothermia may be one of the least frequent causes of death among high latitude peoples (see Steegmann 1983; Beall and Steegmann 2000). Still, selective pressures associated with the cold have been sufficient to produce convergence in body form and physiological processes in modern humans of disparate ancestry (e.g., Sami and Inuit). It is true that, in comparison to cold tolerance of indigenous arctic birds and mammals, the ability of Neandertals to withstand the bitter cold (without substantial cultural buffering) was not very good – but the same holds true for modern humans indigenous to the arctic. A good case for cold-adaptation in the Neandertals can be made when one considers that (1) they exhibit body size and shape convergence with modern cold-adapted people (and I'm confident that further work with the Neandertal genome will reveal a significant degree of convergence in physiological processes as well, such as in mitochondrial genes that control rates of oxidative phosphorylation: see Mishmar et al. 2002), and (2) those convergences were energetically costly. Neandertals had absolutely massive bodies with large surface areas, and accordingly they had very high metabolic rates (see above, and Sorensen and Leonard 2001; Steegmann et al. 2002; Churchill 2006). Furthermore, the combination of short limbs and massive bodies resulted in high costs of locomotion relative to the lighter and more linear body builds of their African ancestors (see Weaver and Steudel-Numbers 2005). As a rule, energetically-costly structures do not arise unless the selective benefits outweigh the reproductive costs involved. If Neandertal body size and shape are not reflecting adaptation to cold, what adaptive benefit did they provide that justified their high energetic cost? The most parsimonious explanation, given the evidence that Neandertals did at times endure very cold climates, is that the morphological similarities between Neandertals and circumpolar modern humans represent a case of convergence and that, like modern peoples of the arctic, Neandertals were cold-adapted.

NOTES

1. It is common to see references to MIS 6 as the "penultimate" glacial, despite there being two glacial episodes (MIS 4 and 2) that succeeded it. This reflects the common understanding that MIS 4, 3 and 2 together represent a single glacial episode (the Weichselian, also known as the Würm, Devensian, or Wisconsin). MIS 3 is simply a protracted period of somewhat warmer conditions (not a full interglacial episode like MIS 5e or 7a). MIS 4, then, is the first maximum of the last glacial, with MIS 2 being the second (commonly called the "last glacial maximum"). To further complicate matters, the Weichselian glacial is also known as the "Pleniglacial," with MIS 4 being the Early (or Lower) Pleniglacial, MIS 3 being the Interpleniglacial, and MIS 2 being the Late (or Upper) Pleniglacial.
2. Bergmann's (1847) rule holds that, within a species or a closely related set of species, larger individuals will tend to be found in colder environments. The generally (but not universally) accepted explanation is that for a given shape, larger individuals will have more heat-producing mass relative to heat-losing surface area, and hence greater thermal constancy. Among humans, ecogeographic patterning in mass follows the expectations of Bergmann's rule more closely than does stature, and among humans and some other mammals (e.g., ursids) variation in mass is a function of variation in both stature and body breadth, such that cold-adapted peoples tend to be short but wide bodied (Ruff 1994). Allen's (1877) rule holds that individuals living in colder environments will tend to have shorter extremities than conspecifics living in warmer climates. Since extremities (including ears and tails) have high surface area-to-volume ratios, shortening of extremities reduces heat loss in cold environments. Adherence to Allen's rule is the likely reason that variation in human stature does not follow Bergmann's rule (Holliday and Ruff 1997).
3. Among cold- and temperate climate-adapted peoples, vasoconstriction is the first physiological response to extremity cooling. After 10–15 minutes, however, blood flow to the extremities resumes, presumably to rewarm the tissues and prevent frostbite. The absence of this response, known as cold-induced vasodilation or the "Hunting response," is believed to underlie the higher frequencies of frostbite seen in tropically-adapted peoples in cold environments (Lewis 1930; Meehan 1955).
4. Under normal (non-hypothermic) circumstances, active transport of sodium and potassium against their chemical gradients (the Na+/K+ pump) consumes about 12% of cellular energy (van de Graaff and Fox 1986). Under the influence of thyroxine the pump works much faster, thus demanding an increase in ATP production (using elevated levels of circulating energy substrates liberated from the liver and adipose tissue by norepinephrine) at 40% efficiency, and also releasing heat as the ATP is used in work – moving Na+ and K+ ions "up hill"

against their solution gradients. Since the cell membranes are permeable to these ions, the last bit of free energy in the process – the potential energy stored in the transported ions – is released as they re-cross the cell membrane and move down the gradient. This situation is analogous to furiously pumping water from the downstream to upstream side of a very leaky dam – not very effective in keeping one's feet dry, but a great way to generate heat!

5. Pearson (2000) also reported similar long bone lengths in all three groups, with the exception that Neandertals had somewhat longer forearm bones. This finding is not supported by a review of the data on brachial indices in Neandertals and circumpolar groups (see Table 5.2).
6. The origins of the Sami and their genetic relationships with European and arctic Asian groups remain a bit cloudy, but recent work (Ingman and Gyllensten 2007) suggests they are derived from a European population from which they split roughly 7600 years BP.
7. The mean radial length for male Sami (Lapps) is listed as 266.8 in Table 5 of Trinkaus 1981. The corresponding mean humeral length and mean brachial index indicates that the original value was 226.8, which is the value used in Table 5.2.
8. It is also important to bear in mind that a ratio of means does not necessarily equal the mean of the individual ratios.
9. Conductivity ($W\ m^{-1}\ K^{-1}$) is a measure of the ability of a material to conduct heat, and is equal to the quantity of heat transmitted through a given thickness of material (normal to the surface of the material) over a given period and due to a given temperature difference. Conductance ($W\ m^{-2}\ K^{-1}$) is a measure of the combined rates of conductive and convective transfer per unit surface area per degree of difference between body core and ambient temperature (see Vogel 2005a). Note also that in the measures of conductivity and conductance given above, K denotes temperature (degrees Kelvin), not the heat exchange mechanism of conduction, as we had previously used it (confusing, I know).
10. An alternative, and oft-cited, explanation is that increased BMR in arctic people is entirely due to the high proportions of protein and fat in traditional diets (e.g., So 1980). Two observations, as pointed out by Leonard et al. (2005), militate against this explanation: traditional arctic diets are not substantially different from western diets in the absolute and relative amounts of protein and fat; and any increase in metabolic rate from the specific dynamic effect of fats and proteins would be gone by the end of the 12-hour fasting period that preceeded the measurement of BMR in most studies.
11. For humans, thermal neutrality involves maintaining skin temperatures within the fairly narrow range of 33 °C (below which we shiver) and 35 °C (above which we sweat). For modern humans, the zone of thermal neutrality (the ambient air temperatures associated with this skin temperature range) is 29–31 °C (Hardy and DuBois 1938). It is assumed that the same values would have applied for Neandertals.
12. Although death from hypothermia does not usually occur until core temperature drops to 32 °C, at 35 °C human thermoregulatory systems themselves begin to fail, muscular coordination is impaired, and the ability of the individual to initiate and sustain actions that may help alleviate the hypothermia is deeply jeopardized (see Danzl and Pozos 1994).
13. For computational simplicity, heat loss estimates were done using a constant core temperature of 37 °C, which imposes a constraint of a constant rate of heat loss at a given ambient temperature. As heat loss progresses, body core temperature decreases which reduces the thermal gradient between body core and body surface. This reduction in thermal gradient reduces the rate of heat flux, and thus slows slightly the rate of continued body cooling. Thus the actual curves depicting time to hypothermia as ambient temperature drops should flatten out more gradually than those depicted in Figure 5.11.

CHAPTER SIX

The Caloric Economy of Pleistocene Europe

Any attempt to explore the energetic budgets of an organism must include some understanding of the environment in which it lives, because the environment determines the amounts and structure of usable energy that the organism can capture, and the amounts of energy that the organism must allocate to thermoregulation, capture of calories, and avoidance of predators and other dangers. However, just as the physical and biotic environments determine many of the limiting factors important to the ecology of an organism, our state of knowledge of key environmental parameters in prehistory sets limits on our ability to comprehend the ecology of prehistoric human populations. The problems inherent in reconstructing Neandertal habitats are manifold. The Neandertals were spatially distributed across the European subcontinent and the western margins of Asia, an area that today encompasses five major biomes (or "formations of plants" in the European parlance) and considerable ecological variation within these biomes. They were also temporally distributed across the late Middle and Late Pleistocene which, as we began to explore in Chapter 5.1, was a time of

Thin on the Ground: Neandertal Biology, Archeology, and Ecology, First Edition. Steven Emilio Churchill.

dramatic climatic oscillations and environmental change in Eurasia. Long stratigraphic sequences at numerous Mousterian sites (e.g., the French sites of Combe Grenal and Pech de l'Azé II) testify to Neandertal persistence in some places despite major changes in the environment. And despite a tendency to abandon the coldest areas during glacial episodes (van Andel et al. 2003; Davies and Gollop 2003; Aiello and Wheeler 2003), a consideration of the distribution of Mousterian sites during various climatic episodes (see below) leads to the conclusion that Neandertals occupied *most* – and possibly *all* – of the humanly habitable environments that Pleistocene Eurasia had to offer. This eurytopism (environmental "generalism," a trait which appears to be shared with all of the later members of the genus *Homo*) provides a variable environmental backdrop against which to explore adaptive dynamics during the Mousterian, but makes difficult efforts to characterize the environmental parameters important to Neandertal ecology. In earlier chapters we tried to partition the Neandertal fossil record into climatic (glacial vs. cold–temperate vs. interglacial) and geographic (Europe vs. the Near East) subsets: if one were to partition further the archeological and fossil records by local habitat (e.g., MIS 6 steppe-tundra vs. MIS 6 steppe, etc.), one would be left with samples inadequate to delineate adaptive patterns in a single type of environment through time, and possibly even across environments at a single point in time. On top of this is the inescapable reality that we are dealing with past environments, and we are therefore subject to the limitations – discussed below – inherent in paleoenvironmental reconstruction. These limitations especially constrain our ability to estimate ecological variables reflecting productivity and biomass structure. Despite these limitations, we can at least make some approximations (that are hopefully "right within an order of magnitude") of the energetic and selectively important components of Neandertal habitats. Very good paleoenvironmental reconstructions and syntheses of European Pleistocene climatic change already exist (e.g., Frenzel et al. 1992; Harrison et al. 1995; van Andel and Tzedakis 1996a, 1996b; Williams et al. 1998; van Andel and Davies 2003), and it is not my intention to review this literature in detail here. Rather, I hope to provide a sketch of the major habitats and climates in which Neandertals found themselves, as a background against which to explore their capture and use of energy.

6.1 ISSUES IN THE RECONSTRUCTION OF PAST ENVIRONMENTS

The emergence of the Neandertals as a morphologically distinct group dates to the relatively mild interstadial climate of MIS 7 (Chapter 2.1). In the more than 200,000 years from their emergence to their disappearance in late MIS 3, the Neandertals endured (as did their ancestors) climatic cycles going from full interglacial to full glacial conditions (Figure 6.1) with attendant large-scale shifts in the distribution and character of plant and animal biotas across Europe and western Asia. While these shifts were likely played out over millennia and hence over tens or hundreds of Neandertal generations (with the exception of MIS 5e and the later part of MIS 3, during which major climatic shifts may have occurred within the lifetimes of individuals: Dansgaard et al. 1993; GRIP 1993; White 1993), they still represent long-term adaptive challenges faced by populations of archaic Eurasians. As Europe cycled between open landscapes of steppe and tundra and closed treescapes of dense broadleaf forests, Neandertals had but two options: do as most mammals and follow their preferred habitats as

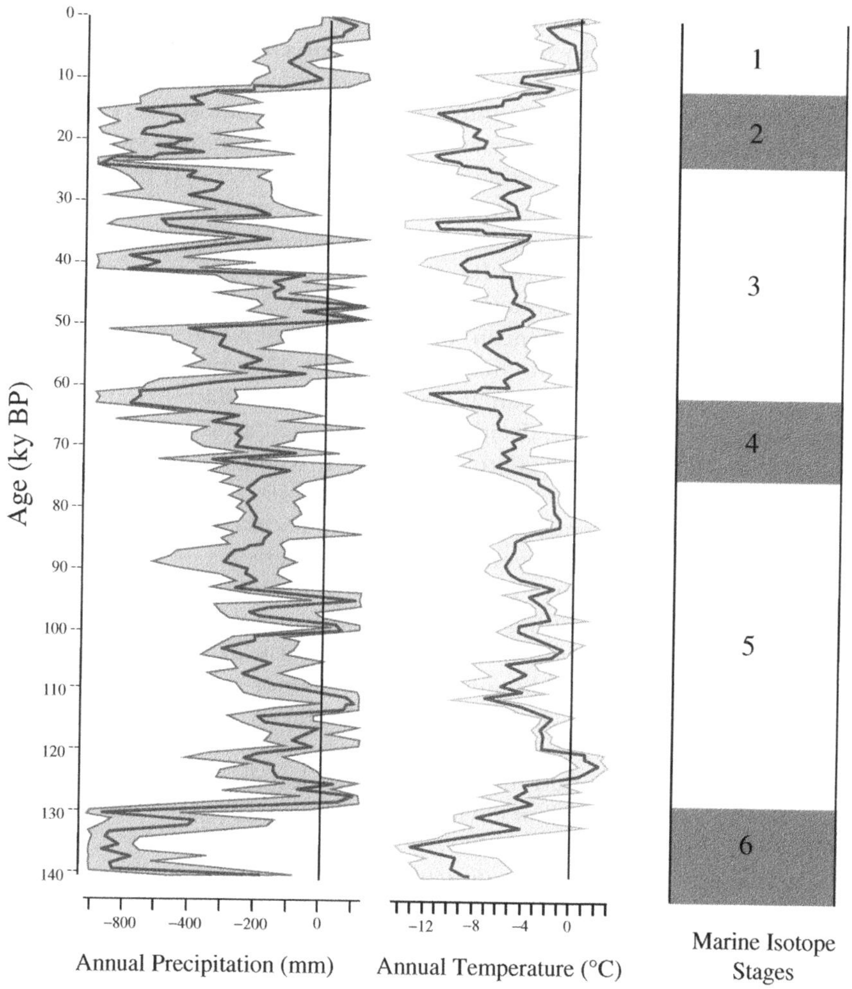

FIGURE 6.1

Reconstruction of rainfall and temperature in eastern France over the last 140 Ka, expressed as deviations from modern values (represented by grey lines), based on pollen samples from Les Echets. Shading indicates confidence limits (computed by simulation) of the estimates. Modern annual total precipitation at Les Echets = 1080 mm; modern mean annual temperature = 11 °C. Redrawn from Figure 3 in Guiot et al. 1989.

ecozones shifted north or south (Hofreiter et al. 2004), thus either abandoning or dying out in regions that had become ecologically unsuitable (Hublin and Roebroeks 2009); or adapt to changing temperatures, altered fauna and flora, and changes in the spatial and temporal distributions of edible resources. The spatial and geographic distribution of Middle Paleolithic sites certainly reveal a pattern of discontinuous occupation of northern areas, which most likely reflects local extinction during climatic downturns (rather than abandonment as they tracked preferred habitats that were shifting southward) (Hublin and Roebroeks 2009). However, this same evidence also reveals a remarkable adaptive flexibility in response to climate change. The geographic distribution of Mousterian sites throughout the climatic fluctuations of MIS 7–3 (see below) shows Neandertals to have been a feature of the fauna of almost every type of biome Europe had to offer (and probably also western Asia, although the association there of modern humans with Mousterian lithic assemblages [Chapter 3.1] precludes drawing any conclusions about Neandertal environmental preferences in the Near East from the distribution of Mousterian sites). Their archeologically-visible presence in steppe (e.g., in MIS 6 deposits at Ariendorf [level 2], Schweinskopf-Karmelenberg [horizon 5], Tönchesberg I, Wannen, Achenheim and La Cotte de St. Brélade: see references in Roebroeks et al. 1992), Mediterranean woodland and semi-desert areas (for example, from MIS 4 and 3 deposits at Guattari [level G0], Kebara [Units XII–VI], Cabezo Gordo and Fate [pollen zone 3]: Grün and Stringer 1991; Valladas et al. 1987; Carrión et al. 2003; Karatsori et al. 2005), deciduous forests (such as MIS 7 and 5e Mousterian assemblages from Maastricht-Belvédère, Ehringsdorf, Weimar, Taubach, Veltheim, and Rabutz, to name a few: references in Roebroeks et al. 1992), and even wetland environments (van Andel and Runnels 2005; see also Karatsori et al. 2005) shows them to have carried further the ancestral genus *Homo* characteristics of wide environmental tolerance and adaptive flexibility. The general (but not total) absence of Mousterian sites above the 45th parallel during cold intervals (van Andel et al. 2003) may indicate that Neandertals were averse to tundra habitats (or it may simply be an aversion to the cold: see Chapter 5.1), although there is evidence to suggest that they could sometimes be found there (Pavlov et al. 2004; Uthmeier et al. 2011). MIS 3 Neandertal remains from Denisova and Okladnikov caves in the Altais of southern Siberia (Krause et al. 2007b) also support the idea that Neandertals were able to colonize tundra or near-tundra environments, or environments with mixed tundra/steppic elements (Derevianko et al. 2005; Shichi et al. 2009).

This adaptive flexibility makes it difficult to make generalizations about the productivity and energy structure of Neandertal habitats (because they were exploiting all of them!). It does, however, provide the opportunity to explore the relationship between variation in ecological parameters and variation in Neandertal behavior across space and time, and to identify different solutions that may have arisen in geographically- or temporally-distant Neandertal populations facing similar adaptive challenges. At present, our ability to explore these relationships is hampered by a far from perfect knowledge of the ecological properties of past ecosystems and by gaps in details (for some regions at least) about the local environments in the proximity of Neandertal sites.

A number of methodological and other issues make impossible the accurate reconstruction of amounts of primary and secondary production and the distribution of edible biomass. First, while it is convenient to talk about biomes (e.g., tundra, temperate deciduous forest, etc.), biome subtypes (e.g., loess tundra, moss-grass tundra, tundra-forest, etc.) or large-scale plant–animal associations (e.g., Mammoth steppe), the use of these broad units conceals an enormous amount of variation in primary and secondary productivity and species composition (both plant and animal), having to do with regional and local variation in insolation, precipitation, soil chemistry and hydrological factors, the nature and intensity of interspecific competition, and historic factors (Colinvaux 1978). The term "tundra," for example, encompasses various types of low arctic and alpine floras (tall shrub-, low shrub-, cottongrass subshrub-, subshrub-, wet sedge-moss-, cushion plant-herb-lichen-, and various forest-tundras) and high arctic floras (wet sedge-moss tundra, cushion plant-lichen, cushion plant-moss and moss-herb semidesert floras) (Bliss and Richards 1982). Within every biome there exist patterns of small- and large-scale variation in the constellations of resident organisms and the structures of communities, such that every biome is in fact a mosaic of ecosystems more-or-less integrated at various hierarchical levels (Moen 1973; Colinvaux 1978; Begon et al. 1996). The scale of this patchwork problem can be appreciated by considering that estimates of net annual primary productivity obtained from different studies of the same biomes can vary by an order of magnitude (see for example Whittaker 1975)! Two of the best proxies available for reconstructing environments local to Mousterian sites – pollen and macromammal remains – provide only the broadest brushstrokes (i.e., allowing reconstruction at the biomic or sub-biomic levels only); pollen because of its highly mobile aeolian nature, macromammals because of their generally wide environmental tolerances (although certain key animal species tend to be associated with a narrow range of habitat conditions and thus may prove useful) (Williams et al. 1998). Animal mobility is also an issue when trying to assess the productivity of Pleistocene environments, since secondary production may occur in one environment and then move to other environments where it is captured and consumed by humans, effectively as "unearned resources" with respect to the local human habitat. Anadromous fish are one prime example, reindeer another (since reindeer productivity derives mainly from tundra resources, yet they may be taken in boreal forest wintering grounds).

Second, we lack appropriate analogs for Pleistocene environments in the modern world (see Hopkins et al. 1982; Guthrie 1990; Huntley and Allen 2003). Tundra was a major feature of the glacial environments of mid-latitude Europe (today it is found only above the Arctic Circle along the coastal margins of the Barents Sea), and it certainly shared with modern arctic tundra key aspects of vegetation (predominance of lichens, mosses, sedges, and dwarf trees) and fauna (musk ox, reindeer, and arctic fox to name a few). However, during glacial maxima tundra elements extended as far south as 45° N latitude, some twenty-one degrees south of extant high latitude tundra (van Andel and Tzedakis 1996b). At this latitude the tundra flora would have benefited from less variation in photoperiod (which ranges from 0–24 hours d^{-1} in the arctic, but only between 8–16 hours d^{-1} in mid-latitude Europe) and from a greater intensity of solar radiation (since the angle of incidence of solar radiation is much higher at lower latitudes), and hence probably experienced considerably greater productivity than does arctic tundra today (Butzer 1971; Stanley 1980; Guthrie 1982, 1990). Additional factors having to do with increased aridity and greater evapotranspiration in Pleistocene compared to modern tundras likely also substantially increased the productivity of the former (paradoxically, increased aridity may improve tundra productivity: see Guthrie 1990), which may have produced a grassland-tundra mosaic "Mammoth steppe" rather than true tundra across a wide swath of Europe. Even under wetter conditions, when permafrost and a short growing season may have constrained the growth of non-tundra elements in moist lowland areas, better drained uplands may have supported grass steppe, resulting in areas of steppe-tundra mosaic unknown in the modern world (van Andel and Tzedakis 1996b). Thus, by some accounts, low latitude Pleistocene tundras may have had primary productivities on a par with modern

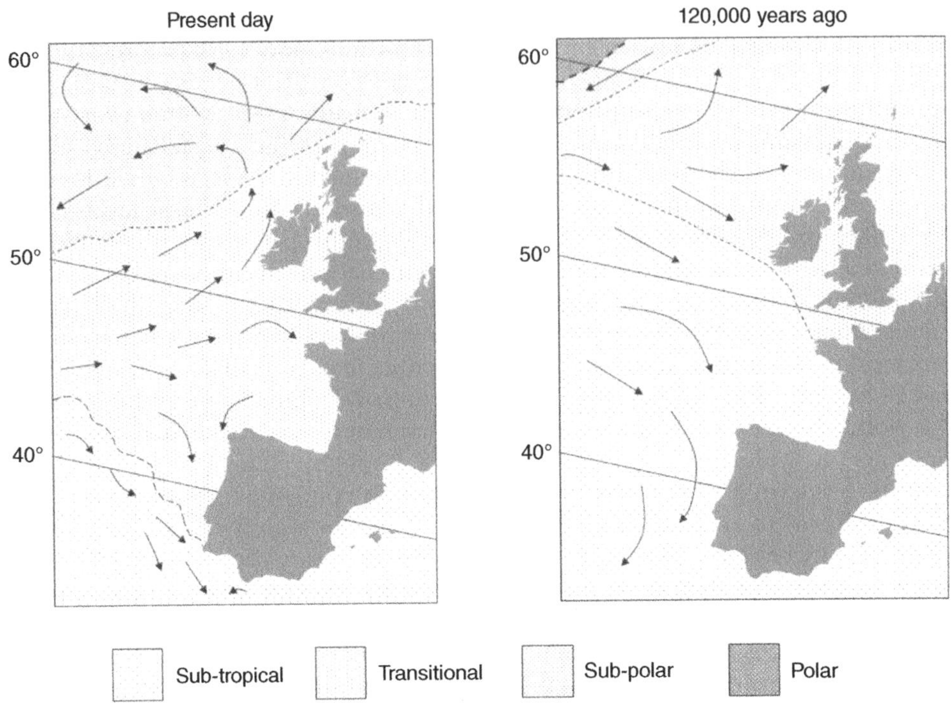

FIGURE 6.2

Maps of the distribution of polar and tropical waters in the north Atlantic today (left) and near the end of MIS 5e (right). Note the more northerly extent of sub-tropical and transitional waters during last interglacial times. Redrawn from Figure 6.2 in Sutcliffe 1985.

temperate grasslands, and would have supported a large standing crop of animal biomass (Butzer 1971).

A similar situation holds for the last interglacial, which was warmer and wetter than the present interglacial. Elevated sea levels (as high as 15 m above present sea level during the peak of MIS 5e: Butzer 1975) resulted in an inundation of the Baltic Sea through the lake region of southern Finland and Karelia, connecting the Baltic and Barents Seas (as the "Eemian Sea": van Andel and Tzedakis 1996b) and creating an island of the northern parts of Scandinavia (Ruddiman and McIntyre 1976). The resulting alterations in ocean currents in the north Atlantic would have brought sub-tropical currents almost as far north as the British Isles, with warm "transitional" (between sub-tropical and sub-polar) waters extending along the northern coast of the Baltic Sea (Sutcliffe 1985) (Figure 6.2). This is in marked contrast to the Holocene pattern, in which sub-tropical waters penetrate north only to about 40 °N latitude (about the level of the Straits of Gibraltar), and the North, Baltic, and Barents Seas contain sub-polar to polar waters (Sutcliffe 1985). Thus, during the peak of the last interglacial the western portion of Europe was wetter (with annual precipitation of 100+ mm above present values in places: Guiot et al. 1989), the central and eastern parts more oceanic (with annual precipitation of 50–300 mm above present values: Frenzel 1973) and the northern coast much more productive than today (Gamble 1999). Again, we are confronted with paleoenvironments – in this case, temperate forests under warmer and wetter conditions than exist today – for which there is no modern analog.

If estimating primary productivity of prehistoric ecosystems is difficult, estimating secondary productivity is no better. Animal production depends upon carrying capacity, which is a function of primary productivity and how that productivity is allocated into edible versus non-edible resources, and also upon the abilities of the resident fauna to locate, ingest, digest, and assimilate the edible portions of the vegetation cover (Moen 1973). Our potential for deriving realistic estimates of animal biomass for given places and times in prehistory is thus hampered by both an inability to accurately estimate primary production and by uncertainty about the (undoubtedly) complex ecological relations between members of prehistoric faunal communities, communities that included now extinct forms and that may have had species associations (e.g., mixed tundra and grassland faunas) not found today. Indeed, recent work on North American Pleistocene faunas indicates that mammalian communities are continually and unpredictably emergent, and that many Late Pleistocene faunal communities have no modern analogs (Graham and Group 1996).

All of this is not to say we shouldn't try to make inferences about the ecological relationships that obtained in Pleistocene Eurasia. Indeed, such efforts afford a rough idea of the environmental conditions and ecological context in which the Neandertals operated and in which the archeological and fossil records accumulated, and provide approximate models that can be refined as the archeological and paleontological records grow, our ability to interpret those records improves, and as we further develop our facility to model past environments (see Huntley and Allen 2003).

6.2 PLEISTOCENE BIOMES OF EUROPE AND WESTERN ASIA

Present-day Europe encompasses five major biomes (Van der Hammen et al. 1971) that are determined largely by temperature and rainfall variation having to do with latitude and continentality (Figure 6.3). Southwestern Europe, south of 45° N latitude and west of 30° E longitude, is dominated by Mediterranean chaparral (maquis) and warm oak forest, while the southeast is dominated by semi-arid steppe and desert vegetations. Mid-latitude Europe preserves the remnants of once-extensive deciduous mixed forest – forest that was likely arrayed as a continuous belt across the western part of Europe prior to large-scale clearing of trees for agriculture. The mixed deciduous forest biome extends between 45° and 55° N latitude in the west, but pinches out progressively from the south (to be replaced by steppic vegetation) east of 30° E longitude as climate becomes less oceanic and more continental. The diminution of the temperate forest belt is complete by the eastern margin of the European subcontinent (the Ural Mountains at 60° E longitude), beyond which to the east the steppe and desert biome is bordered by boreal forest to the north. Boreal coniferous forest begins above 55° N latitude in western continental Europe, while the British Isles, owing to their maritime climate, lack this biome altogether (being covered in deciduous mixed forest). The fifth biome, tundra, is today found only above the Arctic Circle (arctic tundra) and in the high Alps and Pyrenees (alpine or mountain tundra). Maquis covers the Mediterranean coasts of

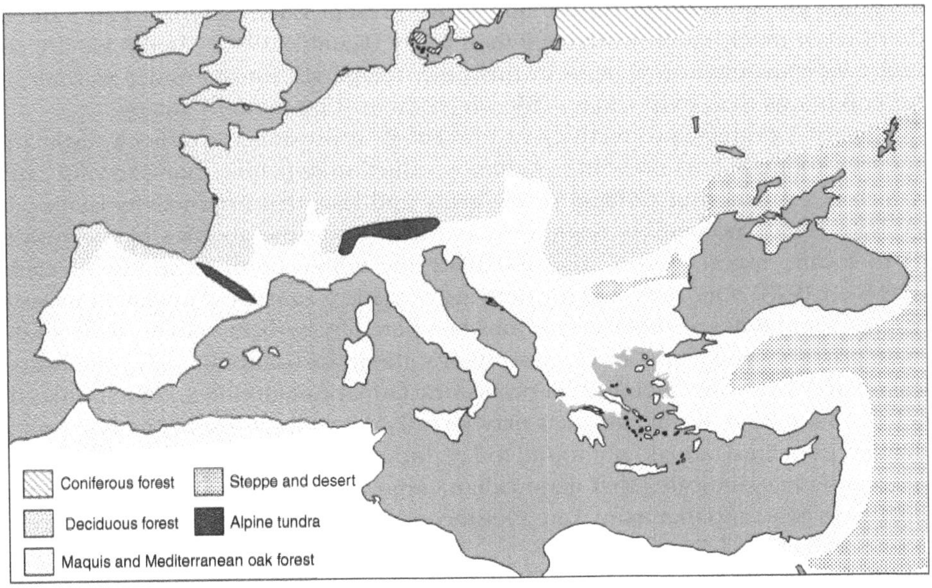

FIGURE 6.3

Distribution of biomes (formations of plants) in modern-day Europe (redrawn from Van der Hammen et al. 1971).

Asia Minor and the Levant while oak forests extend into the highlands of the Kuzey Anadolu Dağlari and Taurus, Caucasus, and Zagros Mountains. The non-mountainous regions of the interior of Asia Minor and the Near East are covered by semi-arid steppe, grading to full desert to the south (Syrian Desert) and east of the Zagros (the Dasht-e Kavir or Salt Desert of Iran).

The climatic oscillations of the later Pleistocene, as reflected in the oxygen isotope record of deep sea cores, ice layer thicknesses in the Greenland Summit ice cap, and terrestrial pollen and malacological records (e.g., Guiot et al. 1989; Reille and de Beaulieu 1990; van Andel and Tzedakis 1996a, 1996b), dramatically altered the vegetation zones of Europe and, to a lesser extent, southwest Asia. These oscillations saw north–south (largely following temperature gradients) and east–west (largely following precipitation gradients) marches of vegetation types across the subcontinent. In general, colder, drier oscillations saw expansion of open country with the spread of tundra environments southward and grasslands westward in Europe and the growth of semi-desert environments in southwest Asia. The warmer, wetter intervals witnessed the development of forests of warmth-loving (thermophilous) deciduous trees across the mid-latitude and northern stretches of Europe, while large areas along the Mediterranean coast (including the Near East) were probably covered in maquis.

Over the long term, regional groups of Neandertals adapted to the alternating faunas and floras that the changing climate brought, and there appear to be some general adaptive responses to climatic extremes (both warm and cold) that are played out across the Middle Paleolithic (e.g., Mellars 1996; Pettitt 2003). Still, we might at the outset expect some adaptive differences in the ways in which these groups responded to similar environmental conditions at different times (for example, the cold maxima of MIS 6 and 4). This is to be expected because no region contained identical environments at disparate points in time, no matter how similar the climate may have been. Each recolonization of an area by a particular vegetation type, whether it be tundra or forest, brought with it a different climax mix of plant and, to a lesser degree, animal species due to a variety of factors: alterations in soil chemistry over time, stochastic factors having to do with the order of succession during recolonization of an area, macroevolutionary changes in the plant and animal biotas, and variation in the rapidity of onset and duration of the climatic episode (Zagwijn 1989). Accordingly, fine-scaled studies of Neandertal adaptation should seek to isolate samples by time and environment but, as noted above, this results in site or fossil sample sizes inadequate for analysis. Despite differences in the regional species composition of biomes at various times, it seems reasonable to assume that there were broad similarities in overall ecological aspect – namely productivity and distribution of biomass – of the major biomes across time. The following is an attempt to characterize (in gross terms at least) the climatic and ecological conditions of the major environments to which the Neandertals adapted. Reviews of climatic reconstructions for the glacial intervals MIS 6 and 4 and the cold–temperate intervals MIS 5d–a and 3 are provided in Chapter 5.1, and thus the following review concerns their ecological characteristics only. Reviews of both climatic and ecological reconstructions for the interglacial episodes MIS 7 and 5e are provided in this chapter.

6.2.1 Interglacial Europe: Marine Isotope Stages 7 and 5e

The marine and terrestrial data for global and European climatic and environmental change is best known for the last interglacial–glacial cycle, spanning MIS 5e (Last Interglacial) to MIS 1 (Holocene). Because the data for earlier periods (of which only MIS 7 and 6 concern us) are less complete it is difficult to compare climatic conditions in MIS 7 (242–186 Ka BP) and 5e (127–118 Ka BP) (dates from Bassinot et al. 1994). Oxygen isotope ratios in most deep sea cores (as evident in the standard reference Equatorial Pacific core V28-238: Shackleton and Opdyke 1973) show stage 7 oceans to have been isotopically heavier than those of 5e, indicating larger ice caps, colder climate, and lower sea levels during the earlier (Penultimate) interglacial. Accordingly, western Europe would have been less oceanic and would have had a steeper latitudinal gradient in vegetation types than seen during the high sea level interglacials of MIS 9, 5e and 1 (Zagwijn 1989). Additionally, MIS 7 encompasses a distinct 10–15 Ka cold spell (MIS 7b) centered on 230 Ka BP (Andrews 1983) in which global ice volumes approached those of glacial conditions. The maxima of the warmer intervals (MIS 7a and 7c) may have approached MIS 5e or 1 (current) global temperatures, but overall were more similar to the later part of MIS 5 (substages d–a).

In contrast to this marine isotopic data, terrestrial climatic proxies indicate that MIS 7 was comparable in temperature and environments to the Eemian (MIS 5e). Floral and faunal remains from MIS 7 sites in northern Europe are indicative of full interglacial conditions: mixed deciduous forest cover and forest-dwelling or thermophilic fauna (see discussions in Roebroeks et al. 1992 and Gamble 1999). The presence of pond tortoise (*Emys orbicularis*) in MIS 7-aged deposits at the German sites of Maastricht-Belvédère and Neumark-Nord even suggest conditions *warmer* than those that obtain today (Gamble 1999). While discrepancies between the global climatic record reflected in deep sea and ice cores versus terrestrial proxies are yet to be resolved (see Webb 1986; Larsen et al. 1995), the similarities in floral and faunal characteristics between MIS 7 and 5e sites suggest that these two intervals can, from the perspective of human forager ecology, be considered together.

Turner and West (1968) proposed a four stage model of interglacial development based on European pollen records, beginning with the pre-temperate zone (phase I) in which open ground gave way to forests dominated by birch (*Betula*) and pine (*Pinus*) along with relicts from the preceding late glacial like juniper (*Juniperus*) and willow (*Salix*). Open country herbs and shrubs remained important in the pollen spectra as pine forests spread northward from their glacial refugia in the south. In the early-temperate zone (phase II), elements of a mixed oak forest (especially oak [*Quercus*], elm [*Ulmus*], ash [*Fraxinus*], and hazel [*Corylus*]) dominated as broadleaf trees replaced conifers and deciduous forests were established across Europe. In the late-temperate zone (phase III), hornbeam (*Carpinus*) and other late-immigrating temperate elements (e.g., fir [*Abies*] and spruce [*Picea*]) entered the mix of species of the established mixed oak forest, generally with a decline of the phase II dominants. Finally, in the post-temperate zone (phase IV) progressive cooling towards the approaching glacial saw further decline and extinction of thermophilous trees, as pine and birch forests once again dominated the landscape. These post-temperate landscapes also saw a thinning of forests (as they transitioned from predominantly deciduous to predominantly coniferous) and expansion of open country heathland (Goudie 1983). This model characterizes the general pattern of floral succession during long warm episodes, but does not account for the dramatic and potentially rapid climatic downturns that punctuated the interglacials (GRIP 1993; White 1993). Although numerous, these cool spikes were probably short lived (the exception being the 10–15 Ka-long cold snap in the otherwise climatically stable penultimate interglacial, marine isotope substage 7b: Andrews 1983), and their effect on the vegetational succession outlined above appears to have been fairly minimal. The 10 Ka-long last interglacial (MIS 5e) appears to fit the general succession model well, as can be seen in the temporal succession of pollen of arboreal taxa in cores from France (Ribains [de Beaulieu and Reille 1992] and Le Bouchet [Reille and de Beaulieu 1990]), England (various: Fig. 6.10 in Sutcliffe 1985), Germany (Geesthacht [Kaiser 1969]) and, less clearly, in Macedonia (Philippi [Van der Hammen et al. 1971]). According to van Andel and Tzedakis (1996a), the interglacial succession differed in northern and southern Europe. In the north, oak and elm gave way to hazel (*Corylus*) and yew (*Taxus*) (although oak remained an important component of the pollen spectrum throughout the interglacial), followed by hornbeam (*Carpinus*), then fir (*Abies*), then spruce (*Picea*), with pine coming to redominate in late MIS 5e. In the south, the oak/elm interval was followed by an expansion of Mediterranean woodland with wild olive (*Olea*) and evergreen oak, after which the hornbeam expansion can be seen (thus a Mediterranean woodland phase replaced the hazel/yew phase farther north).

The longer (56 Ka) penultimate interglacial fits the succession pattern less well. In places – for example the pollen sequence from Marks Tey in England (Jones and Keen 1993) – the overall pattern of succession is borne out (although alder [*Alnus*] played a larger role in the early and late temperate zones, indicative of a more oceanic climate, and hornbeam and spruce did not dominate early and late, respectively, temperate zones as they did in most of the continental MIS 5e sequences). The cold conditions of substage 7b are poorly indicated in the sequence from Marks Tey, with only a moderate increase in grass pollen in an otherwise early temperate flora. In other places (for example the Philippi sequence from Macedonia: Van der Hammen et al. 1971), the pollen record denotes a more variable alternation between oak, pine, and fir, with little development of non-oak deciduous elements. In striking contrast to the Marks Tey core, the Philippi core shows a marked abatement of arboreal pollen and a dominance of open country grasses and herbs in the middle of the MIS 7 sequence, undoubtedly corresponding to the cold conditions of substage 7b.

Paleobotanical, faunal, isotopic, and marine geological records all indicate that continental climate at the peak of MIS 5e was warmer than anything experienced during the Holocene. Oxygen isotope records show the ice volume in MIS 5e to be the smallest of any point, including the present, during the last 730 Ka (Gamble 1986), with beaches 7–15 m above present sea levels at some times (Butzer 1975; Mussi 2001a; Siegert 2001). The oceanic nature of this interglacial is evident by the presence of ivy (*Hedera*) and holly (*Ilex*) far to the east of their current distribution, and by the spread of fir (*Abies*) well north of its current limit (Gamble 1999). It is during MIS 5e times as well that hippopotamus (*H. amphibius*) could be found on the upper floodplain terrace of the Thames at London (Sutcliffe 1985). A Mediterranean climate (warm, dry summers and mild winters) is indicated for much of the continent by the presence of Mediterranean shrubs like Montpelier maple (*Acer monspessulanum*), firethorn (*Pyracantha coccinea*), ivy (*Hedera helix*), holly (*Ilex*), and great fensedge (*Mariscus serratus*), and by the occurrence of taxa now absent in Britain and the Low Countries, including water chestnut (*Trapa natans*), lesser najad (*Najas minor*), and water fern (*Salvinia natans*) (van Andel and Tzedakis 1996a).

Synthesis of available micro- and macro-botanical data for MIS 5e (van Andel and Tzedakis 1996b) produces a picture of a continental Europe (along with the British Isles) that was predominantly covered in mixed deciduous oak forest at the height of the last interglacial (although Scandinavia was covered in mixed coniferous/deciduous forest or boreal forest at this time, it was isolated from the continent by the incursion of the Baltic Sea) (Figure 6.4). To the south, Mediterranean woodland covered the southern reaches of the Iberian peninsula and the coast of France, and was the predominant formation in the Apennine and Balkan peninsulas, the western part of Turkey, and the Mediterranean coastal zone of the Levant (van Andel and Tzedakis 1996b). Judging from the distribution of Mousterian sites (or in the case of the Near East, sites with Neandertal fossils) dating to the last two interglacials, Neandertals had effectively colonized both oak forest and maquis habitats. Again, however, the importance of environmental variability cannot be underestimated. The biogeographic picture portrayed in Figure 6.4 represents a snapshot of one point in time, most likely corresponding to the maximum development of thermophilous forests during the last half of the Quaternary. This snapshot lacks the scale of resolution necessary to evaluate the extent and distribution of ecotonal environments that may have been important, if not critical, to Neandertal adaptation, and it ignores the perhaps marked temporal variation that existed in even short climatic episodes, variation that resulted from climatic fluctuation, natural successions of plant and animal taxa during recolonization

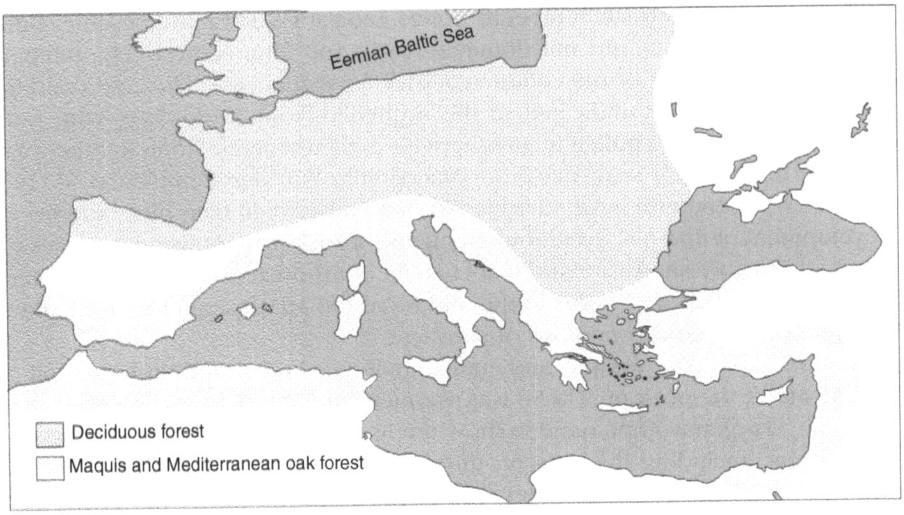

FIGURE 6.4

Distribution of biomes during the peak of the last interglacial, MIS 5e, following van Andel and Tzedakis 1996b. The distribution of biomes is assumed to have been roughly similar during MIS 7.

of an area, temporal changes in soil chemistry and nutrient availability, and chance or historical factors. In what follows, the biotic conditions of the peak of MIS 5e (centered on 125 Ka BP: Berger 1978) are taken to represent one extreme in terms of the ecological conditions in which Neandertals lived, with the full knowledge that conditions were not uniform across Europe, across the interglacial, or between interglacials (with MIS 7 likely being less oceanic and less productive overall than MIS 5e Europe).

6.2.2 Productivity and Edible Resources of Interglacial Environments

In addition to the problems discussed above that impair our ability to reconstruct past habitats and their ecological parameters, we also lack the data needed to derive even moderately reasonable estimates of primary productivity for any given paleohabitat (minimally evapotranspiration [Sharpe 1975], which can in turn be estimated from average monthly temperatures and average number of daylight hours by month [Thornthwaite 1948]). Productivity values for extant deciduous and boreal forests provide some idea of the conditions that likely pertained in interglacial Europe. Representative values for Holocene plant and animal productivities in the biomes that would have predominated the interglacials are shown in Table 6.1. All three formations – deciduous and coniferous forest and woodland (including maquis) – have high net primary productivities (NPP). As discussed further below, they also have large amounts of standing biomass, such that most of the primary production is in woody tissues and therefore inaccessible to herbivores.[1] Woodlands are somewhat better, as reflected in their higher NPP/1° biomass ratios, and accordingly they tend to support higher biomasses of large animals. The forests at the peak of MIS 5e, having developed under

TABLE 6.1 ■ Productivity estimates[a] for forests and woodlands

	NPP $g/m^2/yr$	1° Biomass g/m^2	NPP/1° Biomass	2° Biomass kg/km^2
Temperate deciduous forest				
Mean	1,200	30,000	0.04	700–4000
Range	600–2,500	6,000–60,000		
Coniferous forest				
Mean	800	20,000	0.04	62–3600
Range	400–2,000	6,000–40,000		
Woodland and shrubland				
Mean	700	6,000	0.12	-
Range	250–1,200	2,000–20,000		

[a] Net primary productivity (NPP), primary biomass (1° Biomass), ratio of mean NPP to mean primary biomass (NPP/1° Biomass) and biomass of larger animals (2° Biomass). Plant productivity and biomass estimates from Whittaker 1975 (see also Clarke 1976); animal biomass is for larger mammals, primarily ungulates: estimates based on ungulate data from Russia and USA (Bannikov 1967; Botkin et al. 1981; Crête 1999; see also Redmann 1982).

conditions warmer and wetter than today, were undoubtedly more productive than extant temperate broadleaf forests, but how much more productive?

That MIS 5e was warmer and wetter than today's climate can be seen from the plant taxa registered in northern Eurasian pollen samples that accumulated at this time (Table 6.2). These data suggest mean January temperatures 3.3 ± 3.3 °C above present values, and mean July temperatures 1.6 ± 1.5 °C above today's (with mean annual temperatures about 2.4 ± 2.3 °C above present). Paleoclimatic indicators suggest similar conditions in mid-latitude Europe. According to Van Andel and Tzedakis (1996a), beetle remains at La Grande Pile (France) during MIS 5e suggest mean annual temperatures 0.5–2.0 °C warmer than present (10–12 °C vs. 9.5 °C), with a mean July temperature between 16–18 °C, which is close to contemporary values. While winter temperatures in the north during the peak of MIS 5e may have been 2–8 °C warmer than found in Europe today, mid-latitude Europe may have experienced winter temperatures comparable to those of the Holocene (van Andel and Tzedakis 1996b).

These estimates of mean January and July temperatures can be used to calculate effective temperature (ET: Bailey 1960; Binford 1980), a measure of the temperateness of an environment. ET is based on the mean temperatures (°C) of the warmest (*W*) and coldest (*C*) months, and is thus a proxy measure of the distribution of insolation throughout the year:

$$ET = \frac{18W - 10C}{(W - C) + 8}$$

Using the July temperature estimates from La Grande Pile and making the heuristic assumption that the 2–8 °C higher mean January temperatures indicated for northern Europe applied across the mid- and lower-latitudes as well (resulting in mean January temperatures of 4–10 °C, based on present-day temperatures in Paris) produces ET estimates in the range of 12.4–14.0 for much of MIS 5e Europe. Figure 6.5 illustrates the curvilinear relationship of ET to primary productivity (NPP) from the tropics to

TABLE 6.2 ■ Climatic differences between MIS 5e and present, based on plant taxa represented in pollen cores

	Mean temperature			
Region and plant species or genus	January °C	July °C	Annual °C	Annual precip. (mm)
Denmark				
Lime (*Tilia platyphyllos*)	+2	+1 to 2	+1 to 2	=
Water chestnut (*Trapa natans*)	=	+2	?	?
Northern/central Germany				
Lime (*Tilia tomentosa*)	=	+3	+2 to 3	=
Box (*Buxus sempervirens*)	+1 to 2	+1	+1 to 2	=
Central Poland				
Beech (*Fagus silvatica*)	+3	=	=	=
Lime (*Tilia tomentosa*)	=	+3	+3	=
Lime (*T. platyphyllos*)	+1 to 2	=	+1	+50
Maple (*Acer tataricum*)	=	+1 to 2	=	=
Holly (*Ilex aquifolium*)	+3 to 4	=	=	=
White Russia				
Maple (*Acer tataricum*)	=	+3	=	=
Lime (*Tilia platyphyllos*)	+5 to 6	+3	+4	=
Lime (*T. tomentosa*)	+6	+5	+7	=
Water chestnut (*Trapa* sp.)	?	+1	?	?
Central Russia				
Beech (*Fagus silvatica*)	+10	=	+5 to 6	+
Lime (*Tilia platyphyllos*)	+6 to 9	+2	+4 to 6	+100
Maple (*Acer tataricum*)	=	+2	=	=
Holly (*Ilex aquifolium*)	+9 to 10	=	+3	+100
Water chestnut (*Trapa natans*)	?	=	?	?
Northwest Ukraine				
Lime (*Tilia platyphyllos*)	+2 to 3	=	+1	+50
Western Siberia				
Distribution of conifirs	+4	+3	+3	+100
Central Siberia				
Distribution of permafrost	?	?	+6	?

Data from Table 10 in Frenzel 1973.

the arctic (NPP estimated from evapotranspiration values for 123 localities: data from Table 3-1 in Kelly 1995). ET serves as a crude proxy for the intensity of insolation and the length of the growing season, but it does not take precipitation or nutrient availability into account. Water availability is likely the primary determinant of variation in NPP at any given ET, and the curvilinear nature of the plot in Figure 6.5 is simply reflecting the fact that warm environments tend to be wetter (the extremes occurring in the tropics), and hence doubly productive.

When primary productivity is considered within an ET range more reflective of temperate environments (ca. 10–15) the weak relationship between these two variables is better illustrated (Figure 6.6). Note also that the data used in the construction of Figure 6.5 and Figure 6.6 (from Kelly 1995) have poor representation of

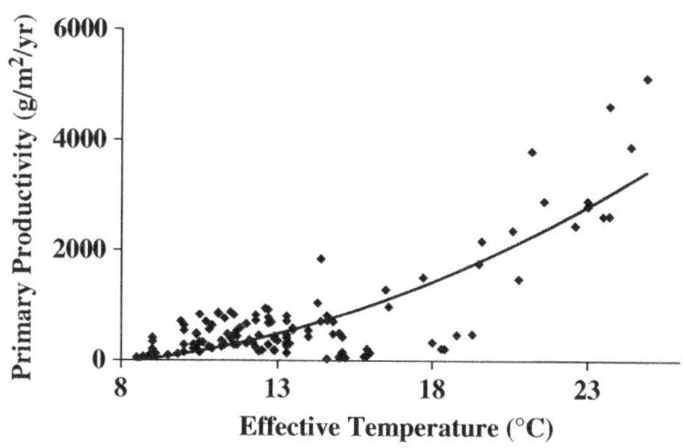

FIGURE 6.5

Effective temperature (°C) and net primary productivity (g m^{-2} yr^{-1}) for localities occupied by 123 historically-known foraging groups (data from Kelly 1995), with a 2nd order polynomial fit to the data ($y = 8.4403x^2 - 72.6x$; $r = 0.8301$).

well-watered temperate forests, which would be expected to have higher productivities than the environments represented, which in turn would weaken even more the ET-NPP relationship.

Nonetheless, effective temperature can be used to derive a *rough* idea of how meat-dependent foragers in interglacial forests might have been. Figure 6.7 shows the relationship between ET and meat consumption across the global range of effective temperatures. Here, meat consumption includes percent contribution to the diet of both fish and terrestrial game (data from Murdock 1967 and Kelly 1995).[2] As can be seen, reliance on meat varies inversely with ET, but only to a point: at ETs above about 15 dependence on meat again increases (although variably). This is largely because more

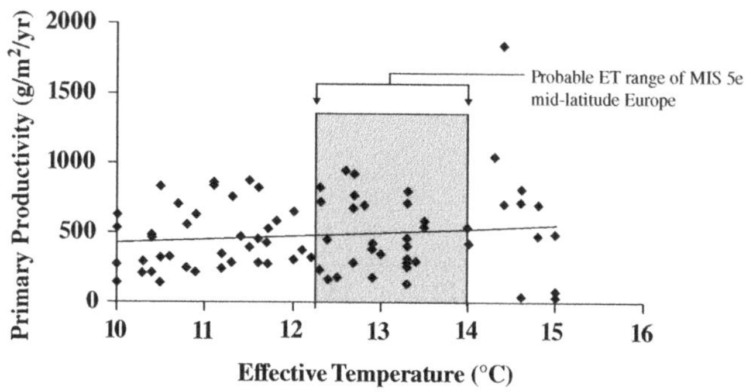

FIGURE 6.6

Effective temperature (°C) and net primary productivity (g m^{-2} yr^{-1}) for localities in the effective temperature range of 10–15 occupied by 79 historically-known foraging groups (data from Kelly 1995), with an ordinary least squares linear regression fit to the data ($y = 26.304x + 163.03$; $r = 0.1312$).

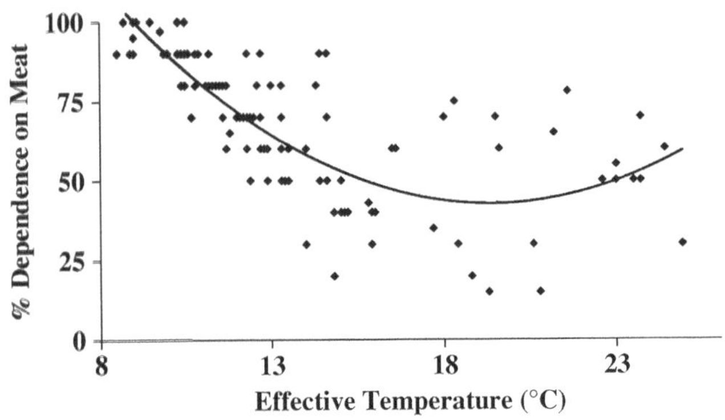

FIGURE 6.7

Effective temperature (°C) and meat consumption (terrestrial game and fish as a percentage of total diet) in 123 historically-known foraging groups (data from Kelly 1995 and Murdock 1967), with a 2nd order polynomial fit to the data ($y = 0.5355x^2 - 20.742x + 243.6$; $r = 0.7737$).

of the productivity is locked away in non-nutritive woody tissues in these high-ET ecosystems, such that they tend to be poorer in edible plant material. Simple linear regression (OLS, $N = 60$, $df = 58$, $r^2 = 0.5042$, $p < .01$), again applied to a narrower range of ETs, predicts that last interglacial foragers would have had to derive roughly 57–70% of their diet from animal sources (Figure 6.8). Since fish appear to have been an inconsistent (and probably insignificant) part of Neandertal diet (Chapter 7), a somewhat better estimate can be derived by using only terrestrial hunting by groups that derive less than 25% of their diet from fishing (see Kelly 1995). We can also expect to

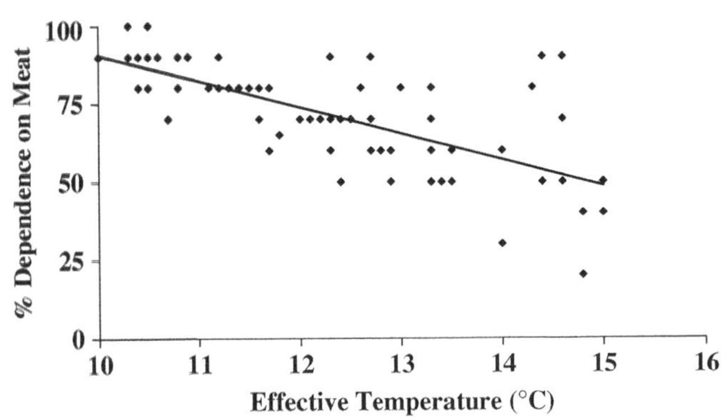

FIGURE 6.8

Effective temperature (°C) and meat consumption (terrestrial game and fish as a percentage of total diet) for localities in the effective temperature range of 10–15 occupied by 60 historically-known foraging groups (data from Kelly 1995 and Murdock 1967), with an ordinary least squares linear regression fit to the data ($y = -8.3799x + 174.41$; $r = 0.7101$).

improve our predictions of meat consumption by including in the regression estimates of primary productivity (Kelly 1995). The resulting multiple regression (with ET and NPP as the independent variables) predicts terrestrial meat consumption among modern foragers fairly well (N = 54, $df = 51$, $R^2 = 0.3520$, $p < .01$: Kelly 1995: 71). Using productivity values for extant temperate forests from Table 6.1 (since NPP is unknown for interglacial forests) and an ET range of 12.4–14.0, we predict about 56–64% dependence on terrestrial meat in forest-dwelling Neandertals, *using modern human foragers as a model*. If MIS 5e forests were more productive on average than extant deciduous forests, the abundance and distribution of edible plantstuffs would, paradoxically, be less favorable for human habitation (see below), and we would expect an even greater dependence on meat (up to ca. 78% using the maximum values for NPP recorded in extant temperate forests).

The question of Neandertal occupation of deciduous forests (Gamble 1986, 1987 vs. Roebroeks et al. 1992) serves as an example of the difficulties, discussed above, inherent in trying to make human–habitat associations when dealing with mosaic ecosystems. Temperate broadleaf forests, although typically having high primary productivity, have edible resource distributions that don't favor human exploitation. First, much of the primary production goes into the formation of tree trunks and branches (Ovington 1965; Odum 1975), and thus the bulk of the plant biomass is unavailable to consumers (except those that can digest cellulose, such as termites). Second, most of the edible production – metabolically active leaves and reproductive organs, as well as the herbivores that consume this production – are found in the forest canopy. Without technological means of accessing those resources, foragers in this environment are left with what edible plant production there is to be gleaned from the understory, the solitary (and generally small) herbivores that roam the forest floor, and social insects and their by-products as a nutritional base. So daunting is this problem of energy distribution that some researchers have wondered whether tropical forests (in which this situation is even more marked than deciduous forests, since less solar radiation penetrates the canopy) were inhabitable by humans before the advent of horticulture (see Hart and Hart 1986; Bailey et al. 1989: that humans *have* long been capable of exploiting tropical forests has recently been shown by the presence of Acheulean and Middle Stone Age sites in the deepest Ituri; Mercader 2002).

Although not as severe in energy structure as tropical forests, northern deciduous forests present similar difficulties.[3] Despite having diverse flora and fauna, most of the edible resources in deciduous forests come in small packages, may require considerable processing before being palatable or digestible, and are often highly seasonal. Historically-known foragers in the eastern temperate woodlands of North America practiced mixed economies of horticulture and gathering, and relied heavily on storage of horticultural products and nuts and berries as winter staples (Driver and Massey 1957; Campbell 1995). The Iroquois, who practiced a blended horticulture/foraging economy in the mixed deciduous/pine forests of New York and Ontario, relied on at least eight varieties of nuts and about 20 types of berries (Parker 1910). While cervids (*Odocoileus virginianus* and *Alces alces*) and bears were economically important resources, these groups also relied on smaller game with high handling costs relative to the calories they return: squirrel, rabbit, raccoon, beaver, fish, turkey, ducks, and other birds (Campbell 1995). For Pleistocene foragers of interglacial Europe, the *potential* foodbag certainly must have contained nuts of oak, beech, walnut, and hazel, and berries of vine creeper (*Vitis* sp.), strawberry (*Fragaria* sp.), raspberry, blackberry and cloudberry (*Rubus* spp.), cranberry, bilberry and whortleberry (*Vaccinum* spp.), and currants and gooseberry (*Ribes* spp.), and edible mushrooms in the form of ceps (*Boletus* sp.), chantarelles (*Chantarellus* sp.), and milk caps (*Lactarius* sp.) (Goudie 1983; Gamble 1986). A host of geophytes (plants that store calories in underground storage organs) would have also been available, including cattails (*Typha latifolia*), bistorts (genus *Polygonum*), arrow heads (*Sagittaria sagittifolia*), dog's tooth violets (*Erythronium dens-canis*), great burdock (*Arctium lappa*), wild carrot (*Daucus carota*), and wild parsnip (*Pastinaca sativa*) (Hardy 2010), and these would have provided an important source of carbohydrates in the Neandertal diet (see Chapters 4.3 and 7.4). The extent to which Neandertals used any of these resources is an open question, however, and stable isotope analysis of human remains from last interglacial (MIS 5e) level 4 at Scladina Cave (Belgium) indicates that the Neandertals there were deriving most of their dietary protein from animals (Bocherens et al. 1999). That Neandertals relied heavily on meat even in productive forest environments (in which we would expect, based on modern human behavior, ca. 36–44% reliance on plant foods) is likely a function of the complexity and relatively low energy returns involved in scheduling and storing seasonal plant foods that come in small packages, and possibly a dietary dependency on key fats in animal tissues (see Chapter 4.3).

For North American temperate forest foragers, survival depended heavily upon (1) scheduling residential or individual movement to capitalize on critical seasonal resources (nuts and berries) that may have short seasons of availability, (2) procurement of certain resources in excess of immediate demand and storage of those resources for winter use, (3) consumption of resources that are costly to collect, capture, or make edible, and thus return few calories per unit mass relative to the calories spent acquiring/consuming them[4], and (4) technology designed to reduce the handling costs of small, costly plant and animal prey, along with storage technology.[5] The pattern of human occupation in Europe over time, along with a lack of archeological signatures for strategies 1–4 in the Lower and Middle Paleolithic, led Gamble (1986) to suggest that successful human exploitation of temperate forest environments did not come until the Mesolithic.

What do we make, then, of the persistence of Neandertals across the continent during interglacial times, when Europe was largely covered with a thick carpet of broadleaf forest? Perhaps the Pleistocene inhabitants of Europe had found ways to cope with inherent difficulties of forest environments (Roebroeks et al. 1992; Gaudzinski-Windheuser and Roebroeks 2011). Another possibility is that during warmer intervals the Neandertals (and their ancestors) were exploiting larger patches of open land or bush cover that may have existed (due to variation in drainage, soil conditions, and biotic factors) within the matrix of primary forest. Such areas may have provided rich pasture and supported large herds of closed country or parkland forms – like red deer – that thrive on open ground but which still require closed habitats in close proximity (Larkin et al. 2001). Thus, though interglacial sites may preserve fauna, pollen, and macrobotanical remains indicative of closed forest, the subsistence strategies of the humans that occupied those sites may have been focused on the myriad ecotones that no doubt existed within an otherwise continuous continental forest. In other words, because of a lack of resolution in the reconstruction of local environments, we may be mistaking a forest-edge adaptation for a forest adaptation (see also Mussi 2001a).

In this regard Gamble (1992) noted that northern European sites tend to be concentrated towards the east and are rare in the west during the last two interglacials (see also Ashton 2002), suggesting that "... significant human habitation in these northern areas during oceanic interglacials, such as 5, required at the very least the dampening effect of continentality, presumably to open up the forests and produce conditions for the biomass more comparable to those in the later intermediate environments of the west" (p. 570). Such a conclusion is certainly supported by the distribution of Middle Paleolithic sites in the Crimea, where fully continental conditions would be expected even during the more oceanic interglacials. Sites here are largely constrained to a roughly 70 by 6 km strip in the Crimean Mountains, lying on an ecotone between the steppe to the north and the forest of the mountains to the south (Marks and Chabai 2001). However, all sites here – no matter the time or environmental conditions – show a pronounced (almost exclusive) bias towards steppe fauna, even when forests were well developed (e.g., during the Hengelo interstadial of MIS 3). Thus, Middle Paleolithic humans (presumably Neandertals) were not exploiting the ecotone faunistically, in that their subsistence efforts appear to have been focused almost exclusively on open country prey. Marks and Chabai (2001) interpret this pattern as reflecting a solution to differential environmental distributions of critical resources: animals on the open steppe; shelter, wood, and flint sources in the forest. Alternatively, they may have needed forest/steppe ecotones to successfully ambush open country prey (see Chapter 8.2). There also exist a few last interglacial-aged Mousterian sites with faunas predominantly indicative of closed forest, such as Maastricht-Belvédère, in the Netherlands (Roebroeks 1988), or Neumark-Nord (Roebroeks et al. 1992) and Taubach (Svoboda et al. 1996) in Germany. Forest- or parkland-loving animals such as fallow, roe, red, and giant deer (*Dama dama*, *Capreolus capreolus*, *Cervus elaphus*, and *Megaceros giganteus*), wisent and aurochs (*Bison priscus*, *Bos primigenius*), wild boar (*Sus scrofa*), and forest elephant and forest rhino (*Elephas antiquus*, *Dicerorhinus kirchbergensis*) have also been recovered in MIS 5e faunal assemblages containing remains of more open country forms or that are uncertainly associated with cultural material (see e.g., Cordy 1988; Roebroeks et al. 1992; Straus 1992; Mussi 2001a). That last interglacial Neandertals were at times exploiting the faunal resources of deciduous forests seems clear, but questions concerning the extent to which they were dependent on forest-edge ecotones, and the factors that were conditioning variation in the relative importance of open versus closed country forms, require further study.

6.2.3 Temperate Interglacial Europe: Marine Isotope Stages 5d–a and 3

Both the latter part of MIS 5 (substages 5d–a) and MIS 3 were marked by rapid fluctuations (on the order of 10,000 years for MIS 5, and on the order of 100–1000 years for MIS 3) between cold stadials and warmer interstadials as climates deteriorated towards the glacial episodes of MIS 4 and 2, respectively. These intervals are distinguished from full interglacial and full glacial intervals by their moderateness: the warmer interstadials were neither warm enough nor long enough to promote significant expansion of deciduous forests (Guiot et al. 1989; Gamble 1999), and the colder stadials were neither cold enough nor long enough to denude Europe of forest biomes. Instead, these temperate interglacial periods saw steeper gradients in vegetation change following

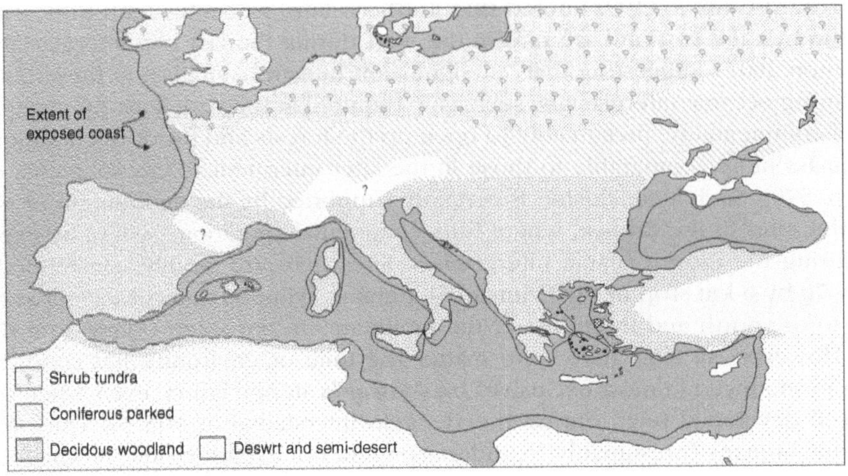

FIGURE 6.9

Approximate extent of major biomes during a warm episode (warm type D/O event) of MIS 3. Based on reconstructions from Huntley and Allen 2003 (Figure 6.12) and van Andel and Tzedakis 1996a (Figure 8).

latitude than characterized either interglacial or glacial episodes, and likely with greater biomic diversity as well (Figure 6.9).

The warmer oscillations of MIS 5 – substages 5c and 5a – correspond to the Brørup (Amersfoort) and Odderade interstadials, respectively, and represent reversals in the expansion of open country habitats that characterized environmental change towards the glacial conditions of MIS 4. In the more maritime western part of the subcontinent these warmer interstadials saw the development of deciduous forests with maquis elements along the north coast of the Mediterranean, forests of thermophilous trees north of the Alps but below 45° N latitude, coniferous forests of pine, spruce, fir, birch, willow (*Salix*) and larch (*Larix*) north to the Baltic Sea and covering the British Isles, and birch forests in Scandinavia (Frenzel 1973; van Andel and Tzedakis 1996a). The coniferous forests of the northern regions dominated in central Europe as well, but thinned towards the east and south, giving way to broken landscapes of steppe with stands of pine, spruce, and birch (Frenzel 1973).

The Interpleniglacial (MIS 3) was colder and less stable than the latter part of MIS 5. Warm phase flora differed from cold phase flora, and as noted by Huntley and Allen (2003), no two warm-type or cold-type D/O phases were vegetationally the same, making it difficult to characterize European floras during this stage. Pollen data (van Andel and Tzedakis 1996a; Huntley and Allen 2003) suggests that the warmer phases of this stage were characterized by a steep north–south vegetational gradient across the continent (Figure 6.9), with shrub tundra (characterized by dwarf birch [*Betula nana*], willow [*Salix*] and Juniper [*Juniperus*]) in the Low Countries and just south of the Baltic giving way to steppe tundra south to the 45th parallel. Initial reconstructions (van Andel and Tzedakis 1996a) had scattered spruce joining the vegetation mix around St. Petersburg, and pine, spruce, and birch woodland covering the middle latitudes from western France to the Russian plain. More recent analyses of a larger number of pollen samples, along with computer modeling, suggest that evergreens were a much less important part of the MIS 3 warm phase mix, being largely restricted to northeastern corner of the continent (Huntley and Allen 2003). Below mid-latitude, Europe was dominated by open country herbaceous cover to the west (temperate grassland or mixed grassland/forest mosaic from Italy westward, with some tundra elements in Iberia) and temperate woodlands to the east (Huntley and Allen 2003). Thermophilous trees seem to have been absent north of the Alps during MIS 3, but are a consistent part of the forest mix south and east of the Alps. Here, mixed evergreen and deciduous forest covered parts of the Mediterranean region and much of central Europe, with regional variation coming in the form of the relative importance of evergreen oak, pine, juniper, hazel, beech, lime and elm in the mix (van Andel and Tzedakis 1996b). Across this region, the Mediterranean woodland was open in character, with highest tree densities in central or southern Italy and northwest Greece. To the east, Mediterranean woodland or shrubland ended, as it does today, in semi-desert and dry steppe conditions as one approaches the interior of Asia Minor and the Near East. The extent to which patches of more open country existed within the forest cover of southeastern Europe is uncertain, as is the degree to which forest cover gave way to steppic habitats during the colder intervals of MIS 3. The colder phases of MIS 3 may have seen Europe largely covered in grassland (temperate grassland to the south and steppe tundra to the north). The Micoquian and Mousterian assemblages of the G-Komplex of Seßelfelsgrotte (southern Germany) date to Interpleniglacial times (48–37 Ka BP) and are associated with botanical and faunal remains (namely mammoth, reindeer, and horse) indicative of mixed steppe and tundra habitat (Richter 2001), indicating that tundra elements extended to mid-latitude Europe during some of the cold phases.

Although global temperatures fluctuated dramatically during the MIS 5d–a and MIS 3 intervals, the response of arboreal taxa was weak, so the conditions described above (van Andel and Tzedakis 1996b), representing as they do a relatively mild period of MIS 3, may well describe the prevailing biogeographic picture for the entirety of the temperate interglacial periods. Pollen from La Grande Pile and Les Eschets (Guiot et al. 1989) from a mild interval in MIS 3, in which the distribution of biomes was as described above, indicate conditions in France cooler (by 4 °C) and drier (by 200–400 mm rain) than today. Beetle assemblages at one of these sites, La Grande Pile, indicate a mean temperature for the warmest month of 8 °C and –5 °C for the coldest month (van Andel and Tzedakis 1996b). These values suggest an effective temperature of 9.2 for eastern France, well below that of the warmer full interglacial periods (12.4–14.0). Estimates of primary productivity for modern analogs to some of the biomes that existed in temperate interglacial Europe can be found in Table 6.1 and Table 6.3. Productivity as measured in modern temperate grasslands is likely to be higher than the productivity of Pleistocene steppe during temperate interglacial and glacial episodes, because the modern analogs are both warmer and wetter. Assuming annual precipitation was reduced by 200–400 mm relative to current levels, the drier conditions would predict large herbivore biomass in the range of 1226–2962 kg km^{-2}, based on the relationship between rainfall and secondary biomass in the drier parts of tropical Africa (Coe et al. 1976). Temperate grasslands, however, experience lower temperatures and less insolation, and consequently reduced productivity: northern temperate grassland in Yellowstone National Park supports only 39% of the large herbivore biomass of the Serengeti

TABLE 6.3 ■ Productivity estimates[a] for open environments

	NPP g/m²/yr	1° Biomass g/m²	NPP/1° Biomass	2° Biomass kg/km²
Arctic and alpine tundra				
Mean	140	600	0.23	17.5–300
Range	10–400	100–3000		
Steppe (temperate)				
Mean	600	1600	0.38	920–5200
Range	200–1500	200–5000		
Desert and semi-desert shrub				
Mean	90	700	0.13	2200
Range	10–250	100–4000		

[a] Net primary productivity (NPP), primary biomass (1° Biomass), ratio of mean NPP to mean primary biomass (NPP/1° Biomass) and biomass of larger animals (2° Biomass). Plant productivity and biomass estimates from Whittaker, 1975; (see also Clarke 1976); animal biomass is for larger mammals, primarily ungulates: estimates based on ungulate data from Russia (Bannikov 1967) and Europe and the US (Guthrie 1968; Klein 1970; Redmann 1982).

(Frank et al. 1998). Taking these considerations into account, it is likely that secondary biomass in temperate interglacial steppe ecosystems was on the order of 500–1200 kg km^{-2}, which overlaps the low end of measured secondary biomass in modern steppe habitats (Bannikov 1967; Redmann 1982). While Pleistocene grasslands were probably less productive than modern analogs, tundra and mixed steppe/tundra above 45° N latitude was probably more productive than extant arctic tundra (as discussed above), and likely supported a greater amount of animal biomass than does modern tundra. Low arctic tundra supports an herbivore biomass on the order of 17.5–300 kg km^{-2} (Guthrie 1968; Klein 1970), and while Pleistocene mid-latitude tundra may have been considerably more productive, it is unlikely that its secondary biomass exceeded that of Pleistocene steppe.

A Neandertal living in maquis and mixed conifer–oak woodland areas, like the ones who frequented Guattari (Latium, Italy) during level G0 times (Grün and Stringer 1991: this level was apparently accumulated by hyenas [Stiner 1994], but the presence of a Neandertal cranium indicates that humans were nearby at the time of formation of the assemblage) or Kebara Cave (Israel) during the formation of Units XII-VI (Valladas et al. 1987), would have foraged in broken woodland and shrub that likely had net primary productivities (NPP) on the order of 250–1200 g m^{-2} y^{-1}, and with relatively dense secondary biomasses (values for secondary production in Mediterranean woodlands are not available, but given their higher proportion of edible biomass to woody structures, their standing crops of animals tend to be higher than those of closed forest environments [see Begon et al. 1996]). Assuming ETs 1–2 degrees above those of mid-latitude France, and using the regression equation above for predicting dependence on terrestrial game from ET and NPP, we can predict that modern foragers living in these environments would derive roughly 69–70% of their diet from meat. Neandertals in the coniferous forests north of the Alps, such as near Wallertheim (probably MIS 5d: Gaudzinski 1995) or Sclayn Cave (Level 1A, dating to the MIS 3 Hengelo interstadial: Gamble 1999), were likely exploiting habitats with NPPs of 400–2000 g m^{-2} y^{-1}, but with most of this productivity going into inedible, woody structures. Once more using the regression above (and an estimated ET of 9.2), this predicts terrestrial meat dependencies on the order of 75–89%. Again, these productivities are based on the responses of modern foragers to environments with similar ETs and NPPs. As discussed in the next chapter, these predictions do not necessarily hold well for Neandertals.

6.2.4 Cold Steppic Europe: Marine Isotope Stages 6 and 4

MIS 6 (186–127 Ka BP:Bassinot et al. 1994) was both more uniformly cold and longer than MIS 4 (71–57 Ka BP: Bassinot et al. 1994) and accordingly the extent of the Fennoscandian ice sheet and the distribution of vegetation zones differed somewhat at the climaxes of these two cold spells. Nevertheless, both intervals were marked by the expansion of open country biomes – tundra, steppe, and semi-desert habitats – and the near exclusion of arboreal taxa across most of Europe and the Near East. While refugia of both deciduous and coniferous trees did persist in sheltered areas in Iberia, Italy, the Balkans, and Turkey during both of these episodes, the predominant picture is of a subcontinent of open landscapes populated in the south by herds of grassland grazers – bison, horse, steppe ass, saiga, and steppe rhino – and in the north by cold-adapted fauna – reindeer, musk ox, woolly rhino, and mammoth.

Throughout much of MIS 6 permanent ice sheets extended as far south as northern Belgium and Germany, while in the British Isles (at that time joined to the continent by lowered sea levels) the glaciers extended almost to London (van Andel and Tzedakis 1996b). Immediately to the south of the ice sheets, forming a nearly continuous strip some 50–100 km wide, could be found unproductive and largely barren polar desert (van Andel and Tzedakis 1996b; see also Bliss and Richards 1982). Extant polar deserts have vascular plant coverage generally less than 2–5%, and few avascular plants (Bliss and Richards 1982), and thus little basis for mammalian or avian production. To the south of the polar desert and extending south to the permanent glaciers of the Alps and nearly to those of the Pyrenees was a wide permafrost zone (Frenzel 1973; van Vliet-Lanoë 1989).

The permafrost zone of mid-latitude MIS 6 Europe supported a discontinuous steppe-tundra vegetation (Figure 6.10) for which no modern analog exists (see discussion above, as well as Stanley 1980; Guthrie 1990; and papers in Hopkins et al. 1982). This unique vegetation mix was dominated by grasses, sedges, chenopods (goosefoot family) and mugwort (*Artemisia vulgaris*), and other asters (van Andel and Tzedakis 1996b). While sedge-grass-moss tundras can be found today in parts of Siberia and Alaska (Bliss and Richards 1982), the presence of xerophytes (chenopods and mugwort) in MIS 6 steppe-tundra indicates conditions drier than that of any modern tundra (Hibbert 1982). Despite the underlying permafrost, the drier conditions and greater insolation (relative to contemporary arctic tundras) produced a cold, dry grassland rich with floral elements today found only in tundra contexts (Hibbert 1982), and with plant and animal productivities more akin to extant grassland than tundra proper (Butzer 1971).

To the south of the Alps and extending to the Mediterranean coast, Europe was blanketed in cold, dry steppe. This grassland was dominated by cold-tolerant and xeric herbs and grasses, including *Artemisia*, and the chenopods eurotia (or winterfat: *Eurotia ceratoides*) and annual kochia (or sand-sward: *Kochia laniflora*) (van Andel

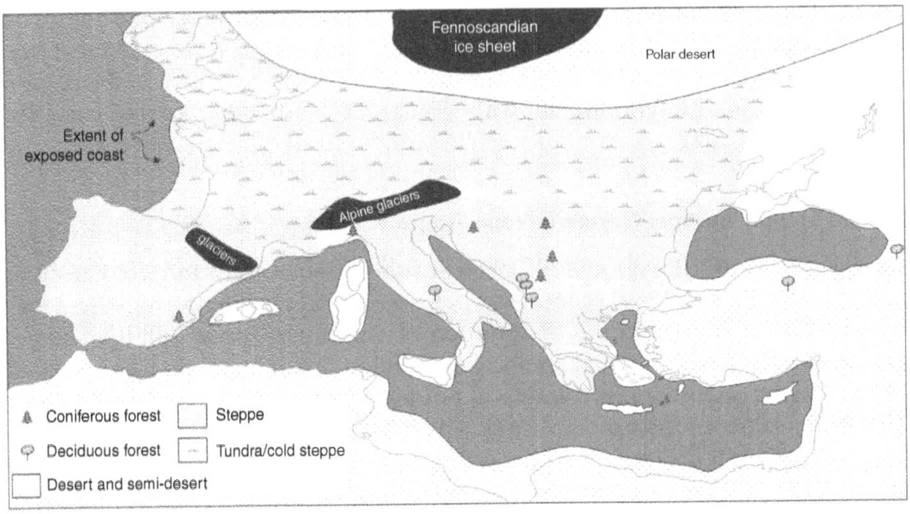

FIGURE 6.10

Approximate extent of major biomes, glacial features (mountain glaciers and Fennoscandian continental ice cap) and coastlines during MIS 4. Coastline position based on –75 m sea level. After van Andel and Tzedakis 1996a.

and Tzedakis 1996b). In southern Iberia, the Apinines, and the western Balkans, sheltered areas of more moderate temperature and greater precipitation provided refugia for stands of temperate trees (van Andel and Tzedakis 1996b). The Levant was covered by semi-desert vegetation, although the uplands of the Taurus Mountains may have provided refugia for temperate trees (van Andel and Tzedakis 1996b).

Similar conditions likely prevailed during MIS 4, at least at the cold climax near the end of the interval (van Andel and Tzedakis 1996b). Although the distribution and general character of the vegetation zones was probably similar during MIS 6 and 4, the Lower Pleniglacial saw greater development of forest refugia in southern areas, undoubtedly due to the milder and wetter conditions that prevailed during this glacial. The seasonal temperatures estimated in Chapter 5.1 produce an estimated effective temperature (ET) range of 8.5 to 9.7 for MIS 6 and 4, comparable to high arctic values of today. Assuming that the productivity of the steppe-tundra of northern Europe was considerably higher than that of modern arctic tundra (it certainly must have been to support Mammoth) and more similar to modern productive grasslands (with NPP between about 200–1500 gm m^{-2} y^{-1}), we can predict (again, based on the behavior of modern foragers) Neandertal dietary dependencies of 71–87% on meat. Unfortunately for foragers in these types of environments, almost all of the primary production in tundra and steppe environments (and presumably in mosaic mixes of these two biome types) is inedible to humans. Thus, despite warmer climates (and higher ETs) and possibly great NPP in the milder MIS 4 cold interval, the dominance of steppe–tundra vegetation during both the penultimate glacial (MIS 6) and the Lower Pleniglacial (MIS 4) likely constrained the Neandertals to a very heavy dependency on animal foods year round (as is the case with modern foragers above the Arctic Circle: see Table 3-1 in Kelly, 1995). Such a conclusion is supported by isotopic studies of five Neandertal fossils dating to this interval, as detailed in the Chapter 7.5. However, this is not to say that plant foods were unimportant. Hardy (2010) notes that geophytes (plants with underground storage organs) would have continued to be available to Neandertals living in a variety of cold climate habitats, including steppe, tundra and evergreen forest. These geophytes include cattails (*Typha latifolia*), bistorts (genus *Polygonum*), dog's tooth violets (*Erythronium dens-canis*), and great burdock (*Arctium lappa*). In addition to being a source of valuable calories, these plants would have provided carbohydrates that would have greatly helped with the digestion and assimilation of proteins in a meat-rich diet (see Chapter 4.3). Starch grains recovered from the dental calculus of Neandertals living in open, steppic habitats (Henry et al. 2011), possibly from the underground storage organ of a water lily, demonstrate that Neandertals were at times consuming these resources.

The review above paints a partial picture of the environmental and ecological backdrop to Neandertal adaptation. We lack data of sufficient detail to allow for the construction of real energy flow (e.g., Bliss and Richards 1982) or economic models (e.g., Keene 1985) that would allow some theoretic basis for evaluation of faunal and (in rare cases) macrobotanical evidence from Mousterian sites. We do have, at least, a broad framework within which to examine variation in Neandertal subsistence strategies. My intent in this chapter was to provide some idea of how modern foragers would be expected to react (in terms of proportions of plants and animals in the diet) to the types of environmental settings in which Neandertals lived. The next chapter is devoted to diet: at what trophic levels were Neandertals feeding, and what specifically were they choosing from the realm of potential plant and animal prey in various environments?

NOTES

1. Some nutritive value can be found in the metabolically active inner bark (secondary phloem and cambium) of woody plants, and thus bark and twigs provide critical winter forage to reindeer, elk [moose] (Guthrie 1990), humans (Campbell 1995), and other animals. It has recently been proposed that consumption of inner bark may have been a seasonally-important component of Neandertal diet (Sandgathe and Hayden 2003).
2. The data used to construct these curves derive from fairly coarse-gained ethnographic accounts of diet, or in some cases on recollections of informants. They are used here with the intent of establishing a broad understanding of how modern human diet relates to ecological variation.
3. Clarke (1976: 457-458) depicts the temperate forests of Holocene Europe as a veritable Garden of Eden:

   "Until the work of Lee and DeVore (1968) there was a general archaeological failure to grasp the dietary actualities of man the omnivore, a failure to comprehend the quantity, quality, stability and variety of highly nutritional plant-foods widely available throughout most of the year in the forests of Temperate Europe and scarcely less abundant in the undegraded Mediterranean woodlands. Figures are misleading and to emphasize one particular plant species is to miss the very significance of the composite abundance and interlocking seasonality of large groups of plant food-species, which meant that there was hardly a month in which a new combination of edible gums, saps, barks, shoots, stems, buds, flowers, fruits, nuts, roots, tubers, rhizomes, corms, bulbs, mosses, seaweeds, water-plants or fungi was not available, waiting immobile, predictable, for the plucking, even signaling their presence with coloured flowers and distinctive foliage; insignificant perhaps species by species but, gathered in diversified bulk, they provided an impressive and stable subsistence basis.

At most seasons of the year, any child over four could gather sufficient to feed itself and others, and in the lean seasons there were always the efficient and nutritious natural storage organs – the roots, seeds, and nuts designed to overwinter without further treatment. It is easy to forget that the steady plucking of four Temperate-forest gatherers could at will pick the edible carcase weight of an adult red deer in less than four hours; whilst, in the Mediterranean park-lands, one worker with a harvesting-knife could gather enough wild seed to produce one kilogram of clean grain with twice the protein value of domestic strains. In three weeks of such work, a family could gather more grain than it could consume in a year. The same Mediterranean woodlands in Italy alone yield some 400 tonnes of Stone Pine nuts in an average year with the protein equivalent of more than 600 tonnes of lean round steak, making these pine kernels the most nutritious of all European nuts, richer even than the mongongo nuts of the !Kung (Howes 1948: mongongo 27% protein; *Pinus pinea* nut 33.9% protein, 48.2% fat). Cooler Mediterranean woods will also yield a tonne of hazel nuts to the hectare, 3 tonnes of chestnuts a hectare and 10–20 tonnes of edible bulbs for each square kilometre of fertile woodland (Howes 1948; Hill 1952).

The temperate oak/hazel forests are no less productive and commonly yield 700–1,000 litres of edible acorns for each mature oak tree, half a tonne of hazel nuts a hectare, 20–50 tonnes of edible bracken root per square kilometre, 5,000–10,000 kilograms of fungi and 13–15 kilograms of blackberries a day in season. These figures are themselves dwarfed by the huge quantity of protein directly available in a wide range of edible herbaceous leaves and plants, consumed as cress, salad, spinach and asparagus meals. Even the less rich northern Boreal forests will regularly yield 137,000–273,000 litres of edible berries rich in sugars and vitamins (Howes 1948; Hill 1952)".

While Clarke's vision of the bounty of temperate forests is appealing, it is also pure fantasy. One wonders why, with so much edible plant productivity "waiting immobile, predictable, for the plucking", human population densities in these environment were so low. Braidwood and Reed (1957) estimated Mesolithic (ca. 7.5 Ka BP) temperate forest population densities to be about 0.049 km^{-2}. Kelly (1995: Table 6-4) lists densities for western North American deciduous forest foragers ranging between 0.42–0.013 km^{-2}. Historic estimates of North American temperate forest-dwelling Powhatan indicate considerably higher densities (0.79 km^{-2}: Turner 1985), but here the practice of corn agriculture likely greatly increased the carrying capacity. These estimates are only marginally higher than (but within the range of) those of arctic groups (0.65–0.002 km^{-2}: Kelly 1995) and well below typical values for agriculturalists (e.g., the early argiculturalists of Iraq [ca. 6.7 Ka BP] had population densities of ca. 10.55 km^{-2}: Braidwood and Reed 1957). Indeed, if temperate forest resources are so abundant and easily gathered, why have most of the human groups known to live in them adopted some form of agriculture? The answer lies in the small package size and high cost of gathering and processing edible goods in deciduous forests, such that however abundant they might be they result in rather low human carrying capacities and somewhat low secondary production overall.

4. Acorns, which served as a staple for many North American groups (Driver and Massey 1957) provide a prime example. After an acorn fall, the nuts must be collected, shelled, ground into a meal, leached in water to remove tannic acid, and then baked or boiled to be made palatable. Post-processing weights of the edible portion of tree nuts may amount to only 50% of the original weight of the nut (Woodroof 1967). To these costs must be added the indirect costs of producing technology to facilitate this processing (in the form of collection containers, grindstones, and watertight containers [usually ceramics] suitable for leaching and boiling) and the costs of gathering firewood. Furthermore, even though oak trees may be abundant within an area, good acorn crops may occur only every three to five years (Dalke

1953), and even during good years only about 10% of the crop is available to human foragers, the rest being lost to other consumers (mammalian and insect), fungus, and incomplete seed development (Gysel 1957). If this ledger doesn't sufficiently underscore the high costs of this single resource, consider also that human consumption of acorns constitutes direct competition with deer, for whom acorn consumption can be a prime determinant of winter fat stores (Moen 1973). Thus, heavy reliance on acorns exacts an indirect cost in production and nutritive value of one of the few large packages of animal calories available, and at a time of year when fat is a critical human resource (Speth 1987).

5. Such technology for one group (Iroquois: Campbell 1995) included tools that helped with accessing arboreal, aquatic, and flying prey (bow and arrow, blowgun, canoe and paddle, fish weirs, harpoons), technology that improved the effectiveness or efficiency of food transport, processing or storing (ceramic pottery, baskets, storage pits), and tools that helped to reduce the costs of searching for prey (traps, canoes). Thus technology was employed to reduce both the high handling costs associated with small but locally abundant resources like nuts and small animals, and to help reduce the high cost of searching for larger, dispersed and mobile prey.

CHAPTER SEVEN

Neandertals as Consumers

Diet is arguably the primary component of an animal's ecology. Energy must be captured before it can be allocated to somatic maintenance and reproduction, and thus the abundance, quality, and distribution of edible resources in the environment is the primary determinant of production in any given species of consumer. Dietary options are themselves constrained by the energy and nutritional requirements of the organism, and by aspects of masticatory morphology, gut anatomy, and organismal physiology that place limits on the ability of the organism to ingest, digest, and assimilate calories and nutrients. An animal's anatomy and physiology proscribes diet (to a large but not complete degree), but by the same token the edible environment and the economic and competitive interactions that take place in that environment are major sources of selection operating on body size, trophic morphology, physiology, and behavior in heterotrophic organisms. Furthermore, variation in abundance and nutritive value of actual and potential foodstuffs through time and across space (the temporal and geographic patchiness of resources) and dynamic interactions with competitors for those

Thin on the Ground: Neandertal Biology, Archeology, and Ecology, First Edition. Steven Emilio Churchill.

resources – both factors operating at multiple time scales, from daily to seasonal to evolutionary – produce the complex set of economic functions that we recognize as subsistence. Humans, as large-brained and behaviorally flexible omnivores, are capable of feeding at all of the trophic levels available to heterotrophs and, indeed, differential reliance on resources at different trophic levels constitutes the primary mechanism by which historically-known foragers adapted to latitudinal variation in plant food abundance (see, for example, Kelly 1995). Our intelligence and behavioral plasticity have also allowed us, as a species, to resolve the myriad scheduling problems (for both dietary items and critical nonfood resources) associated with exteremely patchy environments and, not surprisingly, the study of subsistence organization has received considerable attention in research on the adaptive modalities of prehistoric humans (see Chapter 10.1).

As outlined by Mellars (1996: 194), research on Neandertal diet and subsistence has revolved around a series of interrelated questions. These involve the means by which Neandertals procured animal tissues (the relative importance of hunting versus scavenging); the extent to which Neandertals were resource specialists (that is, were certain prey species predominantly exploited, regardless of means of procurement?); the hunting strategies and methods that were employed (assuming hunting was practiced) and the extent to which these methods involved technology, planning depth, and social organization; the nature of carcass processing and transport (including intensity of utilization and questions of storage); and the extent to which Neandertals seasonally adjusted their animal exploitation patterns.

To this I would add another question, one that is perhaps more basic and arguably fundamental to all the others, and that is, "How variable were Neandertals in trophic ecology?" Modern humans are highly variable in this regard, ranging from tropical foraging groups that were largely operating as primary consumers (upwards of 85% dependence on plant foods) to arctic hunters who were essentially carnivorous, even regularly operating as tertiary consumers (Murdock 1967). This makes it difficult to define the human ecological niche, unless "trophic generalist" is an acceptable characterization. Did the same trophic plasticity apply to Neandertals, or was their niche[1] more narrowly constrained?

Recent isotopic and trace element analyses (reviewed in Section 7.5) reveal that the European Neandertals were deriving most all of their dietary protein from animal sources, suggesting that their diets were dominated by meat and other animal tissues. While this line of research brings new evidence to bear on the reconstruction of paleodiet, these results are harmonious with a pronounced tendency among many who work on subsistence in the Mousterian to see the Neandertals as fundamentally carnivorous. Recent work on plant macrofossils at Mousterian sites (Section 7.3) and microfossils contained in dental calculus (Section 7.4) is showing the Neandertal diet to be more varied than earlier workers might have appreciated. The former position derived in part from the somewhat facile and not entirely correct (see Chapter 6) characterization of European periglacial environments as being devoid of humanly-consumable plants, at least for long stretches of time every year. It also came from the subtle and insidious, if perhaps unavoidable, bias imposed by the taphonomy of cave, rock shelter, and open air sites: animal bones tend to preserve while plant remains do not. Studies of Neandertal subsistence have focused, almost exclusively, on faunal remains simply because this is the archeologically visible remnant of past subsistence dynamics. This exclusivity of research focus, however, lends itself to thinking that there was also a dietary exclusivity on the part of the Neandertals. Clearly there were edible plantstuffs in all of the environments inhabited by Neandertals – more so in some than others – but despite a refinement in our understanding of Neandertal diet from new biogeochemical studies, the question remains: to what extent were these archaic humans exploiting the full spectrum of resources in any given environment? Are the proportions of plant and animal items in the Neandertal diet predictable from knowledge of ecological parameters in the same way that they are for modern foragers, or was Neandertal diet conditioned by other, as yet unidentified, variables?

The review below is intended to address questions of diet (what items were included, what mix of foods from different trophic levels were consumed, and how variable was diet across space and time?) and not questions of subsistence organization (which are discussed in Chapter 10.1). Table 7.1 provides a bestiary of the larger mammals of Pleistocene Europe, included here as a simple enumeration of the potential fauna in the Neandertal foodbag, and as a means of avoiding the tiring repetition of the Linnaean names of various animals in the discussion that follows.

7.1 ANALYSIS OF FOOD RESIDUES: THE MACROMAMMAL COMPONENT OF NEANDERTAL DIET

The mainstay of Middle Paleolithic dietary inference has long been economic analysis of food residues, focusing on macromammal skeletal remains associated with diagnostic tool assemblages. The intent of this approach has, however, seldom been the simple enumeration of the prey species exploited by Neandertals, but rather to further our understanding of their predatory capabilities, niche breadth, and subsistence organization. The road leading from bone assemblages to these higher-level inferences is, however, fraught with obstacles, and questions of Neandertal subsistence behavior remain issues of intense research and debate (see for example Binford 1985, 1989; Chase 1986a, 1988, 1989; Klein 1989b; Stiner 1991c, 1994; Shea 1993, 1998; Marean 1998; Marean and Kim 1998; Patou-Mathis 2000; Daujeard and Moncel 2010; Speth 2012). While the elucidation of diet items from bone assemblages is considerably simpler, it is still not without difficulties. Prime among these is the problem of discerning human- from carnivore-accumulated assemblages or, as is often the case, discerning the relative contribution of these agents to a single assemblage (Webb 1989). Unfortunately, conscientious efforts to delineate the agent of accumulation by thorough analysis of cut and gnaw marks, occurrence of coprolites, animal part representation, and presence of carnivore remains (e.g., as done by Stiner 1994; Enloe et al. 2000; and Valensi 2000) are not the norm, and in many cases faunal assemblages are attributed to human agency on the basis of a few associated stone tools. Careful consideration of Mousterian sites for which a confident diagnosis of anthropogenic accumulation of the contained faunal assemblages can be made reveal some interesting patterns that are played out across the subcontinent. In the review that follows, I draw heavily on Straus (1992) for Iberia, Mellars (1996) and Boyle (2000) for southern France, Patou-Mathis (2000) for northwestern and central Europe, Conard and Prindiville (2000) for the Rhineland of Germany, Stiner (1994) and Mussi (2001a) for Italy, Burke (2000b) for the Crimea, Hoffecker and Cleghorn (2000) on the northwestern Caucasus, Henry (1995) and Shea (1998) for the Levant, and Marean and Kim (1998) for Iran.

TABLE 7.1 ■ A bestiary of larger mammals of the European late Middle and Late Pleistocene (MIS 7-3)

	Inferred habitat preference	Comments
Primates		
Gibraltar "ape" (*Macaca sylvana*)	temperate conditions	
Neandertal (*Homo neanderthalensis*)	eurytopic	
Modern humans (*H. sapiens*)	eurytopic	stage 3 only
Rodentia		
White's squirrel (*Sciurus whitei*)		stage 7 only
Red squirrel (*S. vulgaris*)	temperate deciduous forest and coniferous forest	stages 5e–3
Alpine marmot (*Marmota marmota*)	mountainous meadows	stages 6–3
Bobak marmot (*M. bobak*)		stages 5e–3
European souslik (*Citellus citellus*)	steppe	
Spotted souslik (*C. suslicus*)		stages 4–3
Red-cheeked souslik (*C. major*)		stages 4–3
Beaver (*Castor fiber*)	eurytopic	
European giant beaver (*Trogontherium cuvieri*)	eurytopic?	stage 7 only
Extinct crested porcupine (*Hystrix vinogradovi*)	open ground and hill slopes, drier conditions	
Crested porcupine (*H. cristata*)	open ground and hill slopes, drier conditions	stages 5e–3
Lagomorpha		
Sardinian pika (*Prolagus sardus*)	steppe	stages 6–3, possibly 7
Steppe pika (*Ochotona pusilla*)	brushy valleys	stages 6–3, possibly 7
Rabbit (*Oryctolagus cuniculus*)	meadows, forest edge	
Varying hare (*Lepus timidus*)	coniferous forest, cold conditions	stages 5e–3, possibly 7–6
Brown hare (*L. europaeus*)	open fields	stages 5e–3
Cape hare (*L. capensis*)	fields, woodlands	stages 4–3, possibly 7–5
Carnivora		
Striped hyena (*Hyaena hyaena*)	eurytopic	stages 7–5e
Cave hyena (*Crocuta crocuta spelaea*)	eurytopic	
Lesser scimitar cat (*Homotherium latidens*)		rare throughout this interval
Wild cat (*Felis silvestris*)	woodland/mixed forest, also steppe	
Steppe cat (*F. manul*)	steppe	stages 4–3
Jungle cat (*F. chaus*)	tropical/subtropical forest	small zone in eastern Europe only, stage 5e only
Pardel lynx (*Lynx.pardina*)	woodland?	
Lynx (*L. lynx*)	woodland	stages 5e–3

TABLE 7.1 ■ (*Continued*)

	Inferred habitat preference	Comments
European jaguar (*Panthera gombaszoegensis*)		stage 7 only: also known as the Tuscany lion (*P. toscana*)
Cave lion (*P. leo spelaea*)	eurytopic	
Giant lion (*P. fossilis*)	eurytopic	possibly to stage 7
Leopard (*P. pardus*)	eurytopic	
Wolverine (*Gulo gulo*)	bogs, marshes	
Pine marten (*Martes martes*)	coniferous forest	stage 5e
Beech marten (*M. foina*)	woodland and steppe	stages 4–3, possibly 5
European mink (*Mustela lutreola*)		possibly stages 4–3
Polecat (*M. putorius*)	streams (but otherwise eurytopic)	
Steppe polecat (*M. eversmanni*)	steppe	stages 5e–3
Stoat (*M. erminea*)	deciduous and coniferous forest, also tundra	stages 4–3
Weasel (*M. nivalis*)	forest	
Least weasel (*M. rixosa*)	tundra	stages 4–3
Badger (*Meles meles*)	deciduous forest	
Corsican otter (*Aonyx antiqua*)		
Otter (*Lutra lutra*)		stages 5e–3
Wolf (*Canis lupus*)	eurytopic	
Dhole (*Cuon alpinus*)		
Red fox (*Vulpes vulpes*)	forest, steppe, desert	
Steppe fox (*V. corsac*)	steppe	stages 5e–3
Arctic fox (*Alopex lagopus*)	tundra/cold steppe	stages 6–3
Cave bear (*Ursus spelaeus*)	eurytopic	
Brown bear (*U. arctos*)	forest	
Asiatic black bear (*U. thibetanus*)		stages 7–6
Proboscidea		
Straight-tusked elephant (*Palaeoloxodon antiquus*)	parkland/forest	
Steppe mammoth (*Mammuthus trogontherii*)	steppe and tundra	stage 7 only
Woolly mammoth (*M. primigenius*)	cold steppe and tundra	stages 6–3
Perissodactyla		
Merck's rhino (*Dicerorhinus kirchbergensis*)	parkland/forest	
Steppe rhino (*Stephanorhinus hemitoechus*)	steppe	
Woolly rhino (*Coelodonta antiquitatis*)	cold steppe and tundra	stages 6–3
Horse (*Equus caballus*)	steppe	(= *E. przewalskii*)
Steppe ass (*E. hydruntinus*)	steppe	
Wild ass (*E. hemionus*)	steppe	stages 6–3
Artiodactyla		
Wild boar (*Sus scrofa*)	forest	

(continued)

TABLE 7.1 ■ (Continued)

	Inferred habitat preference	Comments
Hippo (*Hippopotamus amphibius*)	forest, also cool steppe	
Red deer (*Cervus elaphus*)	temperate forests	
Dwarf verticornis deer (*Praemegaceros cazioti*)		Corsica only
Giant deer (*Megaceros giganteus*)	steppe	
Clacton fallow deer (*Dama clactoniana*)	forest	stages 7–6
Fallow deer (*D. dama*)	forest	stages 5e–3
Roe deer (*Capreolus capreolus*)	woodlands, dense brush	
Elk (*Alces alces*)	coniferous forest	stage 5e
Reindeer (*Rangifer tarandus*)	tundra (summer) and coniferous forest (winter), cold conditions	stages 6–3
Saiga antelope (*Saiga tatarica*)	cold steppe	stages 4–3
Chamois (*Rupicapra rupicapra*)	mountain slopes	stages 5e–3
Cave goat (*Myotragus balearicus*)		Baleares only
Musk ox (*Ovibos moschatus*)	tundra	
Sheep (*Ovis* spp.)		stages 5e–3
Ibex (*Capra ibex*)	mountainous areas	stages 6–3
European tahr (*Hemitragus bonali*)	mountainous areas	stages 5e–3
Steppe wisent (*Bison priscus*)	cold steppe	
Woodland wisent (*B. schoetensacki*)	woodland	
Murr buffalo (*Bubalus murrensis*)	river, lake or sea margins	stage 7 only
Aurochs (*Bos primigenius*)	steppe and open woodland	

From Kurtén 1968; Sutcliffe 1985; Jones and Keen 1993; Turner and Antón 1997; Agustí and Antón 2002.

The overwhelming pattern across the entire Neandertal range is of predominant exploitation of certain key species, set against a faunal record that indicates that the Neandertals were *capable* of exploiting most of the herbivorous macromammals in their environment. In other words, when considered in aggregate, Neandertal-accumulated assemblages contain the remains (often with clear indications of human utilization: cut-marks, diaphyseal cracking for marrow extraction, and burning) of a diversity of prey – essentially every moderate-to-large bodied herbivore in the western Eurasian later Pleistocene paleontological record – but the assemblages are nonetheless dominated by a few species of medium-to-large bodied ungulates.

In Iberia, the emphasis appears to have been on red deer, horse, and bovines (since it is difficult to distinguish postcranial elements of aurochs and bison, these remains are generally combined in taxonomic lists as *Bos/Bison*, and are referred to here simply as bovines). This pattern continues as one moves towards the east, as reflected in the Pyrenean site of Gabasa Cave, Spain (although here bovines are rarer and montane caprids more important, although the contribution of non-human predators to this assemblage is not clear: Blasco 1997). Reindeer begin to enter the mix on the eastern side of the Pyrenees (e.g., in Mousterian levels at Cap del Saut, the west cave of Le Portel, France: Gardeisen 1999), appearing quite consistently in southwestern France and remaining a major aspect of the large ungulate mix (variably between levels and sites) throughout western and central Europe. While different levels may demonstrate a predominance of one species (likely reflecting changes in local abundance as climate and environments changed), these four elements (red deer, reindeer, horse, and bovines) constitute the staple prey of Neandertals across most of Europe throughout the later Pleistocene (they also appear to be the preferred prey of humans before and after the Neandertals: see Gamble, 1986). All four of these elements are well-suited to open country habitats, especially horse, steppe bison, and reindeer, and thus all were likely available during colder episodes in which grasslands and tundra prevailed. Red deer and aurochs, as more eurytopic mixed feeders that require some closed cover for winter browse, may have been more minor (but still consistently present) elements of cold steppic or steppe–tundra habitats, but may have replaced the open country forms in economic importance during warmer, more heavily-wooded intervals. It is also worth noting that Mousterian deposits at two sites in southern Iberia – Vanguard and Gorham's Caves (Gibraltar) – have yielded a small number of marine mammal (monk seals, and common and bottlenose dolphins) and fish bones, as well as mollusk shells (Stringer et al. 2008b). The persistence of marine mammals, fish, and shellfish remains through the Mousterian layers at these caves has been taken as evidence that use of marine resources was a regular behavior for the Neandertals there (Stringer et al. 2008b). While these finds do indicate Neandertal exploitation of near-shore coastal resources, the small number of marine animal remains recovered has led to some debate as to their significance in Neandertal diets (Klein and Steele 2008 vs. Stringer et al. 2008a). Interestingly, a Neandertal frontal bone dredged up from the North Sea (Hublin et al. 2009), and presumably representing an individual who lived on the coast of the Netherlands during MIS 3, appears *not* to have been eating aquatic resoruces based on stable isotope analysis (see Section 7.5).

In Italy, the principal prey species appear to have been red deer (far and away the major component of most levels attributed to human accumulation), roe deer, and fallow deer, usually occurring in that order of importance (judging from numbers of identifiable specimens [NISP]) in lithic-rich and carnivore-poor (and hence likely anthropogenic: see Stiner 1994) Mousterian layers. Of secondary but seemingly still significant importance are, variously, ibex, aurochs, horse, and wild boar. Chamois, Merck's rhino, and hippo also occur sporadically (Mussi, 2001a) and even monk seal makes an appearance in Grotta di Sant' Agostino in the form of a single bone with cut-marks from level S2 (Stiner 1994). The remains of carnivores, notably wolf, red fox, wild cat, and brown bear, occur occasionally in layers with abundant lithics (e.g., Grotta di Sant' Agostino levels S0-S3, Grotta Breuil levels B3/4 and Br: Stiner 1994), and although they may derive from carnivore occupation of the cave when the humans were away, the possibility that some smaller carnivores were occasionally eaten cannot be ruled out (see Chapter 9).

As one moves north through the Rhineland and onto the northwestern plains certain large, gregarious and migratory forms – namely mammoth, rhino (Merck's, steppe, and wooly), and ibex – take on additional importance (although horse is the dominant prey over time, and red deer, bison and reindeer retain their importance). Accordingly, there is a greater (although still small) diversity in principle prey species in this area (Patou-Mathis 2000; Conard and Prindiville 2000; see also Gautier 1989). Other species, although they can occasionally numerically dominate assemblages in some levels, make much more sporadic appearances in the Mousterian of western Europe.

These include wild boar (although virtually absent in Spain: Straus 1992), roe and fallow deer (e.g., some 60 individuals of the latter at Wallertheim, Germany: Conard and Prindiville 2000), chamois (e.g., 16 individuals in Level VII of Amalda, Spain: Straus 1992), tahr (which dominates the assemblage at Grotte des Cédres, France: Boyle 2000), and steppe ass. Ibex are rare in Iberia, but fairly abundant in montane regions in western Europe and as one moves north (Straus 1992; Boyle 2000; Patou-Mathis 2000). According to Rolland (1987) alpine bovids (chamois, ibex, Siberian mountain goat) were taken regularly during the Middle Paleolithic across Europe, at the sites of Teshik-Tash (Uzbekistan), Okladnikov and Denisova (Russia), Repolusthöhle and Gudenushöhle (Austria), Drachenhöhle (Germany), Subalyuk (Hungary), Lezetxiki (Spain), l'Observatoire (Monaco), and Bau de l'Aubesier, Lazaret, Baume-Bonne, Les Peyrards, and l'Hortus in France (Gábori 1979 and de Lumley 1969–72 as cited in Rolland 1987). However, the involvement of carnivores in the accumulation of many of these sites is uncertain, and thus the economic importance of alpine bovids remains hazy. Of even rarer occurrence, at least in southwestern France, are megafaunal remains – giant deer, rhino, and mammoth (Mellars 1996). In northern Europe, the occasionals in the faunal assemblages include hippopotamus, forest elephant, elk (moose), musk ox, saiga antelope, and wild sheep (Patou-Mathis 2000).

In eastern Europe, a similar pattern obtains, albeit with different taxa playing the role of dominant prey. In some assemblages, here more than 90% of the identifiable elements belong to a single species, either steppe ass or saiga antelope (Patou-Mathis 2000; *cf.* Burke 2000b on Starosele, Ukraine). Further east, in the northern Caucasus Mountains, bison appears to have been the dominant prey, although saiga, goats, and sheep take on increased importance at higher elevations (Hoffecker and Cleghorn 2000). Species choice in the Near East is a bit more difficult to gauge, in large part because the number of excavated sites drops dramatically as one moves away from the Mediterranean coast. At Kobeh Cave, in the Zagros Mountains of Iran, wild goat/wild sheep (*Capra aegagrus/Ovis orientalis*: the same difficulty in distinguishing postcranial remains as encountered with *Bos/Bison* applies here) dominate the assemblage, with equids (wild ass and *Equus* sp.) finishing a distant second. Wild goat remains are also the most abundant in layer D of Shanidar Cave (Zagros Mountains, Iraq), although tortoise (*Testudo graeca*), wild boar, red and roe deer, and red fox remains were also recovered in lesser abundance (Trinkaus 1983b). Towards the Mediterranean coast, the pattern of emphasis on a narrow range of medium-to-large ungulates seems to erode, and here there appears to be much greater diversity in Mousterian faunas (bearing in mind that some of these assemblages may have been, and indeed probably were, accumulated by anatomically modern humans: Chapter 3.1). Mousterian foragers here tended to rely equally on a more diverse mix of small to moderate-sized, non-migratory ungulates with small home ranges (Bar-Yosef 1992): gazelle, wild goats, ibex, wild sheep, wild boar, Mesopotamian fallow deer, and roe deer. At several coastal sites there seems to have been an emphasis on a few key species, such as the moderate-sized fallow deer and the small-sized gazelle that predominate the MIS 7 site of Misliya Cave (Yeshurun et al. 2007) and the MIS 3 deposits of Kebara Cave (Speth and Tchernov 1998), both in Israel. Larger occasionals occurring in sites from the coast to the Zagros include aurochs, rhino, hippo, and camels (including an as of yet unpublished giant camel, *Camelus moreli*, discovered in 2005 at Hummal in Syria), red deer, hartebeest, and horses (Speth 2012).

7.2 ANALYSIS OF FOOD RESIDUES: THE SMALL ANIMAL DATA

Modern foragers rarely depend entirely on large herbivores, even in cases where meat may comprise most or all of the diet. Boreal forest and arctic groups commonly rely heavily on fish (in part because andramanous fish constitute an important unearned resource for northern terrestrial habitats, allowing appreciably higher population densities than is possible with terrestrial resources alone), small mammals, and birds as either dietary staples, ancillary items in the diet, or seasonally-important resources critical to managing temporal gaps in the availability of other food items (Rasmussen 1931; Birket-Smith 1935; Weyer 1969). It is also true that small animals can be protein- and fat-rich (although some tend to be poor in fat: Speth 1983), and may be a good source of limiting nutrients. In this latter regard, marine resources can be exceptionally good sources of the long-chain polyunsaturated fatty acids (LC-PUFAs) docosahexaenoic acid and arachidonic acid (Broadhurst et al. 2002) and of iodine (1.16 mg and 1.37 mg per kg for mussels and clams, respectively: Tressler 1951), both essential to human development and physiological functioning (see Chapter 4.3). Small animals are also abundant in the landscape – for example, squirrels have somewhat lower biomass densities than whitetail deer in North American temperate forests (1.0 to 5.0 kg ha^{-1} vs. 1.5 to 10.5 kg ha^{-1}: Pianka 1994), but given the size differential (ca. 0.5 kg for squirrel vs. 95 kg for deer, on average: Schmidt and Gilbert 1978; Burt and Grossenheider 1976) squirrels are generally about two orders of magnitude more abundant than deer. In addition, some small animals have high intrinsic rates of reproduction (that is, they tend to be *r*-selected relative to other mammals) and high rates of secondary production,[2] allowing higher rates of sustained predation (although focused predation on a single, *r*-selected resource can lead to dramatic population cycling in both prey and predator: Campbell 1995). Thus, small animals tend to be a more homogeneously distributed (although some, like salmon, can be both spatially and temporally patchy) and reliable resource than larger prey. However, the high handling costs of most small animals (see below and Chapter 3.4, and also Winterhalder and Smith 1981; Stephens and Krebs 1986; Pianka 1994) generally results in foragers passing them by when encountered in favor of continued searching for higher net return items. For this reason, small animals may play more of a role as emergency or short-term resources and as sources of limiting nutrients than as staple items relied upon to satisfy bulk calorie demands. Since small animals generally cost more per calorie captured, intensified reliance on them (as in the late Upper Paleolithic of Europe: Stiner et al. 1999) may signal real stress on the overall resource base.

It is most probable that small animals, both vertebrate and invertebrate, played an important role in Neandertal subsistence. However, the contribution of small animals continues to be poorly known, largely unstudied, and rarely discussed (but see Stiner 1993, 1994; Stiner et al. 1999, 2000; Cochard et al. 2012; Fa et al. 2013). While small animal remains do occur in Mousterian levels – sometimes with great abundance (e.g., the 2200 identifiable rabbit bones from Salpêtre de Pompignan, France: Boyle 2000) – it is generally difficult to distinguish humanly accumulated remains from those introduced to the site through: carnivore activity; burrowing and shelter-seeking behavior of rodents, reptiles and amphibians; faunal rain from the ceiling and walls of caves and shelters (dead birds, bats, owl pellets, etc.); and fluvial or tidal transport (this latter being especially problematic with respect to marine invertebrates and coastal sites).

After a careful taphonomic study of small animal remains at four sites in Latium (Italy), Stiner (1994) concluded that the humanly-generated component of this fauna was relatively small, suggesting to her that human use of small animals did not generally entail transport back to habitation sites. The exception among the Italian sites was Grotta Moscerini, where 47% of the total (large plus small animal) humanly-generated faunal remains are tortoise and shellfish. The data from Moscerini suggests that, at times, small animals were a major part of the diet base. A similar situation is seen in the early Middle Paleolithic layers of Hayonim Cave (Israel), where small game (mainly tortoises, large lizards, and ostrich eggs) make up 52–29% of the total, humanly-generated fauna (Stiner et al. 1999). Cut-marked beaver remains from Taubach (Germany) and Vindija (Croatia) indicate extensive use of this resource during the Eemian Interglacial (Gaudzinski-Windheuser and Roebroeks 2011). With the exception of a few sites like these, we lack the conscientious taphonomic analyses needed to identify the predators and accumulators of small game at most sites. Thus, we have no idea of the degree of variation in small game exploitation by Mousterian foragers, nor how representative Moscerini and Hayonim Caves are of general patterns of Neandertal predation on small animals. These sites do tell us, at least, that at times – and perhaps at *most times* – small animals were an important part of the Neandertal diet base.

The small animal remains from the Latium sites that could be attributed to human agency consisted almost exclusively of tortoise and shellfish. These are both resources that are either sluggish or immobile – effectively sessile animals that allow "hunting by gathering" (Stiner 1994). In the case of shellfish, which have also been recovered in small numbers from two Mousterian sites in Gibraltar (Stringer et al. 2008b), they can also be found in appreciable numbers in shallow water colonies. These small animals are considerably less expensive to capture than faster and more agile game (hares, rabbits, birds, fish) and thus shellfish and tortoise probably ranked higher in the array of potential diet items to Neandertal foragers. However, these slow-moving resources are also slow growing and thus not very resilient to over-harvesting (Stiner et al. 1999), placing limits on the degree to which humans can exploit them as dietary staples.

Tortoise remains, including both the terrestrial genus *Testudo* (probably *T. graeca*, the Mediterranean spur-thighed tortoise) and the fresh and brackish water European pond tortoise (*Emys orbicularis*), have been recovered from several levels at Grotta di Moscerini, most notably from the predominantly humanly-accumulated faunal assemblages from level M6. Here, tortoise makes up 6% of the total identifiable small prey. At least 46% of the identifiable tortoise bones have green bone fractures, and carapace and plastron pieces commonly have impact cones or impact depressions, indicative of human processing and consumption (Stiner 1994). As noted above, *Testudo* also occurs in layer D at Shanidar Cave (Trinkaus 1983b). Remains of terrestrial small game in the early Middle Paleolithic (MIS 7-5) at Hayonim Cave also reflect human selection of "sessile" prey. Here, tortoise and legless lizards comprise 89–97% of the small game remains, with cape hares, birds (predominantly partridges: *Alectoris chukar*, *Perdix perdix*, and *Coturnix coturnix*) and ostrich (*Struthio* sp.) eggs occurring in frequencies of usually less than 5%.

Other terrestrial small game in the Latium caves are generally attributable to carnivore, and not human, consumption (Stiner 1994). Rabbit remains are abundant in some other sites in levels with Mousterian artifacts, such as Salpêtre de Pompignan in France (Boyle 2000) and Cova Beneito and the lower levels of Cova Negra in Spain (Villaverde et al. 1996), but in every case faunal and taphonomic indicators raise the possibility that carnivores – notably lynx, wolf and owl – are responsible for the predominance of rabbit. The scarcity of humanly-modified rabbit and hare bones in Mousterian levels (see, for example, Fa et al. 2013) may reflect, as mentioned by Stiner (1994), human consumption at the point of capture or, alternatively, the lack of utilization of this resource due to the relative expense of exploitation. It is clear that the capture of fast and fleet prey was not beyond the capabilities of Neandertals. For example, cut-marked rabbit remains, along with cut-marked duck and goose bones, occur in levels spanning from MIS 9 to 5 at Bolomor Cave (Spain: Blasco and Fernándes Peris 2009: see also Fa et al. 2013), as do rabbit and hare remains in MIS 5/4 deposits at Les Canalettes (France: Cochard et al. 2012). The relatively rare occurance of humanly-introduced, fast and fleet small game in Middle Paleolithic deposits is thus likely a function of the high handling costs relative to the net caloric return of these types of prey. Davidson (1976) provided capture rate data for European rabbits, with densities of about 5–6.5 animals hectare^{-1} during non-stress conditions, that suggest that without the aid of traps, snares, noxious gases, or poisons, capture rates by humans may top out at about 0.5 rabbits h^{-1}. Given an average caloric return of ca. 1080 kcal rabbit^{-1}, roughly five hours hunting per person (for every member of the social group) per day would be required to subsist on rabbits (Davidson 1976). This low return rate, coupled with the low fat value of rabbits (Speth 1983), makes this resource undesirable as a staple item in the diet. This undesirability is underscored by the accounts of Ojibwa foragers of North America, who experienced periodic food shortages and who expended intense foraging effort after depletion of moose and caribou necessitated a dietary emphasis on rabbits and fish (Rogers and Black 1976, as cited in Gamble, 1986).

The capture costs (known as "handling" or "pursuit" costs in optimal foraging theory) of fast and agile prey like lagomorphs, fish, and birds can be reduced through the use of traps, snares, nets, projectile weapons, and specialized fishing gear (leisters, harpoons, gorges, hooks and line, fish wiers) (Oswalt 1976; see also Churchill 1993; Holliday 1998; Stiner et al. 1999). Technological solutions along these lines are evident in all modern foragers that rely on these difficult to capture and small packages of calories. Intensified exploitation of lagomorphs in the late Upper Paleolithic (LUP) of Europe is undoubtedly a reflection of resource stress related to human population pressure and perhaps, secondarily, the reduced productivity of the last glacial maximum (Stiner et al. 1999; see also Straus and Clark 1986; Villaverde et al. 1996). The LUP technological response to this change in diet structure is clear: there was a revolution in subsistence technology that paralleled the expansion of diet breadth and the intensified exploitation of small prey (see, for example, papers in Peterkin et al. 1993). Mousterian sites, in contrast, are poor in fast and agile small game (at least the securely identified humanly-generated assemblages) and there is no evidence for technologies specifically geared towards their exploitation (see Chapter 3.4). This suggests that Neandertal use of these resources was both at a low level and energetically costly when it did occur.

Shellfish use at coastal Middle Paleolithic sites is also relatively rare, which Rolland (1987) attributes to "neothermal submerging" from elevation in sea level during the Holocene. However, where the continental shelf is steep and thus the present coast and glacial coast are in the same position, shellfish remains can be found in Middle Paleolithic levels (although care must be taken to discriminate between shellfish deposited by wave action during high sea stands and humanly-transported remains: Stiner 1994). This is the case at five Mediterranean locales: the Gibraltar caves; the caves of Balzi Rossi; Moscerini and Breuil caves in Latium; the caves of Puglia (southern Italy); and at Haua Fteah (Cyrenaica, Libya) (references in Stiner 1994: 177). If one ignores assemblages occurring in marine beach deposits or with wave-worn shells, a picture of Middle Paleolithic shellfish exploitation emerges that emphasizes a handful of near-shore (littoral or infralittoral) bivalves and univalves. These include mussels (*Mytilus galloprovincialis* and *M. edulis*) and limpets (*Patella caerulea* and *P. ferruginea*) at Gibraltar and Balzi Rossi, and mussels and clams (*Callista chione* and *Glycymeris* spp.) in Latium and southern Italy (Stiner 1994). One-toothed turbins (*Monodonta turbinata*) are also a common component of these mollusk assemblages (*ibid*). Also occurring at Grotta di Moscerini in low frequencies are cockles (*Cardium tuberculatum* and *Cerastoderma edulis*), scallops (*Pecten jacobeaus*), oysters (*Ostrea edulis*), vongola clams (*Venus gallina*), turret shells (*Turritella* sp.), worm shells (*Vermicularia* ?sp.), sponges (taxon unknown) and possibly the large conch-like univalve *Strombus bubonius* (Stiner 1994).

While it is clear that small animals in general, including shellfish, have high handling costs relative to return rates, I think that the return rates have been grossly underestimated with respect to shellfish (see Bailey 1975; Osborn 1980), leading to an occasionally repeated misconception that shellfish are not worth or are hardly worth collecting (see Gamble 1986; Wilson 1993). Osborn (1980), for example, argues that it would require 52,267 European oysters (*Ostrea edulis*) to equal the caloric return of one red deer, and that about 700 oysters per person would have to be collected, processed and consumed to meet mean daily caloric requirements of a modern human forager. A moment's reflection shows these numbers to be highly exaggerated. At an average edible mass of 75 g per oyster and a caloric value of 748 kcal kg^{-1} (Tressler 1951), humans would have to have a mean metabolic rate of 39,270 kcal d^{-1} to require the 52.5 kg per day supplied by 700 oysters! While the mass and caloric values quoted above are for commercial oysters from deep-water fisheries (but are specific to *O. edulis*) and thus may be higher than values obtained for near-shore oysters, they still serve to illustrate the general problem. Using a mean male/female mass for North American wapiti (red deer) of 309 kg (Schmidt and Gilbert 1978) and assuming that, like caribou (reindeer), about 65% of the animal is edible (Foote 1965), a red deer should on average return about 200.85 kg of edible tissue. Again using caribou as a model (Foote 1965), about 54% of this edible tissue will be in the form of meat, about 31% in viscera, and 15% in fat (with average caloric values of 1190, 1250 and 9000 kcal kg^{-1}, respectively: Keene 1985). Thus we would expect, on average, an adult red deer would provide about 478,062 kcal, the equivalent of 8522 oysters, not 52,000. While this shows that shellfish returns are perhaps not as bleak as some claim, it is still the case that net return rates are lower than those of larger, terrestrial game, and that a lot of shellfish must be collected and processed to provide any kind of sustained contribution to the diet base. Foraging effort does vary by shellfish type, however, with colonial and sessile bivalves like mussels being easier to harvest in abundance (easily scraped off a rock with a digging stick) than the generally individual (but usually larger) gastropods like limpets (Wilson 1993). All things considered, it seems most likely that shellfish played a seasonally important role in the diet of coastal Neandertals, providing a critical source of late winter/early spring calories and fat to foragers who were focused on larger terrestrial herbivores at other times of the year (Stiner 1994).

The extent to which Neandertals used freshwater or marine fish is also uncertain (Rosello Izquierdo and Morales Muniz 2005), but has perhaps been underappreciated. Fish remains have been recovered from a fair number of Acheulean and Mousterian/ Micoquian sites: Vallonet, Lazaret, and Salpêtre de Pompignan (Desse and Desse 1976), Carrière, Vaufrey, Vallière, Frechet and Olha (Le Gall 1992), Grotte XVI (Rigaud et al. 1995), Orgnac and Grotte des Pêcheurs in France (references in Rolland 1987); Tito Bustillo and Cueva Millan in Spain (Le Gall 1992); Figueira Brava Cave, Portugal (Bicho and Haws 2008); Vanguard and Gorham's Caves, Gibraltar (Stringer et al. 2008b); Grotta Maggiore in Italy (Fiore et al. 2004); Vindija Cave in Croatia (Paunovic and Smith 2002); Raj Cave, Poland (Patou-Mathis 2004); and Ust-Kanskaya Cave in Russia (Derevianko et al. 2005). Microwear traces indicative of processing of fish remains have also been found on tools dating to MIS 7-5 at Payre in France (Hardy and Moncel 2011). These sites, however, seem to be the exception rather than the norm. The impact of late 19th and early 20th century archeological methods, in which lack of fine screening of sediments may have precluded the recovery of fish bones, and taphonomic factors (perhaps differentially destroying the small and more lightly mineralized bones of fish) are hard to assess, but fish are rare even in modern excavations in which mammalian and reptilian microfauna are abundantly represented. At Vindija Cave, where fish and amphibian remains with taphonomic indications of human consumption occur in deposits spanning MIS 6-3, the emphasis appears to have been on key species with high nutritive value: trout, pikeperch, and frog (Paunovic and Smith 2002), while the French and Spanish sites listed above show a preponderance of salmonids (mostly trout), carp, and perch (Desse and Desse 1976; Le Gall 1992). As with shellfish, freshwater fish may have played an important role in Neandertal diets as occasional or emergency resources but, based on the totality of evidence, do not seem to have been a major staple during Mousterian times.

7.3 Analysis of Food Residues: Macrobotanical Remains

Plant remains seldom preserve in archeological deposits of Pleistocene age, and so the least can be said about the importance of plant food in the Neandertal diet. Stable isotope and, to a lesser degree, tooth microwear evidence reviewed below (Sections 7.4 and 7.5) suggests that the paucity of macrobotanical remains in Mousterian sites may actually reflect low levels of consumption of these resources and not just plant tissue taphonomy (bucking an old archeological adage: the absence of evidence may, in this case, be evidence of absence!). Certainly plant foods play a reduced role in human diet in cases where seasonality and the nature of primary production (edibility and digestibility of plant growth) limit the range of humanly-consumable plant matter, but even groups that are largely completely carnivorous (such as arctic foragers) consume some small amounts of plant matter. For example, arctic Inuit at Kobuk in the Alaskan interior harvested an estimated 3500 kg of berries in a 5-year period (for a village group of 60). Assuming similar consumption rates in Neandertals – about 11 kg y^{-1} – it's not hard to imagine that seasonal consumption of berries would have had little effect on stable isotopic signatures, but may nonetheless have been an important supplement to an otherwise carnivorous diet (especially since berries tend to be rich in ascorbic acid [Vitamin C], an essential nutrient that is lost from meats during cooking). Along these lines, it's interesting that Rolland (1987) mentions the recovery of hackberries from the Mousterian of Mas-des-Caves (France). Likewise, regular consumption of the gut contents of ruminants (Buck and Stringer 2014), in which mechanical and chemical breakdown of plant cells walls has occurred through rumination and bacterial action, could have been an important source of Vitamin A and other nutrients while still having a negligible effect on stable isotope ratios in Neandertal bone collagen.

Important exceptions to the general situation of unpreserved macrobotanical remains do exist – notably at Kebara Cave in Israel (Lev et al. 2005), but also at Douara Cave (Syria: Matsutani 1987) and Gorham's Cave (Gibraltar: Barton et al. 1999). Charred seeds and fruits from Mousterian levels at Kebara (predominately from units IX-X, which date to end MIS 4–early MIS 3) indicate a diverse complement of plant foods in the Neandertal diet in the Levant, including seeds of grasses (goat grass, oat, and barley), goosefoot, and vetches and lentils, and acorns and pistachio nut (Lev et al. 2005). While nuts are well-represented (and likely correspond to harvesting activity during the fall), the bulk of the plant remains derive from wild pulses (mainly lentil and some vetches), which appear to have been gathered in the early spring. Consumption of nuts (or acorns) of Mt. Tabor Oak (*Quercus ithaburensis*), Atlantic pistachio (*Pistacia atlantica*) and wild almond (*Amygdalus communis*) is also thought to have occurred in the Middle Pleistocene of Israel, based on shell remnants from Gesher Benot Ya'aqov (Goren-Inbar et al. 2002). Nut consumption is also evident in Mediterranean Europe, where some of the plant residues recovered at Gorham's Cave represent stone pine nuts (*Pinus pinea*) (Barton et al. 1999). The limitations inherent in trying to subsist on acorns are discussed in Chapter 6, but these limitations may be less important for the Mt. Tabor Oak, which has acorns with a relatively high nutritional value and low amount of tannins (reducing the amount of processing that must be done to make them edible). A consideration of the nutritional value of Mt. Tabor Oak acorns (Table 7.2) shows that they would have been a good source of energy (in the form of carbohydrates) and a reasonable source of protein as a seasonal (fall) component of the diet. However, their relatively limited representation at Kebara may mean either that Neandertals did not have means of leaching the tannins from the nuts (and they entered the cave as non-food items, hitchhiking along with firewood) or they were processing them in such a way that few of the nutshells survived in the archeological deposits (Lev et al. 2005). Pistachio is much better represented and, given its high nutritive value (in terms of fat and protein: Table 7.2), this nut would have likely been much sought after when in season (late summer and early fall). Fruit residues are rare at Kebara, although wild grape is represented (Lev et al. 2005), while hackberry remains have been recovered at Douara (Matsutani 1987) and, as mentioned above, at the French site of Mas-des Caves (Rolland 1987), and wild olive (*Olea* sp.) has been found at Gorham's Cave (Barton et al. 1999).

Recent work focusing on indirect evidence of plant consumption has also produced some interesting results. Cachel (1997) noted that tools with use-wear indicative of grinding vegetable matter against stone have never been recovered from Mousterian deposits, despite the fact that they are a common occurrence in the earliest Upper Paleolithic assemblages (Straus 1992; de Beaune 1993). Likewise, possible mulling stones, grindstones, and pestles have been recovered from early Upper Paleolithic deposits but

TABLE 7.2 ■ Comparison of nutritional value (per 100 g) of plant foods consumed or potentially consumed by Neandertals, compared with the edible tissues of a reindeer

	kcal	Carbohydrate (g)	Protein (g)	Fat (g)
Legumes[a]	350	60	29	1
Mt. Tabor Oak acorns[a]	270	54	5	5
Pistachio nuts[a]	594	17	19	54
Pine cambium[b]	103	26	2	1
Hardwood cambium[c]	27	6	0	1
Reindeer				
meat	119	0	27	1
fat	900	0	0	100
guts	125	0	15	5

[a] Kilocalorie value from Lev et al. 2005; other nutritional information based on domesticated lentils (*Lens culinaris*), from Ereifej 2004. Wild lentils are the most numerous legume recovered from Kebara cave.

[b] Data for western hemlock (from Sandgathe and Hayden 2003), taken as representative of pines generally.

[c] Data for cottonwood (from Sandgathe and Hayden 2003), taken as representative of hardwoods generally.

are unknown in the Mousterian (de Beaune 1993).[3] However, this appears now not to be the case for at least one category of plant-processing tool: nut crackers. Hammer stones from Gorham's Cave, in association with pine cone residues in ash lenses and sediments, have been interpreted as tools used in processing pine nuts (Barton et al. 1999), and pitted stones from the Acheulean at Gesher Benot Ya'aqov have been likewise interpreted (Goren-Inbar et al. 2002). Analyses of residues and wear patterns on the edges of lithics from Mousterian sites have also produced some evidence of plant processing. The edges of tools from Starosele (Crimea) preserve appreciable quantities of starch grains, as well as raphides, recognizable plant cell remnants, and wear striations referable to abrasion against material with a high silica content (Hardy 1997). The abundance of starch grains on the tools from Starosele suggests that processing and, by inference, consumption of starchy storage organs of plants (e.g., roots and tubers) may have played an important role in Neandertal subsistence. Edible roots also comprise part of the macrobotanical record at Kebara Cave (Lev et al. 2005). It should be noted that plant residues have also been recovered from stone tools at other sites (e.g., La Quina, France: Hardy 2004), but have generally been referable to wood working activities in the context of tool manufacture. Finally, pointed bone tools from the Mousterian at Salzgitter-Lebenstedt have been suggested to represent bark peelers used to get at the edible inner bark of mature trees (Sandgathe and Hayden 2003). The metabolically-active parts of the tree trunk, including the vascular cambium and the secondary phloem, were eaten during lean times by a number of boreal forest foragers in Eurasia and North America (references in Sandgathe and Hayden 2003), and inner bark may have served as an important survival food for the Neandertals during late winter and early spring. Trees are somewhat variable in the nutritive quality of their inner bark, but in general they are not a rich source of macronutrients (Table 7.2).

At present, the direct and indirect evidence for plant food consumption by Neandertals is sparse, and we lack the quantity of sites with good quality data (such as

Kebara) that would allow us to begin to explore how variation in Neandertal plant consumption relates to other dietary and ecological variables. At best, all we can say at present is that, as expected for omnivorous foragers, Neandertals were not adverse to some degree of herbivory, and may have been doing substantial amounts of it at times. Dental microwear studies (reviewed below) also indicate a substantial degree of variation in Neandertal diet, although no clear geographic or temporal patterns in the inferred degree of reliance on plants are evident.

7.4 DENTAL WEAR AND FOOD RESIDUES ON TEETH

The analysis of dental microwear, focusing on striations on the enamel of the occlusal surfaces of teeth, has been an important means of inferring diet in extinct primates (see references in Lalueza et al. 1996). Because Neandertal teeth often exhibit extreme amounts of occlusal wear and dentin exposure, initial attempts to glean dietary information from dental microwear focused on examination of striation patterns on the buccal (vestibular) surfaces of their cheek teeth. More recent work has focused on microwear and macrowear on the occlusal surfaces (on individuals for whom extreme wear has not obliterated the enamel). Added to this is the recent analysis of microfossils of food residues from dental calculus, which has also proved very useful in understanding dietary adaptations during the Middle Paleolithic. We'll first review the buccal wear studies, and then explore the results of wear studies of the occlusal surfaces and analyses of food residues.

Unlike occlusal wear, which results from both tooth-to-tooth attrition and abrasion, buccal striations result predominantly (seemingly entirely) from abrasion by opal phytoliths in vegetal matter and other abrasive adventitious particles (from dirt and grit) in ingested foods (Lalueza Fox and Pérez-Pérez 1993). Buccal striations can be observed and quantified even in worn molars, and comparative analysis (of 153 individuals from ten dietarily diverse groups) has show striation numbers and patterns to be highly consistent from tooth to tooth within individuals, and between-group variation in these features to be much greater than within-group variation (Pérez-Pérez et al. 1994). Within samples, an insignificant amount of variation is attributable to variation in sex or the side, type, or jaw of the tooth analyzed (Lalueza et al. 1996). Additionally, buccal striation patterns appear to develop gradually over time (thus they reflect average diet rather than the last few meals before death), and can be shown to be of dietary rather than taphonomic origin (burial and soil abrasion remove rather than create striae) (Pérez-Pérez et al. 1994).

Lalueza et al. (1996) established baseline data for ten contemporary (extant or recently extinct) human groups of known diet (or for whom dietary inferences could be made with reasonable confidence), against which data for 20 fossil humans were compared. The contemporary groups fell into four broad diet categories, as follows: agriculturalists (strict vegetarian Hindus from Bihar and Orissa, India: 20 individuals); tropical forest foragers (with possibly up to 70% vegetable matter in diet, comprised of 18 Andaman Islanders from the Gulf of Bengala and 9 Sri Lankan Veddah); subtropical arid and mesothermal environment foragers (generally between 60–80% vegetable matter in diet, comprised of 15 Kalahari Bushmen from South Africa, 18 Australian Aborigines and 11 Tasmanians); and carnivorous foragers and pastoralists (arctic or cold-temperate groups whose diet was based mainly on terrestrial and aquatic animals, including 20 Greenland Inuit, 5 Lapplanders from Norway, Finland and Russia, 20 Fueguians from southern Argentina and Chile, and 17 individuals from Vancouver Island, Canada). The major differences between the diet categories followed expectations based on the amount of plant food in the diet: those groups with higher amounts of plant foods, and hence greater exposure to plant phytoliths, have a higher number of striations and a higher proportion of horizontal relative to vertical striations (this latter attribute likely having to do with increased horizontal movements of the mandible during mastication of fibrous plant material: Hinton 1981). This pattern is most pronounced in the complete vegetarians from India, with the tropical and arid/mesothermal foragers being largely similar in numbers and orientations of striations (although the arid/mesothermal groups tend to have slightly higher proportions of vertical striations while the tropical forest groups have slightly higher proportions of horizontal striations). In contrast to these groups, the carnivorous foragers and pastoralists had the fewest number of striations (although some, notably the Inuit, still have fairly high densities of striations) and much higher proportions of vertically oriented striations. The observed striations in these samples likely originated from exogenous abrasives (grit, dirt, ash, bone powder, ancient phytoliths from the soil), and variation in striation density between groups in this category may reflect variation in food cleaning, preparation, or storage (see also El Zaatari et al. 2011). The predominance of vertical striations in all of these groups likely reflects the more vertical mandibular movements associated with chewing meat (Hinton 1981).

Of the fossil humans examined by Lalueza et al. (1996), 11 are either Neandertals or their immediate late Middle Pleistocene ancestors (Figure 7.1). Also included in the analysis was the isolated mandible from Tabun layer C (Tabun C2), of uncertain taxonomic affinity but possibly representing a last interglacial Levantine Neandertal (see Stefan and Trinkaus 1997 vs. Quam and Smith 1998). All of the Neandertals examined fell within the 95% confidence ellipse of the carnivorous foragers when the ratio of vertical to total number of striations (NV/NT) is plotted against the ratio of horizontal to total striations (NH/NT), although many also fell into the 95% equiprobability ellipses of the two mixed diet foraging groups (tropical forest and arid/mesothermal). As can be seen in Figure 7.1, the Neandertals are similar to the meat-dependent foragers in having high proportions of vertical striations and low proportions of horizontal ones. Interestingly, though, many of the Neandertals with low NH/NT ratios also have low NV/NT ratios (but still higher generally than the NH/NT values), and thus fall below the means of the carnivorous foraging groups in their NV/NT ratios. A number of individual Neandertals fall closer to the mean ratio values for the mixed diet groups than the carnivorous groups. Indeed, of the 12 specimens plotted in Figure 7.1, seven were classified as members of the tropical forest groups and three as belonging to the arid/mesothermal groups by a discriminant analysis based on eight microwear variables (Table 8 in Lalueza et al. 1996). Only two specimens, the female from Tabun layer C (Tabun C1) and the child from Devil's Tower (Gibraltar 2), were assigned to the carnivorous forager groups, even though other specimens (notably Amud 1 and Malarnaud) appear "hyper-carnivorous" on the basis of NV/NT and NH/NT ratios. The variation overall in the relative proportions of vertical and horizontal striations and other variables (as reflected in the variation in the discriminant function assignments of the various Neandertals) may reflect geographic or temporal/climatic variation in

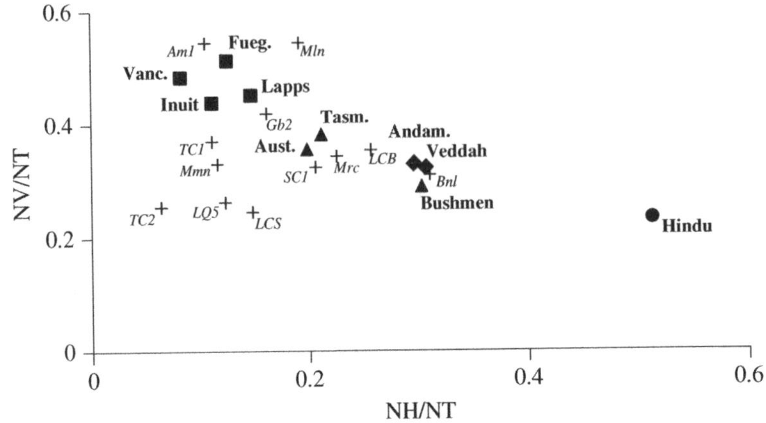

FIGURE 7.1

Dental buccal microwear patterns in Neandertals and modern humans. Ratio of number of vertical to total striations (NV/NT) versus ratio of number of horizontal to total striations (NH/NT) for individual Neandertals (crosses) and mean values for samples of vegetarian agriculturalists (●: Hindu), tropical foragers (◆: Andamanese and Veddah), arid climate foragers (▲: Kalahari Bushmen, Australian Aborigines and Tasmanians) and cold climate foragers (■: Greenland Inuit, Lapplanders, Tierra del Fueguians and Vancouver Islanders). Neandertal specimens include Amud 1 (Am1), Banyoles (Bnl), La Chaise Abri Suard (LCS), La Chaise Bourgeois (LCB), Gibraltar 2 (Gb2), Malarnaud 1 (Mln), Les Pradelles (Marillac: Mrc), Montmuarin 1 (Mmn), La Quina 5 (LQ5), Saint Cesaire 1 (SC1), Tabun C1 (TC1) and Tabun C2 (TC2). All data from Lalueza et al. 1996.

the amount of meat vs. plants in the diet mix. However, no real temporal, climatic, or geographic patterns are evident in the distributions of ratios (Figure 7.1) or the manner in which the discriminant function assigned individuals to groups.

Subsequent analysis, using the same modern human comparators but with a larger sample of fossil hominins and looking only at the total number of buccal striations, more strongly suggests an omnivorous diet in the Neandertals, but perhaps with a high degree of carnivory during some climatic episodes. Pérez-Pérez and colleagues (2003) examined non-occlusal striations in 25 Neandertal teeth from individuals spanning MIS 6 through 3, and found a continuum of striation densities that they interpret as reflecting dietary variability over space and time. Interestingly, Neandertals from colder intervals tend to have a higher density of striations, suggesting greater consumption of vegetal foods during glacial periods. The Neandertals with striation densities most similar to those of highly carnivorous arctic or cold-climate foragers derive from warmer intervals – the reverse of what one would expect based on plant productivity (see Chapter 6). When specimens are divided by climatic conditions and geographic region (Figure 7.2) it can be seen both that meat eating increases (the average density of striations decreases) with warmer intervals and, within climatic intervals, the Neandertals from the presumably more plant-rich Near East appear to be more carnivorous than

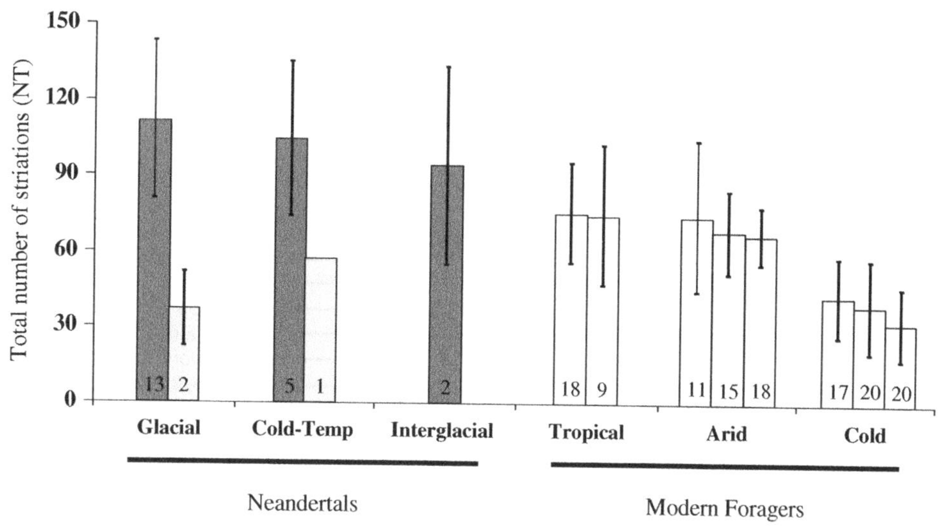

FIGURE 7.2

Mean number of buccal striations in the teeth of Neandertals and modern humans. European Neandertals; ▮; Near Eastern Neandertals: ▯. Whiskers denote ± 1 standard deviation, sample sizes for each sample are provided at the base of each bar. Modern human samples are as in Figure 7.1 (minus Hindu and Lapplanders). All data from Pérez-Pérez et al. 2003.

their European kin. The Near Eastern Neandertals appear to be following the expected trend of decreased carnivory with climatic amelioration, but this result is based on only three individuals. The interpretation of these results is complicated by the finding that the European Neandertals have absolutely high densities of buccal striations – higher even than tropical hunter-gatherers. It is highly unlikely that they were *more* herbivorous than modern tropic foragers, and the observation that Neandertals tend to have relatively low numbers of horizontal striations (Lalueza et al. 1996: see also Figure 7.1) also argues against substantial consumption of plant items. It is possible that the European Neandertals were regularly consuming something that was highly abrasive – either because of contained silicates (although it is probably premature to postulate a Neandertal grazing niche!) or because of adventitious particles.

Ecological patterning of wear fabrics has also been detected in studies of occlusal microwear (El Zaatari et al. 2011) and macrowear (Fiorenza et al. 2011) to Neandertal molars. Using microwear texture analysis on first and second molars, El Zaatari and colleagues (2011) found that Neandertals deriving from sites associated with open country environmental indicators (n = 4; from MIS 5b, 4 and 3 from Spy, Belgium; La Quina and Arcy-sur-Cure, France; and Subalyuk, Hungary) had wear fabrics most similar to those of a sample of six Fueguians from shrub and tundra habitats (Figure 7.3). The Fueguians had largely carnivorous diets, with plant foods comprising <15% of the total diet (El Zaatari 2010; El Zaatari et al. 2011). Eight Neandertals from mixed habitats (with both open and arboreal vegetation indicators) of MIS 6, 6/5e, 5d, 4 and 3 (from St. Césaire, Petit-Puymoyen, Rochelot and La Chaise, France; Vindija, Croatia;

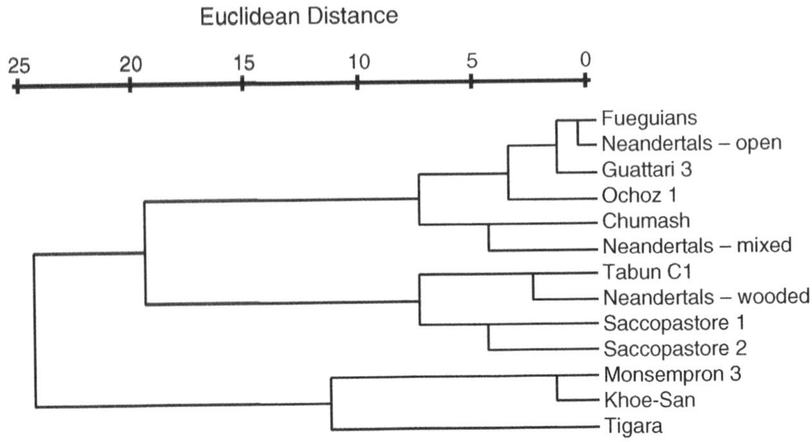

FIGURE 7.3

Hierachical cluster analysis of Neandertals and recent human groups based on occlusal molar microwear analysis (redrawn from El Zaatari et al. 2011).

and Kebara and Tabun, Israel) had wear patterns that clustered with a small sample (n = 13) of Chumash from Santa Cruz Island (California, USA) (Figure 7.3). The Chumash subsisted on a marine diet (fish, shellfish, marine mammals) augmented with terrestrial small game and plant foods, with plants comprising 36–45% of the diet, with perhaps less plant food in the diet in the groups living on Santa Cruz and the other Channel Islands (El Zaatari 2010; El Zaatari et al. 2011). The Neandertals from wooded habitats (n = 4, all from MIS 3, from the sites of Zafarraya and El Sidrón, Spain; Grotta Breuil, Italy; and Amud, Israel) had higher levels of molar occlusal surface complexity and heterogeneity compared to the other Neandertals, suggesting a more varied diet, and perhaps a diet richer in plant foods. Interestingly, despite this suggestion of a more varied diet, the wooded ecosystem Neandertal group still formed a second-order cluster in multivariate space with the other two Neandertal groups, and did not cluster with a small forager sample from a wooded environment (nine South African Khoe-San from a montane forest environment: El Zaatari 2010; El Zaatari et al. 2011) (Figure 7.3). This might suggest that, even for Neandertals with the most varied diets, the proportion of plant foods in the diet was still lower than that of meat (El Zaatari et al. 2011). Thus the occlusal microwear data suggests (1) Neandertal diet was not monotonous across space and time, but varied with ecological conditions, as would be expected for omnivorous, adaptively-flexible foragers, and (2) despite some variation across habitats, the Neandertals' diet may have always been fairly meat dependent (>50% of the diet: see Chapter 6.2).

Studies of occlusal macrowear have also contributed to our understanding of Neandertal diet. Application of a technique known as occlusal fingerprint analysis (which analyzes the three-dimensional topography of wear surfaces) to 19 Neandertals, 12 early modern humans, and 42 recent human foragers detected differences in molar wear patterns between Neandertals from deciduous woodland habitats (MIS 5e Neandertals from Krapina, Croatia) versus those associated with steppe or coniferous forest biomes (MIS 5d-c and MIS 3 specimens from Monsempron and Le Moustier, France, and Vindija, Croatia) versus those from Mediterranean evergreen habitats (MIS 5e and MIS 3 fossils from Tabun and Amud, Israel, and Shanidar, Iraq). Both the temperate woodland and Mediterrean evergreen samples had wear patterns indicative of a mixed diet, potentially containing a substantial amount of plant matter (Fiorenza et al. 2011). The woodland sample did not differ in overall wear patterns from Australian aboriginies ($n = 3$) or Kalahari Khoe-San ($n = 7$), both of which consume a substantial amount of plant material, yet did differ significantly from three samples of foragers with largely meat-based diets (Inuit, $n = 10$; Vancouver Islanders, $n = 15$; Fueguians, $n = 7$). The Mediterrean evergreen Neandertals were significantly different from some recent forager samples but not others, but with no clear ecological pattern to the results (they differed from the meat-eating Inuit and Fueguians but not the Vancouver Islanders, and they differed from the more mixed-diet Khoe-San but not the Australian aboriginies: Fiorenza et al. 2011). The Neandertals from colder steppic or coniferous forest habitats, on the other hand, had wear patterns more similar to modern foragers who consume a largely meat-based diet (Fiorenza et al. 2011), although this Neandertal sample was not significantly different from *any* of the recent foragers, regardless of diet.

The dental wear studies reviewed above produce some contrasting results, which is probably to be expected when one considers a number of factors that underlie this type of research. Sample sizes – both fossil and comparative – tend to be small, and thus statistical power to detect differences between samples tends to be limited. Also, the climatic and ecological conditions associated with each specimen are estimated based on stratigraphic context within the sites from which they were recovered, and the possibility exists that some specimens may be misclassified with respect to their assigned ecological groups. Differences between researchers in how categories are defined and constructed also no doubt adds to differences in the results obtained by different workers. The introduction of exogenous grit, either from food preparation and storage practices or perhaps from environmental conditions, likely also complicates the dental wear picture (see El Zaatari et al. 2011). There is also an effect of the temporal scale across which these dietary indicators develop: occlusal microwear reflects dietary behavior over the last couple of weeks, or even days, of the life of the individual (El Zaatari et al. 2011), whereas buccal microwear and occlusal macrowear are likely to reflect average behavior over many years. Regardless, the combined microwear (buccal and occlusal) and macrowear evidence is consistent overall with relatively high levels of carnivory, yet show that the Neandertal diet varied over space and time, most likely reflecting variation in the availability, digestability, and relative energetic returns of plant foods in different habitats.

Our consideration of the caloric economy of Pleistocene Eurasia in Chapter 6.2 suggested, on the basis of modern human forager ecology, that even in the most plant food-poor environments (open country and boreal forests) Neandertals would still be expected to have derived somewhere between 11–29% of their calories from plants. Analysis of plant microfossils from dental calculus does show that some Neandertals from open country habitats were indeed eating some plant matter (Henry et al. 2011). Starch grains recovered from the teeth of the Spy 1 and 2 Neandertals from Belgium are consistent with the consumption of the underground storage organs of water lilies, grass seeds, and other plants (Henry et al. 2011). The Spy fossils have been directly dated by AMS ^{14}C to ca. 36 Ka BP (Semal et al. 2009), roughly at the climatic downturn at the end of the MIS 3 Hengelo Interstadial. The associated fauna includes horse, reindeer, and red deer, consistent with a steppe habitat (Cordy 1988). As argued by Hardy (2010), geophytes may have been an important source of calories and carbohydrates for Neandertals in all of the climates and environments of Pleistocene Eurasia.

The only other Neandertals thus far sampled for plant microfossils are Shanidar 3 from Iraq and five individuals from El Sidrón, Spain. The age of Shanidar 3 is uncertain, but the partial skeleton derives from layer D1 (Solecki and Solecki 1993), near the top of the Mousterian sequence at Shanidar and slightly below two radiocarbon samples that produced dates of 46.9 ± 1.5 Ka BP and 50.6 ± 3.0 Ka BP (Vogel and Waterbolk 1963). These dates were obtained in the early 1960s using the solid carbon method, and their reliability has been questioned (Olszewski and Dibble 1994), and there has been some suggestion that Shanidar 3 may be as old as 75 Ka BP (Trinkaus 1983b). Palynological samples indicate a warm climate (see Henry et al. 2011), and it thus seems most likely that Shanidar 3 lived during a warm episode, possibly the Glinde Interstadial, of MIS 3. The calculus adherent to the teeth of Shanidar 3 included both starch grains and phytoliths. The starch grains attested to the consumption of grass seeds (from the tribe Triticeae, which includes the wild relatives of wheat, barley, and rye), most likely from wild barley of the genus *Hordeum*, as well as legumes and other plants (Henry et al. 2011). As noted in Chapter 3.5, some of the recovered starch grains showed modifications characteristic of cooking. The phytoliths in the calculus of Shanidar 3 belonged primarily to the fruit of the date palm (genus *Phoenix*), with perhaps some representation of other tree fruits as well (Henry et al. 2011). The El Sidrón sample derives from a 50.6–47.3 Ka BP (MIS 3) woodland habitat in Asturias, and microfossils from four of the five individuals are consistent with consumption of starches, which were possibly cooked. One individual from the site preserved a phytolith from what may have been a grass plant (Hardy et al. 2012). Chemical composition of the calculus in the El Sidrón Neandertals indicated consumption of a variety of carbohydrates, possibly deriving from nuts, grass seeds, and green vegetables, yet failed to detect traces of protein or fats from animal sources (Hardy et al. 2012). This result is consistent with the molar macrowear data that suggest that Neandertals from wooded habitats (the sample of which included an individual from El Sidrón) had higher proportions of plant food in their diets (El Zaatari et al. 2011). The calculus of one individual, a young adult female, had a chemical signal consistent with the consumption of the bitter-tasting, non-food plants yarrow and chamomile, which has been interpreted as indicating medicinal use of plants among the Neandertals (Hardy et al. 2012). Alternatively, these asters may have been introduced into the oral cavity through the consumption of chyme (the partially-digested stomach contents of herbivores), which is an important source of carbohydrates and micronutrients in the diets of arctic foragers, and likely was for Neandertals as well (Buck and Stringer 2014).

7.5 STABLE ISOTOPE AND TRACE ELEMENT ANALYSES

An increasingly important method of investigating diet in archaic humans is the analysis of stable isotopes of carbon (^{13}C) and nitrogen (^{15}N) preserved in the collagen of fossil bone. Because isotopes of an element share the same number of protons they behave identically in chemical reactions, but because they differ in number of neutrons, and thus vary in atomic weight, they vary in their rates of reaction (with heavier isotopes reacting more slowly). Since their chemical reaction rates differ, biological tissues will differentially incorporate different isotopes of an element. For example, plants incorporate ^{12}C preferentially to ^{13}C, such that the ratio of ^{13}C to ^{12}C in plant tissue is reduced relative to that of the atmosphere (Park and Epstein 1960). Additionally, differences between plant species in the photosynthetic pathways they use contribute to variation in the ratios of these two isotopes of carbon. Temperate broad-leafed trees, shrubs, geophytes and some grasses use the Calvin-Benson, or C_3, pathway, which discriminates against the heavier ^{13}C isotope in atmospheric CO_2, and thus these plants have lighter (more negative)[4] $\delta^{13}C$ values, usually between −22‰ and −38‰, with an average value of −26‰. Forest or closed-cover taxa also have lower $\delta^{13}C$ values due to the canopy effect, whereby recycling of ^{13}C-depleted CO_2 further militates against ^{13}C uptake (van der Merwe and Medina 1991). Plants using the Hatch-Slack, or C_4, photosynthetic pathway discriminate less against ^{13}C, and thus they more closely approximate the standard and have less negative $\delta^{13}C$ values. Many open country and dry-adapted forms use the C_4 pathway, including most grasses, chenopods, amaranths, and setarias. A greater proportion of C_4 plants, in conjunction with little or no canopy effect, produces open country floras that have a mean $\delta^{13}C$ value of −12.5‰, with a range between −9‰ and −21‰ (Tieszen 1991). Although cacti and succulents employ crassulacean acid metabolism (CAM) and have values intermediate between C_3 and C_4 plants, these plants were likely of minor importance in Pleistocene European floras, although they may have been part of the semi-desert vegetation that occurred at times in western Asia. These well-documented relationships between stable carbon isotopes and photosynthetic pathways allow for the identification of type of plants (C_3, C_4 or CAM) that animals were eating on the basis of the $\delta^{13}C$ values.

Experimental and field data show that stable carbon isotope ratios in an animal's tissues reflect the values of these ratios in the diet (references in Larsen 1997), regardless of trophic level. That is, open country grazers consuming C_4 grasses will have less negative $\delta^{13}C$ values than closed country browsers, as will carnivores preying on open country herbivores relative to those hunting in closed habitats (there is a 1‰ isotopic fractionation between herbivores and carnivores, such that carnivores have about 1‰ less negative $\delta^{13}C$ than their prey: see references in Fizet et al. 1995). In human bone collagen, the carbon turnover rate (in adults) is greater than ten years (Silberbauer 1979), thus isotopic readings represent an average of dietary intake over ten years. This means that dietary staples overwhelmingly determine $\delta^{13}C$ values, short-term changes are not registered, and seasonal fluctuations are averaged out.

The ratios of two stable isotopes of nitrogen, ^{14}N and ^{15}N, can also be used to make inferences about types of plant matter consumed and the trophic level at which an animal feeds. Atmospheric air has a $\delta^{15}N$ value of 0‰, while the $\delta^{15}N$ value of soils averages about 10‰ (Shearer and Kohl 1989) because the nitrates released during the decomposition of plant detritus are enriched in ^{15}N relative to the atmosphere. Thus plants using nitrogen from the soil will have $\delta^{15}N$ values about 2‰ higher than nitrogen fixing organisms (algae, bacteria, and legumes) (Larsen 1997). Accordingly, nitrogen fixing plants – such as legumes – have $\delta^{15}N$ values close to zero (since they derive their nitrogen from the atmosphere), while the majority of plants derive their nitrogen via nitrogen fixers in the soil, and thus have $\delta^{15}N$ values closer to that of soil.

In addition to indicating something about consumption of nitrogen fixing versus non-nitrogen fixing plants, $\delta^{15}N$ values also reflect the trophic level occupied by higher-order consumers. All herbivores show a trophic level shift of about +3‰ relative to their forage, with the exception of animals that consume leguminous plants, who will experience a weaker shift (they will have lower $\delta^{15}N$ values than consumers of non-leguminous plants) (Katzenberg 1992). Thus, primary consumers eating nonleguminous plants should have bone collagen $\delta^{15}N$ values in the neighborhood of 5–6‰. Carnivores experience again another +3–4‰ shift (Minagawa and Wada 1984; Schoeninger and DeNiro 1984), and organisms that feed at higher trophic levels (tertiary consumers and nursing carnivores) will have further ^{15}N enriched isotopic signatures (Fogel et al. 1989; Larsen 1997).

Stable isotopes of both carbon and nitrogen also can be used to make inferences about the contribution of marine resources to the diet. In coastal areas where C_4 plants are not consumed, less negative values of $\delta^{13}C$ indicate consumption of marine foods, as marine environments tend to be richer in ^{13}C, both because of input of terrestrial organic matter representing both C_3 and C_4 plants, and because of dissolved ^{13}C-rich carbonates in ocean water (Schoeninger and Moore 1992). $\delta^{15}N$ values for terrestrial plants also tend to be lower (by about 4‰) than for marine and even freshwater aquatic plants, and amplifications of relative ^{15}N by trophic level brings some aquatic resources up to 20‰ higher than terrestrial resources, for an average marine enrichment of 10‰ (Schoeninger and DeNiro 1984). Richards (Richards et al. 2001; Richards and Trinkaus 2009), however, notes that European Neandertal isotopic values (see Table 7.3) indicate diets dominated by terrestrial herbivores: the Neandertals thus far studied exhibit neither the elevated $\delta^{15}N$ signal nor more negative $\delta^{13}C$ values that would be characteristic of consumption of aquatic resources. This holds even for the Zeeland Ridges (Netherlands) specimen, which presumably derives from a coastal-living Neandertal (Hublin et al. 2009).

Because stable isotopes do not transmute to other isotopes or elements as do radioactive isotopes, their ratios remain unchanged in bone collagen over geological time. Of considerable use with Holocene human fossil and archeological remains, stable isotope analysis has only relatively recently been applied to older specimens. In general, bone collagen degrades with time, but collagen of sufficient quality for analysis has been extracted thus far from 16 Neandertals (Table 7.3). An additional two individuals, both from Les Pradelles (France), have collagen of more dubious quality, but have nonetheless produced isotopic values similar to those of other Neandertals: (see Table 7.3). Yet another individual, from the French site of Payre, has yielded stable carbon isotope values from dental enamel. The Payre specimen derives from level G at the site, which appears to have been deposited during MIS 8 or 7, making this the oldest Neandertal for which isotopic data has been recovered. The next oldest are two specimens from Scladina Cave, Belgium, which derive from MIS 5 deposits (Bocherens et al. 2005). Several specimens from Les Pradelles in France date to MIS 4, but the bulk of the Neandertal sample derives from MIS 3 (Table 7.3). Since local soil chemistry, temperature, and ground water conditions (affecting rates of diagenesis) can alter bone collagen isotopic ratios (Nelson et al. 1986), results must be interpreted with care. However, the practice of comparing human and non-human faunal material from the same sites and levels – material that has presumably been exposed to the same thermal and chemical environment – allows us to evaluate isotopic ratios of

TABLE 7.3 ■ Bone collagen $\delta^{13}C$ and $\delta^{15}N$ (‰) values of Neandertals

Site/Sample no.	$\delta^{13}C$	$\delta^{15}N$	Age (ky BP)	Source
MIS 8-7				
Payre level G	−14.6	–	–	Ecker et al. 2013
■ MIS 5c-b				
Scladina 4A	−19.9	10.9	ca. 100	Bocherens et al. 1999; Bocherens et al. 2005
Scladina 1B	−21.2	11.8	ca. 90	Bocherens et al. 2001; Bocherens et al. 2005
■ MIS 4				
Les Pradelles 9[a]	−20.2	9.3		Fizet et al. 1995; Beauval et al. 2002
Les Pradelles 10	−19.1	11.6		Fizet et al. 1995; Beauval et al. 2002
Les Pradelles M300	−19.1	11.5		Bocherens et al. 2005; Beauval et al. 2002
Les Pradelles M400	−19.5	11.4		Bocherens et al. 2005; Beauval et al. 2002
Les Pradelles M100[a]	−21.8	8.4		Bocherens et al. 2005; Beauval et al. 2002
■ MIS 3				
Engis 2[b]	−19.6	12.6	ca. 40-35	Bocherens et al. 2001
Jonzac 1	−19.7	11.2	ca. 40-35	Richards et al. 2008c
Feldhofer 1	−21.5	9.0	39.9	Richards and Schmitz 2008
Feldhofer 2	−21.6	7.9	39.2	Richards and Schmitz 2008
Okladnikov	−19.1	12.9	34.2	Krause et al. 2007b
Les Rochers-de-Villeneuve	−19.0	11.6	45.2	Beauval et al. 2006; date from Richards and Trinkaus 2009
Saint-Césaire 1	−19.8	11.4	ca. 36	Bocherens et al. 2005
Spy 572a	−19.8	11.0	ca. 36	Bocherens et al. 2001
Vindija 207	−19.5	10.1	ca. 35–30	Richards et al. 2000; date from Richards and Trinkaus 2009
Vindija 208	−20.5	10.8	32.4	Richards et al. 2000; date from Richards and Trinkaus 2009
Zeeland Ridges	−20.4	10.2	<45	Hublin et al. 2009

[a] Uncertainty remains about the isotopic reliability of the collagen from these specimens (see Bocherens et al. 2005).

[b] Juvenile less than six years old at death: high $\delta 15N$ value may reflect relative recency of milk diet.

various species relative to one another, providing a measure of confidence and interpretability to the values obtained from the human material. In addition, the similarities in results (for both human and faunal fossils between studies and also, for the fauna, with values obtained for similar Holocene forms) across studies (involving analyses done in different laboratories) lends a degree of confidence in the dietary inferences that can be drawn from the obtained values.

Herbivores are characterized by low $\delta^{15}N$ values (with mean values ranging between 7.7–4.5‰), as would be expected of animals consuming a largely non-leguminous diet, and a range of $\delta^{13}C$ values reflective of the mix of open versus closed country forage peculiar to each species (Table 7.4). As would be expected, open country grazers or mixed graze/browse feeders have lower mean $\delta^{13}C$ values than do closed cover forms. In this case, the open country forms (horse, steppe bison, mammoth, reindeer, giant deer, red deer, and wooly rhino) have mean $\delta^{13}C$ values ranging from −19.0 to −21.6‰, while the closed covered browsers (fallow deer) have more negative values (mean −23.0‰). Note, however, that one sample of horses from forested conditions of MIS 5 produced a mean $\delta^{13}C$ signal of −22.4‰ (Table 7.4), which suggests that these animals lived in more forested conditions, where the forage contained a higher proportion of C_3 plants. The other samples of horses, as with the other open country grazers, have less negative values, reflecting a higher proportion of C_4 plants in the diet in open country (but the mean values overall indicate that C_3 plants still comprise the major portion of the diet).

Carnivore $\delta^{15}N$ values vary depending upon the degree of omnivory (or, in the possible case of the cave bear, herbivory)[5] characteristic of the species. Cave bear, long thought to be dedicated herbivores based on dental morphology ("le moins carnivore des Carnivores et le plus ours des Ours," Albert Gaudry quoted in Kurtén 1976), certainly appears to have been highly vegetarian judging from their low $\delta^{15}N$ values (means ranging from 1.4–3.6‰), although at times they may have practiced greater omnivory ($\delta^{15}N$ values of 5.7–9.8‰ have been found in some cave bears from Peştera cu Oase, Romania: Richards et al. 2008b). Differences between studies in $\delta^{13}C$ values for cave bears may reflect habitat-related differences in diet in these eurytopic animals: the cave bears from the last interglacial site of Scladina Cave (Belgium), which likely formed under more closed conditions, have $\delta^{13}C$ values similar to fallow deer (as do the horses from the site: see above); the cave bear from the potentially more open habitats of the interpleniglacial have $\delta^{13}C$ values more similar to steppe-adapted ungulates (Richards et al. 2000). The omnivorous ursids (brown bear) and canids (red and arctic fox) have $\delta^{15}N$ values intermediate between herbivores and strict carnivores, and have $\delta^{13}C$ values indicative of open habitats. Although all these omnivorous carnivores are eurytopic in habit, all of them, especially brown bear and arctic fox, are especially at home in open grassland or tundra environments. The more strictly carnivorous felids (cave lion), hyaenids (cave hyena), and canids (wolf) have, as would be expected, the most elevated levels of $\delta^{15}N$, with means ranging from 9.2–10.3‰ (Table 7.4).

Humans eating a diet of only C_3 plant products (with a mean $\delta^{13}C$ value of −26‰) and herbivores that eat C_3 plants should have bone collagen $\delta^{13}C$ values of about −20.9‰, since bone collagen is enriched in ^{13}C ($\delta^{13}C$ of about +5.1‰) relative to plant tissue or muscle tissue (Silberbauer 1979). Humans consuming a diet based on C_4 plants (with a mean $\delta^{13}C$ value of −12.5‰) and C_4 herbivores should likewise exhibit bone collagen $\delta^{13}C$ values of about −7.4‰. Humans feeding on some mix of C_3 and C_4 resources should have intermediate values of $\delta^{13}C$, depending upon the relative proportions of each type of resource.

$\delta^{15}N$ values in the bone collagen of strict terrestrial foragers (who eschew even freshwater resources) should reflect the relative contribution of plants and animals to the diet. For example, the isotopic value ($\delta^{15}N$) for one omnivore, the domestic

TABLE 7.4 ■ Mean bone collagen $\delta^{13}C$ and $\delta^{15}N$ (‰) values of Pleistocene fauna

	$\delta^{13}C$	$\delta^{15}N$	Source
Open country herbivores			
Horse (*Equus caballus germanicus*) n = 22	−20.6 ± 0.6	5.1 ± 1.4	Fizet et al. 1995
Horse (*E. caballus*) n = 4	−22.4 ± 1.8	5.9 ± 1.2	Bocherens et al. 1999
Horse (*E. caballus*) n = 6	−20.8 ± 0.4	5.7 ± 1.4	Bocherens et al. 2005
Wooly rhinoceros (*Coelodonta antiquitatis*) n = 5	−20.2 ± 0.5	7.0 ± 1.0	Bocherens et al. 2005
Steppe bison (*Bison priscus*) n = 6	−20.0 ± 0.3	6.7 ± 0.9	Fizet et al. 1995
Bos/bison n = 15	−20.4 ± 0.3	6.0 ± 1.0	Bocherens et al. 2005
Giant deer/red deer (*Megaloceros giganteus/Cervus elaphus*) n = 5	−20.2 ± 0.3	6.4 ± 0.9	Bocherens et al. 2005
Reindeer (*Rangifer tarandus*) n = 20	−19.6 ± 0.4	4.7 ± 1.6	Fizet et al. 1995
Reindeer (*R. tarandus*) n = 6	−19.0 ± 0.4	5.5 ± 1.5	Bocherens et al. 2005
Mammoth (*Mammuthus primigenius*) n = 3	−21.6 ± 0.3	8.3 ± 0.4	Bocherens et al. 2005
Closed country herbivores			
Fallow deer (*Dama dama*) n = 10	−23.0 ± 0.8	7.7 ± 0.9	Bocherens et al. 1999
Carnivores			
Cave bear (*Ursus spelaeus*) n = 2	−20.9 ± 0.3	1.4 ± 0.1	Richards et al. 2000
Cave bear (*U. spelaeus*) n = 4	−23.2 ± 0.2	3.6 ± 0.6	Bocherens et al. 1999
Brown bear (*U. arctos*) n = 2	−20.3 ± 2.4	4.8 ± 0.1	Bocherens et al. 1999
Red fox (*Vulpes vulpes*) n = 4	−19.4 ± 0.1	6.5 ± 0.8	Bocherens et al. 1999
Red fox (*V. vulpes*) n = 2	−19.8 ± 0.3	8.4 ± 0.3	Fizet et al. 1995
Arctic fox (*Alopex lagopus*) n = 2	−19.6 ± 0.1	6.3 ± 1.0	Bocherens et al. 1999
Cave lion (*Panthera leo spelaea*) n = 1	−19.2	10.3	Bocherens et al. 1999
Cave hyena (*Crocuta c. spelaea*) n = 1	−19.6	10.3	Fizet et al. 1995
Cave hyena (*C. c. spelaea*) n = 3	−20.5 ± 0.5	8.8 ± 0.9	Bocherens et al. 1999

(*continued*)

TABLE 7.4 ■ (Continued)

	$\delta^{13}C$	$\delta^{15}N$	Source
Hyena (*C. crocuta*) n = 5	−19.3 ± 0.3	9.3 ± 0.5	Bocherens et al. 2005
Wolf (*Canis lupus*) n = 3	−18.6 ± 0.1	10.3 ± 1.5	Fizet et al. 1995
Wolf (*C. lupus*) n = 1	−20.2	7.6	Bocherens et al. 1999
Primates			
Neandertal (*Homo neanderthalensis*) n = 16[a]	−20.0 ± 0.9	11.0 ± 1.2	see Table 7.3

[a] Excluding the MIS 8-7 specimen from Payre and two specimens from Les Pradelles with questionable collagen preservation (see Table 7.3).

dog (8.5–9.5‰: Katzenberg 1992), falls between the means of herbivores (4.7–8.3‰) and obligate carnivores (hyaenids and felids: 8.8–10.3‰) (Table 7.4). Thus humans, as trophic generalists capable of feeding at all levels, should, assuming plant foods are available, have $\delta^{15}N$ ratios intermediate between herbivores and carnivores. As we shall see, this is generally not the case with the Neandertals, regardless of climatic conditions and, presumably, availability of plant resources (see Chapter 6.2).

To date there are reported bone collagen stable isotope values for two individuals from forested conditions during MIS 5, from five individuals from steppe-tundra habitats during MIS 4 (although two of the data points are suspect), and 11 individuals from the cold–temperate climates of MIS 3, ranging from the south of France to the North Sea coast of the Netherlands to the southwest corner of Siberia. The isotopic values derived from these Neandertals, along with values from fauna from the same levels or from near-by sites of the same age, are plotted in Figure 7.4 (with the understanding that isotopic values are not directly comparable between sites or across time, but recognizing that fairly robust patterns emerge in the data regardless of this caveat). In Figure 7.5 and Figure 7.6, the Neandertal values are plotted against (respectively) the herbivore and carnivore values only.

The oldest Neandertals to produce collagen appropriate for isotopic analysis derive from Scladina Cave (Namur, Belgium) (Bocherens et al. 1999), represented by a partial maxilla of a 10-year-old child from layer 4 and an adult pedal first phalanx from layer 3 (originally thought to derive from layer 1, subsequently reassigned to an older level: see Bocherens et al. 2005). Layer 4 is thought to date to MIS 5c at around 100 Ka BP, with layer 3 accumulating in MIS 5b at about 90 Ka BP (references in Bocherens et al. 2005). Proportions of arboreal pollen suggest that the area around Scladina was forested during the deposition of these levels (Cordy and Bastin 1992), and this interpretation is supported by an abundance of fallow deer in layer 4. However, more open country forms, including horse, brown bear, lion, and arctic fox, have also been recovered from layer 4 (this accumulation has no evidence of human agency and is thought to represent a carnivore assemblage). All of the carnivorous and omnivorous carnivores (that is, excluding cave bear) from Scladina have $\delta^{13}C$ values indicative of a diet heavier in ^{13}C than would be expected if they were consuming forest-living browsers in the vicinity

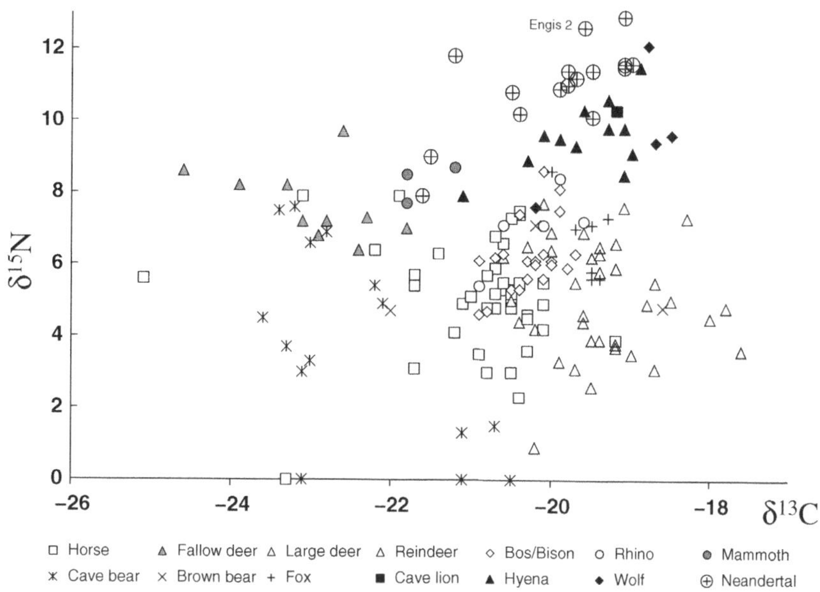

FIGURE 7.4

Stable isotope values for Neandertals and contemporaneous fauna. Two specimens from Les Pradelles of questionable reliability have been omitted (see text), and Engis 2 (indicated) is a juvenile whose isotopic signature may reflect nursing. Data from Fizet et al. 1995; Bocherens et al. 1999, 2001, 2005; Richards et al. 2000, 2008c; Beauval et al. 2002, 2006; Krause et al. 2007b; Richards and Schmitz 2008; Hublin et al. 2009.

of the cave. Bocherens and colleagues (1999) interpret this as a disjunct in carnivore hunting and denning locations, meaning that carnivores residing in the forested habitat of the cave were obtaining their food farther afield, in open country some unspecified distance from their den.

The Scladina Neandertals have high δ^{15}N values – higher than those of most of the carnivores (Tables 7.3 and 7.4 and Figure 7.6), indicating that their dietary protein derived from feeding at a high trophic level. The Scladina specimens also have some of the most negative δ^{13}C values of all the Neandertals (Table 7.3), which no doubt reflects the more forested environments of Belgium during MIS 5. However, their δ^{13}C values – at −19.9‰ and −21.2‰ – are still higher than would be expected based on the δ^{13}C values of the two herbivores sampled from the site – the cover-loving fallow deer (mean $\delta^{13}C = -23.0‰$) and the presumably more open-country horse (mean $\delta^{13}C = -22.4‰$). The strongly negative mean value for the horses from Scladina suggests that they were animals that were feeding in the more heavily forested environments proximal to the cave (Bocherens 2011). Although horses are generally found in open country, they have the ability to exploit a variety of habitats; their habitat selection varies seasonally (with greater use of sheltered temperate or boreal forest habitats in the winter: Pratt et al. 1986; Girard et al. 2013), and forest horses were historically known in Europe and were a feature of some Pleistocene woodlands (Kaagan

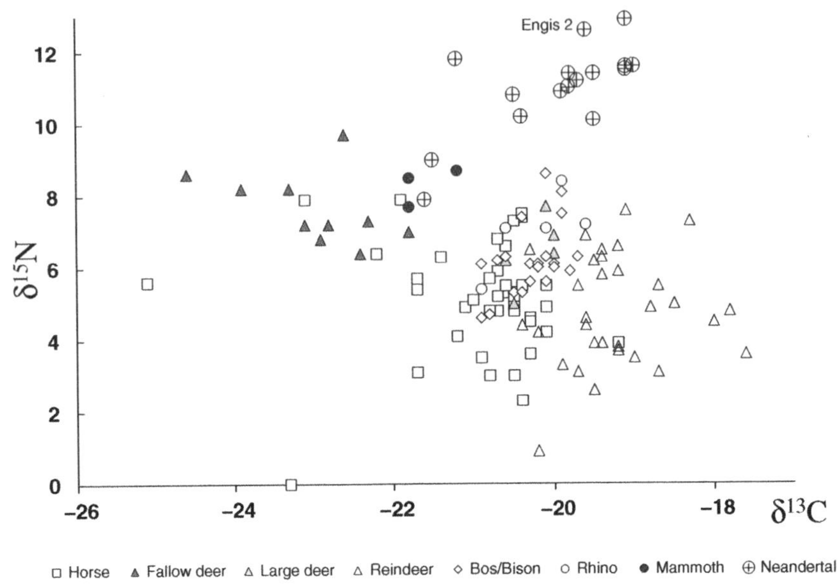

FIGURE 7.5

Stable isotope values for Neandertals and contemporaneous herbivores. Two specimens from Les Pradelles of questionable reliability have been omitted (see text), and Engis 2 (indicated) is a juvenile whose isotopic signature may reflect nursing. Data from Fizet et al. 1995; Bocherens et al. 1999, 2001, 2005; Richards et al. 2000, 2008c; Beauval et al. 2002, 2006; Krause et al. 2007b; Richards and Schmitz 2008; Hublin et al. 2009.

2000). A Neandertal diet equally dependent on fallow deer and horse would predict Neandertal $\delta^{13}C$ values around −21.7‰ – the less negative values observed suggesting greater dependence on open country over closed country prey, as was suggested for the carnivores in layer 4 at Scladina. As discussed in Chapter 3.2, Pettitt (2003) has argued that the low secondary productivity of temperate forests would have necessitated very high mobility among Neandertal groups during interglacial periods (thus putting a premium on lightweight and functionally flexible Levallois-dominated toolkits), the MIS 5 isotopic data from Scladina would certainly support this idea. On the other hand, a single $\delta^{13}C$ determination from tooth enamel of a potentially interglacial (MIS 8–7) Neandertal from level G at Payre (France) of −14.6‰ (Ecker et al. 2013) suggests that the Neandertals there were preying upon forest-dwelling herbivores (Bocherens 2011).

To date, five Neandertals from cold steppic habitats of MIS 4 have been sampled, all of them deriving from Les Pradelles (formerly Marillac) in France (Fizet et al. 1995; Bocherens et al. 2005). Two specimens – a portion of posterior cranial vault of an adult and the maxillary dentition of a young adult – are well-provenenced and derive from layers 9 and 10 at the site. These levels contain Quina-type Charentian Mousterian assemblages and fauna indicative of cold, steppe-tundra conditions, and are believed to date to MIS 4 (Beauval et al. 2002). In these layers reindeer remains were most

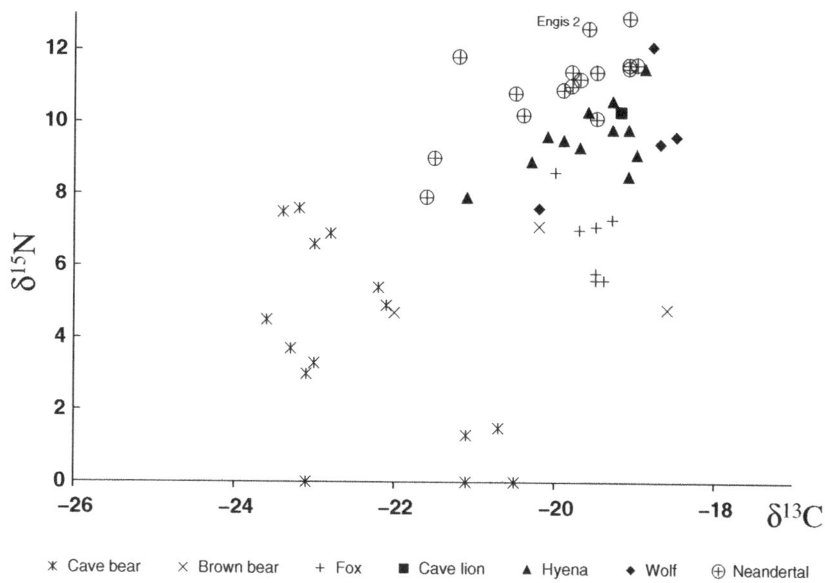

FIGURE 7.6

Stable isotope values for Neandertals and contemporaneous carnivores. Two specimens from Les Pradelles of questionable reliability have been omitted (see text), and Engis 2 (indicated) is a juvenile whose isotopic signature may reflect nursing. Data from Fizet et al. 1995; Bocherens et al. 1999, 2001, 2005; Richards et al. 2000, 2008c; Beauval et al. 2002, 2006; Krause et al. 2007b; Richards and Schmitz 2008; Hublin et al. 2009.

numerous, followed by steppe bison and then horse: cut-marks on some of the faunal elements suggest that the assemblage is partly, if not completely anthropogenic (the occurrence of wolf and hyena in both layers may indicate partial carnivore involvement). Four additional fossils were recovered from disturbed sediments (Beauval et al. 2002), which are considered here with the Layer 9 and 10 materials as deriving from MIS 4 – although some or all of these additional remains may belong to MIS 3.

As with the Neandertals from Scladina, those from Les Pradelles have enriched values of $\delta^{15}N$ (mean of the five specimens = 10.4 ± 1.5‰), higher than the means of the carnivores sampled (Table 7.4) and again indicating a diet coming from the higher trophic levels (at least with respect to protein). Some question exists as to the quality of the collagen extracted from two of the specimens (Fizet et al. 1995; Bocherens et al. 2005) – one found *in situ* in layer 9 and one coming from the disturbed deposits – and if these two specimens are removed from consideration the mean $\delta^{15}N$ actually *goes up* (11.5 ± 0.1‰). For comparison, a single hyena from layer 9 had a $\delta^{15}N$ value of 10.3‰, and three wolves from the same layer also had a mean value of 10.3‰ (± 1.5). In terms of stable carbon isotopes, the Les Pradelles Neandertals have $\delta^{13}C$ values (mean of −19.2 ± 0.2‰ for the three specimens with reliable collagen), which reflect a predominant consumption of open country herbivores. The fauna in levels 9 and 10 are dominated by reindeer (Costamagno et al. 2006), and the observed $\delta^{13}C$ values would be consistent with heavy – but not exclusive – reliance on this resource. The three wolf fossils analyzed from Les Pradelles have $\delta^{13}C$ values indicative of predation focused on reindeer (the wolves have an average $\delta^{13}C$ value 1‰ higher [less negative] than reindeer, which themselves have $\delta^{13}C$ values 0.4–1‰ higher than other open country ungulates. The difference between wolves and reindeer corresponds with the 1‰ isotopic fractionation between herbivores and carnivores as documented in modern food webs [references in Fizet et al. 1995], while the higher $\delta^{13}C$ values in reindeer relative to other herbivores may be a function of high proportions of lichen in the reindeer diet, which are ^{13}C enriched relative to other C_3 plants growing in the same environment [references in Fizet et al. 1995]). The average $\delta^{13}C$ value for the Neandertals is 0.6‰ lighter (more negative) than the mean value for wolves from the site (−18.6 ± 0.1‰). Based solely on the proportions of the three major ungulates (reindeer, bison and horse) in layers 9 and 10 (based on number of identifiable specimens [NISP] from Fizet et al. 1995) and the average $\delta^{13}C$ values for these three species, we can predict the $\delta^{13}C$ values we should see in Neandertals (taking into account also the 1‰ isotope fractionation between herbivores and carnivores), assuming that they were eating these ungulates in the proportions in which they are represented in the faunal assemblages. This is an heuristic assumption and not one expected to be met in reality, since NISP produces imperfect estimates of actual species abundance at a site (Stiner 1994: 69). In layers 9 and 10 reindeer make up 92.0% and 88.6% of the NISP attributed to the three predominant prey taxa, and the proportion of reindeer along with those of bison (4.7% and 8.5%) and horse (3.3% and 2.9%) predicts $\delta^{13}C$ values of −18.7‰ in the Neandertals from both levels. The observed mean value is 0.5‰ lighter, suggesting a somewhat greater emphasis on animals other than reindeer than might be expected based on faunal representation alone.

Isotopic analyses have thus far been conducted on 11 Neandertals from cold-temperate MIS 3 (Table 7.3). As noted above, these specimens come from across Europe (with one specimen, Okladnikov, coming from western Asia). They may also represent a range of climatic conditions: the 45,200 ± 1100 y BP specimen from Les Rochers-de-Villeneuve (Beauval et al. 2006) may derive from the relatively mild Moershoofd D/O event, or from the cold downturn that directly succeeded it; Saint Césaire, at 36,000 y BP (Mercier et al. 1991) may derive from the post-Hengelo climatic downturn; while the two specimens from Vindija (Higham et al. 2006; Richards et al. 2001) likely derive from the ameliorated conditions of the Denekamp D/O event. As with the specimens from MIS 5 and 4, these Neandertals had $\delta^{15}N$ values enriched relative to most carnivores (Table 7.3 and Table 7.4). The mean value (10.8 ± 1.5‰) encompasses a considerable amount of variation from a low of 7.9‰ in one of the Feldhofer specimens to a high of 12.9‰ in the Siberian Neandertal from Okladnikov. The second highest $\delta^{15}N$ value of 12.6‰ in Engis 2 derives from a 5–6-year-old child, and it is possible this high value reflects a late age of weaning in Neandertals (see Chapter 11.2). Since suckling mammals are feeding at one trophic level higher than their mothers, their ^{15}N values are enriched, but continued bone growth and remodeling after weaning decreases the $\delta^{15}N$ signal until an equilibrium (reflecting the post-weaning or adult diet) is reached 1–2 years later. If the Engis child is accurately aged to 5–6 years at death (Tillier 1983), the elevated $\delta^{15}N$ value would be consistent with weaning around age 4 (Bocherens et al. 2001). If the Engis child is removed from consideration, the mean $\delta^{15}N$ value falls only to 10.6 ± 1.4‰ (n = 10). The lowest of the Neandertal $\delta^{15}N$ values (in the Feldhofer, Vindija, and Zeeland Ridges specimens) still falls among

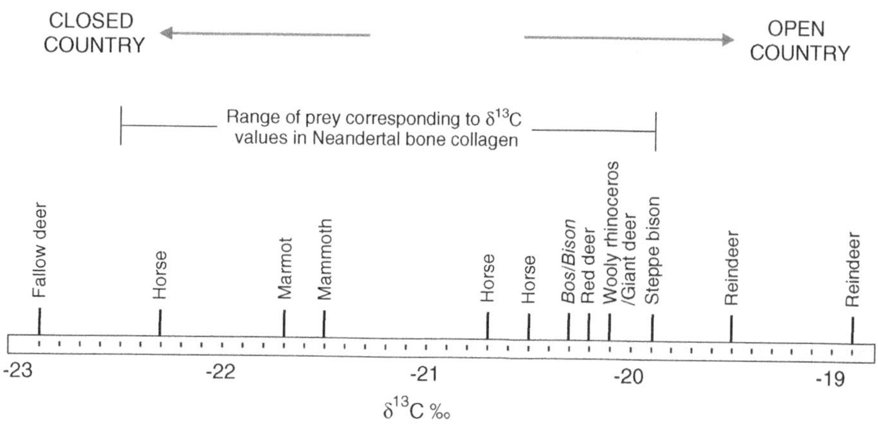

FIGURE 7.7

Sample means of various herbivore species along the $\delta^{13}C$ continuum, based on data from Fizet et al. 1995; Bocherens et al. 1999, and Bocherens et al. 2005. Variation in positions of samples from the same species represents regional and temporal variation in dietary mix of C_3 and C_4 plants (note the position of the sample of horses to the right of the fallow deer, derived from forested interglacial conditions: Bocherens et al. 1999). Dietary range of Neandertals assumes a +1‰ isotope fractionation between primary and secondary consumer.

values observed in carnivores, and still signify a diet in which protein was predominantly obtained from animal and not plant sources.

The MIS 3 Neandertal sample has a mean $\delta^{13}C$ value (-20.1 ± 0.9‰, n = 11) again indicating predominant exploitation of open country prey. Isotope values have not been determined for fauna from all of the sites represented by the MIS 3 sample of Neandertals, but in the context of $\delta^{13}C$ values obtained on the total spectrum of fauna from Feldhofer, Jonzac, Les Pradelles, Scladina, and Vindija it is clear that the relatively isotopically heavy $\delta^{13}C$ values of the Neandertals are most consistent with a predatory focus on grazers (and mixed feeders) from open country habitats (Figure 7.7).

The stable isotope data is consistent (Richards and Trinkaus 2009) and compelling, arguing for a Neandertal niche of substantial carnivory focused on open country prey. That Neandertal $\delta^{15}N$ values are enriched – in every study thus far conducted – relative even to the top-level, obligate carnivores represented in the associated faunas, suggests that the Neandertals themselves were behaving like obligate carnivores (this is not to suggest that Neandertals were physiologically unable to consume plant foods, as are true obligate meat-eaters: MacDonald et al. 1984). Caution is advised in accepting the nitrogen isotope data uncritically, however, as a number of confounding factors may have operated to produce relatively high $\delta^{15}N$ values in the Neandertals. Among the possibilities: (1) Nitrogen isotopes reflect only the trophic level at which protein was acquired, and are mute with respect to the relative contribution of plant versus animal foods to the Neandertal diet. Neandertals may have been obtaining a substantial number of calories from low-protein plant sources, which are not represented in $\delta^{15}N$ values from bone collagen; (2) Neandertals from inland areas were eating more aquatic resources than previously recognized; (3) cooking of meat prior to consumption may have altered its isotopic values; (4) regular bouts of starvation and loss of condition makes the Neandertals appear to have been feeding at a higher trophic level; and (5) Neandertals were preferentially consuming herbivore prey that itself had enriched $\delta^{15}N$ values. We will explore each of these possibilities in turn.

As noted, stable isotope values reflect the source of dietary protein, and the observed $\delta^{15}N$ values in Neandertals do not preclude consumption of plant material. As pointed out by Bocherens et al. (2005), since plant food is poorer in proteins than meat it would be isotopically invisible in humans eating a mixed diet of some plants plus lots of meat, and thus the isotopic data would be consistent with an omnivorous diet with significant plant consumption in Neandertals. The extent to which Neandertals did add plants to the mix, however, remains a question, as the isotopic evidence would suggest that they must have avoided proteinaceous plant foods such as nuts and berries (which comprise much of the humanly-edible primary productivity of temperate woodlands: see Chapter 6.2) and seeds of various grasses (despite microfossil evidence indicating that they did at times eat them: see Section 7.4) for a nearly exclusive focus on protein-poor but carbohydrate-rich plants like geophytes (see Richards and Trinkaus 2009). Another possibility is that the contribution of plant proteins has been masked by other dietary behaviors. Mammoths have high $\delta^{15}N$ values, such that if the Neandertals were consistently eating these proboscidians it might have artificially raised their $\delta^{15}N$ values and masked a certain amount of plant consumption. Recent work on trace elements in bone apatite, however, suggests that this is not the case. While trace element studies have not been widely used with Neandertals, some interesting results have been obtained from the Saint Césaire specimen (Balter et al. 2002; Balter and Simon 2006). Ratios of strontium to calcium (Sr/Ca) and barium to calcium (Br/Ca) in bone mineral become impoverished with increasing trophic position, and thus can be used to infer the trophic level of extinct organisms. In terms of trace element ratios, the Saint Césaire Neandertal appears to have been "quasi-strictly carnivorous," with a Sr/Ca ratio that can only be accounted for by a diet that was 97 wt% meat (Balter et al. 2002: 137). A mixing model using both elemental and isotopic signatures suggests that Saint Césaire's diet was predominantly comprised of bovids and large cervids (58%), followed by horse and rhino (22%), reindeer (13%) and mammoth (7%) (Figure 7.8). Importantly, this dietary reconstruction roughly mirrors that deduced from the zooarcheological remains at the site.

What about the possibility that consumption of fish or other aquatic foods by inland Neandertals produced their elevated $\delta^{15}N$ values? While I did claim that the consumption of fish by Neandertals may be underappreciated, the zooarcheological evidence for Mousterian fish-eating (Section 7.2) comes from a very limited number of sites, and only one of the sites (Vindija) from which Neandertal remains have been isotopically sampled has produced any fish remains (Paunovic and Smith 2002). Oddly, the Vindija Neandertals have among the *lowest* $\delta^{15}N$ values of all the Neandertals thus far studied (Table 7.3), as does the presumably coastal-living Neandertal from Zeeland Ridges. It is also the case that $\delta^{15}N$ enrichment of freshwater ecosystems has been overestimated (perhaps by an undue emphasis on analyses of lacustrine rather than riverine systems), such that freshwater fish consumption would increase $\delta^{15}N$ values to a much lesser degree than previously thought (Drucker and Bocherens 2004). This is supported by the observation that brown bears generally have low $\delta^{15}N$ values (see

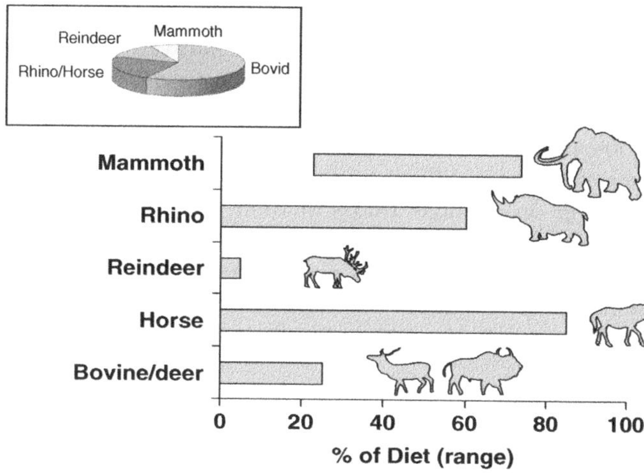

FIGURE 7.8

Range of percentage prey composition in the diet of the Neandertal from Saint Césaire, based on a multisource mixing model applied to ^{15}N and ^{13}C ratios in fossil bone collagen (Bocherens et al. 2005). Inset: a similar approach using trace element ratios (Sr/Ca and Ba/Ca) in the bone mineral of the same specimen produces a different dietary picture (Balter and Simon 2006).

Table 7.4 and Bocherens et al. 2004), despite variable but sometimes high amounts of fish in the diet. Finally, $\delta^{13}C$ values in Neandertal bone collagen have also been argued to indicate that they were deriving their protein from terrestrial herbivores, and eschewing inland aquatic resources (Richards et al. 2001; Richards and Trinkaus 2009). Freshwater fish have more negative $\delta^{13}C$ signatures than do terrestrial animals, and people eating significant quantities of this resource tend to have strongly negative values (e.g., $-23 \pm 1‰$ for Mesolithic people with substantial fish consumption as judged from zooarcheological evidence: Lillie and Richards 2000: but see Drucker and Bocherens 2004). Thus, the overall evidence suggests that the Neandertals so far sampled largely eschewed fish (with the exception of those at Vindija), despite some of the sites (e.g., Jonzac and Les Pradelles) being situated close to rivers or streams, and focused instead on the secondary productivity of terrestrial ecosystems.

The idea that cooking of meat (or plantstuffs) altered the isotopic signature of the food and falsely inflated the Neandertal $\delta^{15}N$ values can likewise be ruled out. Experiments with cooking various foods (reviewed in Drucker and Bocherens 2004) reveal fairly small-scale shifts in isotopic composition, which are negligible relative to the within- and between-group variation seen in bone isotope values. Periodic starvation can probably also be ruled out as an explanation for high $\delta^{15}N$ values in Neandertals. Under conditions of nutritional stress an organism catabolizes its own muscle tissue for protein, and because this tissue is already enriched in $\delta^{15}N$ values relative to the organism's diet, additional enrichment in any newly-formed tissues may occur, making it appear that the organism feeds at a higher trophic level than it actually does (Hobson et al. 1993). Given the evidence for seasonal starvation in Neandertals (reviewed in Chapter 4.3), it is possible that periodic bouts of fasting may explain their elevated stable-nitrogen ratios. However, geese showed only about 1.5‰ enrichment in stable nitrogen in muscle tissue after four weeks of fasting, and it is expected that even less enrichment would be evident in bone collagen, which has a slower isotopic turnover (Hobson et al. 1993). Furthermore, nutritionally-stressed arctic ground squirrels showed no significant difference in blood $\delta^{15}N$ values relative to unstressed squirrels, despite losing on average 18.6% of their body mass over the course of a two-month study period (Ben-David et al. 1999). Thus, it's possible that, like ground squirrels, stable nitrogen enrichment was not a sequelae of starvation and, even if it was, the enrichment in bone collagen would probably not amount to much more than 1‰ or so – which would still place Neandertals amongst meat-eating carnivores in terms of their $\delta^{15}N$ signals.

Finally, as noted above, mammoth bone collagen is enriched in ^{15}N relative to other herbivores (Richards et al. 2000; Bocherens et al. 2001). This likely has to do with greater intake of plants which have symbiotic mycorhiza that take up nitrogen from humus (see Bocherens et al. 2005), and it raises the possibility that the elevated levels of ^{15}N in Neandertal bone collagen may reflect significant consumption of proboscidians. Richards and colleagues (2000) discounted the notion that Neandertal $\delta^{15}N$ values reflect mammoth exploitation over other prey by noting the rarity of proboscidian remains in Mousterian sites (which may hold for southern central European sites dating to MIS 3, but does not necessarily apply to colder intervals or more northerly regions: see Section 7.1). Attempts to infer the proportions of different types of prey in the diet of the Neandertal from Saint Césaire from ^{15}N and ^{13}C values using mathematical mixing models (Drucker and Bocherens 2004; Bocherens et al. 2005) conclude that mammoth *must* have been a significant component of his diet – at a minimum 15–30%, and possibly up to 85–90% – to produce the observed isotopic signature. Similar work using trace elements (see above) also identifies mammoth as a part of Saint Césaire's diet, but a much smaller part (7%: Balter and Simon 2006) (Figure 7.8).

The high $\delta^{15}N$ values observed in Neandertals may also indicate that part of their niche space extended into an even higher trophic level – that of tertiary consumer – through consumption of other carnivores (including, on occasion, one another: see Section 7.6). Predation on other carnivores was not likely to ever have been a major component of subsistence (evidence of humanly-consumed carnivores is rare in Middle Paleolithic sites), but may have been an occasional by-product of technologically-assisted confrontations over carcasses or access to choice shelter locations.

7.6 THE THORNY ISSUE OF CANNIBALISM

No consideration of Neandertal diet would be complete without some discussion of cannibalism. The possibility that Neandertals were eating one another was first raised by Gorjanović-Kramberger (1909) to explain the highly fragmentary nature of the Krapina skeletal assemblage as well as charring of some of the remains. Subsequent workers expanded the argument by highlighting the pattern of bone damage and cutmarks evident on some of the remains, noting for example that the marrow-rich larger long bones were longitudinally split while the marrow-poor small diameter bones were more often intact, and that cranial vault and cervical vertebral remains were consistently damaged, likely as the result of extraction of brain tissue (Tomic-Karovic 1970; Smith 1976; Ullrich 1978). This explanation has not been universally accepted, however, and consideration of taphonomic factors and cut-mark patterns has led others to reject cannibalism as the explanation for the condition of the human fossil assemblage (Trinkaus 1985; Russell 1987b). For Trinkaus (1985), the overall state of preservation of the Krapina material indicated interment shortly after death (whether intentionally by humans or through natural geological processes), and while some of the cut-marks may have been attributable to human activity (whether through cannibalistic defleshing or simply pre-mortuary preparation of bodies), the overall breakage pattern was referable to soil pressure and other post-burial taphonomic processes. Longitudinal fracturing of major long bones, as is common in the Krapina assemblage, also occurs in non-cannibalized assemblages, and may occur with transverse forces produced by sediment compaction. Thus none of the presumed evidence of cannibalism at the site could be taken as *prima facie* evidence of such (Trinkaus 1985). Russell (1987b) attributed the cut-marks on the Krapina bones to human defleshing of partially decomposed remains as part of a mortuary ritual or practice, based on similarities in the position and form of the marks with those found on skeletal remains of recent human groups that ritually deflesh bones preparatory to secondary burial.

Additional human fossil assemblages that may be attributable to cannibalism have been recovered at the early Middle Pleistocene (*Homo heidelbergensis* or *H. antecessor*) site of the Gran Dolina at Atapuerca (Spain: Fernández-Jalvo et al. 1999) and at the Neandertal sites of Vindija, Les Pradelles, Combe Grenal, Macassargues, Zafarraya, Moula-Guercy (references in Defleur et al. 1999) and El Sidrón (Rosas et al. 2006b). Neandertal fossils from a late MIS 5 (120–100 Ka BP) horizon at Moula-Guercy (Ardèche, France) have compelling signs of human modification for consumption. These signs include patterns of cut-marks indicative of filleting of muscle tissue, and anvil striae and conchoidal fractures that indicate directed hammer and anvil percussion of marrow-bearing long bones (Defleur et al. 1999). In addition, the overall similarities in human and non-human faunal skeletal part representations and spatial distributions of bones (reflecting discard patterns) support an economic rather than ritual modification of the human material.

Mounting evidence thus suggests that Neandertals were at times eating one another. That they did so should come as little surprise, since this behavior occurs (or occurred) with some regularity (but in low frequencies) in both archaic and modern humans (see White 1986a, 1986b, 1992b, 2001). Intraspecific killing and cannibalism are also relatively common among carnivores (Fedriani et al. 2000; references in Diedrich and Žák 2006) and, as argued in Chapter 9, Neandertal ecology has to be interpreted in the context of their position on the trophic pyramid (i.e., as carnivores). The unanswered question concerns the conditions under which Neandertals practiced cannibalism, and the economic or social role that the practice played. That Neandertals engaged in cannibalism as a regular economic practice seems unlikely (an economically impossible situation, to which a colleague invoked the metaphor of making a living by taking in one's own washing: Vogel 2001), but they may well have resorted to it during times of starvation.

As noted in Chapter 4.3, temperate and arctic foragers regularly experience calorie shortfalls in late winter and early spring, when body fat and food stores are depleted, plants are dormant, and animals are devoid of fat and in poor nutritional condition.

Predators may elevate hunting and killing frequency during particularly bad winters (Huggard 1993; DelGiudice 1998) or may focus their attention on the more nutritionally rich sex and age classes of prey (Speth 1987). Human hunters often focus on prime age male ungulates at this time of year, since they are likely to be in better condition (having more fat) than pregnant or lactating females, and may eat only the fat rich portions of the carcass of animals in poor conditions (Speth 1987). Marrow fat is one of the last fat stores to be depleted in the body during starvation, and thus even animals in poor condition might be expected to have some nutritional value if the marrow fat can be accessed. The consistent pattern of diaphyseal fracturing of marrow-bearing long bones in the cannibalized Neandertal assemblages suggests that fat may have been a critical resource at the times that cannibalism occurred; that is, that the practitioners were nutritionally stressed and possibly starving. The fracturing of cranial vault bones and cervical vertebrae, likely indicating accessing and consumption of fat-rich brain matter, would also be consistent with this idea. However, Neandertals seem to have been prodigious bone crackers and marrow consumers with all prey and pretty much at all times, and fat may have always been a limiting resource (see Chapter 4.3). Thus it is impossible to tell with current evidence if Neandertal cannibalism was associated with periods of starvation, or if it just occurred periodically (perhaps in the form of exocannibalism, the possible result of violent interactions between neighboring groups) and the victims were processed, with all of the edible portions of marrow bones and brain cases eaten, as were all fauna.

7.7 THE TROPHIC ECOLOGY OF NEANDERTALS

From the multiple lines of evidence reviewed above, a few tentative conclusions about Neandertal trophic ecology can be made. First, diet breadth of Neandertals was narrow in comparison to modern human foragers who occupy (or historically occupied) environments similar to those inhabited by Neandertals, and relative to early modern Europeans of the Upper Paleolithic (Richards et al. 2001; Stiner et al. 1999, 2000; Stiner 2001; Richards and Trinkaus 2009). This is reflected in (1) a seemingly heavy reliance on meat and other animal tissues, regardless of temporal variation in biomic characteristics and the structure of edible resources in the environment (that is, the availability of edible plants); (2) an emphasis on small animal prey (in the few sites in which they are visible in any quantity and clearly attributable to human agency) that was slow-moving and easy to collect, while largely ignoring fast and agile small game such as hares, birds and, with a few notable exceptions, fish; and (3) a predatory emphasis on a few key prey species – horses, bison, aurochs, reindeer and red deer, with wooly rhino and mammoth being variably important additions to this mix. The degree to which plant foods were consumed is difficult to gauge. The isotopic and elemental evidence for European Neandertals suggests that plant items were a minor component of the diet. Dental microwear evidence is more equivocal. When buccal striation patterns (ratios of numbers of horizontal and vertical to total striations) are considered, the European Neandertals look largely carnivorous, consistent with biogeochemical studies. However, in terms of total number of striations alone, European Neandertals look much more omnivorous. Interpreting this latter finding is made difficult by the fact that Neandertals exhibit a much greater number of striations than do modern human foragers whose diets are heavily plant based, and that the total number of striations follow unexpected and difficult-to-explain temporal and geographic patterns (being greater in European than Near Eastern individuals, and being greater during warmer intervals). Occlusal microwear and macrowear reveal ecological patterning, with Neandertals from more productive environments having indicators of greater dietary variation (and likely more plant food consumption), although the overall importance of animal tissues appears to have been high across environments. Macro- and micro-botanical remains from a small number of sites indicate that Neandertals were eating plant foods in some places, but again the quantities and the importance of these items to diet and subsistence remains unclear.

The second conclusion that we can tentatively draw from the various lines of dietary evidence is that the Neandertals' trophic responses to ecological variation appear less flexible than we would expect based on the behavior of modern human hunter-gatherers. The isotopic and elemental data suggest that Neandertal diet was heavily meat-focused under all ecological conditions – from cold glacial episodes in which edible plants would have been scarce to warmer interglacial episodes when plant foods should have been more available. Based on predictions from the responses of modern human foragers to ecological variability (Chapter 6.2), Neandertal dependence on meat should have varied somewhere between a low of about 55% during interglacial intervals to a high of about 90% in some places (evergreen forests) during cold–temperate phases (and close to that in steppe habitats during glacial intervals). Instead, the isotopic and elemental signatures of Neandertal bones suggest dependencies at or above the high-end of the range under all conditions. While the difficulties inherent in trying to infer percentage consumption of plant material from isotopic data was discussed in Section 7.5, it does seem unlikely that Neandertals could have derived as much as 36–45% (based on the occlusal microwear similarity of the mixed-habitat Neandertal with the Chumash, and on predictions for interglacial conditions from global patterns in modern foragers) while somehow managing to avoid proteanceous plants like nuts and berries. Reduced dietary flexibility (relative to modern human foragers) and a greater-than-expected dietary dependency on large mammals may denote in the Neandertals a lack of the technological or organizational means to respond to climatic and ecological variation in the same way that modern humans do (see Kuhn 1995, 2011), or it may reflect nutritional constraints on the diet having to do with the energetic and macronutritional (especially regarding fats important to brain development and maintenance) requirements of producing and maintaining large bodies and big brains (see Chapter 4.3). Clearly, the combined dietary indicators show that there was *some* flexibility to Neandertal dietary ecology (as would be expected of any omnivorous primate). Predictions for Mediterranean habitats during cold–temperate events, in which about 30% of the diet should derive from plants, may have been met, if the plant data from Kebara Cave is indicative of a significant plant component of the diet of the Neandertals there (and further inland as well, given the plant microfossil data from Iraq). Unfortunately, isotopic data have not yet been obtained for Near Eastern Neandertals (despite efforts to do so: Ambrose 1998b). While the macrobotanical evidence is currently too scant to allow for confident generalizations, the preponderance of dietary evidence indicates that outside of the Levant the Neandertal menu was a bit more rigid. Thus, the take-home message is that Neandertal dietary responses to ecological variation weren't inflexible, they just seem to have been less flexible than those of modern human foragers.

This review of the available lines of evidence for Neandertal diet allows us to begin to circumscribe the Neandertal niche (at a minimum for those in Europe, if not for those in the Near East), at least along a couple of dimensions: Neandertals were moderate body-sized omnivores (generalists) with a dietary focus on moderate-to-large bodied, open country herbivores.

NOTES

1. There has been some debate about the existence of a distinct "Neandertal niche" (see, for example, Binford 1985; Soffer 1989a; versus papers in Burke 2000a), but the concept as applied in that context refers to aspects of overall strategy and organization. My intent here is not to address this question, but rather to focus on delineating the staples in the diet, or what might be considered "niche breadth."
2. Smaller animals have higher rates of production thanks to negative allometric scaling of metabolic rate to body mass (such that smaller animals have proportionally higher metabolic rates). A 590 kg cow will consume a ton (English, equal to 909 kg) of hay in 120 days, with an average weight gain of 0.91 kg d^{-1}. The same mass of rabbits (equalling about 300 bunnies) will consume the same amount of hay in 30 days. Both the rabbits and the cow will add the same amount of new cells (109.2 kg) with roughly the same efficiency (both losing about 2,400,000 kcal in heat) from the ingested food, but the rabbits will do so in one-fourth the time (Kleiber 1961).
3. Note, however, that the makers of the Châtelperronian (possibly Neandertals) at Arcy-sur-Cure appear to have been using grind stones to process ochre, and possibly vegetal matter, some 32 Ka BP (de Beaune 2003).
4. The standard reference for $\delta^{13}C$ is based on Cretaceous fossils of *Belemnitella americana* from the PeeDee formation in South Carolina. *B. americana* has a higher $^{13}C/^{12}C$ ratio than most organic substances, and for reference purposes it is assigned a $\delta^{13}C = 0$, which gives most organic samples negative $\delta^{13}C$ values.
5. While it may seem oxymoronic to speak of herbivorous carnivores, it should be kept in mind that "carnivore" is a taxonomic label (designating members of the order Carnivora) and not a dietary categorization. Thus not all carnivores are carnivorous: many (e.g., canids, ursids) are omnivorous, and some (e.g., pandas) are herbivorous. The term "herbivore," on the other hand, *is* a dietary categorization.

CHAPTER EIGHT

Red in Tooth and Claw: Neandertals as Predators

In the last chapter we defined the Neandertal niche as one of omnivory, but with a focus on consumption of moderate-to-large bodied, open country herbivores. This naturally leads to the question of how the Neandertals were obtaining these herbivores – whether by predation, confrontational scavenging, passive scavenging, or some mix of these methods. At one time, much attention was devoted to the question of scavenging as the primary means of meat acquisition by Neandertals, and whether they even had the ability to actively hunt medium- and large-bodied prey (Binford 1985, 1989; Chase 1988, 1989; Stiner 1991c, 1993, 1994; Marean 1998; Speth and Tchernov 1998). Active predation is pretty much taken as a given in recent works (e.g., Hoffecker and Cleghorn 2000; Patou-Mathis 2000; Richards et al. 2000; Costamagno et al. 2006; papers in Burke 2004; Yeshurun et al. 2007), and the role of scavenging in Neandertal subsistence has largely faded from discussion. Scavenging is, however, a component of the calorie-capturing strategies of all large-bodied mammalian carnivores, and it no doubt was a part of Neandertal subsistence – their ability to effectively hunt a range

Thin on the Ground: Neandertal Biology, Archeology, and Ecology, First Edition. Steven Emilio Churchill.

of prey notwithstanding. The debate about hunting versus scavenging is reviewed in Section 8.4, and the nature of scavenging as one of several competitive interactions that occur within carnivore guilds will be explored in Chapter 9. But the focus of the present chapter is on active predation – the means by which Neandertals procured their prey. Mousterian hunting technology was reviewed in Chapter 3.4, so in this chapter we will focus on the zooarcheological and human fossil evidence, to delineate as best we can Neandertal hunting strategies and tactics. This is central to efforts to better understand the energetic costs of foraging, some of the limitations that operated on rates of energy capture, and the overall energetics of Mousterian subsistence.

8.1 NEANDERTAL MORPHOLOGY AND PREDATION

The archeological evidence reviewed in Chapter 3.4 suggests that the major components of the Neandertal hunting arsenal were relatively large, heavy spears, sometimes tipped with stone armatures. As argued in Chapter 3, these appear to have been used as close-range weapons. Aspects of Neandertal upper limb and upper body morphology are also consistent with the idea that predation regularly involved the forceful use of thrusting spears at close range. Before reviewing this evidence, it might be helpful to review why we would expect habitual behavior patterns – such as those involved with the pursuit and capture of game – to be reflected in the skeletal anatomy of the actors themselves.

First, we can assume that natural selection has operated to optimize morphology for important behaviors (such as hunting) performed by a species, within the limitations imposed by genetic, developmental, and functional constraints (Sommerhoff 1950; Smith and Savage 1956; Rudwick 1964; Williams 1966: but see Lauder 1995 for cautions against uncritically adopting this thinking). Trophic apparatus and locomotor systems are central to an organism's fitness, and should be under strong selection for effective design. For predators like lions and Neandertals that grapple with prey larger than their own body size, aspects of postcranial morphology become, in effect, extensions of the trophic apparatus of the animal – for example, highly mobile wrists, digits that spread upon protraction, and sharp, strongly curved claws are all important components of prey capture morphology in lions (Bryant et al. 1996; Russell and Bryant 2001). We might expect, then, that some aspects of Neandertal postcranial morphology were shaped by selection for effective performance in the context of their usual modes of prey acquisition.

Second, we know from a wealth of experimental and comparative studies that living bone tissue is developmentally plastic, that is, it has a certain capacity to adapt to its biomechanical environment (e.g., Goodship et al. 1979; Lanyon et al. 1979; Woo et al. 1981; Lanyon and Rubin 1984). Given this capacity of the mammalian skeleton to dynamically alter, over the course of weeks or months, the material properties and the shape of its bones in response to changes in mechanical loads, biomechanical analysis of the amount and distribution of bone tissue allows us to infer something about the magnitudes and orientations of the loads that were habitually placed upon the skeleton. From this we can make inferences about the behavior patterns that produced those loads (see reviews in Ruff 1992, 2000a). Note, however, that the results of experimental studies, mainly using strain gauge data collected from bones loaded *in vivo*, advocate caution in using this approach, indicating both that bone geometry may not faithfully indicate the orientation of the predominant loads on the bone (Demes et al. 1998, 2001; Lieberman et al. 2003), and that adult mammals may show little ability to adaptively alter bone shape (Pearson and Lieberman 2004). While it is true that morphological response to mechanical strains is reduced in adult bone relative to that of juveniles, and that this reduction may be evident in both the rate and magnitude of response, it is nevertheless true that adult bone is still responsive to habitual loading. For example, postmenopausal women who followed a bilaterally asymmetrical weight lifting program for one year showed a 2.4% increase in bone mineral density in the lifting arm and a 1.4% loss in the non-lifting arm (Kerr et al. 1996). Likewise, a sample of people who started playing tennis as adults (at an average age of 34) still showed three times greater playing arm/non-playing arm asymmetry in bone mineral content than control subjects (Kannus et al. 1995). When bone mineral content was combined with the geometric distribution of cortical bone in the upper limbs to create an index of biomechanical strength in a subsample of these tennis players, individuals who started playing as adults (starting at a mean age of 26) had an average of 11% playing side/non-playing side asymmetry, compared to 4% in a nonathletic control group (Kontulainen et al. 2002). These and other studies (reviewed in Ruff et al. 2006a) indicate that variation in the density, amount, and distribution of cortical bone in adult long bone diaphyses reflects variation in the magnitude, and possibly frequency (see Chapter 10.3), of mechanical loads placed on the skeleton during adulthood, and thus can be interpreted in terms of adult behavior patterns. The issue of how bone geometry relates to the orientation of forces that operated on the skeleton is a bit more complex, and while there does seem to be a general relationship between bone shape and load orientation, experimental studies suggest caution in trying to directly relate the two. As noted by Ruff et al. (2006a), behavioral inferences derived from bone structure alone should be restricted to comparisons of the same element between animals with similar basic mechanical designs (thus comparisons of the geometric properties of a given bone between Neandertals and modern humans is warranted). Also, it is wise to work back and forth between comparative biomechanical and experimental studies as a way of improving our powers of drawing inferences from skeletal morphology (Churchill and Schmitt 2002).

With that said, it is also important to note that different aspects of the skeletal system vary in their ability to adaptively model (change shape) and remodel (change material properties) in response to changes in levels and patterns of stress. The lengths of long bones and the size and morphology of their articular surfaces show limited adaptability, and are generally only responsive to their mechanical environment prior to skeletal maturity (Ruff et al. 1991; Ruff and Runestad 1992; Trinkaus et al. 1994; Lieberman et al. 2001). Long bone diaphyseal cortical tissue, on the other hand, appears to be more plastic throughout life (see above). Thus between-group differences in mean measures of joint size or shape are more likely to reflect the evolutionary-selective history of the group (that is, the genetically-encoded adaptations, favored by natural selection, that promote effective functioning of bone-muscle mechanical systems),[1] as well as perhaps the habitual behavior patterns of juveniles, while studies of long bone shaft morphology provide a finer-grained picture of the habitual behavior patterns of individuals. In both aspects of morphology (less- vs. more-plastic), it is assumed that the upper limb, as an organ of prehension and manipulation, will best reflect genetic and developmentally-plastic adaptation to the mechanical forces associated with the production and use of technology (including hunting technology) and the subduing and processing of prey. In this former category (genetically-encoded features or features that reflect, to a degree, juvenile behavior), we can consider – as relevant to the issue of Neandertal predatory behaviors – scapular glenoid fossa shape, ulnar trochlear notch orientation, aspects of leverage about the shoulder and elbow, and possibly the morphology of the pubic ramus. In the latter category (the more plastic features that better reflect the behavior of the individual) would go aspects of humeral diaphyseal shape, bilateral asymmetry in humeral strength, and aspects of upper limb muscle scar morphology that may reflect something of hunting behavior. We'll look at each of these characters in turn, beginning with the more genetically-canalized features.

8.1.1 Scapular Glenoid Fossa

Neandertals are characterized by scapular glenoid fossae that are narrow relative to their height, which is reflected in a low glenoid index (articular breadth/height: Figure 8.1). While the absolute area of the glenoid fossa is large (thus the joint was able to withstand relatively high magnitude joint reaction forces), the shape of the articular surface has been interpreted as reflecting a joint that was poorly-adapted to withstand dorsally- and ventrally-directed forces at the shoulder which occur during throwing (Churchill and Trinkaus 1990). Throwing is one of the few motions that engenders shoulder joint loads in the transverse plane, brought about by the combined abduction and external rotation of the humerus that occurs during the late cocking phase of throwing (which produces an obligatory posterior translation of the humeral head on the glenoid: Howell et al. 1988). Furthermore, the forces that develop at the shoulder joint during throwing can be quite high, going from 400 N of posterior shear (combined with 1000 N compression) at the beginning of a throw to 75 N of anterior shear (combined with 400 N compression) at the end (Meister 2000).

Although the relative width of the glenoid fossa is variable among and between samples, most modern humans tend to have relatively wide glenoid fossae (Figure 8.1). The Mousterian-associated early modern humans from Skhul and Qafzeh appear to be the exception and are more similar to Neandertals, as judged by two specimens (Figure 8.1: there are also some early modern European specimens, such as Muierii 1 from Romania [Trinkaus 2008a] which have relatively narrow glenoid fossa, as do a few individuals in many of the recent human samples). In previous work, I interpreted the relative width of the glenoid fossa as indicating something about the selective history of Neandertals and modern humans with respect to throwing-based hunting technology (Churchill and Trinkaus 1990; Churchill and Rhodes 2009). The pattern in modern humans was argued to reflect a shared selective history of forceful throwing while hunting with long-range projectile weaponry, in which individuals with wider glenoid fossae were more resistant to degenerative joint disease (osteoarthritis) at the shoulder, and thus able to provision dependents over a greater portion of their life span. Assuming the pain and limited shoulder mobility associated with glenoid fossa osteoarthritis was selectively disadvantageous, one might predict relatively rapid evolution of glenoid fossa shape following on the development of (or first real dependence upon) spearthrower technology and the increased frequency of forceful throwing it entailed. The narrow glenoid fossae in Neandertals and Mousterian-associated modern

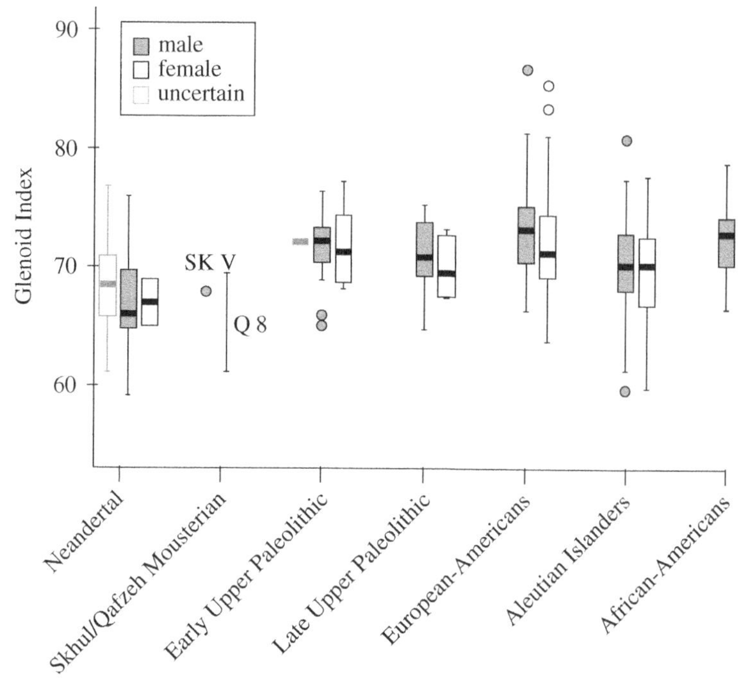

FIGURE 8.1

Median values of the glenoid index (100 * articular breadth/length) in Neandertals (6 males, 2 females, and 8 individuals of unknown sex), Mousterian-associated early modern humans from Skhul and Qafzeh Caves (2 males), early modern Europeans from the Early Upper Paleolithic (>20 Ka BP: 9 males, 6 females, 1 indeterminate) and Late Upper Paleolithic (20–9 Ka BP: 9 males, 4 females), and recent modern human European-Americans (41 males, 42 females), Aleutian Islanders (25 males, 22 females) and African-Americans (25 males, 25 females). For all but the Skhul/Qafzeh sample, the transverse line represents the median value, the box is the interquartile range, the whiskers represent the spread and the circles are outliers. For the two individuals from the Levantine Mousterian, the index for Skhul V is represented by a shaded circle, while the range of estimated values for Qafzeh 8 is represented by whiskers. Data from Churchill and Rhodes 2009.

humans was argued to indicate that forceful throwing was not an important component of Mousterian subsistence systems – at least, not important enough to have brought about an adaptive widening of the shoulder joint (Churchill and Trinkaus 1990; Churchill and Rhodes 2009). Subsequent two-dimensional morphometric analysis of glenoid fossa shape (di Vincenzo et al. 2012) supports the contention that Neandertals have narrow joint surfaces relative to modern humans. However, when seen in phylogenetic context, the morphology in Neandertals appears to represent one point along an evolutionary continuum from very narrow glenoid fossa in australopiths to very wide ones in modern humans (di Vincenzo et al. 2012), and it is very unlikely that this continuum is explained by a gradual increase in the importance of throwing (however, if recent claims that *H. erectus* relied on throwing in hunting [Roach et al. 2013] are correct, then perhaps the observed evolutionary pattern is actually attributable to the evolution of human throwing ability).

The relative width of the glenoid, then, appears to reflect evolutionary-developmental trends in growth rate rather than adaptive morphological responses to mechanical stress at the shoulder joint (seen in this light, the pattern of variation in glenoid fossa shape in Neandertals may reflect varying levels of gene flow from modern humans across space and time: see Chapter 2.6). But this same analysis suggests that another aspect of glenoid fossa shape, the relative development of the upper, or coracoid, portion of the joint surface, may reflect developmentally plastic responses to habitual throwing behavior prior to skeletal maturity (di Vincenzo et al. 2012). Neandertals have a greater dorsoventral development to the coracoid portion of the joint surface, resulting in a more ovoid glenoid fossa that tends to contrast with the more piriform shape seen in modern humans. The piriform shape has been suggested to increase joint stability during lifting and/or to increase the range of humeral rotation with the arm abducted (Bateman 1978; Fahlström 1981; Aiello and Dean 1990; Miles 1997). Throwing, of course, requires humeral rotation with an abducted arm. The glenoid fossae of Upper Paleolithic modern humans and Tierra del Fueguians are more strongly piriform than those of sedentized modern humans from agricultural or industrial economies, and more so in the right than the left limb, which might reflect the employment of throwing-based weapon technology in certain modern human groups but not Neandertals: (di Vincenzo et al. 2012).

8.1.2 Elbow Joint Morphology

Neandertals also differ from modern humans (including Levantine Mousterian and European Upper Paleolithic early modern humans) in two important aspects of elbow morphology. First, Neandertal ulnae have more anteriorly directed trochlear notches (Trinkaus 1983b; Churchill et al. 1996a), resulting from greater dorsoventral height of the olecranon process in Neandertals (Figure 8.2). The more anterior orientation of the joint has been interpreted as reflecting adaptation to habitual loading regimes in which peak loads were incurred with the elbow in partial flexion, as would occur during forceful use of a thrusting spear (Schmitt et al. 2003). The anterior orientation of the trochlea would maximize the effective joint surface area perpendicular to the joint reaction force with the forearm in flexion, increasing the ability of the joint to resist proximodistal shear stress, whereas the more proximoanterior orientation of recent human ulnar trochlear notches would more effectively resist such forces in extension (Trinkaus and Churchill 1988).

Second, Neandertal ulnae have proximodistally longer olecranon processes (Churchill and Rhodes 2006). This feature resulted in greater leverage of the triceps muscle in Neandertals, which would have increased their ability to forcefully extend the forearm at the elbow, but would have substantially decreased their ability to generate angular acceleration of the hand (as needed to accelerate a projectile for a long-distance throw). Even during the earlier stages of throwing, in which humeral internal rotation is providing most of the acceleration (Tullos and King 1973), the Neandertals

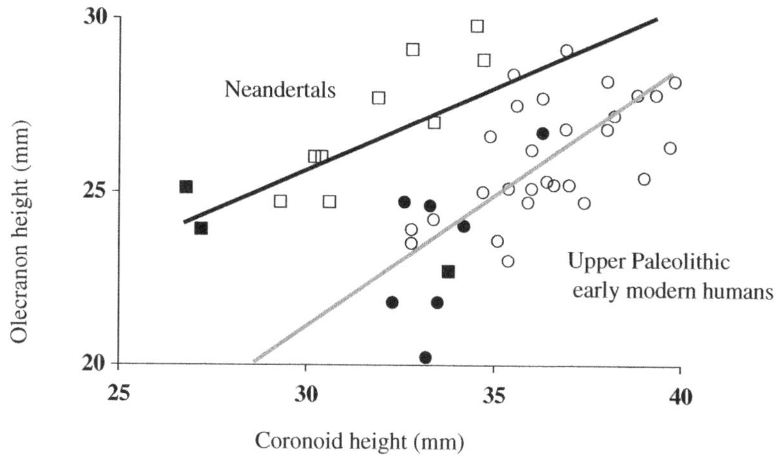

FIGURE 8.2

Ulnar trochlear notch orientation as reflected in the dorsoventral height of the olecranon process relative to that of the coronoid process. Anteriorly-directed trochlear notches have relatively high olecranon processes, as seen in the Neandertals (9 males, □; 3 females, ■); relatively shorter olecranon processes result in anteroproximally oriented trochlear notches, as seen in early modern humans from the European Upper Paleolithic (28 males, ○; 7 females, ●). Ordinary least squares regression lines presented for combined sex Neandertal (__: $y = 0.475x + 11.418$, $r = 0.5721$) and Upper Paleolithic (__: $y = 0.6658 + 1.5791$, $r = 0.6905$) samples. Data from Churchill 1994a.

would have been disadvantaged by their shorter and presumably more massive forearms (which would have simultaneously reduced the speed advantage[2] of the musculoskeletal lever system while increasing the inertial resistance to acceleration of the hand). Amongst modern humans, forearm extension is predominately either passive (that is, largely accomplished by gravity) or rapid (as in throwing or hammering), and mechanical advantage has been sacrificed for speed advantage in extension. In Neandertals, strength in forearm extension was likely part of the overall package of upper body strength required for coping with Ice Age environments at a Mousterian level of technological development: subduing and butchering large prey; breaking branches and saplings for firewood or woodworking; scraping hides and planing wood; and otherwise wrestling with their physical world. Forceful extension at the elbow may also have played an especially important role in close-range hunting with hand-delivered spears. Experimental work on thrusting spear use (Schmitt et al. 2003) suggests that the initial penetration of the spear involves force generated by arm and forearm flexion in the dominant limb – the hunter begins with a relatively extended arm and forearm and uses the powerful humeral and forearm flexors to drive the spear into the target (this holds for underhand thrusting only: see Kortlandt 2002 vs. Churchill 2002). It is likely that, following penetration of the spear, further penetration is accomplished by forceful extension at the elbow, combined with further flexion of the humerus. It may also be that forceful extension of the forearm is necessary for withdrawing the spear for additional stabs (see Churchill 2002). Neandertal arms appear to have been optimized more for the production of force than speed: the greater mechanical advantage of their forearm flexors and extensors would have increased the force with which they could have thrust and withdrawn spears or otherwise grappled with prey (Churchill 2002; Churchill and Rhodes 2006), but would have diminished their ability to accelerate missiles of any sort. However, there are experimental data that suggest that angular acceleration at the elbow contributes relatively little to the kinetic energy imparted to a thrown spear (Maki 2012), and that most of the energy comes from motion at the shoulder. This might mean that the long olecranon processes of Neandertals were effectively unimportant with respect to their throwing ability, although their overall proportions (that is, their having had relatively short and massive arms) would still seem to place them at a disadvantage in accelerating missles.

8.1.3 Superior Pubic Ramus Length

It has also been suggested that the elongated superior pubic rami characteristic of Neandertals (Trinkaus 1976, 1984) played a role in generating and resisting forces in the torso during close-range hunting with thrusting spears (Black 1999). The superior pubic ramus acts as a link between the abdominal muscles and the adductors of the medial compartment of the thigh. During thrusting spear use, forceful rotation of the trunk – generated by the abdominal obliques – may have contributed to the force of the delivery of the spear, and trunk rotation may have also been necessary to resist forces imparted, via the spear, from the defensive behaviors of the prey. The adductor muscles would have played an important role in stabilizing the trunk and transmitting reaction forces to the substrate (Black 1999). Elongation of the pubic ramus increases the area of origination of the adductor muscles and increases the space in the medial compartment of the thigh for their muscle bellies. The concerted firing of the abdominal muscles above and the adductor muscles below would have loaded the pubic ramus in tension while largely protecting it from bending stresses, which may explain in part the relative gracility of this aspect of Neandertal morphology.

Before discussing the more plastic aspects of Neandertal skeletal morphology, I'll note that the overall massiveness and muscularity of Neandertals is consistent with subsistence behaviors that involved regular, close contact with large prey (Churchill and Rhodes 2006). As noted by Vogel (2006: 201), "thrusting spears put gravity to use, working best with heavy bodies that leaned forward over well-planted feet to get sufficient purchase on the ground." Larger, more muscular bodies would have also been better able to take advantage of gravity (assuming an underhand thrusting posture: Figure 8.3) in withdrawing a spear in preparation for a subsequent strike (Churchill 2002).

8.1.4 Humeral Diaphyseal Cross-sectional Geometry

As mentioned above, the amounts and distributions of cortical bone in long bone diaphyses provide a record of the consistent and usual mechanical loads that were experienced by those bones. Geometric analyses of Neandertal humeral cross-sections reveal that (1) their upper limbs were adapted to withstand high-magnitude forces

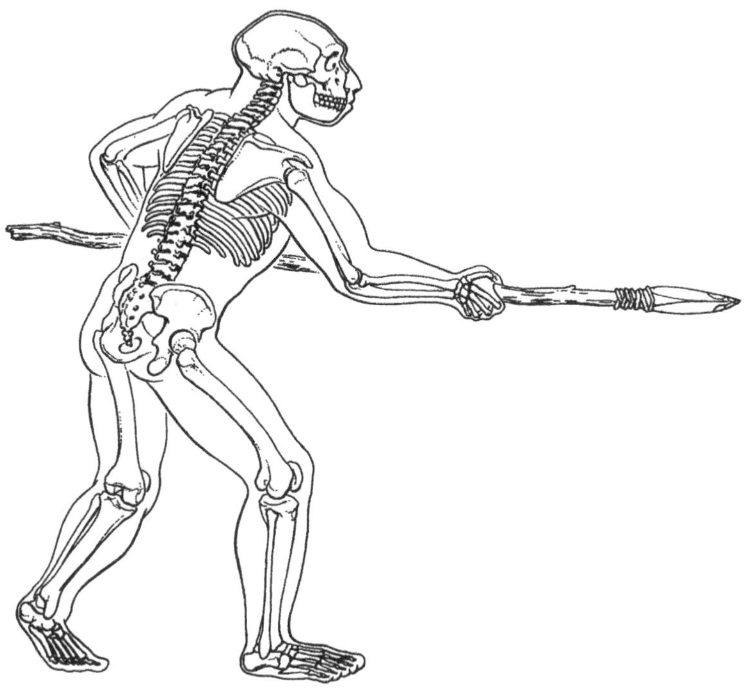

FIGURE 8.3

Neandertal using a thrusting spear. Drawing by Stephen Nash (from Churchill 1998).

relative to most modern humans (that is, their humeri were biomechanically strong); (2) their humeri have their greatest bending strength in the anteroposterior plane, suggesting that they regularly engaged in activities that engendered high-magnitude bending forces in this plane; (3) they had levels of bilateral strength asymmetry (in most cases with the right limb being stronger) well above the levels seen among most recent modern humans.

Consideration of the mechanical strength of bones must be done in the context of overall body size, since the magnitude of the forces experienced by the bone scale with body size (both because larger individuals have more massive body parts and moving them creates greater gravitational and inertial forces, and because longer bones have longer bending moment arms). Measures of the structural contribution of cortical bone tissue to mechanical strength, such as cortical areas (a general measure of the resistance of a bone to compressive forces at a particular point along its length) and second moments of inertia (measures of resistance to bending forces at a particular point along a bone), must be scaled to some mechanically-relevant measure of body size to produce measures of relative strength, or what is commonly referred to as "robusticity" (*sensu* Ruff et al. 1993). Measures of robusticity (also sometimes called "residual strength"), then, should better reflect loads related to activity.[3] When compared to samples of modern humans, it is apparent that Neandertals had robust humeri (Figure 8.4), at least in their right limbs. Figure 8.4 presents sample means for the body size-standardized

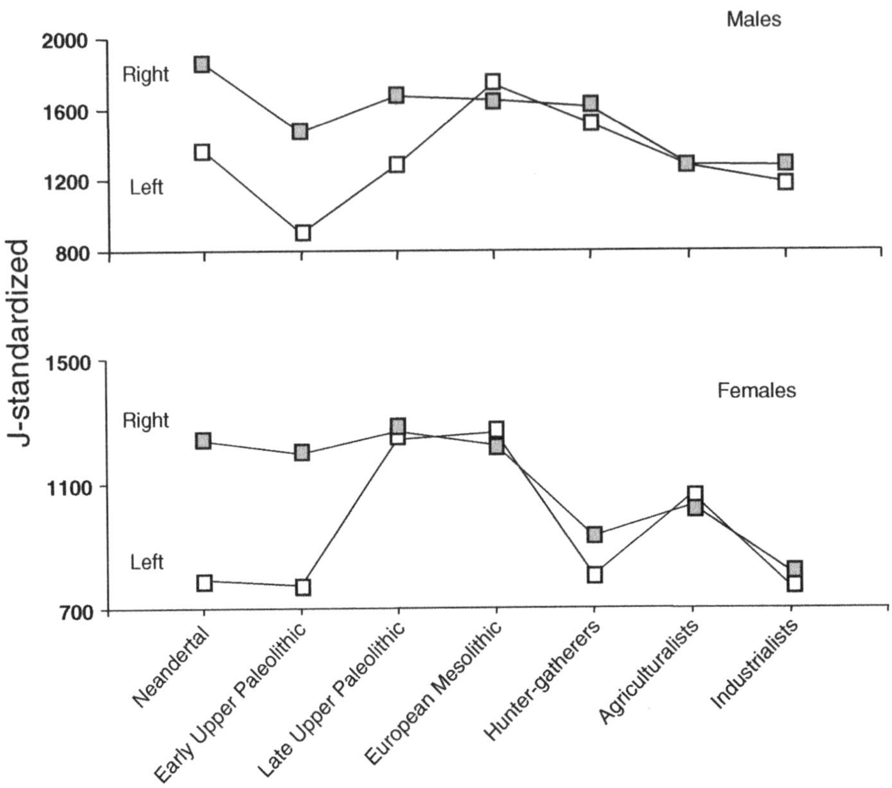

FIGURE 8.4

Average polar moment of inertia (J) standardized to body size in fossil and recent human samples. Recent human samples include hunter-gatherers and pre-European contact agriculturalists from the Georgia Coast (data from Fresia et al. 1990) and autopsy samples of European–Americans ("industrialized": data from Churchill 1994a). Data for European Mesolithic samples is unpublished (collected by the author), all other fossil data from Churchill et al. 1996b, Body size standardization was done using humeral articular length raised to the fourth power (J-standarized = $10^9 * J/HAL^4$).

polar moment of inertia (J), which is a measure of the contribution of bone geometry to generalized bending and torsional strength. Among males (for whom sample sizes are larger), Neandertals have an average right side value that is larger than those of any of the modern human samples (note, however, that Figure 8.4 only presents data for terrestrial hunter-gatherers from the Georgia Coast. Maritime and coastal foragers from Alaska [Aleuts] and Japan [Jomon] have values comparable to the Neandertals: see Churchill 1994a; Churchill et al. 1996b), and a mean left side value that is exceeded only by the Mesolithic and recent hunter-gatherer samples. Similar patterns obtain for other measures of bone strength (Churchill 1994a). For females, the fossil sample sizes are very small but, based on the available data, Neandertal females appear to also have had robust bones in their right arms – comparable to those of females from

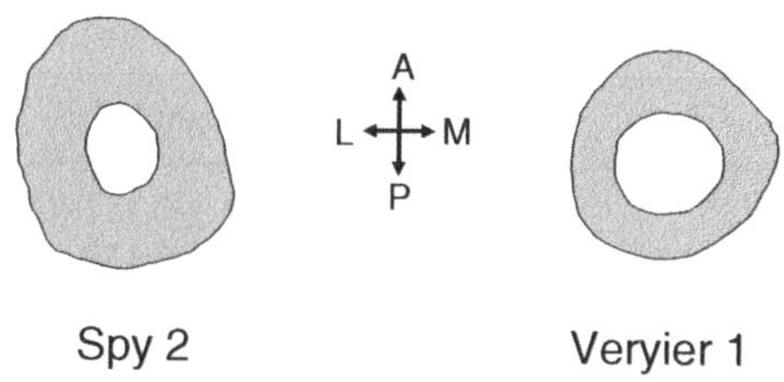

FIGURE 8.5

Representative mid-distal diaphyseal cross-sections of the right humeri of a male Neandertal (Spy 2, left) and a male early modern European from the Late Upper Paleolithic (Veryier 1, right). The sections were reconstructed from external contour molds and radiographs at 35% of bone length from the distal end, are pictured from the distal perspective, and are drawn to the same scale. Note that, in addition to having a greater amount of cortical tissue, the cross-section of the Neandertal humeral shaft is elongated in the anteroposterior direction. A: anterior; P: posterior; M: medial; L: lateral.

the Upper Paleolithic and Mesolithic and greater than seen among any of the recent female samples (Figure 8.4). The mean value for Neandertal female left humeri, on the other hand (or literally, on the other arm), is among the lowest observed in the female samples. The overall pattern suggests that Neandertals were engaged in forceful activities with their upper limbs, at least relative to most modern human groups, but that there was a substantial amount of behavioral asymmetry involved (their right limbs were much more active and engendered higher stresses). It is important to note that the mean values depicted in Figure 8.4 are based on the means of all right and all left humeri. For the females, the means of the two sides are largely represented by different individuals, and thus the picture of bilateral asymmetry is not very trustworthy. Thus, further consideration of asymmetry (below) will be restricted to males, where we can examine average patterns of asymmetry within individuals.

Neandertal humeri also tend to have a greater amount of cortical tissue distributed in the anteroposterior plane (Figure 8.5), and thus appear to be well-adapted to resisting bending stresses in that plane. Second moments of inertia (I) provide measures of the contribution of bone geometry to bending strength in a given plane, and the ratio of the second moment of inertia about the mediolateral axis (I_x: a measure of diaphyseal rigidity to bending in the anteroposterior plane) relative to that about the anteroposterior axis (I_y: a measure of diaphyseal rigidity to bending in the mediolateral plane) serves as a way of comparing the relative strength of a bone to bending in two cardinal planes. Neandertals have high I_x/I_y ratios (Table 8.1), indicative of humeri that are better able to withstand bending stresses in the anteroposterior than mediolateral plane. Experimental work reveals that spear thrusting engenders high magnitude bending moments (with an average magnitude of 733 Nm, but in some cases as high as 1118 Nm) in the anteroposterior plane of the midshaft humerus (Schmitt et al. 2003: in this case, the spears were thrust into a stationary and padded, yet rigid, target, but the forces involved are likely similar to those required to penetrate prey; see Shea et al. 2001; Churchill et al. 2009). Males of the European early Upper Paleolithic also have high I_x/I_y ratios (Table 8.1), which led to the suggestion that they also relied on

TABLE 8.1 ■ Humeral mid-distal diaphyseal shape[a] in fossil and recent males (mean, SD, n)

	Right I_x/I_y	Left I_x/I_y
Neandertals	1.274 ± 0.152 (6)	1.301 ± 0.156 (5)
Early Upper Paleolithic modern humans	1.310 ± 0.178 (5)	1.217 ± 0.112 (6)
Late Upper Paleolithic modern humans	1.160 ± 0.133 (9)	1.121 ± 0.189 (7)
Georgia Coast foragers	1.001 (6)[b]	1.003 (6)[b]
Aleuts (maritime foragers)	1.037 ± 0.135 (19)	1.016 ± 0.780 (16)
Jomon (coastal foragers)	0.897 ± 0.127 (10)	0.891 ± 0.132 (10)
Native Americans (Georgia Coast agriculturalists)	1.125 (11)[b]	1.049 (11)[b]
Native Americans (Pueblo agriculturalists)	0.885 ± 0.152 (16)	0.871 ± 0.135 (19)
European–American (autopsy sample)	1.209 ± 0.105 (19)	1.189 ± 0.171 (20)
African–American (autopsy sample)	1.156 ± 0.150 (14)	1.151 ± 0.157

[a] Ratio of the second moment of inertia (I: mm^4; a measure of diaphyseal rigidity to bending moments in a single plane) in the anteroposterior plane (I_x) to that of the mediolateral plane (I_y) in the humeral diaphysis (at 35% of length from the distal end). Unless otherwise noted, all data from Churchill et al. 1996b.
[b] Ratio calculated using mean I_x and I_y values: data from Fresia et al. 1990.

close-range hunting with thrusting spears (Churchill et al. 1996b). However, consideration of the pattern of right/left asymmetry in diaphyseal cross-sectional shape, as evident in Table 8.1, suggests caution in interpreting the high I_x/I_y ratios in Neandertals. Comparison of right and left side I_x/I_y ratios reveal a small degree of bilateral shape asymmetry within each group, but by and large the mean left side values tend to be very similar to the mean right side values within each group. If humeral diaphyseal cross-sectional shapes were faithfully reflecting the usual mechanical environment of the bones of each limb, we would expect a greater degree of observed asymmetry, since forceful unimanual activities (like throwing and hammering) would be expected to produce very different patterns of stress in the two limbs. The similarities apparent in left and right side cross-sectional shape suggest either that genetically-controlled developmental patterns overprint any activity signal (in which case the cross-sectional shape seen in Neandertals may still reflect adaptation to high levels of humeral bending in the anteroposterior plane, but they reflect genetic adaptation rather than developmentally plastic responses), or, as has been suggested (Demes et al. 1998, 2001; Lieberman et al. 2003), bone cross-sectional shape does not faithfully reflect loading histories. While more work needs to be done to clarify this issue, at present we can only say that humeral diaphyseal shape in Neandertals is consistent with the hypothesis of habitual use of thrusting spears.

Neandertals also exhibit a high degree of humeral bilateral strength asymmetry (Table 8.2: see also Trinkaus et al. 1994), suggesting an emphasis on highly stressful unimanual activities. Throwing is a unimanual activity, and forceful throwing – especially of a relatively heavy projectile – generates very high magnitude torsional moments in the dominant (throwing-side) limb. The throwing-side humeri of professional baseball pitchers, who are only throwing a 90 g missile (but which they are accelerating to 40 m s^{-1} or more), experience on average 92 ± 16 Nm of axial torque, which is estimated to be about 48% of the theoretical maximum torsional strength of

TABLE 8.2 ■ Humeral mid-distal diaphyseal strength asymmetry[a] in fossil and recent males (median, quartile range, range and sample n)

	% Asymmetry CA[b]	% Asymmetry J[c]
Neandertals	16.5	53.1
	12.8–33.2	35.7–77.5
	11.5–47.4 (4)	21.9–98.5 (4)
Early Upper Paleolithic modern humans	13.1	21.7
	0.8–29.6	2.4–68.0
	0.8–29.6 (3)	2.4–68.0 (3)
Late Upper Paleolithic modern humans	31.7	50.9
	28.9–34.0	39.5–97.4
	18.9–50.7 (6)	8.5–101.9 (6)
Aleuts (maritime foragers)	9.5	16.4
	2.6–15.3	6.3–21.8
	0.2–23.9 (24)	1.1–40.1 (24)
Jomon (coastal foragers)	6.7	6.4
	5.1–7.6	3.2–10.7
	0.4–15.0 (10)	1.6–16.7 (10)
Native Americans (Pueblo agriculturalists)	6.8	16.8
	4.0–9.1	8.8–23.6
	0.4–33.4 (14)	5.5–37.8 (14)
European–American (autopsy sample)	5.9	7.5
	3.2–11.6	3.1–21.8
	0.7–26.5 (19)	1.6–41.6 (19)

[a] Percent asymmetry calculated for individuals with both right and left humeri as [(max-min)/min]*100.
[b] CA: cortical area (mm^2: a measure of the thickness of the diaphyseal cortical bone and resistance to compressive and tensile loads).
[c] J: polar moment of inertia (mm^4: a measure of diaphyseal rigidity to bending and torsional moments).

the humerus (Sabick et al. 2004). Not surprisingly, both professional and amateur baseball players occasionally spontaneously spirally fracture their humeri during forceful throws (Tullos and King 1973; Gainor et al. 1980; Pappas et al. 1985; Garth et al. 1988; Linn and Kriegshauser 1991; Sabick et al. 2004). Thus we might expect that forceful throwing of heavy spears (600 g or more: see Chapter 3.4) would stimulate adaptive modeling in the bones of the throwing limb, and would lead to pronounced levels of asymmetry in the upper limbs. Might then the great strength asymmetry of Neandertal humeri reflect throwing-based hunting methods? Not necessarily.

While the use of thrusting spears is a bimanual activity, it in fact generates greater reaction forces and larger bending moments in the trailing limb (the limb holding the proximal end of the spear) than on the leading limb (positioned more distally on the spear) (Schmitt et al. 2003). Based on the results of experimental work using an instrumented spear, the trailing limb – generally the individual's dominant limb – provides most of the muscular force propelling the spear forward and in return experiences most of the reaction forces that develop when the spear hits its target. The leading limb functions more as a guiding limb, although it sometimes (but much less often) experiences large reaction forces. The difference in loading regimes experienced by the two limbs – with a disproportionate amount of stress being experienced by the dominant limb – could certainly account for the observed asymmetry in humeral robusticity seen in Neandertals without having to invoke throwing as the source of asymmetrical loads (Schmitt et al. 2003).

The idea that Neandertal humeral diaphyseal shape and asymmetry reflects habitual use of thrusting spear technology has been challenged by experimental studies of upper limb muscle activity during spear thrusting and other tasks (Shaw et al. 2012). Shaw and colleagues used surface electrodes to look at electromyographic (EMG) activity of the *M. pectoralis major* and the anterior and posterior portions of *M. deltoideus* during simulated spear thrusting and hide scraping tasks. Given that both pectoralis major and the anterior portion of deltoideus are active in forcefully flexing the arm (as would be done when driving a spear into prey), one would expect greater right-side (all test subjects were right-handed) activity in these muscles during spear thrusting. They found that, contrary to expectations, muscle recruitment was greater in the left-than right-side muscles of their subjects during spear thrusting, strongly suggesting that habitual spear thrusting was not the primary cause of the marked right dominant asymmetry in humeral strength seen in Neandertals. Simulated hide scraping, however, did engender greater muscle recruitment on the right side. Given the frequency with which hide scraping must have been performed in the Mousterian (based on modern human analogs from cold environments: see also Chapter 3.5) and the vigorous action required to scrape a hide (which would generate high muscle forces), Shaw and coworkers argued that habitual preparation of skins may have been the behavior responsible for the observed asymmetry in Neandertal humeri.

While the EMG study by Shaw and colleagues (2012) presents some unexpected findings and raises some interesting questions, there are a number of aspects of their conclusions that are problematic. First, their results seem to suggest that the left arm of their right-handed subjects was providing most of the force required to drive the spear into its target. This seems to me to violate common sense, and it is also inconsistent with the empirical observation, discussed above, from a strain gauge-equipped spear showing that most of the propulsive force of spear thrusting was generated by the dominant limb (Schmitt et al. 2003). The suggestion that hide scraping was the activity that produced high degrees of bilateral asymmetry in Neandertal humeral strength is also problematic. Neandertal males have very high levels of strength asymmetry, so the EMG results would suggest that Neandertal males were heavily engaged in hide scraping. Although asymmetry data for Neandertal females is harder to come by, the limited evidence available suggests that their levels of asymmetry were comparable to or less than those of males (Trinkaus et al. 1994; Churchill et al. 1996b), which would seem to indicate (by the argument of Shaw and colleagues) that both sexes engaged in hide scraping. Interestingly, however, humeral asymmetry in J is negligible in late Upper Paleolithic females, and also in females from the Aleutian Islands (see Table 3 in Churchill et al. 1996b). This would seem to suggest, following Shaw et al. (2012), that hide scraping was the sole domain of males in the Upper Paleolithic and among the Aleut (both groups lived in cold conditions, and thus hide scraping would have been frequent). This is certainly inconsistent with ethnographic accounts of Aleut sexual division of labor, which report that females were almost entirely responsible for domestic activities, including construction of clothing (Ransom 1946; Liapunova 1975). Humeral strength asymmetry also appears to have been negligible among female Late

Stone Age foragers of South Africa (based on data from Table 4 in Stock and Pfeiffer 2004), which again would suggest that males were the primary hide preparers in that group as well. Overall, the comparative data on humeral strength asymmetry in various pre-industrialized human groups argue against the conclusion that hide scraping was the osteogenically-important activity that produced the patterns of asymmetry seen among the Neandertals. While the EMG work raises some interesting questions, it does not (at least to me) provide a compelling argument against the hypothesis that habitual thrusting spear use was one of the (if not *the*) principal determinants of humeral diaphyseal shape and asymmetry among Neandertal males.

8.1.5 Entheseal (Muscle Marking) Morphology

Three aspects of upper limb enthesis (muscle attachment site) morphology also suggest that Neandertals were not commonly engaged in forceful throwing. As with diaphyseal cortical bone, hypertrophy of muscle attachment sites is developmentally plastic, reflecting the degree of use of a particular muscle (see Kennedy 1989). The overall pattern of upper limb muscularity and leverage in Neandertals, as discussed above, is consistent with predatory behavior involving close contact with prey, in which muscular effort was critical to hunting success. In the context of this overall hypertrophy and enhanced leverage of the muscles of the Neandertal upper body, one thing that stands out as being anomalous is the small size of the *M. deltoideus*. A number of workers have commented on the relatively small size of the humeral deltoid tuberosity – the insertion site of the deltoid muscle – in Neandertals (Endo 1971; Thoma 1975; Hambücken 1993a; Churchill and Smith 2000b). The narrowness of this muscle scar can be appreciated by standardizing deltoid width by the circumference of the humeral shaft at the same location (Table 8.3) or by the length of the entire humerus (Churchill and Rhodes 2006). As with the leverage of the triceps muscle, the morphology seen in Neandertals may reflect a relative disadvantage in overhand throwing. The primary function of *M. deltoideus* is humeral abduction, but the anterior and posterior fibers are active in elevating a flexed or extended arm (Basmajian 1979). Based on electromyographic studies of overhand throwing, the anterior, middle and posterior fibers of the

TABLE 8.3 ■ Mean deltoid tuberosity width index[a] in fossil and recent humans – mean, SD, n

Neandertals	15.4 ± 3.9 (12)
Skhul/Qafzeh	21.0 ± 3.0 (5)
Early U.P.	24.4 ± 3.1 (14)
Late U.P.	22.5 ± 3.2 (22)
European Americans	23.3 ± 2.4 (45)
African Americans	22.8 ± 3.0 (50)
Aleutian Islanders	23.3 ± 3.0 (47)
Pueblo Native Americans	23.5 ± 3.3 (40)
Peruvian	21.9 ± 2.6 (23)

[a] 100*deltoid tuberosity width/circumference. Both width and circumference taken at 5/12ths humeral maximum length from the proximal end. From Churchill and Rhodes 2006.

deltoid are most active at the point of maximum external rotation, and the anterior and middle fibers are active during the release phase in overhand throwing (Perry and Glousman 1995). Peak activity of all fibers of the deltoid occurs during the early cocking phase of the throw, when the arm is at 95° abduction (Perry and Glousman 1995). The relatively small size of the deltoid muscle, as inferred from its attachment site on the humerus, would indicate Neandertals were comparatively weak in activities involving an elevated arm, such as forceful throwing.

As with the insertion site of the deltoid muscle, the origination of the supinator muscle on the proximal ulna also argues against regular throwing by Neandertals. The supinator muscle fires forcefully during the deceleration phase of throwing to counter hyperpronation of the forearm (Gainor et al. 1980; Zarins et al. 1985), and accordingly the origin of its ulnar head (the ulnar supinator crest) has been found to be hypertrophied in habitual spear-throwing populations in India and the Near East (Kennedy 1983, 2004; Peterson 1998). The supinator index (100 * supinator crest mediolateral diameter/proximal ulnar shaft mediolateral diameter) can be used to quantify the relative hypertrophy of the supinator crest, with the expectation that the relative size of the muscle scar should be highly asymmetric (and larger in the dominant limb) in habitual throwers. The pattern seen among fossil and recent human samples is not entirely clear (Table 8.4), although it is largely consistent with the argument that Mousterian foragers were not regularly using throwing-based, long-range projectile weapons (Figure 8.6). Neandertals and Mousterian Levantine modern human males have low supinator index values for the right (presumably throwing) side relative to Upper Paleolithic and recent

TABLE 8.4 ■ Supinator indices[a] in fossil and recent humans (mean, SD, n)

	Right	Left
Males		
Neandertals	68.6 ± 9.7 (4)	73.4 (1)
Levantine Mousterian	66.4 (1)	67.2 ± 14.9 (2)
EUP	81.9 ± 7.6 (7)	79.7 ± 5.5 (6)
LUP	81.4 ± 5.9 (9)	82.2 ± 6.7 (11)
Euro–Americans	79.0 ± 5.9 (24)	79.4 ± 8.2 (24)
Afro–Americans	74.7 ± 7.8 (24)	76.5 ± 7.8 (25)
Aleutian Islanders	80.5 ± 7.8 (24)	83.4 ± 6.6 (23)
Females		
Neandertals	–	74.6 (1)
Levantine Mousterian	92.1 ± 22.1 (2)	–
EUP	80.5 (1)	84.8 ± 8.8 (2)
LUP	79.3 ± 7.1 (4)	79.2 ± 4.7 (3)
Euro-Americans	80.1 ± 9.6 (13)	77.2 ± 8.5 (13)
Afro-Americans	76.9 ± 6.5 (25)	78.2 ± 8.1 (25)
Aleutian Islanders	85.1 ± 9.3 (13)	89.9 ± 7.8 (14)
Sex indeterminate		
Neandertals	74.3 (1)	–

[a] Supinator index = 100* supinator crest mediolateral diameter/proximal ulnar shaft mediolateral diameter. From Churchill and Rhodes 2009.

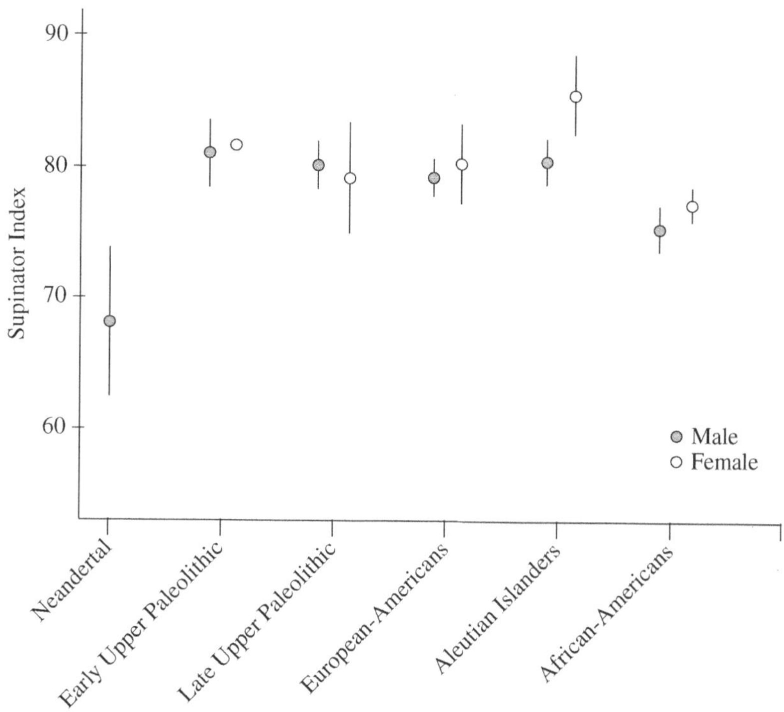

FIGURE 8.6

Mean supinator indices (100* supinator crest mediolateral diameter/proximal ulnar shaft mediolateral diameter) in fossil and recent human samples. Circles denote mean value for each sample, whiskers denote ± 1 SD.

modern humans. Amongst males, the largest mean right-side values were found in the EUP, LUP and Aleutian Island samples. Given that Aleutian Islanders used spearthrowers ("throwing boards") in hunting, the results may be demarking habitual throwers from non-throwers. However, relative supinator size was highly variable across samples in male left-side ulnae and female right- and left-side ulnae (although none of these differences are statistically significant: Churchill and Rhodes 2009).

The relative size of the crest in the right limbs of males is consistent with the expectation that Neandertals and Mousterian modern humans were not habitual throwers while EUP and LUP foragers were. Patterns of bilateral asymmetry in relative supinator size, however, are more difficult to interpret (the expectation of larger size of the crest in the dominant limbs of the throwing groups is often not met: Table 8.4). As with other developmentally plastic traits, the relative size of the supinator crest may be reflecting behaviors other than throwing that involve forceful supination of the forearm (such as, perhaps, hide scraping or woodworking). The small size of the supinator crest in Neandertals may also be a function of the hypertrophy they exhibit in the other supinator of the forearm, *M. biceps brachii*. The morphology of Neandertal radial bicipital tuberosities suggests that they were able to use the biceps muscle for supination over a wider range of pronation-supination than can most modern humans (Trinkaus and Churchill 1988). Perhaps Neandertals relied more heavily on *M. biceps brachii* than *M. supinator* for actions requiring forceful supination. This possibility does not alter the conclusion, however, that supinator crest size in Neandertals suggests that they were not habitual throwers, since the extension of the forearm in the final stages of throwing prevents the use of the biceps as an antagonist to hyperpronation (thus even Neandertals would have had to rely on the supinator muscle in the deceleration phase of throwing). This conclusion, however, rests upon the assumption that spears would have been thrown in a manner resulting in full extension of the forearm in the follow-through phase of the throw (the best but not the only method of throwing a spear: see Miller 1960), and also may not apply to smaller, spherical objects (such as stones) which might be thrown without the elbow moving into full extension. At present, all that we can say is that the small size of the supinator crests suggests either less involvement in habitual spear throwing or biomechanically different patterns of throwing than seen in modern spear-using people.

Finally, there is some negative evidence that supports the hypothesis that Neandertals were not regularly engaged in stressful throwing behaviors. Four male skeletons from the late Upper Paleolithic and Mesolithic exhibit lesions on the dorsomedial corner of the humeral medial epicondyle (Villotte et al. 2010). All four of these cases occur in right side humeri: in three of the individuals there is no corresponding lesion on the left-side humerus; in the fourth individual the state of the left medial epicondyle cannot be evaluated. This particular lesion is rare in archeological samples and recent human populations, and is generally taken as an indicator of throwing-induced stress to the medial collateral ligament (see references in Villotte et al. 2010). The lesion was observed in about 24% of the 17 male Upper Paleolithic and Mesolithic right-side humeri sampled. Similar lesions have not been observed in any Neandertal humeri. While such negative evidence should be viewed tentatively, it does lend support to the idea that Neandertals were not heavily invested in throwing-based hunting technology.

8.2 NEANDERTALS AS CLOSE-RANGE PREDATORS

If I have over-stated the case for close-range hunting with hand-deployed weapons, it is only because I see hunting methods – the actual interface between predator and prey – as being central to a host of subsistence and ecological factors. The assortment of weapons in the hunting toolkit and the number of different ways they can be employed color the way that human hunters view the edible landscape. Over the remainder of this chapter, I will try to make the case that an emphasis on close-range hunting with hand-held spears largely restricted the hunting tactics at the Neandertals' disposal, limited the range of prey sizes that could be profitably targeted, constrained their predatory efforts to a small set of physiographic and habitat characteristics, and kept high the risk of injury, the risk of failure, and the energetic costs of foraging. In the next chapter, our consideration of the limits and realities of close-range hunting will be extended to how they impacted interactions with other carnivores, and I will argue that the limitations of Mousterian hunting technology constrained the degree of social dominance that could be won over other carnivores, and was also a primary determinant of hunting party size. When taken together, these considerations put forward a picture in which Neandertal predatory capabilities – relative to those of other carnivores – fostered a dependency on ecotones and likely severely limited Neandertal population densities and demographic growth.

We know that Neandertals were armed with large, stone-tipped spears, and the morphology of those spears, along with the morphology of the Neandertals who made and used them, indicates that these spears were wielded at fairly close quarters. Beyond that, we don't know how the Neandertals were using their hunting technology – that is, how they were going about positioning themselves within a spear's length of prey, how they were preventing prey from escaping before they could be immobilized or badly wounded, and how they operated to minimize risk of injury to the hunters and maximize the probability of bringing down dinner. The intellectual avenues that lead to answers to these questions are varied, and include experimental studies, considerations of prey ethology, and ethnographic and historic analogy, but each of these avenues has limits and problems. Experimental work with bone and stone weapon tips on spears, darts,[4] and arrows has documented the ability of these weapon systems to deeply penetrate the thoracic and abdominal cavities of a variety of animals, has provided information about their ranges and the most effective ways to deploy them, and has shown the limitations inherent in these technologies (Frison 1986, 1989; Odell and Cowan 1986; Bergman et al. 1988; Shea et al. 2001). While playing an important role in our efforts to infer the functions of Paleolithic artifacts, experimental studies generally elucidate the possible rather than the practical – for example, demonstrating that it is possible to kill a particular type of animal with a particular weapon does not mean that it was practical (that is, that the return benefits outweighed the associated energetic costs and injury risks) for the foragers involved, or that they ever did so. Consideration of the behavior patterns of particular prey species can also shed light on the kinds of opportunities earlier hunters would have had to capture them, as well as the associated difficulties and risks (e.g., Spiess 1979; Binford 1987; Olsen 1989; Blehr 1990; Enloe and David 1997), but can generally only suggest methods that might have fruitfully been employed to capture particular species of prey. A third avenue is the ethnohistoric and military records of spear (and other hunting technology) use (Churchill 1993, 2002; Kortlandt 2002; Villa and Lenoir 2009; Villa and Soriano 2010). Recent hunters employ a wide range of weapon systems and hunting techniques, and thus provide the potential for better understanding the interrelationships among weapon systems, hunting strategies, and prey choice. Data from recent human foragers can also provide information about the practical (as opposed to experimental) limits of various weapons and, ultimately, what the occurrence of different weapons in the archeological record indicates about past human predatory behavior. However, the richness of modern human toolkits – especially with respect to hunting and other subsistence tools – produces its own intellectual challenge. Virtually all historically-known hunter-gatherers had spears in their toolkits, yet there are no such groups which used *only* hand-deployed spears, that is, in the complete absence of projectile weapons, and thus which could be considered analogous to Neandertal hunters (at least with respect to weapon technology). Among all historically- or ethnographically-known hunter-gatherers, the hand-held spear is one of many hunting tools, and thus the ways in which they use this tool may be very different from the ways in which it was used by Mousterian hunters. Nevertheless, I believe that great insights into the nature of close-range hunting with thrusting spears can be had by reference to modern human analogs, and my perspective on Neandertal predation has admittedly been greatly influenced by research into spear use among modern foragers.

Based on patterns in the ethnohistoric record, two things stand out about the use of hand-delivered spears (the discussion that follows derives from an analysis of the ethnohistoric record of hunting among 93 foraging groups, derived primarily from North America but also including South America, Africa, Asia, Indonesia, and Australia: Churchill 1993). First, the use of hand spears is strongly associated with hunting techniques that involve disadvantaging prey. Disadvantaging methods (Figure 8.7), by impeding the prey and preventing its flight (or by taking advantage of prey already in a hindered position), buy the hunter the time and close access necessary to deliver the repeated jabs with the spear needed to kill it. Of 79 reports in the literature of thrusting spear use in terrestrial hunting, 68% of them involved disadvantaging prey in some way, as did 41% of the accounts of hand-thrown spear use (Figure 8.8). The thrusting spear – as used by modern foragers – is primarily a dispatching tool, used to put an animal to death after it has been placed in a disadvantaged position. Among modern hunters, disadvantaging with thrusting spears is strongly associated with cooperative drive techniques (that drive animals into hindrances like bogs, bodies of water, snow drifts, or steep-walled box canyons) and use of domesticated animals (mostly dogs) or other technology (boats and snowshoes) that allows the hunter to be unimpeded despite deep water or deep snow. In some cases, one form of hunting technology is used to disadvantage the animal before hand spears are used to deliver the *coup de grace*. These include snares, traps, or nets that are used to ensnare or entangle prey, or other weapon systems that wound prey and immobilize it either through traumatic shock or with projectile-delivered toxins. Neandertals may have been using cooperative driving techniques (see Kuhn and Stiner 2006), but it is probably safe to assume that they lacked dogs, projectile weapon systems, and transportation technology. In only 14 cases in the modern record were hand-delivered spears employed *without* the use of cooperative drives, dogs, or accessory technology (thus these cases are probably most similar to what obtained in the Mousterian). In these circumstances, the hunters required only suitable landscape features with which to disadvantage the animals, most commonly snow drifts, but also valleys and swamps. Some reported cases also involved hunters taking advantage of hibernating bears naturally "trapped" in their dens.

Hand-delivered spears are associated less frequently with ambush and pursuit hunting (the two tactics together accounting for 29% of thrusting spear cases and 37% of the reported cases of spear throwing). Reported cases of ambushing with hand-held spears also involved either suitable terrain or cooperative hunting involving the driving of animals toward concealed hunters. Most ambushing spear hunters used trees, rocks, and other physiographic features for concealment while waiting for the animal to come within effective weapon range. Cases of pursuit hunting with thrusting spears involved either the use of horses (to close the gap between mounted hunter and fleeing prey), dogs (to catch the animal and then hold it at bay), or hot and dry environments where endurance pursuit tactics could be employed. Only one case of encounter hunting with hand-held spears, involving the taking of peccaries in the neotropics, was noted.

The methods most strongly associated with the use of hand-delivered spears among modern foragers – disadvantaging and ambushing – are heavily terrain-dependent. Disadvantaging is highly reliant on physiographic features for success: without the

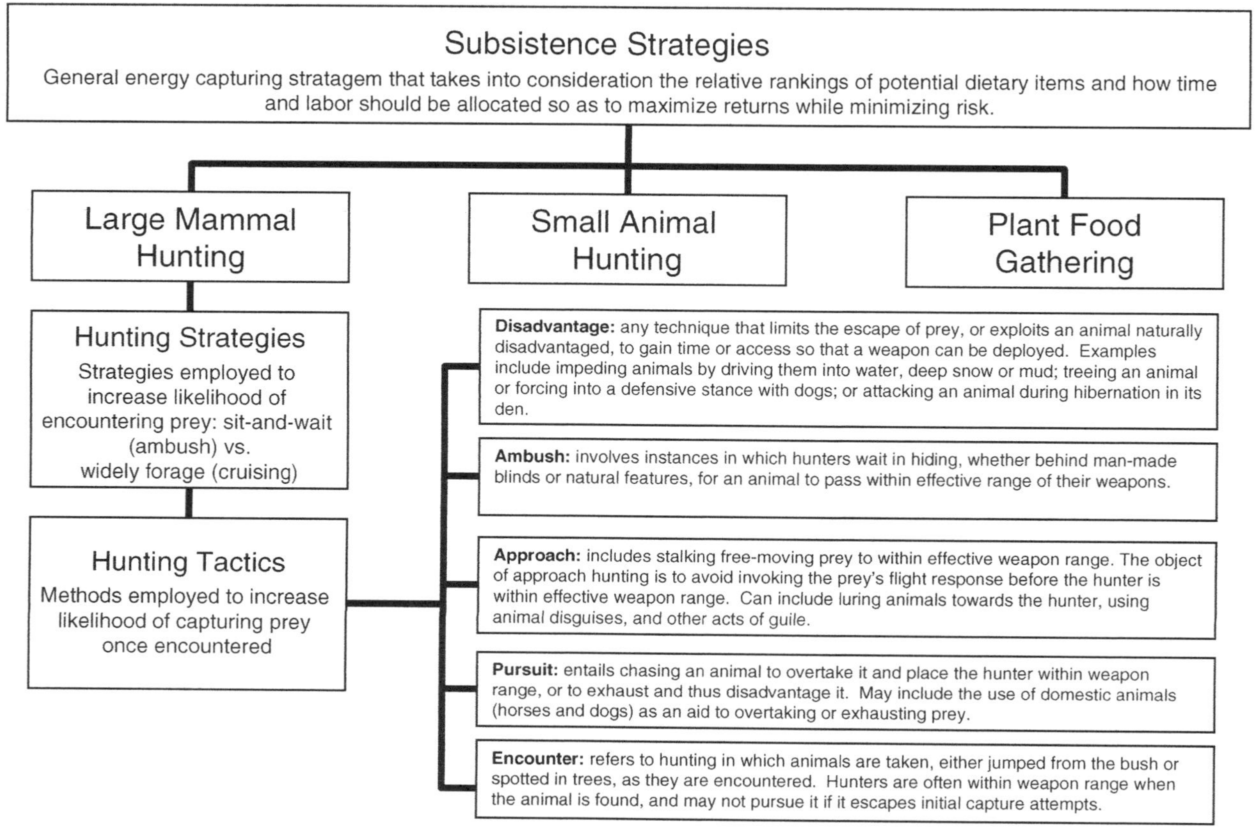

FIGURE 8.7

Hierarchical configuration of subsistence strategies, hunting strategies, and hunting tactics (as applied to larger prey). As defined here, hunting strategies concern the action plans aimed at finding prey, while hunting tactics include the means by which a predator captures prey once it is found. The choice of strategy and tactic are largely but not completely independent, as some hunting tactics are strategy limited (for example, encounter hunting can only occur within the strategy of wide foraging). Note that the sit-and-wait strategy is sometimes called ambush hunting (e.g., Greene 1986), but should not be confused with the tactic of ambushing.

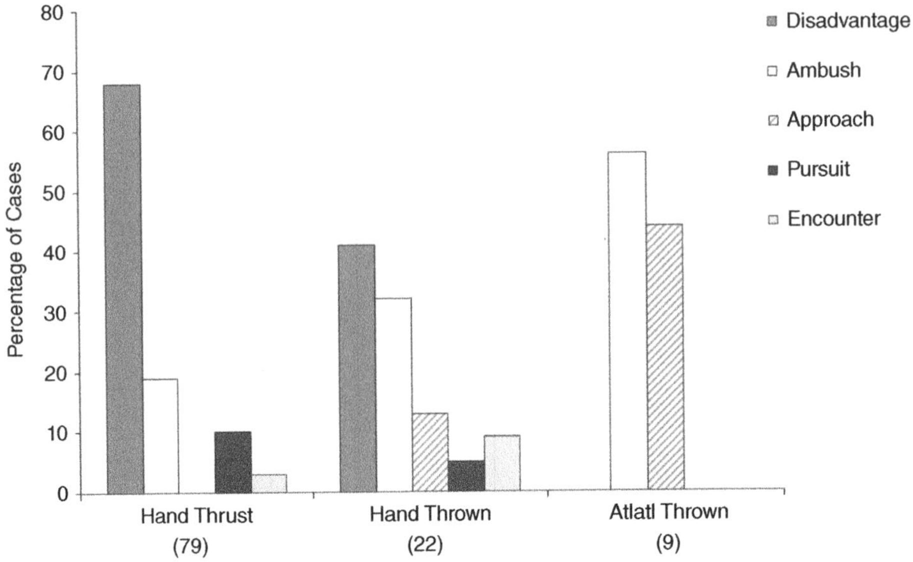

FIGURE 8.8

Hunting tactics associated with hand spears (thrust and thrown) and spearthrower (atlatl) delivered darts. Numbers in parentheses represent the number of reported cases of each type of weapon system in the ethnohistoric literature (Churchill 1993).

use of dogs or true long-range projectile weaponry, it requires natural or man-made traps such as narrow arroyos, box canyons, corrals, swamps, snow drifts, or bodies of water. Cooperative hunting techniques may be very important in moving prey into disadvantaged locations, but even then success relies on some form of impediment that animals can be driven into. Even ambushing, though used much less often, requires physical features suitable for concealing hunters until the animal is within effective range of the weapon – within 5–10 m for hand-delivered spears. These terrain features must not only be suitable for disadvantaging or ambushing, they also must be located in or near places where desirable game animals occur. The terrain-dependence of ambush hunting can also be seen among carnivores: lions, who combine a sit-and-wait strategy with ambush tactics (lions also stalk prey, which is analogous to approach tactics), favor hunting locations with characteristics that facilitate concealment and ambushing (e.g., woodland vegetation and erosion embankments) over more open locations with higher prey densities (Grant et al. 2005).

The terrain-dependency of hunting with hand spears becomes even more apparent in the complete absence of accounts in the ethnohistoric record of approach hunting with thrusting spears, and in the very limited contexts in which hand thrown spears were used in approach hunting (Figure 8.8). The ability of a hunter in open country to stalk prey to within striking distance is limited only by their skill and guile, and it is possible that seasoned and skillful hunters could elude the vigilant watch of grazing herbivores to get within a striking distance of 10 m or so. But even if they managed this difficult feat, they still must be able to either launch a spear or close any remaining distance and effectively wound the target before the quarry flees, within the context of prey reaction times and flight responses that are measured in fractions of a second. Again, lions provide a useful analogy. Hunting by stalking (comparable to approach hunting) is more successful in taller grasses where the lions can get closer to their quarry, but it is also much more successful with smaller prey (ca. 60% success rate with steenbok, duikers, and other small mammals) than larger prey (e.g., less than 20% with wildebeest, zebra, kudu, and waterbuck, about 40% for buffalo) (Funston et al. 2001). Lions have the advantage over human hunters of being raptorial – that is, being able to strike and hold prey to prevent it from fleeing. Seizing prey is no doubt less effective with larger size classes, and lions more often resort to chasing prey in these cases (with better success rates – about 35% for medium-sized wildebeest and 55% for buffalo: Funston et al. 2001). For human hunters with little ability to deter the flight of open country prey, it is hard to imagine that they could obtain even the 15–20% success rates that lions enjoy with medium-sized ungulates.

It should be noted that there are a few (3) reported cases of approach hunting with hand-thrown spears: the Tiwi (Melville Island, Australia) and Tasmanians are known to have hand thrown *very thin and light* spears (unlike the massive javelins of the Pleistocene) at wallabies and kangaroo in conjunction with approach hunting tactics. These cases are curious exceptions to an otherwise robust pattern of hand spears being used exclusively with disadvantage, ambush, and pursuit hunting. These cases show that hand-propelled spears *can* be used as long-range projectiles with approach hunting – although it was only the Tasmanians who threw spears long distances (30–40 m) (Roth 1890); the Tiwi were reported to have approached prey very closely before throwing (Goodale 1971). An effective distance of 30–40 meters greatly reduces the difficulty of stalking prey to within weapon range while avoiding detection, and makes open country hunting possible. This is the effective range of spearthrower-propelled darts, and not surprisingly we see a strong association between the use of spearthrowers and open country approach hunting among modern foragers (Figure 8.8). The Tasmanian case raises the possibility that, prior to the development of other forms of projectile weaponry (which could perhaps deliver a projectile with either greater velocity or greater accuracy) humans were hand launching spears over long distances, and doing so to effectively take open country prey – consistent with some interpretations of the Schöningen spears (Thieme 1997, 1999; Rieder 2001, 2003). Still, the preponderance of evidence indicates that Neandertals were not regularly engaged in throwing, and the majority of the modern human evidence reveals that most foragers do not use hand spears as long-range weapons. The Tasmanian case remains an intriguing outlier, perhaps related to unique circumstances involving historical contingency, prey size and behavior, and environmental characteristics that make this combination of hunting technology and method practical (there is also the possibly that the historic account of this hunting behavior is wrong).

It is only with pursuit techniques that the hand-spear hunter is freed somewhat from terrain constraints. However, among modern humans this tactic usually entails the use of domestic animals – either horses that carry their rider to within weapon distance (e.g., horse-mounted buffalo hunters of the North American historic period) or dogs that run an animal to exhaustion. In a few cases, humans were reported to keep prey moving at a disadvantageous speed (respiration and energetic efficiency are coupled with gait in quadrupeds but not bipedal humans, so by keeping prey moving at a certain speed humans can eventually run it to exhaustion: see Carrier 1984). However, this practice among modern humans is limited to hot, arid environments (where overheating helps to more rapidly bring the prey to exhaustion), and chasing animals to collapse is not the most energy-efficient means of hunting, especially for short-limbed and massive Neandertals (Steudel-Numbers and Tilkens 2004; Steudel-Numbers et al. 2007; see also Raichlen et al. 2011), and would have been even less so in the colder climates of Pleistocene Europe.

All of this suggests that Neandertals were most commonly using disadvantaging and ambushing tactics when hunting, which further suggests that they were constrained to hunting in areas with physiographic characteristics suitable to these methods. The zooarcheological and biogeochemical evidence (reviewed in Chapter 7) clearly points to a dietary (and presumably predatory) preference for open country herbivores. Evidence from kill sites also tells us that Neandertals were capable of hunting in open environments (Gaudzinski 1995; Gaudzinski and Roebroeks 2000). Our consideration of hunting technology and tactics suggests that Neandertals were limited to taking them at ecotones – the places where the grasslands met woodland or forest cover or abutted rivers or other water bodies, and which afforded the Neandertals ample cover for concealment or natural traps for disadvantaging game. Not surprisingly, a number of the Middle or Lower Paleolithic sites with evidence of hunting in open environments are found in association with such physiographic features (see Marks and Chabai 2001; Villa and Soriano 2010: 7–9), and topographic features appear to have been a component of hunting in other environments as well (Rendu 2010: 1808).

The second notable aspect of the way that modern foragers use hand-delivered spears (the first being terrain-dependence) is that they are predominantly directed against larger prey (Figure 8.9). Mean body weights of animals hunted with hand

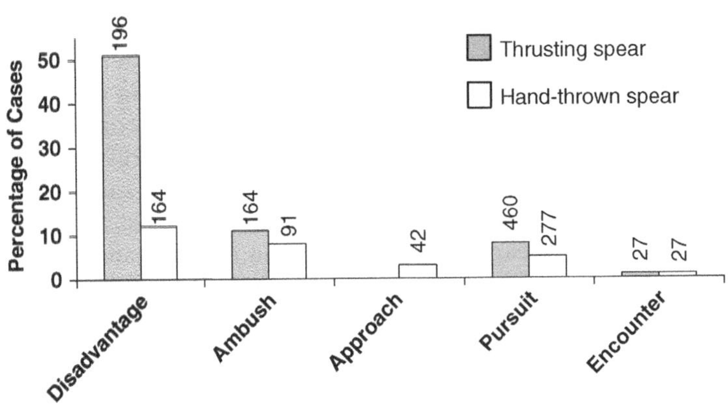

FIGURE 8.9

Hunting tactics and median prey body sizes associated with hand spears. Based on a subset of the cases illustrated in Figure 9.8, for which ethnohistoric accounts provided information about both the hunting tactics employed and the type of game pursued (n = 65 cases). Numbers above the bars represent median prey body sizes (as estimated by Churchill, 1993) associated with each form of spear use and each hunting tactic.

spears have high variances, which does reflect a degree of versatility to this weapon system. These large variances are in part a function of the inclusion of some cases of hunting megaherbivores (elephants and hippo), which makes the median values (as reported in Figure 8.9) a bit more informative, but they also reflect the fact that hand-delivered spears can be used to kill any sized animal once it has been hindered from fleeing. Regardless, modern foragers tend to direct this weapon at larger prey (mean size of prey taken by hand spears = 324.0 kg, SE = 104.3 kg [median = 160.2 kg], n = 48 cases; mean size of prey taken by bow and arrow = 151.6 kg, SE = 17.2 kg [median = 118.8 kg], n = 72 cases; difference significant at p = 0.001: Churchill 1993). This may be because spears are the best tool for dispatching large animals, or because the hunting techniques required to take game with hand spears generally work better with larger prey. Disadvantaging is likely more successful with larger prey, since smaller, fleeter animals can more readily escape traps and impediments (agility generally scales inversely with mass: Alexander et al. 1979; Schmidt-Nielsen 1984; Calder 1996). Disadvantaging allows a spear to be applied repeatedly against an animal incapable of escaping, so even the largest animals (e.g., elephants: see Frison 1986) can be dispatched once disadvantaged. Ambushing is probably also more effective with larger, slower prey, whose slower reaction times allow the hunter greater opportunity to place a well-delivered spear thrust or to launch the spear before the animal moves away. Similarly, large game may be the focus of pursuit hunting, as the physics of respiratory and heat exchange physiology make larger animals easier to exhaust, at which point spears can be used to kill them.

Reduced effectiveness of hand spears with smaller game may thus represent an additional limitation to hunters armed only with this weapon. Modern hunters must often employ additional technology (usually other weapons, boats, or snowshoes) in order to disadvantage an animal or to exploit one already disadvantaged. When cases of disadvantaging involving accessory technology are compared with cases where spears alone were used, a significant difference in body weight is observed (mean body weight of prey taken by hand spears and disadvantaging with the use of accessory technology = 340.5 ± 118.0 kg [median 152.5 kg], n = 55 cases; mean body weight of prey taken by hand spears and disadvantaging without other technology = 894.6 ± 438.0 kg [median 309.0 kg], n = 14 cases; one-tailed p value = 0.0170: Churchill 1993). Disadvantaging without the aid of other technology, as was undoubtedly practiced in the Mousterian, is directed at much larger prey. Average prey size tends to go down from the Middle to Upper Paleolithic in most places (see for example Fa et al. 2013), although this is more likely to reflect population density-driven increases in diet breadth (Stiner et al. 1999) than changes in hunting technology per se (although the two are clearly related: projectile weapons reduce the handling costs associated with smaller game, and thus are part of the picture of diet breadth expansion).

The large-animal focus of hunting with hand spears is interesting in light of the prey preferences of Neandertals, as judged from faunal remains at Mousterian sites and from the isotopic and elemental signatures of Neandertal fossils (Chapter 7). The dietary staples appear to have been moderate-to-large sized ungulates, with an emphasis on reindeer, red deer, horse, and bovines (bison and aurochs). Megaherbivores – mammoth and rhino – appear to have also been important dietary items, at least in some places and at some times. The faunal record of the Mousterian also includes smaller prey, and there's nothing to say that Neandertals were incapable of taking smaller herbivores – we're simply noting that the preferred prey items tend to be towards the larger end of the herbivore body size spectrum. The smallest of the regularly-occurring items are reindeer (average mass of 86 kg: all of the mass averages reported here are for combined sexes, averaged across their geographic range, and are from FA Smith et al. 2003) and red deer (167 kg), which may have been somewhat more difficult for Neandertals to hunt than were larger prey. Both of these species are sexually dimorphic and thus show a considerable range in adult body sizes – from 80 kg female to 150 kg male reindeer and from 70 kg female to 250 kg male red deer (Gamble 1986). Along these lines it would be interesting to see if Neandertals preferentially took larger members of each species. The average masses of the other common prey species are larger, from horse (mean mass of 250 kg) to bison and aurochs (500–900 kg) to rhino (1600–2900 kg) to mammoth (average about 5485 kg, based on estimates from Christiansen 2004). For reasons having to do with the scaling of muscle cross-sectional areas (and thus force production) and bone strength with body size, larger animals are much less agile (as size increases, mass increases disproportionately relative to the ability of the muscles to generate force to accelerate that mass, and relative to the strength of the supporting skeleton to withstand the forces generated by accelerating and decelerating that mass) and, all things being equal, a 250 kg horse should only be about half as agile as a 125 kg red deer, while a 500 kg bison would only be about as fourth as agile as the red deer. Of course, the reduced agility of the animals – which facilitates getting them into disadvantaged positions or striking them in an ambush before they can flee – comes with a cost: their larger size makes them more dangerous, and increases greatly the risk of injury and death for the hunter.

One other thing is interesting about Neandertal prey preferences: there is an emphasis on prey species that use defensive postures or direct attack as anti-predator defenses. Prey species have evolved numerous anti-predator tactics that operate along the predator behavioral continuum of searching for, catching, subduing, and killing prey. Most herbivores adopt methods that reduce the likelihood of being detected (crypsis) or of evading predation once detected (through speed of evasion, safety in numbers, or mobbing predators) (Caro et al. 2004). Larger herbivores, however, sometimes use active defense – either by taking a defensive posture and holding their ground or by aggressively attacking interloping carnivores. Extant African elephants and rhinos actively defend against predators – resulting in documented cases of lion deaths (Hayward and Kerley 2005) – and it is reasonable to assume that their European Pleistocene sister taxa did as well. The behavior of aurochs is unknown, and their extant descendants are too far removed from the evolutionary pressures of regular predation to be reliable indicators of how they might have reacted to large carnivores. However, some extant bovines are known for their aggressiveness and their use of defensive stances against predators, the African buffalo being perhaps the most notorious (Hayward and Kerley 2005). Bison are also known to stand their ground or directly attack predators (Caro et al. 2004), and it would be surprising if aurochs – given their large body size and impressive horns – did not also occasionally adopt this defensive strategy. Finally, red deer and male horses are also known to use active defense tactics against large carnivores (Caro et al. 2004, Hayward and Kerley 2005). Since Neandertals were probably least able to cope with fleeing prey (lacking both long-range weapons with which to strike an escaping animal and the running speed to catch them), the natural defensive behaviors of some of the larger herbivores may have actually made them more vulnerable to human predation. Of course, large size, aggressive demeanors, and defensive anti-predator tactics also make the prey that much more dangerous to hunt.

8.3 PREY SIZE, HUNTING "PACK" SIZE, AND RISK OF INJURY TO NEANDERTAL HUNTERS

Neandertals, with adult body sizes on the order of 60–85 kg, were regularly hunting prey whose body size was an order of magnitude larger than theirs. Most carnivores kill and consume prey that is smaller than them, and with few exceptions predation on larger animals (that is, larger than the carnivore) is limited to socially cooperative hunters (Gittleman 1985). At a few sites, such as Jonzac (Britton et al. 2011), Mauran, and Les Pradelles (Costamagno et al. 2006; Rendu et al. 2012) in France or possibly Salzgitter-Lebenstedt (Gaudzinski and Roebroeks 2000) in Germany, the faunal evidence is consistent with the killing of multiple cervids or bovids at a single time, strongly implying communal harvesting of herd animals by socially cooperative Neandertal hunters. If Neandertals could work together to kill multiple reindeer or bison at a single time, they certainly could have worked together to bring down a single prey animal that was too big for a solitary hunter to handle. Even with social hunting, though, only a handful of carnivores are able to regularly prey on the largest size classes of herbivores (such as wildebeest or red deer – never mind megafauna!). These include the socially-hunting lion, spotted hyena, and wolf, and the more solitary cougar and tiger (Gittleman 1985; Kunkel et al. 1999). While these predators are able to feed on the largest class of herbivores, disparities in body weight between predator and prey seldom exceed two- or three-fold. One notable exception is elephant predation by lions in northern Botswana (Joubert 2006), where group-hunting lions have been regularly observed to attack and kill juvenile elephants who may be as much as 10 times the weight of an individual lion. An empirical relationship has been documented between prey body mass and the total mass of social carnivores (individual mass times the number of predators in the hunting party), supporting the claim that social hunting provides carnivores a flexible system of preying on larger animals (Earle 1987). Given prey in the size range of horse to aurochs (roughly 250–1000 kg), and an average adult male Neandertal body mass of about 78 kg, this relationship would predict average hunting party sizes on the order of 4 to 15 hunters. While the high end of this range may eclipse the average social group size of Neandertals clans (see Chapter 11), these estimates do suggest that large game hunting by Neandertals was necessarily social, and probably required more than a few hunters. The use of females and even children as drivers (or as living "scarecrows" who were strategically positioned to prevent prey from deflecting from the drive path) may have increased the effective pack size of a hunting party, without unduly exposing the smaller (less massive) members of the social group to risk of injury (cf. Kuhn and Stiner 2006). By exercising disadvantaging techniques, hunters may have relaxed somewhat the need for additional hands, but success rate likely increased and risk of injury likely decreased as the size of the hunting "pack" increased.

On the topic of injury, there is no doubt that close-range hunting of large game was a dangerous business. Neandertal fossils exhibit high frequencies of healed trauma, and the patterning of these injuries – occurring with highest frequencies in the head and neck, and thus being similar in overall pattern to a group of modern humans who regularly have close contact with large and irate herbivores, namely rodeo riders – is consistent with the idea that Neandertal hunters had to pay the wages of regular combat with ill-tempered herbivores (Berger and Trinkaus 1995). Subsequent consideration of trauma patterns in early modern Europeans (Trinkaus 2012) shows them to be similar to the Neandertals, which warrants caution in ascribing the observed pattern to close-range hunting (since these modern humans appear to have been armed with projectile weaponry). The ways in which evidence of non-fatal trauma accumulates in the skeleton (which may bias towards a greater frequency of healed traumatic lesions in the upper body), a highly mobile life style (which may select against individuals with lower limb trauma), and interpersonal violence may all also contribute to the observed pattern of upper body trauma in Neandertals and early modern Europeans, although prey-induced trauma may still be a part of the picture (Trinkaus 2012). There are also at least two Neandertal males that evince partially or fully healed traumatic lesions caused by lithic edges – Shanidar 3 from Iraq (Trinkaus 1982; Solecki 1992) and St. Césaire 1 from France (Zollikofer et al. 2002). These lesions are generally interpreted as evidence of interpersonal violence between Neandertals (or possibly between Neandertals and modern humans: see Churchill et al. 2009), and they may well be. However, we are not able with current methods to rule out accidental causes of these lesions, and it may be that one or both of them occurred during close-quarter efforts to subdue prey animals (during which one hunter accidentally impaled another) (cf. Geist 1981). Regardless, there is little doubt that Neandertal predatory methods routinely placed hunters at risk of personal injury, and this was probably a factor that contributed significantly to high mortality rates for prime-aged adults (Trinkaus 1995a).

8.4 NEANDERTAL HUNTING IN ECOLOGICAL CONTEXT

On the basis of zooarcheological evidence, Neandertals appear to have been proficient hunters capable of bringing down a variety of game; tailoring their predatory strategies to the behavioral characteristics of their prey and, to an extent, the physiographic features of the hunting terrain; hunting the prime-aged, adult members of their prey species; and acting cooperatively to either harvest multiple animals at a time or to take megafaunal prey (Stiner 1994; Gaudzinski 1995; Burke 2000a; Gaudzinski and Roebroeks 2000; Steele 2004; Adler et al. 2006a, 2006b; Costamagno et al. 2006; Yeshurun et al. 2007; Rendu 2010; Britton et al. 2011; Delagnes and Rendu 2011; Rendu et al. 2012; Niven et al. 2012; Speth 2012). In their overall predatory capabilities, the Neandertals may have differed little from the early modern humans that succeed them, at least in some places (Grayson and Delpech 2002; Adler et al. 2006a, 2006b). We have come a long way from the debate, which dominated much of the discussion about Neandertal hunting behavior in the last few decades of the previous millennium, about whether Neandertals were "opportunistic" hunters (in which they hunted opportunities in which they could capture game given a limited skill set and limited technology) or were capable of the "specialized" hunting (targeted hunting and mass harvesting of key prey species) that characterized the predatory behavior of Upper Paleolithic-associated modern humans. A brief review of this debate is in order, since it has had a lasting influence on how research into Neandertal hunting and subsistence organization is conceptualized and structured, and since this issue is still occasionaly used to contextualize research on Neandertal predation (for example, Rivals et al. 2004; see also Speth 2012). A clear understanding of Neandertal predatory capabilities along a "generalized-specialized" continuum of prey selection is also important to our knowledge of the subsistence strategies the Neandertals were employing (see Chapter 10).

While the Neandertals were generally seen as capable of directed hunting of small-to-moderate sized herbivores, such as reindeer (Chase 1986a, 1988; Binford 1984b, 1985, 1991), wild goats, and sheep (Marean 1998), animal part representation and mortality profiles in Mousterian faunal assemblages from the French sites of Combe Grenal and Grotte Vaufrey, and patterns of carnivore damage on human modified (cut-marked) bones in these assemblages, led Binford (1982b, 1985, 1987, 1989) to argue that scavenging of large animals was an important component of archaic human meat procurement (see also Binford 1984b for a comparable argument concerning MSA hominins from Klasies River Mouth, South Africa).[5] Binford suggested that Neandertals may have lacked the behavioral or technological sophistication needed to regularly hunt the full range of available prey species, and that Neandertal subsistence may have been dependent upon scavenging for success (Binford 1984b, 1985). Stiner (1990a, 1991a, 1991b, 1991c, 1993, 1994; Stiner and Kuhn 1992) presented zooarcheological evidence from the Italian Mousterian sites of Grotta Guattari and Grotta dei Moscerini that reflected a reduction in the frequency of scavenging and an increase in hunting during the Mousterian, rather than at the Middle to Upper Paleolithic transition (as usually assumed), which she felt reflected adaptive responses to changing ecological conditions (a similar late Mousterian shift is seen in the fauna at Mujina Pećina, Croatia: Miracle 2005). Thus, by Stiner's view, Neandertals shared with modern humans a well-developed degree of adaptive flexibility, and may better be characterized as "opportunistic," rather than "obligate," scavengers. Such a conclusion should not be surprising, because modern foragers engage in variable amounts of scavenging (O'Connell et al. 1988), and because scavenging is a regular component of the subsistence behavior of all carnivores. Nonetheless, Binford's claims of obligate scavenging, coupled with an inferred hunting focus on individual animals (based on a general lack of evidence for cooperative game drives and mass kills: see below) suggestive of opportunistic hunting practices in the Mousterian of some regions (Freeman 1973; Binford 1982b; Straus 1983; Straus and Clark 1986), melded into a general view of Neandertal subsistence behavior as lacking planning depth, tactical behavior, and long-term curation of subsistence technology (Binford 1982b, 1989: see also Simek 1987 for corroborating evidence from spatial patterning of bones and artifacts at two Mousterian sites). By this view, true logistical foraging behavior was a fully modern human trait that arose at the time of, or after, the appearance of modern humans (Binford 1985) (see Chapter 10).

That Neandertal subsistence behavior differed from that of early modern humans was by no means the consensus view, and numerous arguments were made that subsistence organization and hunting behavior differed little between the Middle and Upper Paleolithic (e.g., Chase 1986a, 1988, 1989; Farizy and David 1989; Patou 1989; Patou-Mathis 1996; see also Jelinek et al. 1989 for purported evidence of mass hunting in the French Mousterian). Faunal assemblages dominated by prime-aged individuals (which contrast with the attritional mortality profiles generated by carnivores, and thus are generally taken to indicate human hunting: Stiner 1990b) were also put forward as evidence of purposeful hunting of targeted resources by Neandertals (Baryshnikov et al. 1996; Blasco 1997; Fernandez and Legendre 2003; Steele 2004; Yeshurun et al. 2007). Towards the close of the millenium, Marean (1998; Marean and Kim 1998) argued that the original pattern of faunal part representation upon which Binford had based his arguments owed to taphonomic processes and collector bias, not to hominin behavior. By refitting long bone diaphyseal shaft fragments from small faunal assemblages from Kobeh (Iran) and Die Kelders (South Africa) Caves, Marean showed the assemblages to be dominated by limb elements (characteristic of hominin access to meaty portions of a carcass and, by inference, hunting) and not foot and axial skeletal parts (characteristic of hominin access to carcasses after they have been ravished by carnivores) as claimed by Binford and others. Marean's conclusions have not been universally accepted (see Klein and Cruz-Uribe 1998; Stiner, 1998), but evidence has accumulated to show that Neandertals had early access to carcasses (implying hunting) in a variety of contexts (Speth and Tchernov 1998, 2001; Hoffecker and Cleghorn 2000; Patou-Mathis 2000; Steele 2002; Rabinovich and Hovers 2004; Adler et al. 2006a, 2006b; Stiner 2006; Yeshurun et al. 2007). The issue of scavenging versus hunting has thus become, as Mary Stiner (2006) puts it, an "old horse," and the idea that Neandertals obtained much if not all of the meat in their diet by active hunting, in which they were able to target prime-aged individuals, has become largely accepted.

As has been stressed elsewhere in this volume, what remains to be determined is how variability in Neandertal hunting behavior and subsistence strategies overall related to variation in ecological and environmental parameters, and whether the Neandertals show the same sort of adaptive flexibility (and employed the same broad adaptive strategies) as do modern human foragers (Kuhn 2011). Along these lines, John Speth (2012: 9) has enumerated a number of factors that likely interact to dictate hunting strategy and prey choice among foragers, including what prey was available and in what numbers, the condition and reproductive states of the available prey at any given time of year, the size of the hunting parties that could be mustered, the distance between hunting areas and basecamp, the food and hide needs of the hunter, the technology at the hunter's disposal, the context of the hunt (subsistence versus hide needs versus social needs), the regional competitive landscape, and the larger social landscape. To this list, for Neandertals, I would also add the physiographic topography as it related to the distribution and availability of prey. As noted by Speth (2012), the interaction of the variables produces a continuum of hunting patterns, from hunters who take prey more or less in proportion to their availability in the landscape (so-called "generalized" hunters), to hunters who largely take prey according to availability but modify their hunting strategies based on situational and organizational constraints, to hunters that target very specific prey sets (so called "specialized" hunters). Neandertal hunting behavior no doubt varied in predictable ways along this continuum in relation to variation in the factors identified by Speth, and identifying ways in which Neandertal responses may have differed from those expected of modern humans may serve as a potentially fruitful way of exploring adaptive (and/or cognitive)[6] differences between groups. Along these lines, I wish to make abundantly clear that I have no reason to doubt the capabilities of Neandertals to hunt a variety of prey in a variety of contexts, and as I hope I have demonstrated, I simply think that the nature of their weapon technology introduced an important constraint on their hunting behavior, which in turn had a significant impact on their subsistence ecology.[7]

NOTES

1. Note that the selection responsible for between-group differences in joint morphology may have occurred in one sex only, while the result (the specific morphological feature) will appear in the skeletons of both sexes (since the favored alleles will be passed to offspring of both sexes in subsequent generations).
2. A lever is a rigid beam that rotates at a fixed point, the fulcrum. We tend to think of levers as force multipliers – we apply a force to one end and a greater force is exerted on some object by the other end of the lever. But levers only operate this way when they have a mechanical advantage greater than 1. For a given lever, the force that is applied to one end (the in-force: F_i) times the length of the in-lever arm (L_i: also known as the "power arm," the portion of the beam between the point of application of the in-force and the fulcrum) is equal to the force the lever exerts on its target (the out-force: F_o) times the out-lever arm (L_o: also known as the "load arm," the distance from the fulcrum to the point of application of the out-force). Since $F_iL_i = F_oL_o$, $F_o = F_iL_i/L_o$. The mechanical advantage (or "leverage") of the system, then, is defined by the ratio L_i/L_o – when the ratio is greater than 1, force is multiplied, when less than 1, force is reduced by the lever. Levers also either multiply or diminish the velocity of their moving ends. The velocity that is applied to one end (the in-velocity: V_i) times the length of the out-lever (L_0) is equal to the out-velocity (V_o) times the length of the in-lever (L_i). Since $V_iL_o = V_oL_i$, $V_o = V_iL_o/L_i$, and thus the speed advantage of a system $= L_o/L_i$. The speed advantage of a lever system, then, is the inverse of its mechanical advantage – a single lever system cannot be constructed so as to amplify both force and speed. Most bone-muscle levers in the vertebrate skeleton have mechanical advantages less than 1, which is disadvantageous for exerting force on the environment, but advantageous for both providing our moving parts an ample range of motion and for accelerating parts of our body during locomotion and manipulation.
3. It should be borne in mind that while measures of robusticity are measures of relative strength, they are not free of size effects (see Jungers and Burr 1994; Jungers et al. 1998), and thus robusticity comparisons across groups of different mean body sizes should be done with caution.
4. "Dart" is used to denote a spear propelled by a spearthrower (also known as an atlatl, *propulseur*, throwing board, or woomera), and to distinguish this weapon system from hand-deployed spears.
5. Three indicators are often taken to signal scavenging in humanly-generated faunal assemblages: an over-abundance of low utility skeletal parts (especially heads and feet, since these are the commonly unconsumed portions of a carcass that remain after carnivores have picked it over); attritional prey mortality profiles (characterized by an abundance of young and old individuals, as these are the individuals that non-human carnivores target in hunting); and evidence of human processing (cut-marks) that overlie carnivore chew marks (indicating that humans had late access to the carcasses).
6. Since this discussion has featured the work of John Speth, I cannot help but hear his voice in my head as I tentatively suggest that behavioral differences between Neandertals and modern humans might reflect cognitive differences. I sat beside John at a conference on Neandertals several years ago. He had recently published an article entitled "News flash: negative evidence convicts Neanderthals of gross mental incompetence" (Speth 2004), and

every time a speaker would say something that in the least manner impugned the intellectual ability of the Neandertals, John would lean over to me and emphatically whisper "NEWS FLASH!"

7. Might the lack of true long-range projectile weapons in the Neandertal toolkit reflect a lack of innovation, and thus a cognitive limitiation relative to modern humans? Not necessarily. I suspect that Neandertals employed close-range weapons because it is what their ancestors did and because it worked, and there was little incentive to invent new ways of doing things. It is not clear why Middle Stone Age modern humans did invest in projectile technology, although it is possible that temporary increases in population density in parts of Africa created pressure to expand diet breadth, thus favoring the adoption of projectile technology as a means of reducing the handling costs of certain prey items (see Chapter 12).

CHAPTER NINE

In the Company of Killers: Neandertals as Carnivores

The reviews presented in the last three chapters create a distinct picture of Neandertal ecology, and of the energy structure of their environments. Most of the humanly-consumable calories in the Pleistocene environments of Europe, if not the Near East, were to be found at the higher trophic levels. Whether Europe was covered with broadleaf deciduous forests or open steppe-tundra, most of the primary productivity was in plant structures that cannot be assimilated in the human digestive tract: even in the best of these types of environments, modern human foragers would depend heavily (≥50%) on meat to satisfy their energy and nutritional needs. Apparently lacking technological and behavioral strategies for harvesting, detoxifying and otherwise processing various small packages of plant foods, Neandertals may have found the European pantry even barer. On top of this was the problem of seasonality of plant production, which demands an overwintering strategy of some sort. While there are a number of tactics that can be used in concert to effect such a strategy (such as storage of plant and animal foods, or seasonally reducing energetic expenditures), increased

Thin on the Ground: Neandertal Biology, Archeology, and Ecology, First Edition. Steven Emilio Churchill.

consumption of animal tissues is one obvious solution to seasonal cessation of plant productivity, and one that was no doubt used by the Neandertals. It comes as little surprise, then, that the various lines of evidence at our disposal for deducing Paleolithic diet all suggest (some more strongly than others) that Neandertals had a significant footprint in the higher levels of the trophic pyramid.

The coordinates of the Neandertals in trophic space are of great importance, because when it comes to food webs there is very little room at the top: carnivores require high prey densities just to break even (Oksanen and Oksanen 2000). Based on global patterns of secondary to tertiary production across a large range of carnivore body sizes, Carbone and Gittleman (2002) estimate that it takes a standing crop of about 10,000 kg of herbivores to support 90 kg of carnivore mass, irrespective of the size of the carnivores (in fact, larger carnivores, because they have higher costs of transport, may fare a bit worse: Carbone et al. 2007). Based on rough estimates of secondary productivity (Chapter 6), the Pleistocene biome with the greatest herbivore production – steppe – would have only supported about 45 kg km^{-2} of carnivore mass – roughly one Neandertal per two square kilometers, or one lion per five square kilometers. But acquiring one's slice of the herbivore pie is not easily done, both because of the dangers inherent in the act of predation itself (as discussed in Chapter 8.3) and because the limited resources at higher trophic levels are hotly contested. For a Neandertal, feeding on herbivores meant direct competition with large-bodied carnivores that were aggressive, powerful, and built for killing.

Modern human foragers deal with competition at the higher trophic levels in a number of ways. In areas of reasonable plant productivity, one mechanism for reducing competition with carnivores is by ramping up the plant component of the diet – an option that may not have been available to Neandertals. Modern humans also use social and technological means to gain an edge in competitive interactions with carnivores, again (as we'll explore below) this option may not have been open to Mousterian foragers, or at least not as easy for them to exercise. For these reasons it becomes apparent that Neandertal subsistence ecology must be viewed within the context of competition within the Eurasian Pleistocene large-bodied carnivore guild (cf. Stiner 1991c, 1994; Turner 1992; Mussi and Palombo 2001; Estévez 2004). The nature of the relationship of Neandertals (and early modern humans) to other carnivores is also important with respect to competition for space (i.e., shelter sites), and the possible demographic effect that this competition had on either side (e.g., Mussi 2001b; Grayson and Delpech 2001 vs. Wolverton 2006).

Competitive interactions within guilds of carnivores have long been a focus of study by ecologists and mammalogists, but generally with a focus on exploitative competition – the sometimes direct but usually indirect competition that occurs between different species with overlapping resource bases. A number of more recent studies, however, have signaled the importance of interference competition within carnivores – involving direct and usually antagonistic interactions between guild members – as key determinants in carnivore ecology and demography. In the remainder of this chapter, we explore the implications of both types of competition to Neandertal ecology, with an eye towards defining the niche space occupied by Neandertals *vis-à-vis* other carnivores, assessing habitat use constraints experienced by Neandertals because of the presence of other large-bodied carnivores, and estimating population densities of the major players in the Eurasian Pleistocene large-bodied carnivore guild, Neandertals included.

9.1 LARGE-BODIED CARNIVORES OF THE EURASIAN LATE PLEISTOCENE

The first task in exploring the ecologically-significant interactions that may have defined the relationship of Neandertals to other carnivores is to identify the salient actors. By modern standards, Pleistocene Europe (and western Asia) had a rich carnivore fauna, but it is not apparent that every species of carnivore identified in the fossil record was of equal significance in the competitive interactions experienced by Neandertals. Who, then, were the important players?

Fossils from late Middle and Late Pleistocene faunal assemblages reveal a carnivore community that was mixed in several dimensions. First, the carnivore guild across this time contained a combination of extinct and extant species (as well as forms that today are locally extinct but that survive in other geographic regions). In this regard, the Late Pleistocene carnivore community represented the end stage of the transition from the Middle Pleistocene archaic fauna of the Galerian European Land Mammal Age (ELMA) to the modern faunas of the Late Pleistocene and Holocene Aurelian ELMA (Mussi and Palombo 2001). Few of the archaic carnivores survived the transition to the Aurelian ELMA, roughly around the beginning of MIS 9 (ca. 330 Ka BP), with only the European jaguar (*Panthera gombaszoegensis*: Turner and Antón 1997, also called the Tuscany lion, *P. toscana,* by Kurtén 1968), the archaic, giant lion (*P. fossilis*: Mussi and Palombo 2001), and the lesser scimitar cat (*Homotherium latidens*: Reumer et al. 2003) appearing in late Middle and Late Pleistocene contexts. The two species of *Panthera* do not appear to have survived past MIS 7 (Kurtén 1968; Mussi and Palombo 2001), while the saber-toothed cat appears to have existed – albeit in relatively low numbers (judging from its extremely rare occurrence in fossil deposits post-dating 500 Ka BP) – until the terminus of MIS 3 (based on a ca. 28 Ka BP specimen from the North Sea: Reumer et al. 2003). There is one archaic carnivore species, though, that is ubiquitous across Europe until late in MIS 3: the cave bear (*Ursus spelaeus*). However, as discussed in Chapter 7.5, the cave bear is taxonomically a carnivore but dietarily an herbivore (Kurtén 1976; Richards et al. 2000: although at times more omnivorous/carnivorous: Richards et al. 2008b), and thus does not merit inclusion in the large-bodied carnivore guild.[1] So, while there was certainly a temporally-dynamic quality to the nature of the European large-bodied carnivore guild (and all of the Eurasian mammalian guilds, for that matter) during the time of the Neandertals, by the Late Pleistocene (beginning with MIS 5e) the Eurasian carnivore community had largely settled into a modern configuration, with only a few archaic taxa in the mix.

Second, the Late Pleistocene Eurasian carnivore guild included a mix of northern hemisphere holarctic forms (e.g., wolf, red fox, lynx, brown bear, badger, and wolverine), including some predominantly Asiatic species (dhole and Asiatic black bear), a sprinkling of smaller arctic taxa (arctic fox, least weasel), and a healthy representation of species which today are generally associated with African faunal communities (lion, leopard, spotted and striped hyena). This latter, African component appears to have comprised the nexus of the competitive interactions that mattered to the Neandertals.

Of the 30 or so carnivore species that are represented in fossil assemblages of the European later Pleistocene, only a handful preyed on moderate- to large-bodied ungulates, and thus can be considered major resource competitors with the Neandertals. Smaller carnivores (mean body masses ≤ 20 kg) can reasonably be excluded from the guild occupied by the Neandertals (cf. Lewis and Werdelin 2007 for earlier members of the genus *Homo*). Prey size selection is heavily dependent on carnivore body size (Rosenzweig 1968; Gittleman 1985) and, in general, small carnivores have diets dependent on some combination of small mammals and reptiles, birds, invertebrates, and vegetable matter (e.g., Roper 1994) – items that did not figure heavily in the Neandertal diet (although they apparently did in the diets of early modern Europeans: Stiner et al. 2000; Richards et al. 2001; Stiner 2001). A few of the larger members of this small-bodied carnivore guild – notably lynx and wolverine – are known to consume and even actively hunt moderate and large ungulates, and thus they may merit inclusion in the large-bodied carnivore guild. Given the nature of both dietary flexibility and intraguild competition in carnivores, however, we can probably safely ignore these mid-sized hunters: they may contribute to a significant portion of moderate–large ungulate mortality in some places in the modern world but, given the wealth of larger carnivores in paleoecosystems, I think they were unlikely to have done so in Pleistocene Eurasia.

The diet of a given carnivore species is usually variable across space, and carnivores use dietary and predatory flexibility as a way of adjusting to variation in local abundances of potential food items and in the dynamics of intraguild competition in any given area. Mid-sized carnivores like lynx and wolverine are no exception. In the Jura Mountains of France and Switzerland, lynx focus on small-bodied cervids (roe deer) and bovids (chamois), and these two prey items together constitute about 91% of their quarry (Jobin et al. 2000). The remainder of the diet is comprised of smaller mammals, including several small-bodied carnivores. Farther north, larger ungulates enter the dietary mix. In the Białowieża Forest of Poland, for example, red deer constitute 61% of the ungulates killed by lynx (over a 7-year study period: Jędrzejewski et al. 1993). Lynx can be a significant source of mortality for red deer in some areas (accounting for 9% of their natural mortality in the Białowieża Forest), but direct their predation at fawns rather than adults (Jędrzejewski et al. 1993), which might be expected given the relatively small body size of lynx (combined sex average = 11.3 kg: Gittleman 1985). Wolverines, like lynx, exhibit considerable regional variation in their diet and, like lynx, these relatively diminutive carnivores (combined sex average = 11.6 kg: Gittleman 1985) are known to prey upon moderate- and large-bodied ungulates, including caribou, moose, and sheep (Myhre and Myrberget 1975; Landa et al. 1997; Lofroth et al. 2007). While wolverines have been known to attack and kill moose (references in Myhre and Myrberget 1975), the majority of their larger ungulate intake comes from scavenging winter-killed animals or carcasses produced by larger carnivores (Lofroth et al. 2007). Medium- and small-sized mammals like ground squirrels, marmots, hares, tree squirrels, porcupines, and rodents, as well as ground-nesting birds, become increasingly important in the diet as larger mammals become unavailable (see Lofroth et al. 2007). This last point, and indeed the dietary flexibility of these mid-sized carnivores generally, takes on increased significance when it is noted that most modern studies have been done in contexts with an impoverished large-bodied carnivore guild – with some cases involving a complete lack of larger carnivores (Landa et al. 1997). Given the nature of intraguild interference competition and the importance of predator body size to the outcome of competitive interactions (see below), it seems likely that in a situation involving at least six carnivore species of 35+ kg body mass (as in Pleistocene Europe), these mid-sized predators would not have been able to compete for a share of the large ungulate productivity and would have relied much more heavily on small game in the diet.

Having dismissed the smaller carnivores, I think we can also discount bears as major players in the Eurasian Late Pleistocene large-bodied carnivore guild. There is no doubt that bears were present: both brown and cave bear were visible components of the carnivore fauna during the time of the Neandertals, and up through MIS 6 the Asiatic black bear was also present in Europe. There also is no doubt that bears are large and aggressive, and that they are quite capable of dominating carnivores smaller than them in hostile interactions (for example, wolves eventually lose most of their carcasses to brown bears in Yellowstone National Park: Downey 2006). But, as with the mid-sized carnivores discussed above, it is important to take into consideration the dietary flexibility of ursids. I have already argued that cave bears, by virtue of their vegetarian habit, should be excluded from consideration.[2] Brown bears and Asiatic black bears are omnivorous, and their diet is based primarily on plant food (see, for example, Bocherens et al. 2004; Wolverton 2006). The component of the diet deriving from higher trophic levels includes a mix of fish, insects, honey, and ungulates. In certain places and at certain times, brown bear predation on ungulates may be substantial – for instance, brown bears may kill upwards of six caribou calves per day during the three week calving period in northeastern Alaska (Young and McCabe 1997). In Yellowstone National Park in the United States, brown bears consume a substantial amount of ungulate meat, almost exclusively obtained by driving wolves off of red deer (or, less frequently, bison or moose) carcasses (Gunther and Smith 2004). Brown bears are an order of magnitude larger than wolves (average combined sex body mass of brown bears = 298.9 kg; wolves = 33.2 kg: Gittleman 1985), and thus an adult grizzly will almost always prevail in usurping a kill from even a moderate-sized pack of wolves. Still, doing so isn't easy (and may not even be possible for juvenile bears, despite being two- to three-times the body mass of the wolves), and female bears sometimes lose cubs to wolves during aggressive interactions (Gunther and Smith 2004). I suspect that both brown and black bears of the Pleistocene opted out of confrontations with social groups of lions (who may have averaged about half the size of an adult brown bear), and probably even with packs of hyenas and Neandertals (at about a fourth of the brown bear's size). Given their dietary flexibility, the usually low levels of ungulate meat in their diet, and the injury risks of confrontational scavenging, Pleistocene bears more than likely focused on the vegetable and small animal components of their diet.

With the exclusion of bears and smaller carnivores, our cast of predators is reduced to three felids (lion, leopard, and the lesser scimitar cat), two canids (wolf and dhole), two hyaenids (spotted and striped hyena) and one hominin (Neandertal). The lesser scimitar cat (*Homotherium*) is a rare occurrence in the later Quaternary European record, there being only a handful of fossils recovered thus far that post-date 500 KaBP (see Reumer et al. 2003), which has led to the interpretation that this carnivore existed in low abundances (relative to other carnivores, especially lion) throughout its range (Kurtén 1968). Still, despite its low visibility in the fossil record, *Homotherium* must have maintained sufficient population densities to remain demographically viable for more than a half million years, meaning it must have been a consistent (even if not overly abundant) feature of the carnivore community during the time of the Neandertals.[3] With the stage thus set, the only change that takes place in the *dramatis personae* is the exit, by MIS 5e, of the striped hyena from the theater (Kurtén 1968: with a more recent, possibly Neolithic, recolonization of Eurasia from an African refugium; see Rohland et al. 2005).

In the next section, we review the basic ecological characteristics of the carnivores with which the Neandertals competed. The niche dimensions of Pleistocene carnivores are, of course, unknown, but reasonable inferences can be made from consideration of living conspecifics or congeners (in the case of lions, leopards, hyenas, wolves and dholes), from paleontological and functional morphological studies (for *Homotherium*), and from archeological inference and analogy to modern human foragers (for Neandertals). Some of the more salient niche dimensions of the core carnivores of the Eurasian Late Pleistocene large-bodied carnivore guild are provided in Table 9.1, and brief summaries of each taxon are provided below.

9.2 THE MEMBERS OF THE EURASIAN PLEISTOCENE LARGE-BODIED CARNIVORE GUILD

9.2.1 *Homotherium latidens*

Much of our knowledge of the morphology, behavior and ecology of the lesser scimitar cat, *Homotherium latidens*, comes from key fossil sites in France and Spain, but inferences have also been generated through comparisons with its North American Pleistocene sister species, *H. serum*. This European machairodont (saber-toothed) cat, at 150–230 kg (based on the size of *H. serum*: Anyonge 1993), was within the size range of modern African lions (130–260 kg: Creel 2001), but smaller than European cave lions (see below, and Antón et al. 2005). It differed from lions, however, in having postcranial morphology indicative of a greater degree of cursorial pursuit and less ability at grappling with prey (Anyonge 1996; Arribas and Palmqvist 1999; Antón et al. 2005), suggesting cheetah-like hunting tactics (stalking followed by pursuits of less than 1 km) and a preference for open country (see Lewis 1997; Antón et al. 2005). Also like the cheetah, *Homotherium* had a relatively large brain and an enlarged optic center (Rawn-Schatzinger 1992), suggesting a degree of sociality (see also Antón et al. 2005)[4], along with diurnal activity patterns.

The dental morphology of *H. latidens* suggests (1) a specialized ability for killing relatively large prey or thick-skinned prey (Antón et al. 2005; see also the review of debates about the function of saber canine teeth in Turner and Antón 1997), and (2) a hypercarnivorous dietary strategy (Arribas and Palmqvist 1999). The long and narrow, blade-like carnassial teeth of *H. latidens* imply specialization for stripping flesh quickly and efficiently from a carcass, and little if any bone processing for in-bone nutrients. The long-standing idea that machairodont cats were inefficient at removing meat from carcasses (see review in Antón et al. 2005), and thus left abundant scavengable remains for hominins and other carnivores, appears to be in error, certainly for the homotheres and probably for the "dirk toothed" cats, like *Megantereon* and *Smilodon*, as well. The incisors of the lesser scimitar cat were somewhat procumbent, and proficient for tearing flesh from carcasses (Arribas and Palmqvist 1999): elephant bones from a *H. serum* den in Texas bear tooth marks indicating that the cats were able to clean the bones of meat (Marean and Ehrhardt 1995).

TABLE 9.1 ■ Niche dimensions of Eurasian Late Pleistocene large-bodied carnivores

	Body mass (kg)	Social behavior	Habitat preferences	Hunting methods	Feeding strategy	Prey body size preferences	Probable preferred prey in Pleistocene Eurasian contexts
Lesser scimitar cat	150–230	social?	open> woodland	stalking, coursing	hypercarnivore +	large	large bovines and cervids; juvenile mammoths and rhinos
Cave lion	160–325	social	open> woodland	ambush	flesh, some bone	190–550 kg (possibly larger)	horse, bovines, giant deer
Leopard	45–90	solitary	woodland	ambush	flesh, some bone	20–80 kg	roe and fallow deer, wild boar, gazelles, ovines; juveniles of large forest herbivores
Spotted hyena	65–90	social	open> woodland	coursing	bone cracking	55–185 kg (possibly larger)	saiga, reindeer, horse, bison (smaller individuals), red and fallow deer, juvenile elephants
Striped hyena	30–50	solitary	all	opportunistic	bone cracking	smaller herbivores	saiga, gazelles, roe and fallow deer, wild boar, perhaps ibex and mountain goat
Wolf	45	social	all	coursing	hypercarnivore -	100–300 kg (possibly larger)	horse, bison, moderate and large cervids (reindeer, red and giant deer); flexibly including smaller (saiga, roe deer) and larger (scavenged megafauna) prey
Dhole	20–40	social	woodland	coursing	hypercarnivore +	30–175 kg	reindeer, roe and fallow deer, wild boar, ovines.
Neandertal	60–85	social	woodland/ open ecotone?	ambush	all	80–500 kg	reindeer, red deer, horse, bison, aurochs; varying amounts of megafauna

Although smaller and more gracile than cave lions (Antón et al. 2005), *H. latidens* appears to have directed its predation towards megaherbivores, as it (and its sister species *H. serum*) is often found in association with remains of elephants (e.g., Meade 1961), and because its killing teeth and cranionuchal morphology appear to be adapted to predation on larger animals (Hooijer 1962; Turner and Antón 1997; Antón et al. 2005). However, predation was almost certainly directed at juvenile members of the largest herbivores, just as modern lions do when hunting elephants (Joubert 2006). Thus, as noted by Antón et al. (2005: 1295), "the key advantage of the scimitar-tooth adaptation was probably not so much the access to prey significantly larger than lions can take but the higher efficiency in the killing of animals within the upper range of the lion's prey spectrum." *H. latidens*, then, appears to have been oriented towards prey in the range of large bovines (bison, aurochs) and cervids (giant deer) up to juvenile rhinos and elephants (perhaps up to several metric tons in mass), most likely taken through social hunting in open environments.

9.2.2 *Panthera leo spelaea*

Cave lions were broadly distributed and relatively abundant in Late Pleistocene Eurasia (Kurtén 1968; Nagel et al. 2003; Diedrich 2007; Rothschild and Diedrich 2012). While considered by most to be a regionally- and temporally-defined subspecies of the extant lion (references in Sommer and Benecke 2006: but placed by some into its own species, *P. spelaea*: Sotnikova and Nikolskiy), the cave lion was larger – by about 25% – than its extant African conspecifics (Anyonge 1993; Kurtén 1985). Given a body mass range of about 130–260 kg for living African lions (Creel 2001), this would suggest a size range between about 160 and 325 kg for the Pleistocene form. Note that the cave lion's sister species – the North American Pleistocene lion *P. atrox* (also considered by some to be a subspecies of *P. leo*) – may have weighed as much as 500 kg (estimated body size range 344–523 kg: Anyonge 1993), suggesting that some European cave lions may have attained similar size (equivalent to the mass of a good-sized adult male polar bear!).

Inferences about the behavior and ecology of cave lions come mainly from consideration of extant lions, although there have been suggestions that cave lions differed from their modern analogs in some critical ways. Specifically, the lack of a mane in the Pleistocene forms[5] has suggested to some that they may have lacked the pride structure of extant lions, instead existing as breeding pairs or in very small prides (Guthrie 1990). In prides with few females, males more often assist in hunting, leading Guthrie (1990) to suggest that Pleistocene lion pairs hunted together over large expanses of territory. This picture would imply relatively low lion densities, which thus has important implications for competitive (especially aggressive) interactions between lions and other carnivores – including Neandertals – as discussed below. The connection between small-to-nonexistent manes and social structure may not, however, be so clear. Variation in mane size in extant lions is not strongly associated with variation in pride size, and patterns of body size and canine size dimorphism in cave lions suggest that they had a breeding structure similar to that of modern lions, in which a male or a coalition of males reproductively monopolized a pride of females (Turner 1984; Yamaguchi et al. 2004). The lack of a mane in cave lions may simply reflect a thermoregulatory constraint for large male cats during the summer months (Nagel et al. 2003).

If the social structure of cave lions was like that of extant African lions, they would have lived in prides of up to a dozen adults with offspring. These prides would have had from one to three adult males in residence, who would have been considerably larger than the females (perhaps as much as two times larger). Despite the greater size of the males, much of the prey procured by the pride would have been taken by females in cooperative hunts. Small bachelor bands – males without a pride – would also have existed, and the males in these groups would probably also have hunted cooperatively, but would likely have directed their predation towards prey size classes larger than those normally taken by females (Turner and Antón 1997).

Based on analogy with living lions, the cave lion was likely most active at night, and probably hunted by ambush (stalking prey under the cover of tall grass or bushes, followed by a short rush from cover). Lions thus prefer to hunt under cover of darkness, and in areas with some bush cover or tall grass (Funston et al. 2001; Grant et al. 2005). Despite a need for a modicum of cover for ambush hunting (a limitation that can be overcome to a degree through coordinated group hunting: see Antón et al. 2005), lions tend to direct predation towards open country grazers, and thus feed off the secondary productivity of high-turnover, high-secondary biomass grasslands (e.g., Schaller 1972; Mills and Shenk 1992). Interestingly, patterns of entheseal reactions on cave lion postcranial remains suggest a greater emphasis on pursuit hunting, suggesting that the Pleistocene lions may have engaged in more chasing of prey and less ambush hunting than do their modern counterparts (Rothschild and Diedrich 2012).

The importance of body size in lions relative to the size of their prey was discussed in Chapter 4.4 (see also Chapter 8.3), and given the larger size of Pleistocene lions their prey preferences may have run more towards the larger herbivores (and juvenile megaherbivores) than do those of modern lions. Modern African lions preferentially prey on herbivores in the body size range of 190–550 kg, with a most preferred prey weight of 350 kg (Hayward and Kerley 2005). It is important to bear in mind that lions are flexible in dietary habits, and capable of adapting their predatory strategies as ecological conditions vary: although African lions prefer zebra, wildebeest, gemsbok, buffalo, and giraffe (Hayward and Kerley 2005), in places they rely heavily on prey that is smaller than preferred (warthogs and impala: Hayward and Kerley 2005; Antón et al. 2005) or larger than preferred (juvenile elephants: Joubert 2006). Based on the size of Pleistocene European lions and the predatory preferences of their African analogs, it is likely that cave lions focused on horses, larger deer (*Megaloceros* and possibly red deer) and bovines (bison and aurochs) (Turner and Antón 1997). Evidence for lion predation on a 700–800 kg prime-aged male archaic bison (*Bison priscus*), determined from the nature of the carnivore-induced damage to the carcass, can be seen in a mummified carcass, dubbed "Blue Babe," from Alaska (Guthrie 1990). Given that a considerable amount of secondary biomass was likely to be found in megaherbivores (rhinos and elephants) in European Pleistocene steppe habitats (see below), and given the large size of cave lions, it is possible that their diet also regularly included the juveniles of these open country giants (to the extent that competition with lesser scimitar cats allowed).

9.2.3 *Panthera pardus*

Leopard fossils are well represented in later Pleistocene deposits in Eurasia (Sommer and Benecke 2006), the species having taken up residence in more northerly latitudes sometime during MIS 9 (Wolsan 1993). Extant leopards are solitary and secretive[6] (Bailey 1993), and there's no reason to suspect that their Eurasian Pleistocene conspecifics were not also. Thus this large felid contrasts with lions and the lesser scimitar cats in terms of social structure, and probably also in body size and in important aspects of habitat preference and prey preferences.

As far as felids go, leopards are relatively variable in body mass, showing considerable variation between sexes and within sexes across their geographic range (Bailey 1993). Adult leopards vary in mass between 20 and 90 kg (Mills and Harvey 2001; Stuart and Stuart 2000), although adult males larger than 70 kg may be relatively rare (Bailey 1993). Based on available fossils, European fossil leopards were as large as the largest extant specimens, so adult males were probably in the 70–90 kg range (Turner and Antón 1997). Given an average body mass dimorphism of about 65% (see Table 5.1 in Bailey 1993), adult females were probably in the range of 45–60 kg. Based on limb proportions and other aspects of morphology in fossil and extant leopards (Lewis 1997), and on the behavior of extant conspecifics, it is reasonable to infer that Eurasian Pleistocene leopards were solitary ambush hunters with a marked preference for more closed, woodland environments.

At one-half to one-seventh the size of other felids roaming the landscape, the preferred prey size of Pleistocene leopards would have been correspondingly smaller than that of the larger sympatric cats. In Africa, extant leopards have a very diverse diet, and while they regularly take game weighing between 20 and 80 kg, they significantly prefer prey within the 10–40 kg mass range (with a most preferred mass of about 25 kg: Hayward et al. 2006). Leopards tend to focus their hunting on smaller, bush-loving antelopes (impala, bushbuck, duikers), the young of larger antelopes (such as kudu and wildebeest), and on small game such as baboons, porcupines, and warthogs (Hayward et al. 2006; Turner and Antón 1997). Indian leopards also focus on medium-sized prey (31–175 kg) such as bushpig and chital, muntjac, and sambar deer, with a mean prey size of 37.6 kg (Karanth and Sunquist 2000; Hayward et al. 2006). Thus, the most likely items in Pleistocene Eurasian leopard diets would have been smaller, cover-loving herbivores like roe and fallow deer and wild boar, and the young of larger prey such as red deer, moose, forest bison, and aurochs. In western Asia, gazelles would have entered the diet mix, as would goats and sheep in more mountainous areas.

9.2.4 *Crocuta crocuta spelaea*

Spotted hyenas are well-represented in Late Pleistocene deposits in Europe and the Near East (see references in Diedrich and Žák 2006), no doubt thanks to their proclivity to den in caves and rock shelters. The Eurasian spotted hyena is generally known as the "cave hyena," and (as we saw with the cave lion) its exact taxonomic status (whether a subspecies of the spotted hyena, *C. crocuta spelaea* or its own distinct species, *C. spelaea*) remains a matter of debate (see Rohland et al. 2005). But whether a subspecies or a closely-related sister species, the Eurasian Pleistocene spotted hyena appears to have been morphologically and behaviorally similar to extant African spotted hyenas (albeit with some slight morphological differences: references in Rohland et al. 2005).

Like modern spotted hyenas, the cave hyena was a social carnivore (based on abundances and age distributions of fossils in cave deposits: see Diedrich and Žák 2006), living in clans that probably (like extant spotted hyenas) had an average size of about 25 individuals (but perhaps sometimes reaching as many as 80 individuals: Kruuk 1972; Mills 1990). Although clans may be large in extant spotted hyenas, hunting is generally conducted by a smaller subset of clan adults: hunting may be undertaken as a solitary endeavor, but usually the harassment of ungulate herds and ensuing pursuit of target animals involves the cooperative action of 2–6 individuals (Mills 1990: in some cases, however, as many as 20 animals may participate in a hunt: Kruuk 1972). Hyenas are primarily nocturnal hunters, and take prey by chasing it, sometimes over distances of 1–3 km (Kruuk 1972).

Cave hyenas are generally considered to have been larger and more robust than extant spotted hyenas (Diedrich and Žák 2006), with interglacial individuals larger than extant African hyenas and animals coming from glacial periods even larger still (Klein and Scott 1989). However, reliable estimates of the average body mass of Pleistocene hyenas are not readily available in the literature. My own observations of fossil and recent hyena cranial material suggest that the Pleistocene forms were in the range of 10–25% larger, on average, than extant spotted hyenas from southern Africa. A small sample of spotted hyenas from the Kalahari produced mean mass values of 59 kg for males and 71 kg for females (Mills 1990), which would put the cave hyenas roughly in the range of 65–90 kg.

As coursing hunters, hyenas prefer more open environments. Their tastes are catholic, they are able to take down a wide range of prey body sizes, and they show little marked preference for particular prey species (Hayward 2006). Despite the eclectic nature of hyena predation, they do prefer medium-to-large ungulates (preferred prey body size range 56–182 kg, mode = 102 kg: Hayward 2006; cf. Mills 1990). In open habitats of Pleistocene Europe, their preferred prey likely included saiga antelope, reindeer, and horse, and perhaps smaller individuals from bison herds (see Diedrich and Žák 2006). In more closed environments cervids, notably red deer and fallow deer, probably formed a significant portion of the diet (e.g., Stiner 1991a on the hyena-accumulated fauna from Grotta Guattari). Spotted hyena in Africa have been observed taking prey as large as adult buffalo (Pienaar 1969), and juveniles of giraffes, rhinos and, occasionally, elephants (Hirst 1969; Mills 1990; Berger and Cunningham 1994). In Pleistocene Europe, cave hyena may have regularly preyed on juvenile wooly rhinos and scavenged the remains of adults (perhaps secured through confrontations with *Homotherium* or Neandertals), as suggested by an abundance of rhino remains in hyena dens in central Europe (Diedrich and Žák 2006).

Spotted (and presumably cave) hyenas are active hunters who in most contexts may secure the majority of their caloric needs (corresponding to about 3.8–4.9 kg of meat per day per adult: Henschel and Tilson 1988) through predation. They are, however, also confrontational scavengers, and in contexts in which large carcasses are available through the action of other carnivores they may obtain up to a third of their dietary need through kleptoparasitism (Kruuk 1972). Theft of carcasses works both ways, however, and hyenas may lose up to 20% of their kills to prides of larger, more socially-dominant lions (Kruuk 1972). Accordingly, hyenas are prone to disarticulate carcasses and transport portions back to den sites where they can be consumed in relative safety, and where nutrients within the shafts of long bones or within braincases can be accessed with the aid of their stout, bone-crunching teeth and powerful masticatory apparatus.

9.2.5 *Hyaena hyaena*

Striped hyenas occur in the Eurasian fossil record of the Middle Pleistocene, but appear to have gone locally extinct by MIS 5e times (Kurtén 1968). Their distribution in historic times through southwest and southern Asia appears to be the result of recolonization of the Asian landmass from an African refugium sometime after the last interglacial, and possibly as recently as Neolithic times (Kurtén 1965; Rohland et al. 2005).

Nocturnal and generally solitary in their foraging behavior (Kruuk 1976), striped hyenas consume a diet largely composed of the scavenged remains of large and moderate-sized herbivores, and hunted or scavenged carcasses of smaller herbivores, as well as small mammals, invertebrates, and plant foods (Kruuk 1976). Their relatively small body size (26–41 kg: Mills and Hofer 1998; Yom-Tov 2003), combined with their solitary foraging (or with a single partner: Rieger 1979), may limit their ability to take larger game, as their observed predatory behavior is largely confined to prey about the size of goats and sheep (Kruuk 1976; Leakey et al. 1999). Although European Pleistocene forms were likely larger than extant striped hyenas (if the size differences in their cousins the spotted/cave hyenas are any guide, then Ice Age "stripies" may have been about 29–51 kg), it is still likely that they limited their predation to smaller herbivores (saiga antelope, gazelles, roe and fallow deer, wild boar, and perhaps ibex and sheep in more mountainous areas). As with extant striped hyenas, their Pleistocene conspecifics probably preferred more open country but which had some light bush cover, although they no doubt could do well in more closed country, including even full temperate forest conditions. As nonconfrontational scavengers of the kills made by larger predators (again, based on the behavior of extant striped hyena), Pleistocene striped hyenas probably removed portions of carcasses to safer denning locations (Leakey et al. 1999), where they could work on extracting nutrients from within long bones or brain cases in relative safety.

9.2.6 *Canis lupus*

The wolf was a ubiquitous feature of faunal communities during Neandertal times (Kurtén 1968), and is the only member of the later Pleistocene Eurasian large-bodied carnivore guild (as considered here) that is still extant in Europe. Modern wolves are geographically widespread, eurytopic, and highly variable in body size and hunting ecology (see Jolicoeur 1959; Mech 1970; Kunkel et al. 1999, 2004; Jędrzejewski et al. 2007), and their behavioral flexibility makes them highly adaptable to circumstances. Accordingly, wolf prey preferences, pack sizes, and other aspects of ecology vary as a function of the constitution of the local herbivore guild, the abundance of various prey animals within that guild, and the local mix of other carnivore competitors (including avian competitors: Vucetich et al. 2004).

Wolves vary geographically in mean body size, but fossil evidence suggests that Pleistocene wolves were coeval in size to modern Alaskan wolves (JA Leonard et al. 2007). This claim is based on consideration of Pleistocene fossils of North American *C. lupus*, noting that genetic studies suggest that North American Ice Age wolves were more closely related to European wolves than to extant North American wolves (JA Leonard et al. 2007; extant wolves in Europe, on the other hand, appear to be descendants of local Pleistocene wolves: Sommer and Benecke 2005). Thus, European Pleistocene wolves likely had an average body mass of about 45 kg (Macdonald 1984;

Palmqvist et al. 1999), and did not attain the impressive sizes seen in fossils of the North American dire wolf (*C. dirus*, in which subspecies mass averages range between 59 and 68 kg: Anyonge and Roman 2006). Reasonable assessments of body mass are important, because prey size preferences have been shown to vary positively with body mass in wolves (references in Larivière and Crête 1993).

Wolves are social predators that hunt in packs usually composed of a monogamous breeding pair (the alpha pair) and their offspring (Van Valkenburgh and Sacco 2002), although all male, single male/single female, large fission/fusion, and polygynous groups can also be found (Mech and Boitani 2003). Pack sizes vary between about 4 to 18 animals, as a function of prey density, body size of available prey (Jędrzejewski et al. 2007), and the intensity of competition with scavenging ravens (Vucetich et al. 2004). As social hunters, wolves are able to bring down quarry an order of magnitude larger than them, and prefer large over smaller prey (references in Anyonge and Roman 2006). The North American dire wolf (*C. dirus*) likely had a preferred prey size of 100–300 kg, but could handle prey up to 600 kg (Van Valkenburgh and Koepfli 1993; Van Valkenburgh and Hertel 1998), although recent upward revisions of estimated body mass of dire wolves suggests that typical prey sizes may have been somewhat greater still (Anyonge and Roman 2006). The Pleistocene grey wolf, while being smaller than the dire wolf, also apparently regularly consumed larger prey: based on stable isotope evidence and considerations of body size and jaw biomechanics, North American (and presumably European) Ice Age grey wolves enjoyed a varied diet, but one that regularly included horse and bison, and at times muskoxen and mammoths (the latter likely scavenged) (JA Leonard et al. 2007). Given that extant holarctic wolves regularly take the largest extant cervid (moose) and large bovines (bison), the diet of European Pleistocene wolves probably also included giant deer and red deer, as well as smaller cervids (reindeer, roe and fallow deer) and bovids (saiga antelope and ibex).

Wolves hunt by coursing, and prefer hunting habitats where terrain features or vegetation provide some cover for stalking or using ambush to initiate a chase, but which otherwise favor pursuit of prey (flatter and generally open areas: Kunkel and Pletscher 2001). Despite having an advantage in numbers, wolves often have difficulty defending carcasses from other carnivores (bears) and scavenging corvids, and where they are sympatric with bears, wolves are routinely displaced from their kills by them (Downey 2006). In the rich carnivore guild of Pleistocene Eurasia, wolves would probably have found it difficult to defend carcasses from lions, hyenas, lesser scimitar cats, and possibly Neandertals. Not surprisingly, wolves have evolved a hypercarnivorous feeding strategy (van Valkenburgh 1991) that allows them to feed quickly at a kill, although they do include some amount of bone and within-bone tissues in their diet (Palmqvist et al. 1999). However, North American Pleistocene wolves exhibit much higher frequencies of tooth breakage than do their extant counterparts (11% versus 2–7% in geographic populations of modern wolves: JA Leonard et al. 2007), suggesting a greater amount of bone processing in the Pleistocene forms, perhaps in response to greater competition for carcasses in the richer Ice Age carnivore guilds.

9.2.7 *Cuon alpinus*

Dholes (or cuons) are coursing pack hunters that today survive only in central and east Asia. Although found predominantly in tropical forests and scrublands in the south, dholes are eurytopic carnivores that are equally at home in steppe and alpine habitats (and are found, for instance, on upland plains in Tibet and in the Ural Mountains of Russia: Fox 1984). In the Pleistocene, their range covered all of Eurasia, and while they were never an abundant feature of European faunas during the later Pleistocene, their consistent presence – at least in low numbers – is recorded at a handful of sites from the Middle Pleistocene onwards. Archaic members of the species appear in early Pleistocene fossil assemblages in Europe and western Asia, but the species appears to have gone locally extinct by late Würm (MIS 3) times (Kurtén 1968).

Extant dholes vary in adult body mass between about 10 and 20 kg (Rabinowitz and Walker 1991; Farlow and Pianka 2002; Palmqvist et al. 2002), but craniodental dimensions of European Pleistocene dholes suggest they were larger, and almost the size of grey wolves (Kurtén 1968), suggesting that 20–40 kg might be a reasonable body size range estimate for the species. Cuons are often seen out foraging solitarily, but they also commonly hunt socially, forming parties of up to 10 animals or more (Cohen et al. 1978; Karanth and Sunquist 2000). Although smaller in body size, their social behavior gives them a competitive edge over leopards in antagonistic encounters (they are the socially-dominant carnivore in the match up: see Section 9.3), although leopards do occasionally kill and eat dholes, as do tigers (Karanth and Sunquist 2000). Limited evidence suggests that where dholes overlap with leopards and tigers, they form larger hunting parties (Rabinowitz and Walker 1991). Among modern cuons there is no evidence that they seek out "competition refuges" (that is, portions of their habitat with lowered probability of encountering tigers or leopards: Karanth and Sunquist 2000), which is a strategy commonly employed by African wild dogs and cheetahs in avoiding lions and hyenas (Durant 1998). Nevertheless, competition refuges may have played an important role in cuon ecology in Pleistocene Eurasia, given the richness of the carnivore guild and its dominance by large social felids and hyaenids, which may have both kept dhole population densities low and confined them to more heavily wooded habitats.

Like wolves, pack-hunting dholes are capable of bringing down large prey, and in southern Asia they are known to take gaur (or seladang), which have adult body sizes in the range of 700–1500 kg (Cohen 1978; Cohen et al. 1978; Karanth and Sunquist 2000). While they are capable of bringing down large prey, their diet is focused on moderate-sized herbivores (mainly cervids), including hog deer, sambar, chital, swamp deer, and muntjack (Cohen 1978; Cohen et al. 1978), and one study in India (Karanth and Sunquist 2000) reported a preferred prey size of 31–175 kg and a mean prey size of 43 kg. In Russia, dholes are known to take reindeer, wild goats, and wild sheep (Cohen 1978), and these prey items may have also been a component of the diet of their Pleistocene European counterparts. However, the existence of larger social carnivores in open environments may have constrained cuons to hunting in either more closed habitats (forcing a greater dependence of cover-loving prey such as roe and fallow deer and wild boar) or more mountainous regions (in which goats and sheep may have dominated the diet).

9.3 COMPETITION WITHIN THE CARNIVORE GUILD

Now that we have identified the important players, we can begin to explore the nature of the competitive interactions that probably existed between them. The eight core members of the guild overlapped greatly in the prey species they exploited, and thus they undoubtedly brought considerable competitive pressure to bear upon one another. Intraguild competition can be a potent evolutionary agent among carnivores (e.g., Jones 1997; Linnell and Strand 2000; Creel 2001; Kok and Nel 2004; Yamaguchi et al. 2004), and thus rival carnivores most likely greatly shaped the Neandertal predatory niche. In addition to partitioning and divergence of niches as a result of exploitative competition, carnivore behavior and demography is also shaped by interference competition within the guild. As we will see, interference competition may have been a major determinant of Neandertal habitat use, mortality, and demography.

9.3.1 Exploitation Competition

Competition for limited resources (exploitation competition), whether direct or indirect, is the most studied and best understood form of competition between carnivores. There is no question that resources are quite limited at the highest trophic levels: starvation has been shown to be a significant cause of mortality in many carnivores, including lions, spotted hyenas, wolves, wild dogs, and lynx (Eloff 1980; Van Valkenburgh 1991; Watts and Holekamp 2009). Interest in exploitative competition derives from the early appreciation, following the principle of competitive exclusion (Gause 1934; Hardin 1960), that a struggle for resources among sympatric species leads to niche divergence, often accompanied by character displacement in trophic and/or locomotor structures (Grant 1972, 1994). Character displacement is well documented among sympatric members of carnivore guilds (Dayan et al. 1990; van Valkenburgh 1991; Jones 1997; Dayan and Simberloff 2005), with one good example being seen in the teeth of two sympatric cursorial carnivores from Africa: wild dogs and spotted hyenas. In this case, a degree of niche partitioning has occurred *within carcasses*, that is, in terms of the tissues that each carnivore can extract from a carcass. Wild dogs are easily displaced from kills by larger carnivores, and thus have evolved a feeding strategy focused on stripping as much soft tissue from a carcass as quickly as possible before being dislodged from their kill. Their dentition, which includes elongated and blade-like carnassials, reflects this strategy of hypercarnivory (*sensu* van Valkenburgh 1991, denoting a diet of flesh and few large bones). Spotted hyenas, on the other hand, often acquire carcasses by kleptoparasitism after some or most of the soft tissues have been consumed. While hyenas can and do consume flesh, they are also able to crack relatively large bones to extract marrow fat and brain tissue, and thus can find nutritional value even in carcasses that have been stripped of flesh by other carnivores. This ability to consume nutrients encased in hard tissues ("durophagy") is reflected in their dentition, in the form of stout, conical premolars and thick, stout carnassials.

Niche space is multidimensional, and thus there are many avenues available to carnivores to ease competition with rivals (Lewis and Werdelin 2007). Exploitative competition can be alleviated (at least somewhat) by attaining distance from near neighbors (in competitive space) along any number of axes: degree of omnivory versus carnivory; prey species preferences; prey body size preferences; habitat preferences; and carcass processing abilities (degree of flesh versus bone consumption). Naturally, physiological and morphological constraints may curtail the ability of a given species to move, evolutionarily, along any particular axis. Canids, for example, have digestive physiologies conducive to omnivory and thus can reduce competition under some circumstances by increasing consumption of plant stuffs, whereas the digestive physiology of felids does not afford them this option. Other behavioral aspects of niche divergence – such as activity patterns (diurnality versus nocturnality), social group size and grouping behavior, prey sex and age preferences, carcass transport or caching strategies, preferences for particular types of hunting grounds, and hunting methods – may serve to separate carnivore species in time or space (and thus reduce the chance of aggressive encounters) while not necessarily reducing exploitative competition. For example, a greater emphasis on cursorial hunting in cheetahs may allow them to stalk and capture prey successfully in the open, and thus reduce their rate of encounter with lions, who prefer slightly greater cover for ambushing prey (Grant et al. 2005). Thus, this difference in hunting method may reduce interference competition between the two species (to the great benefit of the cheetah), but leaves the two carnivores vying for a share of the same secondary productivity (and thus subject to exploitation competition). This distinction is important to efforts to understand the ecological pressures that drove shifts in Neandertal predatory behavior, such as the emergence of a prime age prey focus in the later (MIS 4 and 3) Middle Paleolithic (Stiner 1991c, 1994; Gaudzinski and Roebroeks 2000).

Based on the behavior and ecology of their extant relatives, Ice Age carnivores were undoubtedly very flexible in their prey preferences and diet, providing them abundant opportunity to adapt to temporal and regional variation in the composition of the herbivore guild, the relative abundance of various members of that guild, and the competitive pressures exerted by fellow members of the carnivore guild. Despite this adaptive plasticity, it is also apparent that there existed considerable dietary overlap between members of the guild, just as there does in modern carnivore guilds (e.g., Kunkel et al. 1999; Karanth and Sunquist 2000; Hayward 2006). The richness of the Pleistocene Eurasian guild likely exacerbated the problem of exploitative competition for the various member species. In particular, competition was likely particularly intense for the high secondary productivity of open landscapes (steppe and tundra) that was stored in moderate- and large-bodied herbivores such as reindeer, horse, bison, aurochs, and giant deer, as these prey formed the core of the diet for at least five large carnivores (cave lions, lesser scimitar cats, cave hyenas, wolves, and Neandertals). The predators in this subset of the guild may have found some degree of niche separation by specializing on specific size classes of prey (a form of character displacement: Jones 1997), which may have been especially important to *Homotherium* (and possibly cave lions) given their greater ability to capitalize on productivity in the largest herbivore size classes (elephants and rhinos). Still, distancing oneself from one's competitors along a prey body size preference axis only does so much to relieve competitive pressure. Despite a marked difference in preferred prey size, extant lions and hyenas still experience a 58.6% actual prey species overlap and a 68.8% preferred prey species overlap (Hayward 2006). In open environments of Pleistocene Eurasia, where herbivore species richness was relatively low, the overlap in diet items between carnivore species may have been greater still.

The flexibility inherent in carnivore trophic strategies, combined with the multidimensional nature of niche space, makes it difficult to make confident inferences about the exact trophic ecology of Neandertals at any particular place or time. Neandertals may have had, thanks to material culture, a greater potential to move along the multiple dimensions that define niches, such as prey size preferences and carcass feeding

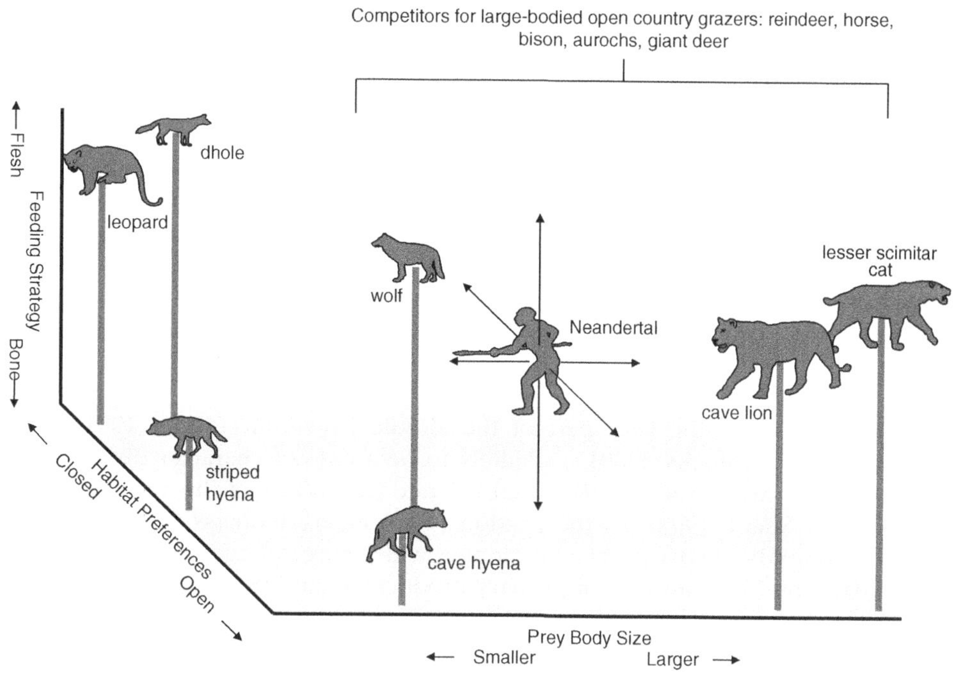

FIGURE 9.1

Positions of the members of the Eurasian Pleistocene large-bodied carnivore guild along three niche space dimensions: prey body size, habitat preference and feeding strategy (the latter denoted by the length of the stalk supporting each critter). The figures represent the overall position of each species on these axes, with due acknowledgement that great variation exists within the trophic ecology of carnivore species. Because striped hyenas and wolves operate equally effectively in closed and open habitats, they are depicted as falling in the middle of the habitat preference range. As discussed in the text, Neandertals may have had a greater ability to adjust their position along these axes – as indicated by arrows – through cultural means.

strategies (Figure 9.1). For example, in situations involving carcasses that were too large to transport whole back to living sites, stone tools would have allowed them to remove flesh fairly rapidly and reduce losses to other carnivores, effectively allowing them to exploit a hypercarnivorous (*sensu* van Valkenburgh 1991) feeding strategy like that of dholes, wolves, and scimitar cats. At the same time, stone hammers and anvils would have allowed them to access nutrient-rich tissues inside the long bones and brain cases of even the largest mammals, allowing them to operate like bone crunching hyaenids. This latter ability would have allowed them to more effectively process carcasses they had brought back to living sites, and to more effectively exploit scavengable remains (for example, by processing megaherbivore bones that were too large even for hyenas to crack). Thus, Neandertals could operate along the entire flesh eating-bone cracking continuum, and may have been able to colonize open niche space in the form of large-animal within-bone tissues. Hunting technology and flexible social hunting strategies may also have allowed Neandertals to go for a fairly broad range of prey body sizes (for instance, being able to kill megaherbivores that are outside of the prey size range of cave hyenas, who were of similar body size as the Neandertals) and to effectively take both closed country and open country forms. The ability of Neandertals to hunt in open country was probably limited (see Chapter 8, and below), but they were undoubtedly able to take open country prey along open country/woodland ecotones. Faunal and geochemical evidence (Chapter 7) indicates that they were able to exploit cover-loving prey, despite a dietary emphasis on open country animals. This sort of behavioral flexibility no doubt allowed the Neandertals to adjust their trophic strategies around the niche space occupied by other members of the guild, and thus to modulate – to a degree – the effects of what was probably fairly intense competitive pressure.

It is important to note that the emergence of a prime-aged adult prey preference among Neandertals in the later part of the Middle Paleolithic (Stiner 1991c, 1994; Gaudzinski and Roebroeks 2000) is unlikely to be a case of character displacement. Prey populations are sensitive to the total kill rate (regardless of the age class of victims: Mills and Shenk 1992), so the portion of secondary biomass that can support carnivores cannot be increased by exploiting a wider range of age classes: the size of the herbivore "pie" is a function of primary productivity and vegetation structure, not the predatory behavior of secondary consumers (see Carbone and Gittleman 2002). In fact, prey populations can support fewer carnivores when predation is aimed at adults rather than juveniles (Mills and Shenk 1992: see also Grenier 1979). Thus, increased predation pressure on adult ungulates would have intensified, rather than ameliorated, exploitative competition within the carnivore guild, and likely would have exaggerated interference competition between guild members.

9.3.2 Interference Competition

While exploitation competition has a long history in carnivore studies, increasing attention has been paid in the last decade to the powerful influence that interference competition has on carnivore ecology and demography (e.g., Palomares and Caro 1999; Linnell and Strand 2000; Fedriani et al. 2000; Creel 2001; Gunther and Smith 2004; Carbone et al. 2005b; Donadio and Buskirk 2006; Downey 2006). Interference competition – involving direct antagonistic interaction between species – can take the form of kleptoparasitism (theft of carcasses), exclusion from preferred habitats, and intraguild predation (Creel 2001). The adverse effects of interference competition can be substantial, so much so that inverse density relationships have been documented in a number of sympatric carnivores, including lions and spotted hyenas (Linnell and Strand 2000; but see Creel 2001), wild dogs and cheetahs (and with both lions and spotted hyenas: Linnell and Strand 2000), and among numerous pairs of sympatric canid species (Fedriani et al. 2000; Linnell and Strand 2000).

The negative effects of interference competition, however, are not symmetrically shared among all members of a given guild. The larger, more aggressive ("behaviorally-dominant" or "socially-dominant") species generally prevail in intraguild interactions. Large carnivore guilds annually remove about 15% of all available prey, but the dominant species in the guild may monopolize upwards of two-thirds of that prey (10% of all available prey), leaving the lower-ranked species to compete for the remainder

(Karanth et al. 2004). Kleptoparasitism and intraguild predation are also highly asymmetrical, with the dominant species being more effective at killing or harassing competitors, and usually winning in confrontational interactions over carcasses (although smaller, less dominant species can score some victories by attacking and killing juveniles of the larger, socially-dominant species: Palomares and Caro 1999; Gunther and Smith 2004). While confrontational scavenging tends to be highly asymmetric (for example, lions and hyenas regularly drive wild dogs from their kills; the reverse has seldom been observed), in some cases dominant species provide scavenging opportunities for subordinate species by killing larger prey that they would otherwise find unattainable (as has been documented between coyotes and wolves: Creel 2001). As a result of these disparities, socially-dominant species enjoy larger ranges, more varied diets and higher densities than their sympatric competitors (Fedriani et al. 2000).

There are two counterintuitive aspects of interference competition that make it particularly damaging to lower-ranked, nondominant species. First, the intensity of this form of competition usually increases as prey density increases, contra the situation with exploitation competition. Higher prey densities generally mean high densities of the dominant species, and thus greater potential for antagonistic interactions. The situation is exacerbated if prey size tends to be large, hunting is costly, and the competition is highly asymmetric (as when a great size disparity exists between the dominant and nondominant carnivore) (Creel 2001). Carcasses are more valuable than live prey, because the injury risk and energetic cost of capture have already been paid. Furthermore, the larger the prey size and the greater the risk and cost of capturing them, the greater the net value of carcasses and the greater the payoff for commandeering them from your competitor. For these reasons, it pays larger carnivores to challenge less dominant species whenever the cost and risk of doing so is lower than that of hunting, which is usually the case when great size disparities exist between dominant and nondominant species. For the nondominant member in such interactions, the net return rate from hunting may vary inversely with prey density: wild dogs, for example, make more kills at higher prey densities but are driven off from a greater proportion of carcasses, such that the energetic cost of foraging actually increases with prey density (Creel 2001). Furthermore, intraguild interactions can reduce the carrying capacity for the nondominant species even when there is minimal dietary overlap and thus little exploitative competitive pressure, simply by excluding the nondominant from preferred habitats (Linnell and Strand 2000). For these reasons, smaller members of the guild may actually survive better in areas of low prey density (as has been documented for cheetahs, wild dogs, and arctic foxes: Linnell and Strand 2000).

Second, and again contra the situation with exploitation competition, the intensity of interference competition is not relaxed at low densities of the nondominant species. Carnivores appear to be demographically more sensitive to changes in mortality than to changes in reproduction (Linnell and Strand 2000). Intraguild predation has been documented in 97 carnivore species pairs (Linnell and Strand 2000), and can account for as much as 89% of the mortality in the victim (nondominant) species (Donadio and Buskirk 2006). Among cheetahs, for example, cub mortality can be as high as 95%, of which most (73%) is due to lions (Linnell and Strand 2000). This mortality slows the population growth rate and increases growth rate variability at low population densities, and thus increases the probability of extinction of the nondominant species (Linnell and Strand 2000) (Figure 9.2).

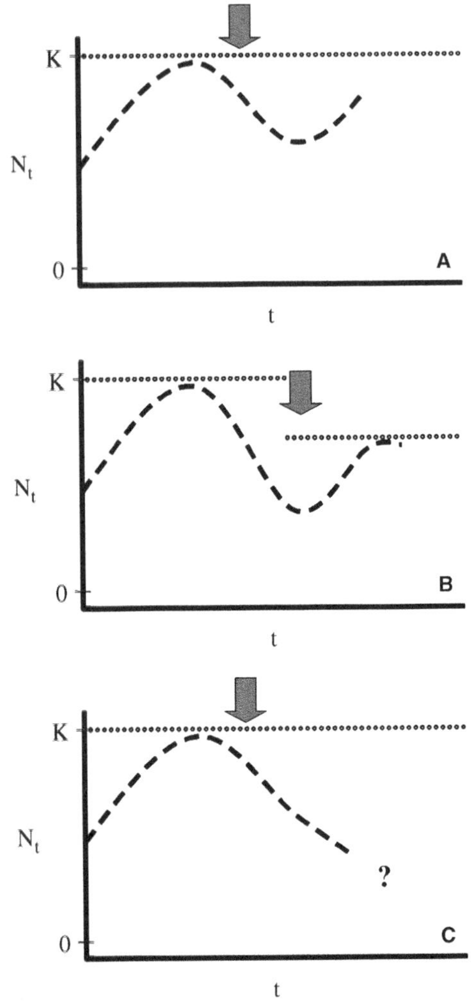

FIGURE 9.2

Possible effects of exploitation competition (panels A and B) and interference competition (panel C) on population growth rate in carnivores following a perturbation (⬇) in population density. A. If the carrying capacity (K) remains unchanged, the relaxation in exploitative competition caused by the reduction in population size (N_t) will allow for rapid population growth over time (t), until the population again reaches the carrying capacity. B. If the perturbation is associated with a reduction in carrying capacity (perhaps by allowing a competitor to grow in numbers), population growth will still occur, but only to the new carrying capacity. C. With interference competition, higher population densities result in increased mortality from larger predators, in turn resulting in reduced and more variable growth rate, increasing the possibility of extinction. Redrawn from Linnell and Strand 2000.

One final aspect of interference competition deserves mention. The Eurasian Pleistocene large-bodied mammalian carnivore guild was particularly rich, and with eight competitors for dietary resources, plus two large and aggressive ursids who were competitors for space and, to some small degree, prey, the guild was much richer than any carnivore community in the modern world. Competition for resources was probably intense, and life was probably difficult even for the larger, more socially-dominant members of the guild. Crania of cave lions, for example, have higher frequencies of bite damage than do modern lions (Rothschild and Diedrich 2012), most likely from aggressive encounters with cave bear, hyenas, and possibly other lions. Thus interference competition was likely to be a potent source of mortality for all the members of the guild, regardless of rank (although certainly the less dominant species would have fared much worse), and was most certainly an important component of the demography of all the carnivores, including Neandertals, in the community.

9.4 NEANDERTALS WERE NOT THE SOCIALLY-DOMINANT MEMBERS OF THE CARNIVORE GUILD

So where did Neandertals fall in the dominance hierarchy of large-bodied carnivores? This can be difficult to know with any certainty, but I think a good case can be made that, wherever they fell, they *weren't* the socially-dominant members of their guilds.

In terms of body size, the Neandertals fell in the middle of the guild (Table 9.1), with cave lions and lesser scimitar cats being appreciably larger, cave hyenas and leopards being about their same size, and wolves, dholes, and striped hyenas being smaller. However, what a carnivore lacks in size can usually be offset by numbers, which gives an edge to social predators. Earle (1987) described social carnivores as having a "flexible body mass," demonstrating that smaller carnivores were able to take prey that would otherwise be above their size range through cooperative foraging, and that flexibility in the size of hunting parties allowed smaller carnivores a way of adjusting their net body mass to local conditions. Carnivores also employ social behavior as a strategy for dealing with intraguild interference competition. Dholes form larger foraging parties in areas where they overlap with tigers and leopards (Rabinowitz and Walker 1991), larger packs of wild dogs are better able to defend carcasses from spotted hyenas (Carbone et al. 1997), wolves are better able to defend carcasses from kleptoparastic crows at larger pack sizes (Vucetich et al. 2004)[7], and even among lions defense of carcasses, and cubs, from spotted hyenas may be a factor favoring social hunting among females (see Packer et al. 1990; Palomares and Caro 1999). Indeed, species that form groups can defend against and even kill competitors up to 12 times their own body mass, as occasionally demonstrated by African wild dogs killing adult lions or dholes killing adult tigers (Palomares and Caro 1999). Solitary predators are generally ranked low within the dominance hierarchy: for example, dholes are socially-dominant over leopards, despite being smaller in individual body mass (Karanth and Sunquist 2000: although leopards do sometimes kill and eat dholes, most likely in instances where leopards encounter a solitary dhole).

When group mass (*sensu* Earle 1987) is taken into consideration, Neandertals still fall in the middle of the pack (Table 9.2). Based on body size considerations alone, larger groups of Neandertals may have been able to defend or capture carcasses from small groups (and solitary individuals) of lions, scimitar cats, and hyenas, as well as

TABLE 9.2 ■ Mass, hunting group size and "group mass" of principle members of the Eurasian Pleistocene large-bodied carnivore guild (rank ordered from greatest to least group mass)

Predator	Body mass (kg)[a]	Group size[b]	Group mass (kg)
Cave lion	243	3–5[c]	729–1215
Lesser scimitar cat	190	3–5[d]	570–950
Cave hyena	78	3–12	234–936
Neandertal	78	3–10[e]	234–780
Wolf	45	6–10	270–450
Dhole	30	8	240
Leopard	68	1	68
Striped hyena	40	1	40

[a] Middle of range of body masses provided in Table 9.1.
[b] Group sizes based on observations of extant analogs (and rounded to nearest whole number), from Table 1 in Earle 1987, unless otherwise noted. Where multiple values were provided for a given species, the range of reported group sizes is used here.
[c] Packer et al. (1990) report a modal group size of 5 for lions.
[d] Group size based on lions, following arguments (see text above) that *Homotherium* was a social predator.
[e] Neandertal hunting party sizes are of course unknown. Values chosen reflect the probable range of possibilities, with the highest value representing the possibility (as expressed by Kuhn and Stiner 2006) that the entire social group hunted together.

from the smaller carnivores. At smaller group sizes, Neandertals would undoubtedly have lost out in confrontations with the large social felids, and may well have had difficulty with all but the solitary leopard and striped hyena.

But what role might hunting technology have played in establishing pecking order within the carnivore guild? Certainly, without spears Neandertals would have fared worse than striped hyenas. But how much of an advantage does weapon technology, at the level of Mousterian thrusting spears, confer upon its possessors? Modern humans can and do kill the largest carnivores with hand-held spears (polar bears in the arctic, lions in Africa), and by analogy Neandertals would have been able to as well. However, the modern human cases generally entail a significant component of mobbing behavior. Among the pastoralist Maasai of east Africa, for example, lion hunting is generally conducted in one of two contexts, both of which entail mobbing solitary lions with large groups of armed people (Hazzah 2006). The first involves retaliatory killing of lions that have depredated livestock, and is undertaken by an entire community as it seeks out and kills a specific "problem" lion (Hazzah 2006). The second context is the ritual killing of lions, which is done by groups of 10–20 warriors armed with spears and shields who isolate a single lion before attacking it (Lichtenfeld 2005). Lions are occasionally killed by individual warriors, generally on an opportunistic basis, but even as a group endeavor lion hunting is regarded as extremely dangerous. In traditional Inuit societies polar bears were attacked and killed with hand spears, but usually with the aid of sled dogs that surrounded and occupied the bear (even when Inuit began using rifles to kill bears, they continued to employ dogs to make the practice safer: Ekblaw 1928; Manning 1944). Thus, while polar bear hunting may involve a single human hunter against a dangerous carnivore that is an order of magnitude larger, it also involved the mobbing behavior of 6–10 wolf-sized canids. Despite the ability to confront predators with large social groups, and despite being armed at times even with firearms, humans still regularly fall victim to lions and other carnivores in Africa (Treves and Naughton-Treves 1999; Treves and Palmqvist 2007; Kushnir et al. 2010), leading to what Eaton (1979) has called, at best, a situation of co-dominance between lions and people. I suspect that Neandertals may have taken on lions, scimitar cats, and hyenas – either to usurp a carcass or to defend one from theft – when numbers were on their side. Given the pride and pack hunting of these carnivores, I also suspect that most times the Neandertals were on the losing side of these interactions.

Another way to determine social dominance patterns within a carnivore guild is to explore evidence for intraguild predation, that is, to see who is eating whom. Intraguild predation is highly asymmetrical (see for example Table 2 in Palomares and Caro 1999) and often (but not always: see below) results in consumption of the nondominant carnivore by its dominant competitor (Fedriani et al. 2000). Intraguild predation is difficult to assess in the archeological and paleontological records, however, for a number of reasons. First, the caves and rock shelters that have produced most of these records were intermittently occupied by Neandertals, hyenas, wolves, lions, bears, and other carnivores (Straus 1982; Mussi 2001b; Beauval et al. 2005), which all potentially or actually contributed bones to the fossil assemblages we find there today. Discerning the major agent of bone accumulation in any particular stratigraphic layer is complicated and can be fraught with difficulty but, in general, the faunal remains that are from Mousterian layers which are rich in artifacts and poor in carnivore remains, that have evidence of human agency (such as cut-marks or burning), and that exhibit anatomical part representation and mortality profiles attributable to human hunting are inferred to have been primarily accumulated by Neandertals. Faunal remains from layers that are carnivore-rich and stone tool-poor, with high frequencies of carnivore coprolites, and that exhibit evidence of carnivore agency (such as gnawed or gastric acid-etched bones) or evidence of denning (such as juvenile or adult skeletons found in complete or partial articulation), are inferred to be carnivore-accumulated assemblages. Thus, the occurrence of carnivores in Mousterian faunal assemblages is not *prima facie* evidence of human agency, and in most cases the carnivores found in these sites entered the assemblages without the help of Neandertals. Human modification of carnivore bones (e.g., cut-marks) can provide important information, but as a general rule the carnivores found in Mousterian levels bear no such traces (e.g., Miracle 2005). These factors make the occurrence of carnivore bones in Mousterian sites difficult to interpret with respect to patterns of intraguild predation.

A second complication is that even clear cases of bone modification constitute only indirect evidence of intraguild predation. Neandertal remains with signs of carnivore gnawing, for example, may represent cases of non-carnivore-induced mortality, with subsequent scavenging of human corpses by carnivores. Secondary involvement of carnivores would have been facilitated in cases in which mortuary practices (or a total lack thereof) left corpses above ground and accessible to scavengers such as spotted and striped hyenas: thus a significant degree of carnivore modification to Neandertal bones may reflect a low frequency of burial (Gargett 1989, 1999) more than it does predation by other members of the carnivore guild. Our understanding of the extent to which carnivores were a source of mortality for the Neandertals would be improved by a systematic examination of traumatic, antemortem lesions for evidence of carnivore attacks on living Neandertals. Along these lines, it is interesting that documented carnivore attacks on modern humans sometime result in carnivore-induced amputations of limbs (Avis 1999; Abuabara 2006; Emami et al. 2012): two Neandertals (Shanidar 1 from Iraq and Krapina 180 from Croatia: Trinkaus and Zimmerman 1982; Gardner and Smith 2006) evince amputations (Shanidar) or possible amputations (Krapina) in their upper limbs.

Added to these problems is the fact that intraguild predation does not necessarily lead to consumption (precluding the possibility of bone modification) or transport back to denning sites (reducing the chance that a victim of intraguild predation would enter the paleontological/archeological record). In most documented cases of intraguild predation, spotted hyenas, leopards, and wolves were observed to consume carnivores that they killed, and hyenas in particular appear prone to consume anything that they kill (and are often cannibalistic in cases of intraspecific predation: references in Diedrich and Žák 2006). Lions, on the other hand, do not consume the carnivore species they kill in most cases (based on data provided in the appendix of Palomares and Caro 1999). And even in cases where lions may consume a rival carnivore, they rarely transport bones away from death sites. Thus, lion-induced mortality in Neandertals may be archeologically and paleontologically invisible. Finally, while intraguild predation is asymmetrical, the dominant members of the guild do sometimes fall victim to subordinate carnivores. For example, despite being the socially-dominant member of their guild, lions experience about 8% cub mortality from hyenas and leopards, and even adult lions are sometimes killed by packs of wild dogs (Palomares and Caro 1999). Adult female lions may also occasionally fall prey to a clan of hyenas, and their skeletons are sometimes found in Pleistocene hyena dens (Diedrich 2007). This adds to the complications inherent in interpreting evidence of intraguild predation in the Pleistocene.

These caveats make the evidence at best suggestive, and at the risk of going beyond any conclusions the data can reasonably bear, I wish to simply note that the available evidence is consistent with the argument that Neandertals were not the socially-dominant member of their guild. At a number of sites, surface modifications on Neandertal remains attest to the involvement of hyenas (or other carnivores) as taphonomic agents. Cave hyenas appear to have been an important agent of bone accumulation at the site of Grotta Guattari (Italy) (Stiner 1991a), and were likely responsible for the presence of at least one of the two Neandertals from the site (White and Toth 1991). Carnivore gnawing, most likely attributable to hyenas, can also be found on a Neandertal femur from Les Rochers-de-Villeneuve (France) (Beauval et al. 2005), on the bones of the juvenile from Teshik-Tash (Uzbekistan) (Movius 1953), and possibly on some of the postcranial remains from Krapina (Croatia) (Russell 1987a). A carnivore-gnawed Neandertal femur was recovered from a hyena den at the French site of Grotte de la Tour (Puymerail et al. 2012). To this we can add a tibial diaphysis from the Middle Paleolithic of Gruta da Oliveira (Portugal), which bears puncture and gnawing marks from a small carnivore (possibly a canid), but which occurs in a cave that also preserves evidence of hyena denning (Trinkaus et al. 2007b), and the remains of a Neandertal accumulated by hyenas at Rochelot (France) (Tournepiche et al. 1996). Similar gnawing damage, consistent with either hyena or wolf, has been observed on Neandertal remains from the Greek site of Kalamakia (Harvati et al. 2013), and a fragmentary Neandertal femur from Zafarraya, Spain, has damage consistent with hyena gnawing

(Barroso Ruiz and Hublin 1994). In addition, Neandertal teeth have been recovered at Les Pradelles (France) with evidence of gastric etching, most probably from being swallowed and subsequently regurgitated by hyenas (Costamagno et al. 2003), and two Neandertal teeth were recovered from a hyena den at Cueva del Camino in Spain (Arsuaga et al. 2010). Finally, a maxillary premolar of a juvenile Neandertal or early modern human has also been found in a carnivore den at the Iranian site of Wezmeh Cave (Trinkaus et al. 2007a), and carnivore-marked Neandertal remains, in which the type of carnivore that caused the damage is unknown, have been recovered from Valdegoba (Quam et al. 2001) and Cova Negra (Arsuaga et al. 2007) caves in Spain. Clearly, hyenas were at times feeding on Neandertals, although again it can't be known if these cases represent predation or scavenging. It should be noted that similar evidence comes from African and Asian sites involving hominins that likely shared a similar level of technological and social development as the Neandertals. Carnivores are implicated in the taphonomy (and possibly mortality) of at least three African Middle Pleistocene archaic humans (attributed to *Homo heidelbergensis*, *H. rhodesiensis*, or *H. helmei*): those from Florisbad and Hoedjiespunt in South Africa and Kabwe in Zambia (Tappen 1987; Berger and Parkington 1995). Hyena predation on Middle Pleistocene Africans is also indicated by the recovery of human hair from a brown hyena coprolite from 195–257 Ka BP deposits at Gladysvale cave, South Africa (Backwell et al. 2009). And in eastern Asia, most if not all of the *H. erectus* individuals represented at Zhoukoudian (China) were accumulated by hyenas (in this case, the extinct giant hyena *Pachycrocuta brevirostris*: Boaz et al. 2000). Again, these data suggest (cautiously) that archaic humans, armed only with close-range weapons and likely foraging in relatively small groups, fell prey to other members of the carnivore guild with sufficient frequency as to be detectable in the archeological and paleontological records.[8]

Evidence for Neandertal predation upon other members of the large-bodied carnivore guild (again, with the exclusion of ursids) is hard to come by. An isolated lion phalanx with cut-marks has been recovered from Mousterian contexts at Caverna delle Fate (Italy) (Valensi and Psathi 2004), and cut-marks have been found on a lion radius in the Mousterian F2 level of Le Portel (France) (Gardeisen 1999). Le Portel has also produced wolf remains with cut-marks (Gardeisen 1999). Although other examples may exist, they have not received much attention in the literature.

The fairly common occurrence of Neandertals and other archaic humans in hyena assemblages, and the rarity of Neandertal-modified large carnivore remains, tentatively supports the claim that Neandertals were not the behaviorally-dominant carnivores in the Eurasian later Pleistocene guild. Based on considerations of their body size, probable social group size (relative to the probable sizes of the groups of their carnivore competitors), and the taphonomic data, it is reasonable to see the Neandertals as falling somewhere in the middle of the guild with respect to dominance rank. Cave lions, lesser scimitar cats, and cave hyenas were likely dominant to the Neandertals in intraguild interactions, while the Neandertals were probably dominant in interactions with wolves, dholes, leopards, and striped hyenas. The position of Neandertals *vis-à-vis* cave lions is supported by analysis of Mousterian cave sites in Italy (Mussi 2001b), which show that Neandertals tended to avoid habitation sites that were also found suitable by lions (and bears). However, the site settlement data suggest a greater parity between Neandertals and hyenas: Italian Neandertals did inhabit sites where they had a distinct probability of encountering hyenas, wolves, and leopards (Mussi 2001b).

Although modern humans are not the focus of this book, the position of the Cro-magnon in the carnivore dominance hierarchy deserves some mention, if only to draw into sharp relief the situation that held for the Neandertals. For the same reasons that applied to Neandertals (discussed above), early modern Europeans would have also had to maintain a position in the highest level of the trophic pyramid to subsist in Pleistocene Europe and, like Neandertals, they would have found themselves facing exploitative and interference competition with a rich and well-established community of large carnivores. I think, however, that unlike the Neandertals, modern humans *were* the dominant member of the guild. The Cro-Magnon may have had two things going for them that allowed them to prevail in interference interactions: true long-range projectile weaponry (Brooks et al. 2005; Shea 2006; Churchill and Rhodes 2009)[9], and perhaps larger social networks that allowed them to effectively exploit coalitionary behavior to outnumber and eradicate competitors (as noted by Shea [2006], low population density results in a high cost to the maintenance of coalitions, and thus humans may not have been able to exploit this competitive tactic below certain density thresholds). Regardless of what exact advantages were possessed by early modern Europeans, the end result was clear: eradication of six of the seven members of the large-bodied carnivore guild (as discussed here) within 25 Ka of the arrival of modern humans (Figure 9.3). Modern humans arrived in Europe roughly 45–40 Ka BP cal, and within about 10–15 Ka Neandertals and the lesser scimitar cat were gone (Reumer et al. 2003). By about 25–24 Ka BP cave hyenas and dhole were locally extinct (Kurtén 1968; Estévez 2004), and by 18 Ka BP the cave lion and leopard had also vanished (Estévez 2004; Sommer and Benecke 2006). Wolves remain the only member of the guild that has not been extirpated, although it has not been for lack of human effort in places. The rapid demise of most of the large carnivores after the Cro-Magnon arrived in Europe suggests to me that modern humans had a very different relationship with their competitors than did the Neandertals (one of dominance and active suppression, as opposed to subjugation and acquiescence), and that the extinction of the Neandertals can be seen in the broader context of modern human invaders who were intolerant of competitors for shares of secondary productivity.

9.5 Neandertal Ecology in the Context of Competition within the Carnivore Guild

An idea of the two forms of competition – exploitative and interference – that were operating within the Eurasian Pleistocene large-bodied carnivore guild provides a general context for understanding Neandertal predatory strategies and ecology. As discussed above, moderately-large open country grazers – reindeer, horse, bison, aurochs, and giant deer – probably formed the core of the diet for five of the eight members of the guild, and competition over these herbivores was likely intense. That Neandertals used their great behavioral flexibility – in terms of being able to exploit productivity of both closed and open habitats, being able to handle prey across a fairly wide range of body sizes, and being able with the aid of technology to employ carcass processing strategies that ran the gambit from hypercarnivory to bone crunching– to ameliorate the effects of this competitive pressure seems certain. However, competition within the carnivore guild, especially interference competition, probably had two profound effects on Neandertal ecology and demography: their subordinate status to some of

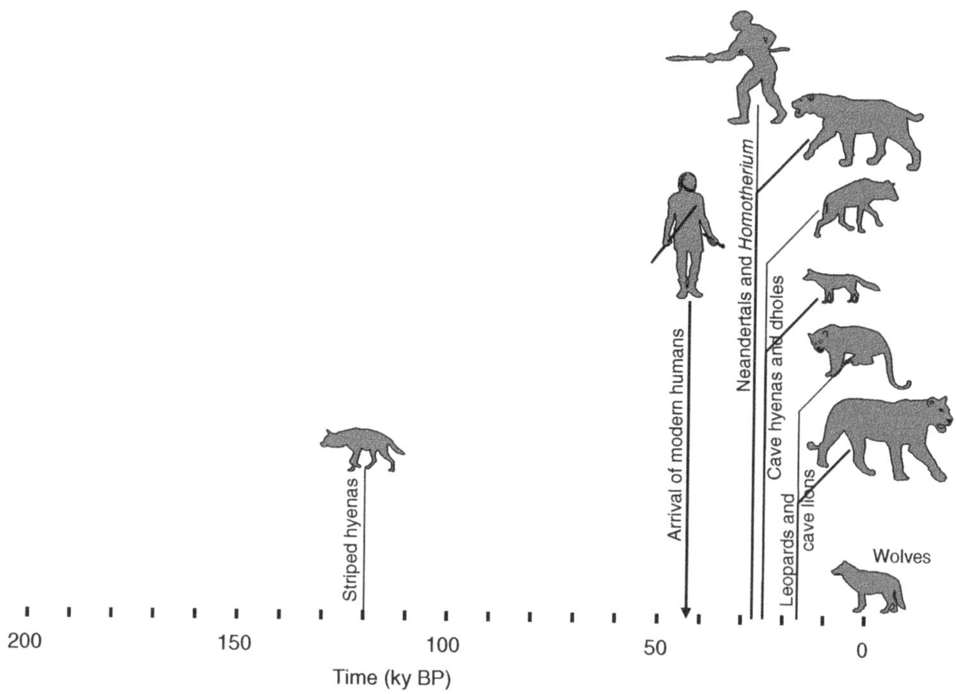

FIGURE 9.3

Timing of extinction events in the Eurasian large-bodied carnivore guild in the late Pleistocene. After the establishment of the guild by late MIS 7 times, the only change prior to the appearance of modern humans was the disappearance of striped hyenas by end-MIS 5 times. Within 25 Ka of the first appearance of modern humans in Europe, six of the seven remaining members of the guild had gone extinct (with only wolves still extant in Europe). Extinction dates after Kurtén 1968; Reumer et al. 2003; Estévez 2004; Sommer and Benecke 2006.

the larger social carnivores probably contributed to what was already a limited ability to hunt in open habitats, and likely also functioned to keep their population density low across Eurasia.

In Chapter 8, I argued that the close-range nature of Neandertal hunting technology limited their ability to hunt in open country, which suggests that the open country herbivores that formed the core of Neandertal diet (Chapter 7) were hunted in more closed habitats, along ecotones, or in mixed habitat contexts (brush-grass mosaics). The presence of cave lions, lesser scimitar cats, and cave hyenas probably also served to exclude Neandertals from hunting in open environments.

Habitat partitioning has long been recognized as a means by which carnivores avoid competition (Farlow and Pianka 2002), and this no doubt was an important component of the competitive dynamics within the Eurasian Pleistocene guild. At the French site of Payre, for example, there is stable oxygen isotopic evidence to suggest that Neandertals were preying on horse, rhino, and large bovids in the valley floor below the site, while wolves were focusing on tahr and chamois on the slopes above (Ecker et al. 2013). Larger carnivores can monopolize areas of high prey density (Creel 2001), which during glacial and cold–temperate intervals would mainly have been open, steppic environments. However, interference competition is also more severe in open habitats. Among wild dogs (a good example of a low ranking member of a relatively rich carnivore guild) the frequency of kleptoparasitism from hyenas varies by a factor of 43 across study sites (Creel 2001); much of this variation can be explained in terms of variation in habitat structure. On the open plains of the Serengeti, hyenas were attendant at 85% of wild dog kills (references in Carbone et al. 2005b), while in the closed habitats of the Selous Game Reserve hyenas were able to locate only 14% of wild dog kills (Creel 2001).

Open habitats favor contest competition because carcasses can be located more easily (Creel 2001) and because the hunting activities of a given species are more readily apparent to competitors. In the open environments of Africa, both hyenas and vultures follow the hunts of wild dogs, and congregate rapidly once a kill is made (Creel 2001). Also, in open environments dropping vultures can be seen from kilometers away (certainly not the case in closed habitats), and because carcasses are more easily found in these habitats, open environments support higher densities of avian scavengers (Creel 2001). Pleistocene Eurasia had its share of scavenging vultures and crows (M. Pavia as cited in Mussi and Palombo 2001), which would certainly have alerted lions and hyenas to the hunting activities of Neandertals (see also Vucetich et al. 2004).[10] Closed habitats provide a measure of relief from interference competition: in the closed habitats of Africa, hyenas do not locate wild dog kills as often, do not converge on their kills as rapidly, congregate in smaller numbers than do hyenas in open country (Creel 2001), and generally scavenge less (Kruuk 1972). Thus both issues of hunting technology (close-range weaponry) and interference competition with other carnivores would have favored the exploitation of more closed conditions by hunting Neandertals. The zooarcheological and isotopic records of Neandertal diet (Chapter 7) clearly attest to an emphasis on open country herbivores. The picture that emerges from the considerations above is one in which Neandertals exploited these animals along edges (closed/open ecotones) which provided cover for hiding as well as concealment from competitors, or hunted individuals that were seeking shelter or resources in denser stands of vegetation (the modern analogs of all of the open country Pleistocene forms can at times be found in heavy bush cover or forest, sometimes as part of seasonal movement patterns and sometimes having to do with individual foraging and thermoregulatory considerations). Relative abundances of faunal elements in Middle and Upper Paleolithic sites in Europe also support (albeit weakly) a Neandertal preference for hunting in areas of closed cover: "as regards the taxa over-represented on Neanderthal sites it would appear that the dominant pattern is one of preferential association with taxa that are indicative of warmer and to some extent closed habitats" (Stewart 2004: 183–184). Bear in mind, however, that while exploitation of ecotones may have eased interference competition it would not have relieved it entirely: both lions (Funston et al. 2001; Grant et al. 2005) and wolves (Kunkel and Pletscher 2001) also prefer to hunt at closed cover/open country ecotones.

The second major effect of competition within the carnivore guild was on Neandertal population density. Studies of large-bodied mammalian carnivore guilds suggest that predators annually remove about 15% of the prey biomass (references in Karanth et al.

2004), but that the energy captured is highly unevenly distributed. In at least some contexts, the socially-dominant member of the guild may crop approximately 10% of the herbivore biomass each year, leaving the other members of the guild to compete for the remaining 5% (Karanth et al. 2004). As a general rule, it takes a standing crop of about 10,000 kg of prey to support about 90 kg of carnivore biomass (irrespective of carnivore body size: Carbone and Gittleman 2002), thus population density of a given carnivore is a function of both the biomass of its prey and its own body size (the ratio of carnivore numbers to prey biomass scales as the reciprocal of carnivore mass: Carbone and Gittleman 2002). It can also easily be seen that the behavior of the dominant member of the guild constrains the population densities that can be attained by the subdominant members, who are competing for a much smaller share of the secondary biomass.

The importance of social dominance rank can be seen in the relative biomass of five members of the modern African large-bodied carnivore guild: lions, hyenas, leopards, cheetahs, and wild dogs (Table 9.3).[11] The average observed density of lions (per 10,000 kg biomass of their preferred prey) is more than four times greater than would be predicted based on general predator biomass to prey biomass scaling relationships, reflecting their effective monopolization of secondary productivity. Hyena density (and biomass) is also greater than expected (although not so dramatically: only about 1.4 times greater than expected), which suggests that hyenas do an effective job of competing for the portion of herbivore productivity not controlled by lions. Cheetahs and wild dogs, on the other hand, are less abundant than expected (0.73 and 0.51 as many as expected, respectively: Table 9.3), reflecting their low rank and disadvantage in competition within the guild. Leopards, surprisingly, are 1.7 times more numerous than expected (Table 9.3), which may be a sign of the effectiveness of habitat and prey body size partitioning as a means of reducing competition (both exploitative and interference) with the more socially-dominant members of the guild.

In modern African ecosystems, lions comprise almost half (49%) of the total biomass of these five carnivore species (based on observed density and mass values provided in

TABLE 9.3 ■ Predicted versus observed population density (per 10,000 kg of prey) for African carnivores

Predator	Body size (kg)[a]	Predicted density[b]	Observed density[c]	% deviation from predicted[d]
Lion	142.0	0.4898	2.150	+339
Hyena	58.6	1.2404	1.755	+41
Leopard	46.5	1.5814	2.720	+72
Cheetah	50.0	1.4654	1.070	−27
Wild dog	25.0	3.0342	1.557	−49

[a] From Carbone and Gittleman 2002.

[b] Number per 10,000 kg of prey, predicted from carnivore mass (M) as $89.1M^{-1.05}$ (Carbone and Gittleman 2002).

[c] Average observed number per 10,000 kg of prey, as determined in Table 1 of Carbone and Gittleman 2002. Values given here were provided by Chris Carbone (personal communication, December 2007) and differ from values published in Carbone and Gittleman 2002.

[d] 100*(observed-predicted)/predicted.

Table 9.3), followed by leopard and hyena (20% and 16%, respectively). Cheetah and wild dog – the perennial underdogs in intraguild interference competition – together comprise only 15% of the biomass of the large carnivore guild (9% and 6%, respectively). The Eurasian Pleistocene carnivore guild was richer (eight member species), and thus biomass must have been approportioned differently than among the African carnivores: the larger, socially-dominant members of the guild may not have been able to monopolize as much of the trophic level, and the smallest, least-dominant members may have been particularly thin on the ground. If we assume that Neandertals, as middle-ranking members of the guild, constituted 10–15% of the productivity at the third trophic level, what might this suggest about Neandertal population density?[12]

Herbivore biomass is of course the major determining factor for carnivore biomass, and as noted above as a general rule it takes 10,000 kg of herbivore biomass to support about 90 kg of carnivore biomass (Carbone and Gittleman 2002), or 1:111 predator/prey biomass ratio. Large-bodied African carnivores do a bit better than the global, pan-mammalian average, and enjoy a 125.6/10,000 kg (1:80) predator/prey biomass ratio (data provided by Chris Carbone). Secondary biomass estimates are provided for various Eurasian Pleistocene biomes in Chapter 6, and from these we can get a rough idea of the overall density of carnivores, including Neandertals. Of course, the dynamics of competition within the carnivore guild and the behavioral and ecological responses of the guild members to that competition (namely, the ways in which various guild members were partitioning niches), as well as the great variation in predator/prey density relationships across space in the modern world, make it impossible to estimate the density of particular species with any precision. Nonetheless, such an exercise provides a rough idea of the limitations imposed on Neandertal population size by their dietary focus on animal tissues and by competition with their fellow meat eaters.

Secondary productivity was highest in open country, steppic ecosystems (including mixed steppe/tundra), and these habitats supported the highest carnivore biomass. Judging from modern ecosystems, secondary productivity in open landscapes likely varied by an order of magnitude across the expanses of Europe and western Asia but, as discussed in Chapter 6, the overall productivity of Pleistocene steppe habitats was likely less than that of analogous Eurasian and North American ecosystems today (because of colder and drier conditions) while, paradoxically, that of midlatitude Pleistocene tundra was likely higher than that of modern low arctic tundra (see also Guthrie 1982, 1990). Using the secondary biomass estimates for steppe from Chapter 6 (500–1200 kg km^{-2}), this ecosystem would have supported about 4.5–15 kg of carnivore per square kilometer (the lower value being based on a secondary biomass of 500 kg km^{-2} and using the pan-mammalian 1:111 predator/prey biomass ratio; the higher value calculated from a secondary biomass of 1200 kg km^{-2} and using the African carnivore 1:80 predator/prey biomass ratio). The scaling of herbivore body size and population density among modern fauna provides us with an idea of how secondary biomass was approportioned across herbivore taxa: habitat-specific (Damuth 1981) and habitat-, diet-, and geographic location-specific (Damuth 1993) equations were used to predict the population density (per km^2) of the major large-bodied herbivores that occupied Eurasian Pleistocene steppe. The product of the population density and the average mass provides a measure of the biomass (kg km^{-2}) of each species and allows the determination of the percentage of the secondary biomass each species represents. These percentages were then used to estimate the biomass of each taxon given a total

TABLE 9.4 ■ Estimated herbivore biomass for late Pleistocene steppe/tundra

Herbivore	Body mass (kg)[a]	Biomass (kg km^{-2})[b]
Steppe & wooly mammoth	5484[c]	137
Steppe rhino	2000	124
Wooly rhino	2897[d]	128
Horse	250	101
Steppe & wild ass	230	99
Giant deer	700[d]	112
Reindeer	86	91
Saiga antelope	29	83
Musk ox	313	103
Steppe wisent	500	108
Aurochs	900	114

[a] Unless otherwise noted, masses are from FA Smith et al. 2003 (and for extant forms, represent combined male and female masses averaged across the species' geographic range).
[b] Average of estimates from high arctic tundra and temperate grassland prediction equations from Damuth 1993 and Eurasian/North American open country grazers equation from Damuth 1993, and adjusted to a total ecosystem secondary biomass of 1200 kg km^{-2} (see text for details).
[c] Average of estimated masses of *Mammuthus primigenius* from Christiansen 2004 (5484 ± 2086, $n = 22$).
[d] Mass estimate from Brook and Bowman 2004.

secondary biomass of 1200 kg km^{-2} (Table 9.4 and Figure 9.4). This exercise suggests that the greatest animal productivity in the open grasslands was to be found in equids (wild asses, steppe asses, and horses), moderately-large bovines (steppe wisent and aurochs), and rhinoceroses (steppe and wooly rhino). The abundance of musk oxen, reindeer, and saiga antelope was somewhat lower but was also probably more regionally and climatically variable (the arctic forms dropping out as tundra vegetation gave way to grasses, the antelope declining where the vegetation mix included more tundra flora). Given the prey preferences of modern analogs (discussed above), cave lions probably focused their predation on the most abundant and most productive ungulates (equids and bovines) while occasionally taking larger animals, while hyenas may have focused on the smaller size classes among the herbivore guild, with some dietary overlap with lions at the higher end of their preferred prey body size range (Figure 9.4: see also Feranec et al. 2010; Bocherens 2011; Dusseldorp 2013). As discussed above, the lesser scimitar cat probably focused its predation on megaherbivores, and likely capitalized on the great productivity of rhinos. As with modern wolves, Pleistocene wolves were likely catholic in prey selection, taking a range of prey from small antelope to highland-loving goats (Stiner 2004; Ecker et al. 2013) to large cervids and bovids (Feranec et al. 2010). Thus, there was probably little in the way of specializing on certain types or sizes of prey that the Neandertals could have done to lessen the competition they incurred from neighboring carnivores (see for example Stiner 2004). Dietary overlap between Neandertals, lions, scimitar cats, hyenas, and wolves would have increased the benefit of interspecific killing (through the elimination of resource competitors), as would the great size differential between the large cats on the one hand and Neandertals and hyenas on the other (see Donadio and Buskirk 2006).

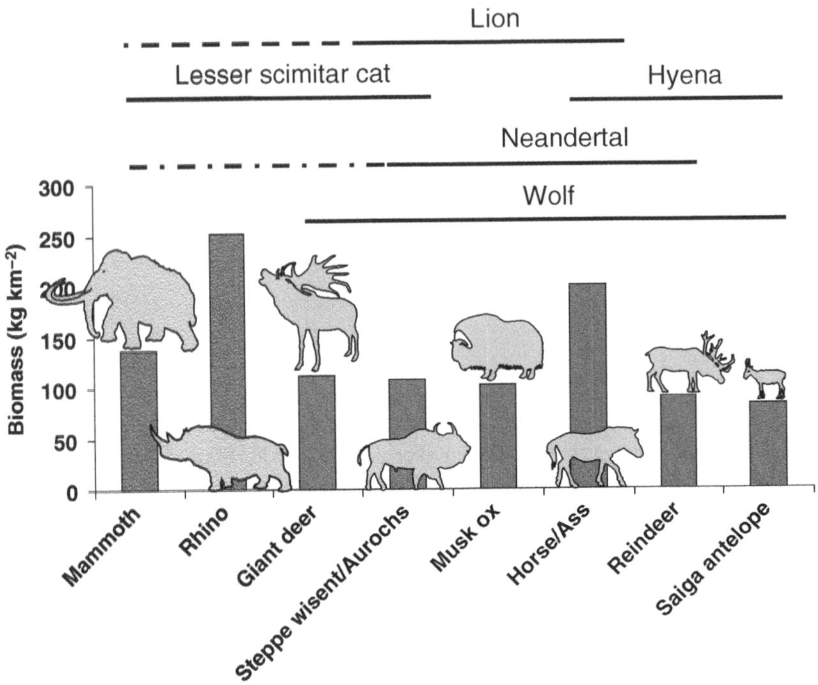

FIGURE 9.4

Estimated biomass of grazing herbivores in Pleistocene steppe/tundra ecosystem, based on values in Table 9.4. The likely preferred prey of the five major carnivores feeding off of open country productivity are shown above the graph. Both cave lions and Neandertals probably preferred moderately-large sized ungulates, but were able to exploit prey of larger size classes.

If Neandertals managed to hold 15% of the biomass at their trophic level (roughly equivalent to spotted hyenas in the modern African large carnivore guild), and assuming an average adult body mass of 72 kg, secondary productivity of steppe/tundra ecosystems would have only supported Neandertal population densities in the range of 0.009–0.031 Neandertals km^{-2}. However, 15% is probably an overgenerous estimate, given the richness of the Pleistocene Eurasian carnivore guild and the probable rank of the Neandertals in that guild. If the dominant members of the guild (cave lions and lesser scimitar cats) together comprised 50% of the large carnivore biomass (a conservative estimate: lions today comprise almost half of the biomass in the African large carnivore guild), and the lowest ranking members (leopards, dholes, and striped hyenas) together comprised 20% of the biomass (higher than might be expected based on rank alone, but taking into account the effects of habitat partitioning: recall that leopards alone account for 20% of the biomass in the modern African guild), the mid-ranking members of the guild (spotted hyenas, Neandertals, and wolves) would together only comprise about 30% of the large carnivore biomass. Assuming Neandertals were able to hold on to 10% of the production at their trophic level, open ecosystems of Europe

TABLE 9.5 ■ Estimated population densities of Neandertals in open and closed habitats, and observed population densities in modern carnivores

	Population density (individuals km^{-2})
Lion	0.008–0.385
Spotted hyena	0.006–1.842
Cheetah	0.006–0.078
Leopard	0.005–0.370
Wolf	0.005–0.042
Dhole	0.013–0.030
Wild dog	0.001–0.015
Neandertal (open country)	0.006–0.031
Neandertal (closed country)	0.009–0.111

Carnivore data from Carbone and Gittleman 2002.

may have supported Neandertal population densities in the range of 0.006–0.021 individuals km^{-2}. As can be seen in Table 9.5, population densities of carnivores are highly variable, and even the large, socially-dominant members of their guilds live at low population densities in some places. Still, if the Neandertals fell in the middle of the guild hierarchy (as argued above), they likely lived at densities that today are more characteristic of lower ranked carnivores (cheetahs, wild dogs, wolves, and dholes) than high-ranking ones (lions and spotted hyenas).

Based on interglacial conditions (during MIS 7 and 5e) that were potentially warmer and wetter than obtain today, Eurasian temperate deciduous forests may have had relatively high secondary productivity (secondary biomass of 700–4000 kg km^{-2}: Chapter 6). Such environments would have supported a greater carnivore biomass, on the order of 6.3–50.2 kg km^{-2} (the lower value being based on a secondary biomass of 700 kg km^{-2} and using the pan-mammalian 1:111 carnivore/prey ratio; the higher value calculated from a secondary biomass of 4000 kg km^{-2} and using the African carnivore 1:80 predator/prey ratio). Neandertals may thus have fared better during interglacial episodes, perhaps with population densities on the order of 0.009 individuals km^{-2} (assuming Neandertals comprised 10% of the carnivore biomass, and based on a secondary biomass of 700 kg km^{-2}) to 0.111 individuals km^{-2} (if Neandertals comprised 16% of the biomass at their trophic level, and secondary biomass = 4000 kg km^{-2}). In addition, a considerable amount of the secondary biomass of interglacial deciduous forests may have been in the form of megaherbivores (elephant, rhino, and hippo: Table 9.6), the adults of which may have been somewhat more available to Neandertals using ambush and disadvantaging hunting techniques (Churchill 1993) than they were to even the largest social carnivores. In western France, the stable isotope data suggest that Neandertals were eating more megafauna than were hyenas (Bocherens 2011), whereas zooarcheological indicator from the same area suggest the reverse (Dusseldorp 2013: see Dusseldorp 2011 on why these two sources of information might produce conflicting conclusions). But even if the Neandertals were able to shift to a greater megafaunal hunting focus at times, it may have relaxed somewhat the degree of exploitative competition they experienced, but would probably have exacerbated the problem of defending large, valuable carcasses from kleptoparasitism.

TABLE 9.6 ■ Estimated herbivore biomass for late Pleistocene temperate deciduous forest

Herbivore	Body mass (kg)[a]	Biomass (kg km^{-2})[b]
Straight-tusked elephant	7839[c]	748
Merck's rhino	1600[d]	556
Wild boar	117	343
Hippopotamus	1418	585
Red deer	167	522
Fallow deer	56	299
Woodland wisent	500	448
Aurochs	900	499

[a] Unless otherwise noted, masses are from FA Smith et al. 2003 (and for extant forms, represent combined male and female masses averaged across the species' geographic range).
[b] Average of estimates from mixed temperate forest equation from Damuth 1993 and Eurasian/North American closed habitat (forest) herbivore-browser, open country herbivore-browser, and open country herbivore-grazer equations from Damuth 1993, and adjusted to a total ecosystem secondary biomass of 4000 kg km^{-2} (see text for details).
[c] Average of estimated masses of *Palaeoloxodon (Elephas) antiquus* from Christiansen 2004 (7839 ± 3328, n = 10).
[d] Mass estimate from Brook and Bowman 2004.

Consideration of the place of the Neandertals in the Eurasian large mammal carnivore guild suggests that they existed at fairly low population densities, the implications of which will be explored in greater detail later (Chapter 12). Studies of carnivore ecology also stand to inform discussion of Neandertal mobility patterns (including aspects of home range size, mobility strategies as they relate to prey density and patchiness, and how variation in snow cover may have impacted Neandertal mobility and predatory behavior), which brings us to the subject of the next chapter.

NOTES

1. The term "guild" derives from medieval associations of tradesmen, and it is used in ecology to denote groups of species that are not necessarily taxonomically related but that use the same kinds of resources in similar ways. Thus cave bears, which were by all means large-bodied carnivores, by virtue of their hypocarnivory do not merit inclusion in the large-bodied carnivore guild. Neandertals, by virtue of their carnivory, can be considered members of this guild, despite not being carnivores (taxonomically). Thus I hope I will be granted some leniency for occasionally referring to Neandertals as carnivores.
2. Cave bears are sometimes found in association with ungulate remains, presumably accumulated by the bears, and interpreted as reflecting at least a modicum of hunting behavior in these largest ursids (see references in Estévez 2004). Still, the morphological and biogeochemical evidence indicates that cave bears were almost exclusively vegetarian.
3. Biogeographic considerations allow us to rule out the possibility that *Homotherium* existed only in isolated refugia: large-bodied cats would require distributions over an area > 180,000 km^2 (the size of the Italian peninsula) to persist over many millennia (O'Regan et al. 2002).
4. Antón and colleagues (2005) make an argument for social living and social predation in homotheres (particularly *H. latidens*) on the basis of their functional morphology and on

presumed competitive interactions within the carnivore guild. Indirect support for this inference comes from the recovery of 30 *H. serum* individuals – representing adults, juveniles, and cubs – from a single den site (Friesenhahn Cave) in Texas (Turner and Antón 1997), noting of course that such associations in fossil assemblages do not necessarily mean group living (McCall et al. 2003).

5. As revealed in cave paintings and sculpted figurines from the Upper Paleolithic: there are dozens of representations of lions in sites across western Europe, some with small ruffs but none with a distinct mane. One ca. 30 Ka BP painting of a pair from Chauvet Cave in France depicts a clear male (with a visible scrotum) who lacks a mane (Clottes 2003).
6. An advertisement in South Africa noted that if you get attacked by a leopard you should count yourself lucky: few people ever get to see a leopard!
7. Although ravens may not seem like much of a threat, their small size, high mobility, and large group sizes make it difficult for lone wolves or small packs to defend against their thievery, and it has been estimated that wolves lose 2–4 kg of meat per day per wolf to these scavengers (Vucetich et al. 2004).
8. Although carnivore-damaged modern human remains are much more rare, early modern humans were not entirely immune to predation (or scavenging), as witnessed by the possibly hyena accumulated Gravettian-aged human humerus from Eel Point (England) (Schulting et al. 2005).
9. Consideration of the morphology of the ca. 30 Ka BP modern human scapula from Peştera Muierii (Romania) has led Trinkaus (2008a) to suggest that projectile weapons post-date the Aurignacian. Consistent with this inference is the occurrence of the remains of the *earliest* western Asian and European modern humans in carnivore assemblages (Trinkaus et al. 2007a; cf. Trinkaus et al. 2000). If so, it makes the rapid disappearance of the large-bodied carnivores (including the Neandertals) all the more dramatic.
10. To make matters worse, some ungulates are sensitive to the vocalizations of scavenging birds, and may use their cues to avoid predation (as has been documented for moose with raven calls: Berger 1999). Thus, troublesome birds may have made hunting in open country that much more difficult.
11. I thank Chris Carbone for sharing his data on carnivore density. Note that the data discussed here were revised since the publication of Carbone and Gittleman 2002.
12. The decision to exclude brown bears (and cave bears) from consideration in the large mammal carnivore guild comes into play here – if one were to include them (see arguments in Richards et al. 2008b about the predatory behavior of male brown bears and the occasional carnivory of cave bears), secondary productivity would then be spread among a richer guild comprised of nine or ten members, and thus most of the guild members (Neandertals included) would have fared worse.

CHAPTER TEN

The Cost of Living in Ice Age Europe

It is time to return once more to the matter of Neandertal energetics. In Chapters 4 and 5 we developed some estimates of the caloric cost of basal metabolism in Neandertals, and in this chapter we'll focus on the activity costs – the caloric costs of finding and procuring food, water, shelter, raw materials for tools, and fuel – that were added on to the Neandertal caloric budget.

Estimating total energy expenditure (TEE: kcal d^{-1}, also known as daily energy expenditure) in Neandertals requires some assumptions about their activity levels, and thus their caloric expenditures above basal metabolism (Sorensen and Leonard 2001; Steegmann et al. 2002; Churchill 2006). The standard approach[1] is to apply a physical activity level (PAL) multiplier, derived from energetic studies of modern foragers, to basal metabolic rate (BMR)[2] to determine total daily caloric expenditure. As a general guideline for assessing human energy expenditure, the World Health Organization considers PALs of 1.55–1.77 for men and 1.56–1.63 for women as constituting light activity, 1.78–2.09 (men) and 1.64–1.81 (women) as moderate activity, and greater

Thin on the Ground: Neandertal Biology, Archeology, and Ecology, First Edition. Steven Emilio Churchill.

TABLE 10.1 ■ Physical activity level (PAL) in modern foragers and pastoralists

Group	Males	Females	Combined	Reference
Ache (neotropical)	2.2	1.9	–	Leonard and Robertson 1992
Hiwi (neotropical)	–	–	1.7	references in Jenike 2001
!Kung (Africa)	1.7	1.5	–	Leonard and Robertson 1992
Turkana pastoralists (Africa)	1.3	1.3	–	references in Jenike 2001
Gunwinggu (Australia)	–	–	1.5	references in Jenike 2001
Groote Islanders (Australia)	–	–	1.9	references in Jenike 2001
Igloolik Inuit (arctic)	2.2	1.8	–	Godin and Shephard 1973; Panter-Brick 2002
Evenki herders (arctic)	1.8	1.6	–	Katzmarzyk et al. 1994
Keto (arctic)	1.7	1.5	–	Katzmarzyk et al. 1994

PAL = Total (daily) energy expenditure/basal metabolic rate.

than 2.09 (men) and 1.81 (women) as representing heavy activity (FAO/WHO/UNU 1985: see Leonard 2003 for additional empirical support of these values). Most workers agree that Neandertal activity levels fell at the high end of this range, but there is little consensus as to how high. Values of PAL are highly variable across modern hunter-gatherers (and relatively active pastoralists: Table 10.1), and provide only general guidelines as to the most appropriate value to apply to Neandertal energy budgets. Sorensen and Leonard (2001) used PALs of 2–3 in estimating Neandertal TEE, based on data from the World Health Organization showing that PALs can reach these levels in some people in developing countries. Elsewhere (Churchill 2006), I have used PALs of 2.0–2.5 with Neandertals, based on direct measurement of calorie consumption in Inuit that still practice a traditional foraging lifestyle (see Shephard and Rode 1996; Steegmann et al. 2002), or values of 2.5 for males and 1.8 for females (Froehle and Churchill 2009) based on sex-specific estimates of Inuit energetic expenditures. How appropriate are these PAL estimates?

The preponderance of anatomical evidence from the Neandertal skeleton, reviewed in Section 10.3, indicates that they led highly active lifestyles. From the stoutness of their bones, to the large size and heightened rugosity of the muscle scars on those bones, to their great thoracic dimensions, skeletal indicators suggest activity levels that may have been extreme relative even to modern arctic foragers (see Churchill 1994a; Franciscus and Churchill 2002; Churchill and Rhodes 2006). In this chapter, we will review the evidence for Neandertal activity levels in an effort to better estimate the caloric costs of subsistence and other activities in their energy budgets. Since mobility (movement on the landscape) represents the major expenditure in forager energy budgets, most of this chapter will be devoted to determining the extent of Neandertal travel across the landscape. Among recent foragers, the overall organization of subsistence (the strategies by which geographic and temporal patchiness in resources are managed and risk is minimized) is the main determinant of how mobility is partitioned (that is, camp moves versus individual movement) and managed, and so we will begin with a review of Neandertal subsistence strategies. Following from this, we'll review what we know about Neandertal home range size and what we can infer about their actual levels of mobility within their territories. This will put us in a good position to estimate the energetic cost of locomotion as part of the activity budget. After a final consideration of energy expenses related to domestic activities, we should be able to evaluate the reasonableness of PAL multipliers that have been suggested for use with the Neandertals.

10.1 SUBSISTENCE ORGANIZATION AND MOBILITY

Moving about the landscape accounts for the largest activity expenditure in forager energy budgets. Quantitative data from extant hunter-gatherers are rare, but it has been estimated that female foragers in the Kalahari desert (!Kung) walk about 2400 km y^{-1} (references in Panter-Brick 2002), with some daily distances of 40–60 km during migratory residential moves (Bentley 1985). It has also been estimated that among neo-tropical foragers of Paraguay (Ache) males and females spend about 34% and 16%, respectively, of their waking time walking (averaged across both active foraging days and rest days in camp: Hill et al. 1985; Hurtado et al. 1985). Across a wider but still tropical/subtropical sample of hunter-gatherer groups, males average between 2–7 hours per day in foraging pursuits (again averaged across foraging and rest days), while females average between 1–5 hours per day foraging (Waguespack 2005). The caloric cost of walking varies by terrain: walking on a level surface consumes about 0.54 kcal per kilogram of body mass per kilometer (Brody 1945), while ascending a steep slope may cost as much as 6.9 kcal kg^{-1} km^{-1} (Lusk 1931).[3] Assuming that most foraging effort involves movement, it's clear that even in fairly productive environments foragers must expend a substantial portion of their energy budgets on mobility. In less productive temperate, arctic, or periglacial environments the costs would be higher still. And while the focus here is on mobility related to subsistence, it should also be noted that reproductive effort may necessitate considerable movement on the landscape, especially (as was likely the case for Neandertals) when population density is low and social groups are widely dispersed on the landscape (MacDonald and Hewlett 1999).

Movement on the landscape is the means by which foragers solve problems of uneven distribution of resources (patchiness) and local biodepletion, and is thus the key element in hunter-gatherer subsistence systems. Not surprisingly, both residential and logistical mobility are highly variable within and between groups of foragers, and there is little doubt that Neandertal mobility patterns were variable across geography and time (on seasonal, multiyear, and generational time scales) (cf. Burke et al. 2008). Variation in mobility is conditioned by overall productivity and the ecological structure of edible resources, by the amount of terrestrial game in the diet, and by the density and patchiness of that game, by the availability of water and other non-food necessities (such as raw material for tools, firewood, suitable camp locations, and shelter materials: Binford 1982a), and by labor availability and demographic parameters (such as the ratio of producers to consumers). Mobility costs are also in part a function of the overall subsistence strategies that are in play, which determine the ways in which the calories spent on mobility are partitioned between residential (camp) moves and logistical forays (for purposes of capturing calories or gathering information) (Kelly 1983).

For foragers living in areas with a relatively homogeneous distribution of resources (as tends to be characteristic of tropical and subtropical zones) residential mobility – picking up and moving camp – may be the primary means by which groups deal with biodepletion and "map on" to specific resource patches or seasonally available foods. As foragers hunt and gather resources in the environment around the residential camp, they begin to deplete those resources, necessitating trips further and further from camp. At some point, most likely when the average return rate of the area within the foraging radius of camp (usually about 10 km) drops below the average return rate for the environment as a whole, it pays to move camp to a new area where resources have had a chance to regenerate since the last time the group lived there. In some settings, residential mobility may be very high, with camp moves on nearly a daily basis. Among the Ache, for example, most female labor goes into moving camp (usually every day), with some foraging conducted while the camp is being carried from one spot to another. This high residential mobility, made possible by female labor, seems to minimize local biodepletion and thus serves to keep male hunting return rates high (Hurtado et al. 1985; Hill et al. 1985). Frequent residential moves fall at one end of a continuum of hunter-gatherer mobility strategies, and groups that employ this strategy have been described as "foragers"[4] (*sensu* Binford 1980) or are said to practice circulating mobility (Wallace and Shea 2006). Tropical and subtropical hunter-gatherers that fall towards this end of the continuum employ a system in which consumers are generally moved to resources, residential camps are provisioned with resources gleaned within a few hours walk from camp, material goods are kept to a minimum, and camp is moved when local foraging return rates drop below the overall environmental average (when calories in the area around camp become depleted) (Binford 1980; Kelly 1983, 1995).

In environments where resources are more patchily distributed, both temporally and spatially, residential moves may be fewer and logistical mobility (movement of individuals or task groups away from residential camps) higher. In general, as productivity decreases, spatial and temporal patchiness of edible resources increases. When resources are highly clumped and those clumps are widely dispersed in space, greater mobility is required to exploit the resource patches. On the other hand, when resources are temporally clumped, the best strategy is generally to harvest resources in bulk when possible, and store them for consumption during the periods of resource scarcity. Storage of bulk-harvested resources is often the key to surviving seasonal drop-offs in productivity (the "overwintering" problem), and also constrains residential movement during at least part of the year (because it may be difficult or impossible to move the stores). Thus, foragers in these patchy environments tend to use logistical mobility as the means of dealing with spatial patchiness (sending smaller labor parties on forays away from the residential camp to harvest resource patches, and to transport those resources in bulk back to the camp for consumption and storage). Thus, at the other end of the continuum from circulating "foragers" lie "logistical collectors" (*sensu* Binford 1980), or hunter-gatherers who might be described as practicing a radiating mobility strategy (Wallace and Shea 2006). In these types of site settlement systems residential moves may be few and logistical mobility may be very high (Kelly 1983). Most arctic and some temperate hunter-gatherers employed this type of system, whereby edible resources tend to be harvested in bulk at specialized and often seasonal procurement locales, and these resources are moved to consumers at residential camps

(Binford 1980; Kelly 1983, 1995). Because resources are generally collected at some distance from the residential camp, local biodepletion tends to be less of a concern for logistical collectors (although it may be an issue with respect to other resources, such as firewood); however, biodepletion does tend to occur across the larger territory over long time periods (8–10 years or more), requiring shifting of territory areas over long time scales (Binford 1983).

The issue of the extent to which Neandertals employed these two site settlement systems (logistical collecting versus foraging) ranks among the great unresolved questions about Neandertal adaptive strategies. If we unpack the question a bit, we see that there are several deeper questions embedded in this issue:

1. Were the Neandertals capable of employing a logistical, radiating system? As practiced by recent cold-temperate and arctic foragers, these systems require considerable planning depth, and they tend to be supported by relatively complex subsistence and transportation technology. If the Neandertals weren't capable of enacting these complex strategies, then why weren't they? Might this imply some cognitive limitation (such as a lack of planning depth), or some other sort of constraint? Perhaps a lack of transportation technology capable of reducing the energetic cost of transporting bulk items back to camp (as recent foragers do with boats or dog sleds, for example) made radiating systems unworkable, or perhaps Neandertal social groups were just too small to accommodate the labor demands of such a system (the production of task-specific parties that could be sent out to gather information or harvest resources) (see Kuhn and Stiner 2006).
2. If the Neandertals weren't capable of working a logistical system, how were they managing? Since modern foragers in cold–temperate/cold environments use radiating systems to deal with the conflicting problems generated by the combination of temporal and spatial patchiness of resources, how were the Neandertals solving these problems without using radiating mobility?
3. If Neandertals were using logistical systems, even occasionally, how were they doing so without the benefit of transportation technology? As noted above, all the historically-known forager groups that fall close to the logistical collector end of the continuum (that is, those that use complicated radiating systems which are heavily dependent on logistical mobility) rely on transportation technology that can be called upon to reduce the cost of moving about the landscape in pursuit of information[5] or of moving foodstuffs in bulk back to camp, or both. If the Neandertals were practicing radiating mobility, what limitations did a lack of transportation technology impose on their subsistence strategies, and how did these play out with respect to the relative economic (energetic) return rates of different resource options?
4. If Neandertals were capable of using both circulating and radiating site settlement systems (that is, they were able to operate anywhere along the forager-logistical collector continuum), were they adjusting their mobility strategies in response to ecological circumstances in the same way that modern human foragers do? In other words, might we expect that Neandertals were behaving more like "foragers," or employing mixed forager/logistical collector (circulating/radiating mobility) strategies, in contexts in which resources were likely to be more homogeneously

distributed (such as temperate forests of Interglacial episodes, or the maquis and mixed woodland of the Near East), and behaving more like logistical collectors in more patchy environments (such as steppe and tundra of colder episodes)?

As we discussed in Chapter 3.2, Middle Paleolithic subsistence technology differs from that of recent modern humans in that it does not become more complex in less productive environments, which means that Neandertals were solving the extractive foraging problems posed by those environments by different (and, as of yet, unknown) means. These are the same environments in which we might expect to see more logistically-organized subsistence systems – might there also be some disjunct between expectations (based on recent hunter-gatherers) and practice with respect to variation in Neandertal subsistence behavior?

So, what can we say about Neandertal site settlement systems? As with every aspect of Neandertal ecology we have considered, the evidence concerning subsistence strategies is difficult to interpret, and a number of divergent opinions have been expressed. Site size, faunal indicators of season of site occupation, inferences about site function (residential camp vs. hunting camp, etc), lithic indicators of hunting intensity, and lithic raw material procurement distances have all been brought to bear on the question. These indicators suggest to some that Neandertals were, at least at times, using a circulating system with high residential and low logistical mobility (Marks 1989; Mellars 1996; Çep and Waiblinger 2001; Zilhão 2001; Martínez-Moreno et al. 2004; Gaudzinski-Windheuser and Roebroeks 2011). However, these same indicators have been interpreted as indicating that, at certain times and places, Neandertals might have been employing radiating mobility, involving few residential moves and high logistical mobility (Lieberman 1993; Lieberman and Shea 1994; Patou-Mathis 2000; Marks and Chabai 2001; Wallace and Shea 2006). Some workers see evidence of seasonal alternation between circulating and radiating strategies (Richter 2001), while others argue that Neandertal subsistence systems were complex, temporally variable, or impossible to discern based on current evidence (see Lieberman 1998; Hovers 2001; Adler and Tushabramishvili 2004; Rivals et al. 2009). The interpretation of evidence having to do with site settlement systems is complicated and problematic, and generally involves considerable taphonomic analysis to qualify the reliability and meaningfulness of the data (see, for example, Gaudzinski-Windheuser and Kindler 2012). Much work has gone into this line of research (e.g., papers in Conard and Wendorf 1996; Conard 2001, 2004; Conard and Delagnes 2010), and a thorough review of the literature would in itself require a book-length effort. My own take on the issue, having waded through most of this voluminous literature, is that Neandertals were capable of behaving logistically, but, by-and-large, they did not operate that way. The evidence suggests to me that their subsistence systems were more often operating near the circulating mobility, "forager" end of the continuum, and that even when they employed a greater amount of logistical mobility, they at best attained systems which were somewhere in the middle of the continuum, involving a mix of residential and logistical mobility. My thinking on this matter has been influenced heavily by the following observations:

1. Evidence for bulk harvesting of resources is rare. There are many Mousterian or Micoquian sites with faunal assemblages that indicate monospecific prey choice or a heavy emphasis on a single species (see list in Gaudzinski-Windheuser and

Kindler 2012). These sites tend to be associated with open country conditions (where one might expect to see logistical systems in operation), and the prey tend to be herd-forming ungulates (such as reindeer or bison) that are represented in appreciable numbers (minimum number of individuals sometimes greater than 100) (Gaudzinski-Windheuser and Kindler 2012). These sites might be signaling bulk harvesting of gregarious ungulates, which would only make sense in the context of meat storage (and which thus implies a certain commitment to a radiating mobility strategy, at least for the colder part of the year). However, these sites still represent a very small fraction of known Middle Paleolithic sites, and at only a few of these sites can a reasonable argument for bulk animal procurement be made. One taphonomic difficulty inherent in interpreting faunal assemblages involves differentiating palimpsests (accumulations over many episodes) from "one-off," single event episodes. There is evidence at some of these sites, such as Mauran in France (where remains of at least 136 bison were recovered from a single Mousterian level) and Wallertheim in Germany (at least 52 bison in level B1), that the fauna were accumulated over multiple individual hunting episodes rather than a single mass kill event (Farizy and David 1992; Gaudzinski 1995). Even at the sites with the best evidence of mass kills, such as Les Pradelles in France (33 reindeer in level 9 and 22 in level 10) and Salzgitter-Lebenstedt in Germany (with at least 86 reindeer), it is not clear how many hunting episodes underlie the assemblages or how many animals were killed at any single time (Costamagno et al. 2006; Gaudzinski and Roebroeks 2000). It is possible that at small social group sizes, Neandertal hunters did not need to kill many animals to provide adequate winter stores for the group. There are a couple of approaches by which we can estimate the storage need, assuming a winter lean period of about three months (the late winter/early spring months of February, March and April: Rowley-Conwy 1984) and using reindeer as our unit of currency. At an average live weight of about 100 kg (Binford 1978), a reindeer in good condition would provide on average about 35 kg of meat and about 10 kg of tallow and marrow fat (Foote 1965) suitable for storage, for a caloric value of approximately 131,650 kcal (using caloric values of reindeer tissue from Keene 1985). If we further assume a daily metabolic need of about 2240 kcal per adult Neandertal (estimated using the low end of the range of estimated basal metabolic rates, averaged for males and females, with a low-activity PAL multiplier of 1.4), then an adult Neandertal would need about 202,000 kcal, or the equivalent of one and a half reindeer, on which to overwinter. Since caloric needs for juveniles were likely similar to those of adults (based on observations on arctic modern humans: see Table 4.8), a Neandertal social group of 10–15 would have needed 15–23 reindeer-sized animals for winter stores. Alternatively, we could use consumption values estimated for Nunamiut Eskimos (Binford 1978) of 1.2–1.3 kg live weight of animal tissue per person per day. Neandertals were probably about 7–14% more massive than Eskimos (Chapter 4.1), which might bump the need up to about 1.3–1.5 reindeer per season, or again somewhere in the vicinity of 13–23 reindeer-sized animals to supply the social group for the winter. It is also possible that food storage was augmented by some level of continued active hunting throughout the winter, reducing the number of animals that needed to be taken in the fall. However, the viability of this strategy is questionable, as winter snow cover would have limited the mobility of hunters (and greatly increased the energetic cost of locomotion: see Section 10.4), and the nutritional quality of the prey would continue to decline over the course of the winter (Speth 1987). Still, at small social group sizes, a Neandertal group practicing a radiating strategy might need to kill only 10–25 reindeer-sized animals to put on stores for the winter. Some of the monospecific or single species-dominated faunal assemblages dating to the Middle Paleolithic (Gaudzinski-Windheuser and Kindler 2012) might represent this sort of "specialized" hunting in the context of a radiating mobility settlement system but, again, such sites constitute a fairly small part of the record of Neandertal subsistence behavior.

2. Evidence of storage during the Middle Paleolithic is rare. Large pits filled with bones, interpreted as storage pits that made use of natural refrigeration, are a common feature of later Upper Paleolithic sites in the colder portions of northern, central, and eastern Europe (Soffer 1989b; Hoffecker 2005). Pit features are extremely rare in the Middle Paleolithic (Stiner et al. 2000; Marean and Assefa 1999). Three pits, two of which may have had stone caps, have been documented in the Mousterian levels of the French cave sites of Combe Grenal (Bordes 1972), Le Moustier (Peyrony 1930), and La Quina (see Mellars 1996). While Bordes (1972) thought these pits had a funerary function (representing burials in which the bones had completely disintegrated), Jelinek and Debénath (as claimed by Mellars 1996: 307) have suggested that they were used for food storage. Unlike similar storage pits in the Upper Paleolithic, however, the three Mousterian pits lack animal bones. It is, of course, also possible that the Neandertals were using means of preserving meat that have not left an archeological signature, such as allowing carcasses to freeze on the ground surface, or smoking or drying meat. Here, the preservation of a natural cast of what appears to have been a wooden tripod erected over a hearth at the Abric Romaní, Spain (Castro-Curel and Carbonell 1995) is of interest, as it may indicate meat storage by smoking. The possibility of meat drying as a storage method has not really been explored, and for this the "meat drying index," which operates under the assumption that the drying of meat places constraints on butchery practices and selection of body parts and that these constraints play out in frequencies of elements preserved in the faunal record (Friesen 2001), may be a useful tool for looking for signatures of storage. In the only case thus far in which this index has been applied to a Middle Paleolithic assemblage (the red deer-dominated fauna of Gatzarria Cave, France), the faunal data fit a model of animal part selection and transport on the basis of marrow fat utility rather than drying needs (Ready 2013). While the application of the meat drying index to other Middle Paleolithic faunal assemblages might turn up some cases indicative of intensive meat storage, at present the Middle Paleolithic evidence for storage is weak. Along these lines, the inference of annual starvation in Neandertals (from age patterning of dental enamel hypoplasias: see Chapter 4.3) might be taken as indirect evidence against storage, and severe calorie shortfalls every winter. However, it should be noted that Inuit, who do practice storage and who generally use very logistical, radiating mobility strategies, also exhibit similar patterns of hypoplasias (Guatelli-Steinberg et al. 2004), presumably reflecting a recurring pattern of starvation between the exhaustion of the winter stores and the return of ecological production.

3. Duration of site occupation appears to have been variable, but patterns of variation do not seem to follow expectations based on a model of radiating mobility.

For Neandertals following a circulating system, we would primarily expect to find residential sites which were occupied for relatively short periods (Binford 1980). One would also expect to find short occupation sites in a radiating mobility system although, rather than being residential camps, these sites relate to the activity of specialized task groups out on logistical forays away from the residential site. Some of these sites may represent overnight (or multiple-night) field camps, or stations where information was gathered or hunting took place (Binford 1980). Thus, some Mousterian sites that appear to represent a short period of occupation (or multiple short-term stays), such as Les Pradelles in western France (Costamagno et al. 2006), may represent special activity sites (in this case, a hunting camp) within a logistical system. However, in this sort of system we would also expect to find residential camps with evidence of longer occupation. So how long were Neandertals tending to live at residential sites? As with other aspects of their behavior, duration of site occupation varied. So the real question is: does the variation follow ecological variation in predictable ways, that is, with a greater occurrence of long duration occupation sites in colder, less productive periods and areas? The qualified answer appears to be "no." One difficulty in addressing this question involves determining the length of occupation represented by a single layer at a site. In general, the recognition of short-term occupations is relatively straightforward, as they tend to have low densities of lithic remains and small quantities of flaking waste, to preserve fauna killed in a single season (as judged by the age and sex structure of the prey samples, or by cementochronological analysis of teeth in the faunal sample), to have evidence of carnivore modification of faunal remains (meaning the bones left behind by the hominins were still fresh and greasy, and thus attractive to carnivores), to preserve more discrete activity areas within the site, to be relatively spatially noncomplex, and to show little investment in site modification (Kuhn 1995; Mellars 1996; Costamagno et al. 2006; Vaquero et al. 2001a). The identification of long-duration residential sites is somewhat more tricky, since most layers in cave sites are palimpsests, representing debris accumulated over multiple episodes of site use. Thus with a thick accumulation of lithic and faunal remains, in which the fauna were killed across multiple seasons, it can be difficult or impossible to differentiate between ephemeral visits spanning multiple seasons versus a single, multiseason stay. Under these conditions, the presence of site modifications, such as structures, stone pavements, and large hearths, or other signals of investment in the site (such as clearing of debris to the site margins), may be the most telling signals of long-term residential camp use (Mellars 1996). Short-term residential sites are relatively common, and have been documented in the Middle Paleolithic of the Near East (Hovers 2001; Wallace and Shea 2006), and eastern (Marks and Chabai 2001), central (Conard and Prindiville 2000; Çep and Waiblinger 2001), and western Europe (Straus 1982; Villaverde et al. 1996; Mora et al. 2004; Vallverdú et al. 2005; Rivals et al. 2009; Moncel 2011). Some sites, or specific layers within sites, have yielded evidence consistent with longer, multimonth or even multiseason occupations, such as the abri Canalettes in the south of France (Meignen et al. 2001), Sant' Agostino in central Italy (Kuhn 1991), or even level Ja at l'Abric Romaní near the Mediterranean coast of Spain (Vaquero et al. 2001a). Mellars (1996) notes that there are many sites in the Perigord region of France that seem to reflect episodes of relatively intense occupation, and at least two sites in the Levant – Kebara Cave in

Israel (Bar-Yosef et al. 1992) and Tor Faraj rockshelter in Jordan (Henry 1998) – may preserve signs of multiseason occupation. What is notable about all of these sites, however, is that they occur along the southern margin of the Neandertal range where, even during colder episodes, we would least expect to see radiating mobility patterns. The Mousterian levels of Trou Magrite in southern Belgium, by way of contrast, record "sporadic and ephemeral" occupations (Straus and Otte 1995: 851) in association with cold climate, open country fauna dating to MIS 5b. This appears to be the usual condition for more northerly Middle Paleolithic sites. Gamble (1999) notes that Middle Paleolithic sites in the northern parts of Europe tend to produce excavated sections containing small collections of artifacts, such that sites have to be extensively excavated to produce assemblages suitable for analysis. Likewise, Roebroeks and Tuffreau (1999: 128) characterize Neandertal residential behavior in northwestern Europe as "short term, episodic and highly mobile." Monospecific or single species-dominated faunal assemblages, which might signal specialized hunting and mass killing of herd-forming prey as part of a logistical strategy, are more prevalent in the northern reaches of Europe (Gaudzinski-Windheuser and Kindler 2012). One wonders, then, where are the long-duration residential sites that were being provisioned by these hunting locations?

4. Transportation technology appears to have been lacking in the Middle Paleolithic. Among historically-known foragers, logistical collecting systems are supported by transportation technology, such as snow mobiles and motorboats in the modern world, or dog sleds and canoes in earlier times (Binford 1978; EA Smith 1991). Such technology reduces the energetic and time costs of transporting calories in bulk from collection sites to residential camps, and may be a requisite of the long-distance logistical systems employed by many arctic groups. What might logistical provisioning look like without the benefit of transportation technology? The 13–23 reindeer needed by our hypothetical group (see point 1, above) would produce about 585–1035 kg of storable meat and fat that would need to be transported back to camp. The number of trips required to get these stores to the residential camp is, of course, a function of the size of the average load. Nepalese male porters carry loads that average 93% of their own body mass (but at great energetic cost: Bastien et al. 2005). Assuming similar burden-carrying prowess among the Neandertals, 8–14 person-trips would be needed to move the stores to camp. If the size of the task party were small, and a few individuals had to be left behind at the kill site to defend the unmoved stores from scavenging by other carnivores (see point 7, below), perhaps as many as five round trips would be required to transport the goods. Keep in mind that, while experienced porters are able to carry very heavy loads (Bastien et al. 2005), they do not do so rapidly. In the Nepalese case, the porters alternate 15 seconds of walking with 45 seconds of rest when moving uphill. Thus, if any appreciable distances were involved, logistical movement of goods to camp would have likely taken many weeks of arduous travel. These problems would be exacerbated at larger group sizes, since the number of producers in the group is always a fraction of the total group membership (and thus the total need). Of course, one solution would be to employ other members of the social group as porters. But, unless the residential camp is tethered to some resources (which seems unlikely),[6] it would make more sense to move the residential site to the vicinity of

the kill site if the entire social group is being called upon to move to there anyway. Consideration of intrasite spatial patterning suggested to Gaudzinski-Windheuser and Kindler (2012) that kill and butchery sites might have frequently transformed into temporal residential camps, suggesting a strategy of logistical mobility to locate and procure prey *en masse*, followed by a residential move to the area of the kill site. While such a strategy employs components of a radiating mobility system, it also relies on residential mobility to move consumers to goods, and thus at best it represents a mixed mobility strategy.

5. Middle Paleolithic subsistence technology seems geared more towards maintainability than reliability. When survival depends on accumulating sufficient stores for overwintering, the technology used to capture those stores tends to be optimized for reliability (Bleed 1986). These systems tend to be specialized for specific resource sets so that they are highly effective, to be made of durable components, to have redundant components in parallel (such that if part of the tool fails, the overall tool will continue to perform), and to be made such that they can be repaired quickly and easily in the field (so that harvesting of the resource is minimally interrupted by equipment failure). These design considerations are part of the reason that, among recent foragers, technological complexity varies inversely with productivity (Oswalt 1976; Bleed 1986). Reliable technology is at a premium in colder environments, because the logistical collecting systems in use there demand extractive tools that will continue to perform effectively until adequate amounts of a given resource have been harvested. As discussed in Chapter 3.2, Middle Paleolithic subsistence technology does seem to vary in complexity across time or space, which is not what we would expect if the Neandertals were at times practicing radiating mobility strategies. The one interesting exception concerns the use of bone points, which have been documented at a few sites (Montet-White 1996; Gaudzinski 1999; Villa and d'Errico 2001). As discussed in Chapter 3.4, bone and antler is more durable than brittle stone for use as weapon armatures, but produces edges that are not as sharp as those of stone. The loss in killing effectiveness of bone points relative to stone points must be offset by some other benefit of using bone, and it is most likely that it is bone's durability, in the context of a demand for reliable technology, that provides that benefit (see Guthrie 1983). The recovery of bone points at Salzgitter-Lebenstedt in Germany (Gaudzinski 1999), an open-air site with the remains of 86 reindeer (Gaudzinski and Roebroeks 2000), may testify to the use of (more) reliable technology to mass harvest reindeer in the presumably cold, open steppic environments of early MIS 3 northern Europe. Still, bone points (and potential mass kill sites) are the exception rather than the rule in the Middle Paleolithic. Instead, Neandertals employed subsistence toolkits seemingly optimized for maintainability (Bleed 1986). In more productive environments, hunter-gatherers are likely to encounter a greater diversity of extractive foraging tasks, and thus they need subsistence tools kits that are versatile rather than specialized. In these contexts, tools may wear quickly or be prone to damage (because they are being used for a variety of tasks), and thus there is a premium on simple gear that is relatively easy to service in the field (Bleed 1986). Middle Paleolithic toolkits appear relatively simple and maintainable compared to the more diversified and specialized technology of the Upper Paleolithic. However, Shea (1997) raises a very good argument with

respect to hafting of stone tips onto spears. A simple wooden spear, he notes, is a versatile tool, which can be employed as a digging stick, probe, lever, and beater (for dislodging fruit or nuts from trees): by adding a lithic point, the spear becomes more effective as a weapon but loses its functionality in other contexts. Thus, Shea argues, composite technology may reflect a move towards more reliable subsistence technology in the context of subsistence strategies designed to deal with seasonal environments (by focusing on the procurement of large game) (Shea 1997). While I agree that hafting of spear points represents a shift along the maintainable-reliable continuum, it seems to me more likely to be a reflection of the increased importance of large-bodied mammalian prey in the subsistence systems of people living in Ice Age Eurasian than a marker of logistical collecting *per se*.

6. Mass killing of herd animals with close-range hunting technology is tricky. Among historically – or archeologically – known arctic hunters, the dispatching of large numbers of reindeer with thrusting spears (lances) is supported either by boat technology or stone retaining walls and large stone carins. In the former case, drivers force the animals into water, where hunters in kayaks spear them while they are swimming; in the latter case, drivers move the animals into ever-narrowing courses that are defined by large (human-sized) cairns and stone walls, ending at cul-de-sacs where the hunters lie waiting (Spiess 1979; Blehr 1990; Friesen 2013). Both strategies require drivers, and it is possible that Neandertals were using natural terrain features and engaging their entire social groups in such hunts (using adult females and children as drivers: Kuhn and Stiner 2006). But again, if the entire social group is moving to a kill site, we are looking at something that is not a fully-logistical subsistence system.
7. Managing carnivores at mass kill sites was likely difficult. As noted in Chapter 9.3, interference competition within carnivore guilds can be fierce. As discussed in Chapter 9, a dead animal is more valuable to a carnivore than a living one, because the injury-risk and energetic cost of capturing it have already been paid. This makes kleptoparasitism – the usurping of carcasses by more socially-dominant carnivores – a strategy with high payoffs, despite the risk of injury that may accompany the aggressive interactions with the original owners of the carcass. If a single carcass is valuable, one or two dozen of them would have probably been very attractive to the large social carnivores with which the Neandertals had to compete. If the Neandertals were not the socially-dominant member of the carnivore guild, as argued in Chapter 9, it seems unlikely that they would have been able to defend carcasses against theft, especially if transport issues (see point 4 above) necessitated a lengthy period over which the carcasses had to remain in the field.

These considerations lead me to doubt that the Neandertals were employing full-blown radiating mobility strategies, even occasionally. They were no doubt capable of logistical organization (that is, flexibly varying the relative amounts of logistical and residential mobility in ways that best fit the resource structure of the environments in which they found themselves: see Daujeard and Moncel 2010; Moncel and Rivals 2011), and I suspect that across most of their geographic and temporal range they were employing mixed strategies that were heavily dependent on residential moves (circulating mobility) but that also made good advantage of logistical mobility to provision residential bases. Along these lines, it is important to note that cold–temperate environments do not necessarily demand logistical strategies, and thus it's not unreasonable to think that Neandertal subsistence/mobility systems differed markedly from the highly logistical systems *normally* seen among modern high latitude hunter-gatherers. At least two groups of Inuit – the Netsilik (Kelly 1983) and the Copper "Eskimo" (Stefansson 1913; Jenness 1922, 1959) – lived historically in contexts in which bulk procurement of terrestrial game for winter storage was not feasible, but edible resources (in the form of seals hunted through the sea ice) were homogenously (yet sparsely) distributed and huntable throughout the winter. Combined with three season exploitation of terrestrial resources (caribou, and salmon taken from inland streams), these groups were able to manage in the arctic with a circulating mobility pattern. The Netsilik and Copper Eskimo, then, may be more germane to understanding Neandertal mobility patterns than are other, more logistically organized, modern foragers. While it certainly wasn't the case that Neandertals across Europe were able to use marine resources during the winter, there may have been some opportunity to capitalize on natural winter mortality (winter-kill). Increased wintertime mortality has been documented in red deer (Coulson et al. 1997; Loison et al. 1999), reindeer (Skogland 1990), roe deer (Gaillard et al. 1993), and ibex (Toigo et al. 1997). Winter mortality primarily involves juveniles (up to 40% mortality in some cases: Skogland 1990), but affects adults as well: approximately 40% of an entire population of roe deer died during one particularly bad winter in Sweden (Gaillard et al. 1993). Much winter-kill is attributable to starvation, and accordingly the carcasses of these animals were likely in too poor of condition (too fat depleted) for consumption (Speth 1987; 1989). But ungulates can also die from cold exposure, and the carcasses of these individuals may have had sufficient fat stores to be nutritive. Thus, it is possible that by smoking and drying what reserves they could during the fall (small-scale storage), combined with some hunting (early in the winter, while animals were still in good condition), some scavenging of winter-killed carcasses, and a great deal of "hunkering down," the Neandertals might have been able to overwinter without food storage, and to use residential mobility to deal with local biodepletion.

10.2 HOME RANGE SIZE

How large were the territories exploited by Neandertal social groups? Regardless of the type of settlement system in play, larger territories require greater mobility to exploit the resources contained therein. While good ethnographic data on logistical mobility are difficult to come by, a few recent hunter-gatherer groups can be called upon to illustrate the relationship between mobility levels and territory size (using data from Table 1 in Kelly 1983). The G/wi of southern Africa occupy a home range of 782 km^2, which they cover on average with 11 residential moves of 25 km per year. The average round trip distance covered in a logistical foray away from camp is 14 km. Data on the number of logistical forays per year are, unfortunately, lacking, but as hunter-gatherers falling towards the "forager" end of the site settlement continuum (similar to the San: Binford 1980), the number of such trips per year was not likely to be great. At the other end of the spectrum lie the Nunamiut, who occupy a territory of 12,850 km^2 in central Alaska. The Nunamiut make, on average, 10 residential moves of 69.5 km per year. Logistical mobility is again difficult to gauge, but Binford (1976) recorded 47 trips, with a mean round trip distance of 80.9 ± 81.4 km, in a two-week period of high mobility in April of 1971, which gives an idea of the kinds of mobility employed by these logistical collectors. Thus, knowing something of Neandertal home range size will give us some idea of the relative amounts of movement required in their subsistence rounds.

As a first approach to estimating home range size in Neandertals, it seems reasonable to think that we could simply estimate it using data derived from modern foragers, combined with paleoenvironmental inferences about Pleistocene Eurasia. Two important considerations, however, limit the extent to which we can apply such a strategy. First, overall mobility is likely dependent in part on the type of site settlement system in operation. Because Neandertals might have been employing systems that differed in key ways from those of modern humans in similar environmental contexts (as discussed in Section 10.1), their home range size-to-productivity relationships may not have varied in the same way as do those of recent foragers. Second, modern human foragers at high latitudes tend to rely heavily on marine rather than terrestrial resources (since productivity in the oceans increases, up to a point, with latitude), which fosters sedentism along rivers and coasts and results in very different mobility strategies than those employed by foragers dependent on terrestrial resources (Kelly 1983). While Neandertals appear to have at times been exploiting coastal and riverine resources (see Chapter 7), there is no evidence to suggest that they did so intensively, and certainly not to the degree necessary to put up stores for overwintering or to support a sedentary existence at the water's edge. This necessarily constrains comparisons to modern groups that were primarily dependent on terrestrial resources but, as noted above these groups practiced highly logistically organized subsistence economies. The Inuit groups that don't practice logistical strategies (Netsilik and Copper Eskimo) manage thanks to marine resources, and thus their mobility levels are not representative of what might have obtained for terrestrially-dependent foragers practicing a circulating mobility or mixed circulating/radiating mobility strategy at high latitudes. As a final consideration, relationships between resource structure and residential mobility, or percent dependence on terrestrial game and home range size (see Kelly 1983) among modern foragers are highly dependent upon population density and subsistence strategy, and cannot be assumed to have been similar in Mousterian foragers. For these reasons, we need to look elsewhere to estimate territory sizes for Neandertals. Studies of how lithic raw materials moved across Pleistocene landscapes, and models derived from carnivore ecology, present two avenues to explore this issue.

10.2.1 Lithic Raw Material Movement

The movement of stone tools across the Paleolithic landscape provides one increasingly important source of information about the movement patterns and territory sizes of the hominins who made and used them (see references in Kuhn 1995). Since flint (and other raw material) can generally be "sourced" (that is, its location of orgin can be determined), the distance a piece of discarded flint has traveled from where it was quarried tells us something about human movement on the landscape. As with pretty much everything we've examined thus far, interpreting the evidence is not entirely straightforward. As noted by Michael Shott (comment on Brantingham 2006: 453), "the distribution and abundance of tool stone across the landscape do not simply reflect ancient hunter-gatherer land use any more than my frequent-flier milage reflects my commuting distance to work." Indeed, if the procurement of high-quality flint was organized logistically (that is, special trips were made to collect flint and transport it back to residential camps), the movement of flint might have little relationship to the size of the area exploited for subsistence, or to the movement of Neandertals within that area. Such a situation has been argued to have obtained during MIS 3 times in the Salento peninsula of southern Italy, where good-quality flint is scarce and flint was imported from the north (Milliken 2007; Spinapolice 2012). Additionally, trade of raw material between social groups, as well as long-distance visiting or individual migration events, would also complicate the interpretation of raw material movement. At Champ Grand, an open air site in the Loire valley of central France, flints have been recovered from sources as far as 250 km to the north (as the crow flies, perhaps as much as 400 km by foot if following the river valleys) and 180 km to the south (again, straight line distance) (Slimak and Giraud 2007). While it should be noted that 99% of the lithic assemblage from the site was produced on local material (mainly from a source about 15 km away), the exotic flints derive from a catchment that encompasses much of central France (see Figure 2 in Slimak and Giraud 2007). Fitting a simple ellipse to the distribution of quarry sources for the exotic flint at Champ Grand produces a territory estimate of 54,000 km^2, which comes close to the largest home ranges recorded for modern foragers (in arctic and subarctic groups: Kelly 1983) and which is unlikely to reflect the actual home range of a single Neandertal social group.

The use of lithic raw material distributions to estimate the scale of landscape use in Neandertals requires the assumption that lithic raw material procurement was embedded in subsistence mobility (Binford 1979), and this is generally considered to have been the case in the Middle Paleolithic (Kuhn 1995; Mellars 1996). Thus, while we need to be attuned to the possibility that rare exotic flints may be informing us about extraordinary movement or possibly intergroup exchanges, and that in some cases lithic procurement may have been scheduled separately from subsistence movement, by-and-large the distance across which flint traveled in the Middle Paleolithic gives us an idea of both Neandertal home range sizes and how mobility was partitioned within those ranges. Unfortunately, raw material transport patterns have not been extensively studied across the Neandertal range, and much of what we know comes from work centered on a few locations. Based on raw material movement as well as intensity of flint knapping evident at sites in Aquitaine (southwestern France), Geneste (1988b) has identified three mobility zones surrounding residential sites. First is the local zone (zone proche), encompassing the area within a 5-km radius from the site. Between 60–98% of the raw material used at Aquitaine sites derives from within this zone, which likely reflects the daily foraging radius away from residential camps. Extending beyond the local zone out to about 15–20 km from camp lay the intermediate zone (zone intermédiaire: Geneste 1988b), from which 5–20% of raw material was procured. Outside of Aquitaine, Neandertals were possibly deriving a greater proportion of raw material from this intermediate zone, as witnessed by sites in central France (Fernandes et al. 2008; Moncel 2011), northeastern Spain (Mora et al. 2004; Fernández-Laso et al. 2011), Crimea (Marks and Chabai 2001; Burke 2006), and the Levant (Hovers 2001). Finally, a distant zone (zone éloignée) extended out a variable distance beyond the intermediate zone. Generally, less than 5% of the raw material at Aquitaine sites, and usually only about 1–2%, derived from this zone (Geneste 1988b). The limits of this distant zone tend to be found at about 100 km from any given site in western Europe, but out to about 200–300 km in central Europe (Féblot-Augustins 1993). Again, however, the difficulty of distinguishing exotic flints traded across social groups or carried in during cases of extraordinary migration from flints collected during more ordinary movement in the home range rears its head.

Overall, the preponderance of local and near-local flint in Mousterian assemblages suggests moderately-sized territories (Féblot-Augustins 1993; Mellars 1996; Fernandes et al. 2008; Fernández-Laso et al. 2011). In MIS 5a unit 1 of Baume-Vallée (south central France), raw material movement suggests a potential territory of 6644 km^2, while the same indicators suggest a potential maximum territory area of 5024 km^2 at the nearby MIS 3 site of Le Rond de Saint-Arcons (Fernandes et al. 2008). A larger territory – 13,000 km^2 – has been suggested for Neandertals of the Aquitaine basin based on raw materials in MIS 7–6 levels at Vaufrey and other sites (Geneste 1988a; Féblot-Augustins 1993). At the (likely MIS 3) eastern European site of Karabi Tamchin in Crimea, the majority of flint was harvested from 25 km away, implying an approximate territory of only 1962 km^2 (Burke 2006). However, Burke (2006) proposes that the Crimean Neandertal population overall (estimated to be between 175–300 individuals) needed the entire 27,000 km^2 Crimean peninsula to harvest their mobile primary prey (*E. hydruntinus*), and that the entire peninsula was within the "knowable" range of individual Neandertal foragers. In central and north central Europe, where raw material transport distances are typically larger, flint movement generally suggests territory sizes on the order of 10,000 km^2 (for example, at Raj Cave [Poland] and Kůlna [Czech Republic]: Féblot-Augustins 1993). Even the largest of these estimates are moderate in the context of more recent, comparable hunter-gatherer groups (Table 10.2). Nunamiut Eskimo and Crow Indian groups, both of which relied primarily on hunting of terrestrial game (87% and 80% of total diet, respectively), exploited territories up to 12,850–63,700 km^2 (Kelly 1983). Other northern latitude groups that relied more heavily on marine resources, such as the Baffinland Inuit (25,000 km^2) and Netsilingmiut (6000 km^2), utilized somewhat smaller but still sizeable territories. The lithic data suggest territory sizes that were large relative to those of recent foragers living in tropical, subtropical, and even temperate areas (ETs ranging between 25 and 10: Table 10.2), but small relative to hunter-gatherers from arctic and subarctic environments (ET less than 10, as would have obtained during all but interglacial episodes: see Chapter 6.2).

10.2.2 Carnivore Models for Estimating Neandertal Mobility

Ecological relationships between secondary productivity and range size in large-bodied carnivores provide a second avenue by which we might estimate Neandertal home range sizes. The logic behind this approach is fairly straightforward: first, across the carnivores, home range size varies positively with metabolic need (and thus body size) and with the proportion of meat in the diet (Gittleman and Harvey 1982). Second, in both lions and wolves it has been shown that territory sizes and core areas vary inversely with the encounter rates of ungulates (Gittleman and Harvey 1982; Jędrzejewski et al. 2007). Third, there is empirical data across a range of carnivore body sizes showing that it takes a standing crop of 10,000 kg of prey to support 90 kg of carnivore, regardless of the body size of the carnivore (Carbone and Gittleman 2002), which means that carnivore biomass density (kgs per unit area) is related to prey biomass density. Based on these last two observations, there must exist detectable relationships between the

TABLE 10.2 ■ Home range size of recent foragers, organized by effective temperature (ET)

Group	ET (°C)	Home range size (km^2)
Penan	24.9	861
Mbuti	23.7	465[a]
Semang	23.7	2475
Vedda	23.0	41
Andamanese (Onge)	21.9	40
Aeta	21.2	3265
Siriono	20.6	780
Birhor	19.7	130
G/wi	19.3	782
Dobe Ju/'hoansi	18.8	1380
Hadza	17.7	2520
Aranda	15.9	260
Walapai	15.1	588
Maidu	14.8	1855
Shasta	14.0	3255
Nez Pierce	13.3	2000
Crow	13.0	61880
Micmac	12.7	3100
Nootka	12.6	371
Gulf Salish	12.6	631
Washo	12.3	2327
Twana	12.3	211
Ainu	12.3	171
Puyallup-Nisqually	12.3	191
Klamath	12.2	1058
Ainu	12.0	171
Southern Kwakiutl	11.6	727
Montagnais	11.6	2700
Quinault	11.5	110
Makah	11.3	190
Haida	11.1	923
S. Tlingit	10.9	1953
Mistassini Cree	10.8	3385
Bella Coola	10.5	625
N. Tlingit	10.0	2500
Nunamiut	9.8	12850[a]
Baffinland Inuit	9.1[b]	25000
Netsilik	9.5	6000

Home range = total area exploited per year. Data from Kelly (1983, 1995).
[a] Using midpoint of reported range of territory sizes.
[b] Using ET reported for Copper Inuit.

aggregate mass of a carnivore group, secondary biomass in their prey species, and their territory size. Because Neandertals were heavily meat-dependent (Chapter 7), they should be expected to adhere to the same ecological relationships as do other carnivores.

Steele (1996) used empirical data on carnivore group size/territory size relationships (Grant et al. 1992) to predict home range size for a Neandertal social group of 25 and for small populations of 125 and 300. Steele's estimates are reported as home range diameters, but equate with territory sizes (assuming a circular territory) of 834 km^2 for the social group, 3409 km^2 for the population of size 125, and 7356 km^2 for the population of 300 Neandertals. These estimates seem unreasonably small based on the inferences drawn from studies of raw material movement. In this case, part of the problem undoubtedly lies in the fact that the carnivore data span several orders of magnitude in body size (from least weasels to lions and tigers), but perhaps the larger problem is that the data are uncorrected for carnivore group size and secondary biomass density. This illustrates some of the difficulties inherent in trying to use ecological relationships in carnivores to predict range size in Neandertals. As with modern human foragers, data on actual mobility in carnivores (such as day range sizes) are scarce. Despite a couple of efforts to empirically delineate the determinants of carnivore mobility (Carbone et al. 2005a; Gittleman and Harvey 1982), the resulting models cannot be applied directly to estimate Neandertal home range for a number of reasons: they require input variables (such as prey density and prey group size) which are highly variable across space and which, for Pleistocene Eurasia, would themselves be highly estimated (see Chapter 6); they generally involve proportional equations but lack proportionality constants, such that direct estimates are not obtainable; and the results are generally applicable only at a broad interspecific scale (since the empirical relationships are determined on carnivores ranging in size from dwarf mongoose to grizzly bear). This latter concern is perhaps the most troubling, since the large intraspecific variation in mobility observed in some of the larger-bodied species (Gittleman and Harvey 1982; Höner et al. 2005; Jędrzejewski et al. 2007) causes the observed interspecific relationships to break down at the family taxonomic level and below. As with humans, mobility is a variable that can be adjusted by carnivores to adapt to local ecological factors and, as with humans, this variability makes it difficult to characterize the behavior of any given species, even within a given biome.

These considerations warrant caution in trying to use carnivores to estimate Neandertal mobility, and suggest that efforts to do so should use data from carnivores of body size close to Neandertals (that is, large-bodied carnivores rather than the full taxonomic range of carnivores) and should proceed from primary data on ecological variables, pack size and aggregate group mass, and territory size (rather than trying to use existing empirical relationships reported in the literature). Chris Walker and I (2014) have thus endeavored to construct a predictive model for Neandertals using data on northern grey wolves (*Canis lupus*). Grey wolves make a good model for Neandertal mobility for many reasons (see Arcadi 2006): (1) they have a varied diet but prey predominantly on large ungulates, and among the extant members of the genus *Canis* they occupy a niche that is most similar to that of the Neandertals; (2) they occupy northern hemisphere environments that are in many ways similar to those occupied by Neandertals; (2) they range widely and are capable of endurance locomotion; (4) they are social carnivores that regularly engage in cooperative hunting; (5) and they have wide

TABLE 10.3 ■ Wolf packs used in the development of a home range size predictive model for Neandertals

Country	Region	Avg. adult mass (kg)	Mean territory (km^2)	Latitude	No. packs
Canada	East-central Ontario	27.8	224	50 °N	2
Canada	Southwestern Quebec	28	351	47 °N	14
Canada	Yukon	36	849	62 °N	28
Poland	Bialowieza Forest	33.4	201	52.5 °N	4
Romania	Brasov	33.4	164	45.5 °N	2
USA	North-central Minnesota	33.2	116	47.75 °N	4
USA	Northeastern Minnesota	30.8	243	48 °N	11
USA	Northwestern Minnesota	33.2	344	48.5 °N	8
USA	Northwestern Alaska	44.4	1868	66 °N	6
USA	South-central Alaska	44.4	1193	63 °N	25

Data from Walker and Churchill 2014.

habitat tolerances and flexible social structure (see Chapter 9.2 for basic information on wolf ecology). Grey wolf territory size is highly variable, ranging from just a few square kilometers to tens of thousands of square kilometers (US Fish & Wildlife Service 2008). Within the territories, grey wolves operate as endurance hunters, traveling up to 200 km d^{-1} (although averaging 20–30 km), and up to 20 km at sustained trotting speeds averaging 8.7 km h^{-1} while actively pursuing prey (Mech 1970, 1994; Mech and Korb 1978; Arcadi 2006). We thus used data on wolf territory size (km^2), group aggregate mass (kg), prey biomass (kg km^{-2}), latitude (°N), average winter temperature (°K), and average winter precipitation (mm) on 104 wolf packs from ten locations across North America and Europe (Table 10.3). Because Neandertals and wolves differ in body size, pack size cannot be used as a direct control for the effects of consumer demand on territory size (namely, that a greater mass of carnivores requires a greater mass of prey to support it, such that if prey biomass is held constant, territory size is expected to covary positively with pack size). Thus we used aggregate mass of the social group as our measure of consumer demand. This was calculated by multiplying pack size by the average mass of mature wolves for the region from which the data derived (masses ranged between 27.8 kg for wolves in East-central Ontario and 44.4 kg for those in Alaska: Table 10.3). Prey biomass may not be the best measure of the caloric richness of a wolf's environment, as it is secondary productivity in preferred prey species, rather than secondary biomass, which determines wolf territory size (Jędrzejewski et al. 2007). Holding the ungulate standing crop constant, Jędrzejewski et al. (2007) report an almost seven-fold increase in territory size in wolves at 60°N relative to those at 40°N, which they note is consistent with theoretical arguments (Carbone and Gittleman 2002) that predator populations are sustained by prey productivity rates rather than prey biomass. Although primary productivity decreased during colder climatic cycles in the Pleistocene, attendant changes in plant biomass structure (specifically, a reduction in allocation of primary productivity to inedible supportive structures, as reflected by increases in the NPP/1° biomass ratio: Chapter 6) as woodland gave way to open steppe would have increased animal production, and thus Neandertal territory size and mobility (as well as those of the other members of the large-bodied carnivore guild) would likely have decreased during colder, more open conditions (cf. Pettitt 2003). Thus it might be more appropriate to use secondary productivity, but such data do not appear to exist for modern wolves.

Analysis revealed that most of the ecological variables were highly intercorrelated (and thus redundant with respect to the model), and backwards elimination regression revealed the data best fit a two variable model constructed with aggregate mass (AM, in kg) and average winter temperature (WT, in °K):

$$\text{Territory Size (km}^2) = 5258.7 + 3.65\text{AM} - 20.91\text{WT} \pm 205$$

We sought to model Neandertal home range sizes for social groups between about 10–30 individuals. To estimate aggregate mass values, we used male and female adult Neandertal mass averages (78 and 66 kg, respectively: Chapter 4) and one-half the mean male/female body size (= 36 kg) to represent the size of a juvenile Neandertal. Neandertal aggregate mass values of 500 kg, 1000 kg, and 1500 kg were used as sample estimates for input into the model. The number of Neandertals represented by these aggregate masses would naturally vary depending on the age and sex composition of the group, but if one makes some simplifying assumptions (dependency ratios varied between 1.25–3.00 [Binford 2001], all dependents were juveniles [see Caspari and Lee 2004], and sex ratios among producers were equal) these values should roughly correspond with groups of 10–11 (3–4 adults, 6–8 children), 19–23 (6–8 adults, 11–17 children), and 29–33 (8–13 adults, 16–25 children) individuals, respectively. We also used the range of average winter temperature values representing an MIS 3 warm interval and the last glacial maximum (LGM) as estimated by the Stage 3 project (ftp://ftp.essc.psu.edu/pub/emsei/pollard/Stage3/). While the Neandertals were obviously extinct by the LGM, we used LGM temperature estimates as broadly representative of climatic conditions endured by Neandertals during the preceding glacial maxima (MIS 4 and 6: see Chapter 5.1). We were interested in modeling Neandertal home range sizes in the regions for which estimates based on lithic raw material movement exist: initially (Walker and Churchill 2014), we modeled home ranges sizes for southern France and the Crimean peninsula, but the results also apply to the conditions that would have obtained in north central Europe.

When applied to Neandertals, the wolf model suggests that Neandertals required and maintained moderately-sized territories (ca. 1400–5400 km^2: Table 10.4). As with the estimates from lithic raw material dispersal (see above), these home ranges are large relative to those of most tropical foragers using predominantly circulating settlement systems, and are large relative even to those of most mid-latitude groups for whom data are available (Table 10.2), yet small compared to arctic and subarctic groups. It is interesting that home range estimates derived from grey wolf ecology (combined with some generous assumptions about Neandertal ecology) are fairly concordant with estimates derived independently from lithic raw material movement distances, at least for south central France and the Crimea (but not necessarily the Aquitaine basin or north central Europe: see below). The home range sizes estimated by our model imply territories with radii on the order of 20–40 km, which is also consistent with the distances that the majority of flint at Middle Paleolithic sites was procured (see above). This seems to lend support to the assumption that underlies the use of raw material movement to estimate territory size, that is, that raw material procurement was embedded in the subsistence

TABLE 10.4 ■ Neandertal home range sizes predicted from wolf model

	MIS 3		MIS 6 & 4	
Southern France	0 °C	–8 °C	–4 °C	–8 °C
Raw material est. (km^2)	5024		?	
500 kg group				
Wolf model est. (km^2)	**1375**	**1543**	**1459**	**1543**
Deviation from raw material est. (%)	–73	–69	–	–
Implied population density (person km^{-2})	0.007–0.008	0.006–0.007	0.007–0.008	0.006–0.007
1000 kg group				
Wolf model est. (km^2)	**3200**	**3368**	**3284**	**3368**
Deviation from raw material est. (%)	–36	–33	–	–
Implied population density (person km^{-2})	0.006–0.007	0.006–0.007	0.006–0.007	0.006–0.007
1500 kg group				
Wolf model est. (km^2)	**5025**	**5193**	**5109**	**5193**
Deviation from raw material est. (%)	0	+3	–	–
Implied population density (person km^{-2})	0.006–0.007	0.006	0.006–0.007	0.006
■ **Crimea**	0 °C	–4 °C	–4 °C	–20 °C
Raw material est. (km^2)	1962		?	
500 kg group				
Wolf model est. (km^2)	**1375**	**1459**	**1459**	**1794**
Deviation from raw material est. (%)	–30	–26	–	–
Implied population density (person km^{-2})	0.007–0.008	0.007–0.008	0.007–0.008	0.006
1000 kg group				
Wolf model est. (km^2)	**3200**	**3284**	**3284**	**3619**
Deviation from raw material est. (%)	+63	+67	–	–
Implied population density (person km^{-2})	0.006–0.007	0.006–0.007	0.006–0.007	0.005–0.006
1500 kg group				
Wolf model est. (km^2)	**5025**	**5109**	**5109**	**5444**
Deviation from raw material est. (%)	+156	+160	–	–
Implied population density (person km^{-2})	0.006–0.007	0.006–0.007	0.006–0.007	0.005–0.006
Poland	–4 °C	–8 °C	–8 °C	–20 °C
Raw material est. (km^2)	?		10000	
500 kg group				
Wolf model est. (km^2)	**1459**	**1543**	**1543**	**1794**
Deviation from raw material est. (%)	–	–	–85	–82
Implied population density (person km^{-2})	0.007–0.008	0.006–0.007	0.006–0.007	0.006
1000 kg group				
Wolf model est. (km^2)	**3284**	**3368**	**3368**	**3619**
Deviation from raw material est. (%)	–	–	–66	–64
Implied population density (person km^{-2})	0.006–0.007	0.006–0.007	0.006–0.007	0.005–0.006
1500 kg group				
Wolf model est. (km^2)	**5109**	**5193**	**5193**	**5444**
Deviation from raw material est. (%)	–	–	–48	–46
Implied population density (person km^{-2})	0.006–0.007	0.006	0.006	0.005–0.006

Estimates from the wolf model (**in bold**) are provided for the high and low mean winter temperature ranges for each area as estimated from climate maps generated by the Stage 3 project. Implied population densities are based on groups sizes of 10–11 (corresponding to the 500 kg group), 19–23 (1000 kg group), and 29–33 (1500 kg group) individuals. Raw material territory size estimates are from Fernandes et al. 2008 for southern France, Burke 2006 for the Crimea, and Féblot-Augustins 1993 for Poland.

rounds of Neandertal groups (see above). The preponderance of local flint at Mousterian sites is likely to be a reflection of the tendency of Neandertal groups to collect flint during hunting or gathering trips within the "foraging radius" of the sites (Mellars 1996; Féblot-Augustins 1999), while the less frequent occurrence of exotic flint likely reflects the movement of some curated tools, made at distant locations from raw material that was locally-collected there, during residential moves or logistical excursions. While it appears that at some times and in some places the Neandertals scheduled raw material procurement separately from subsistence rounds (Milliken 2007; Spinapolice 2012), which might suggest a more complex, radiating mobility strategy, they were probably generally not doing so.

There is likely no simple relationship between home range size and settlement system, but the moderate-sized territories suggested by both estimation methods do have some implications for Neandertal subsistence dynamics. Return-trip movement across a territory with a 20–40 km radius, whether from a residential camp positioned near the center or from one positioned near the margin, was unlikely to be within day-trip distance. Average day-trip distances among modern foragers (on foot) vary by sex and group, ranging between about 6–17 km for females and 12–25 km for males, and generally occur at walking speeds between 2.2–2.6 km h^{-1} (Binford 2001) and 4.6 km h^{-1} (Kramer 2004b) (the shorter limbs of Neandertals would have only minimally affected their optimal walking speeds: Steudel-Numbers and Tilkens 2004). Consequently, logistical forays away from camp, if taken, may have required travel times of up to a week if moving across the home range diameter and back. At the smaller end of the home range size estimates, full exploitation of the territory could likely be accomplished via day trips or single overnight excursions away from a residential camp, and thus a settlement system based on limited residential moves (a radiating mobility system) may have sufficed to ameliorate problems of biodepletion and spatial patchiness of resources. At larger territory sizes, however, residential moves likely became increasingly necessary to position foragers closer to temporally-limited resource patches or to limit logistical mobility while uniformly exploiting the home range (and thus avoiding problems of biodepletion). A similar relationship between territory size and residential moves would not be expected to hold for modern foragers with transportation technology: in the absence of such technology, Neandertals undoubtedly faced energetic and burden-carrying constraints on logistical foraging radii, which may have forced a heavy reliance on camp moves as a personnel positioning strategy (see Verpoorte 2006; Macdonald et al. 2009; and Section 10.4).

In no cases did the wolf model predict home ranges of the sizes estimated from lithic movement in the Aquitaine basin during MIS 7–6 (13,000 km^2: Geneste 1988a) or north central Europe during MIS 4–3 (10,000 km^2: Féblot-Augustins 1993). Based on the climatic reconstructions of the Stage 3 project, the southern part of Poland (in the vicinity of Raj Cave, from which the latter estimate is partly based) would have experienced mean winter temperature in the range of –4 to –8 °C during an MIS 3 warm interval and –8 to –20 °C during glacial episodes. These WT values overlap those of the other two regions, and thus produce overlapping home range size estimates (Table 10.4), the largest of which (corresponding to a Neandertal social group of 29–33 individuals and a WT of –20 °C) is only half the size of the lithic estimate. Might the wolf-based estimates be too small? The wolf data derive in part from places where wolves have effectively no competition from other carnivores (although some of the data comes from places where wolves compete with brown bears), and thus they may be able to subsist on smaller territories than would be possible under conditions of heavier competition. Given the argument that Neandertals were not the socially-dominant members of the carnivore guild (Chapter 9.4), their subsistence demands may have required greater space per unit body mass than do those of modern-day wolves. In Chapter 9.5 we made some back-of-the-envelope calculations of Neandertal biomass density that were corrected for exploitative and interference competition with other carnivores (that is, that assumed the Neandertals were mid-ranking members of the guild). The territory sizes and aggregate masses reported in Table 10.4 correspond fairly well with the population density predictions from Chapter 9.5. The estimates in Table 10.4 range from 0.005–0.008 individuals km^{-2}, and thus fall at the low end of the range of estimates for cold intervals (0.006–0.021 individuals km^{-2}) from Chapter 9.5. Thus the wolf model actually produces estimates of home range size ***higher than expected*** based on our consideration of the secondary productivity of Pleistocene environments and carnivore competitive ecology. This unexpected finding is likely to be due to differences in productivity in the Pleistocene versus today. Our estimates of Neandertal population density were based on estimated, cold interval open country secondary biomasses ranging between 500–1200 kg km^{-2}, whereas the observed secondary biomasses for the modern wolves ranged between 63–634 kg km^{-2}. For reasons discussed in Chapter 6.2, some Pleistocene ecosystems were more productive than their modern analogs, and thus there was likely a different relationship between mean winter temperature and secondary biomass in the Neandertal versus modern worlds. The combined effects of differences in competitive dynamics and differences in secondary productivity appear to be cancelling one another out to some degree, although it appears the wolf model may still be overestimating slightly the Neandertal home range sizes. Nevertheless, neither approach suggests home ranges on the order of 10,000–13,000 km^2. In the case of the Aquitaine basin, Geneste (1988a) has suggested that the flint at Vaufrey accumulated as a palimpsest of occupations over a geologic period during which groups exploiting different, smaller territories were converging on the cave site (that is, the 13,000 km^2 indicated by flint movement represents the cumulative area, whereas any particular social group exploited a smaller territory within that area). The home range sizes estimated here would suggest that a similar situation may have applied in north central Europe.

The territory sizes estimated by the wolf model and the lithic studies, and the general concordance between these two independent estimates, seem to be consistent with settlement systems that may have relied more heavily on residential than logistical movement (that is, a circulating mobility strategy), in territories that, by virtue of their large size, demanded fairly high mobility levels of their occupants. Although high mobility levels cannot be confirmed by reference to the remains of the Neandertals themselves, the morphology of their lower limbs is certainly consistent with their having had elevated levels of movement about the landscape.

10.3 PALEONTOLOGICAL REFLECTIONS OF NEANDERTAL MOBILITY

As reviewed in Chapter 8.1, we expect skeletal morphology to reflect habitual behavior, both because of natural selection for mechanically competent and efficient skeletal design, and because of developmentally-plastic responses of bone tissue to mechanical stimuli during the lifetime of the individual. Neandertals, like modern humans, were efficient bipeds, and while their bodies differed in proportions from those of most modern humans (for reasons having to do with cold adaptation and force production: see Chapter 5.3), the overall configuration of their pelves and lower limbs reflect adaptation to bipedal locomotion (see references in Trinkaus and Hilton 1996). However, based on the reasoning described in Chapter 8, the amount and distribution of cortical tissue in the shafts of their lower limb bones tell us something about the magnitudes and orientations of loads those bones habitually experienced, and by inference something about Neandertal mobility levels and patterns. Accordingly, most comparative functional morphological studies of Neandertal mobility have focused on structural (geometric) analysis of the shafts of lower limb long bones. Orientations of key joint surfaces in the lower limb, which reflect more the activity patterns during skeletal growth (i.e., childhood) than during adulthood, have also entered into the discussion.

Geometric analyses of lower limb long bone (femoral and tibial) cross-sections have been used to compare European Mousterian Neandertals with Upper Paleolithic modern humans (Lovejoy and Trinkaus 1980; Trinkaus and Ruff 1989a, 1989b; Ruff et al. 1993; Trinkaus 1997), Levantine Mousterian Neandertals with Levantine Mousterian modern humans (Endo and Kimura 1970; Trinkaus et al. 1998a), and the possibly Châtelperronian-associated St. Césaire Neandertal with early Upper Paleolithic modern humans (Trinkaus et al. 1998b, 1999a). All of these studies have produced consistent results: regardless of cultural context, Neandertals tend to have lower limb bones that are robust (having elevated measures of resistance to compressive, tensile, bending, and torsional loads) relative to early modern humans when long bone length is used to body size-standardize the strength measures. However, relative to measures of body mass (such as femoral head diameter, or mass estimated from long bone length and bi-iliac breadth) there are no significant differences between groups (Trinkaus and Ruff 1989a, 1989b; Trinkaus 1997; Trinkaus et al. 1998a). Body mass is arguably the most mechanically relevant measure of body size when considering lower limb remains (Ruff et al. 1993; Ruff 2000a), since it is body mass under the acceleration of gravity, and the acceleration and deceleration of body mass during locomotion, that produce the predominant forces on the lower limb (also, the forces generated by lower limb muscles to accelerate and decelerate the body during locomotion must be proportional to mass). Thus, when the greater breadth of the trunk (and hence greater mass for a given stature) of Neandertals is taken into consideration, the apparent between-group differences in lower limb robusticity disappear. This suggests (but see below) that mobility levels overall were similar between Neandertals and early modern humans, across each of the comparative contexts. This result holds for comparison of lower limb bone strength between Levantine Neandertals and early modern humans as well (Trinkaus et al. 1998a), which does not support the idea of significant differences in mobility levels between these groups (see below). Similar adult mobility levels (but not necessarily subadult levels: see below) in Levantine Mousterian groups is further indicated by similarities between them in femoral histology (Abbott et al. 1996), talar proportions (Rhoads and Trinkaus 1977), and pedal phalangeal dimensions (Trinkaus and Hilton 1996).

Differences in the relative strength of the lower limbs of Upper Paleolithic modern humans (Late Pleistocene) and recent (Holocene) humans have typically been argued to reflect differences in mobility levels between groups (Ruff et al. 1984; Bridges 1985, 1989; Holt 1999, 2003; Stock and Pfeiffer 2001; Shackelford 2007), and these inferences have generally accorded very well with ethnohistoric or archeological evidence of mobility in these groups (although some aspects of lower limb robusticity may be better reflectors of mobility than others, and climatic adaptation may introduce additional variation: Stock 2006).[7] Thus the similarity in lower limb bone strength in Neandertals and early modern humans would suggest similarities in their locomotor levels. However, it is not entirely certain that differences in mobility levels between individuals or groups *should* be reflected in the robusticity of the lower limb. There is evidence that the critical stimulus that triggers adaptive responses in bone is the magnitude of the strain engendered during loading, not the number of loading cycles. For instance, Rubin and Lanyon (1984) found that a small number of loading cycles (36) that produce strain levels above a certain threshold were sufficient to stimulate adaptive modeling and remodeling, while an increase beyond 36 (up to 1800 cycles) had no appreciable effect on further osteogenic activity. Numerical models (Whalen et al. 1988; Beaupré et al. 1990a, 1990b) and numerous clinical/experimental studies (reviewed in Skerry and Lanyon 1995) support the claim that, with loads of sufficient magnitude, only a small number of cycles is needed to reach a saturated osteogenic response. This has important implications for analyses of between-group variation in lower limb robusticity: if lower limb loads produced during walking are of sufficient magnitude to either prevent bone loss or stimulate bone formation, a saturated osteogenic response would be achieved within the first hundred meters of locomotion (within 36 loading cycles, or strides), and thus one would not predict a difference in bone mass or distribution in the lower limbs of human groups differing in mobility – humans that walk one kilometer a day should have the same relative bone strength (robusticity) as people who walk 40 kilometers a day.

Bone biologists disagree about the primary 'mechanical stimulus' in osteoregulation, and fatigue damage, strain energy density, strain rate, strain gradients, and abnormal strain distributions have all been implicated (see Martin and Burr 1989 for a review). Of these, the only mechanism that would be consistent with the observed patterning in human skeletal material would be fatigue damage, since microfractures may develop from a few cycles of high magnitude loading or, more importantly, from many cycles of lower magnitude load. Thus, mobility differences between groups may result in different amounts of fatigue damage and, hence, different modeling and remodeling responses. However, at physiological loading levels during walking, roughly 10^8 loading cycles would be required to bring a femur to failure through fatigue (King and Evans 1967), and as noted by Currey (1984: 82): "a person walking at a brisk 100 steps a minute, 10 hours a day, would take 9 years to reach this number of cycles on one leg bone." If the risk of fatigue damage from walking is so low, we might reasonably wonder if even the highest foraging or residential mobility levels among humans would ever be sufficient to stimulate appositional modeling and increased lower limb diaphyseal robusticity. Along these lines, it is instructive that Westcott (2006), using samples of nine North American groups that varied in mobility levels, failed to detect a significant correlation between femoral midshaft robusticity (as determined by external shaft dimensions, rather than cross-sectional geometry) and levels of terrestrial logistical mobility. Likewise, Carlson and coworkers (2007) found that highly mobile foragers from southern Australia had femoral shafts that were gracile relative to samples of recent humans from agricultural and industrial economies (it is important to note here, however, that the comparative samples were drawn from populations that likely had high frequencies of individuals who engaged in manual labor, and who thus probably experienced substantial non-locomotor loads on their skeletons).

If a mechanism other than fatigue damage provides the primary osteogenic signal for bone modeling, the conventional interpretation of variation in human lower limb skeletal robusticity would still require re-evaluation (for example, if strain rate is the primary osteogenic signal, then between-group differences in femoral robusticity may be reflecting differences in load magnitude related to burden carrying, not distance traveled). An additional confound exists in the fact that the sources of biomechanically-significant loads in the lower limb may have varied between Neandertal individuals in important yet unknowable ways. Among recent foragers, males may cover absolutely more ground than females, but may do so while carrying no appreciable weight, whereas females may travel less far but may regularly carry heavy burdens (Hilton and Greaves 2004; Kramer 2004a). Thus a given level of robusticity may (if multiple factors, like fatigue damage and strain rate, serve as significant osteogenic stimuli) reflect a large number of cycles of low magnitude loading in males and a smaller number of cycles of higher magnitude loads in females. While this remains an open issue, it is important to bear in mind that the lack of difference in lower limb robusticity, and by inference mobility levels, between Neandertals and early modern humans may have more to do with the inadequacy of structural analysis for detecting mobility differences than with actual locomotor level parity between groups.

Despite similarities in levels of femoral and tibial robusticity, differences in cross-sectional *shape* of these bones suggest that important differences existed in the orientation of loads on the lower limb, and thus possibly *patterns* of locomotion between Neandertals and early modern humans (of both the Levantine Mousterian and the European Upper Paleolithic). Early modern human femora have mid-shaft cross-sections that are tear-drop shaped with large pilasters on the dorsal surface (Figure 10.1). This results in bones with large second moments of inertia in the anteroposterior plane, which are thus resistant to the kinds of bending stresses incurred by the femur during bipedal locomotion. As can be seen from Figure 10.1, Neandertal femoral shafts tend to lack a pilaster and to be sub-circular in section (i.e., having nearly equal second moments of inertia in both the anteroposterior and mediolateral planes), reflecting greater mediolateral expansion of the shaft (Trinkaus and Ruff 1989a, 1989b; Trinkaus et al. 1998a, 1999b). In all of the comparisons, Neandertal femora are generally more resistant to mediolaterally-directed bending stresses, thus showing a greater resistance to bending moments in all directions, relative to the anteroposteriorly reinforced femora of early modern humans. This pattern is reflected as well, albeit not as markedly, in the tibia (Lovejoy and Trinkaus 1980; Trinkaus and Ruff 1989a, 1989b) and in the pedal phalanges (Trinkaus and Hilton 1996). The greater mediolateral reinforcement of Neandertal femora can also be seen in the ratio of their midshaft cross-sectional anteroposterior to mediolateral second moments of inertia (I_x and I_y, respectively). Second moments of inertia quantify the contribution of bone geometry to bending strength in a given plane, and the I_x/I_y ratios observed in Neandertals (which are generally less

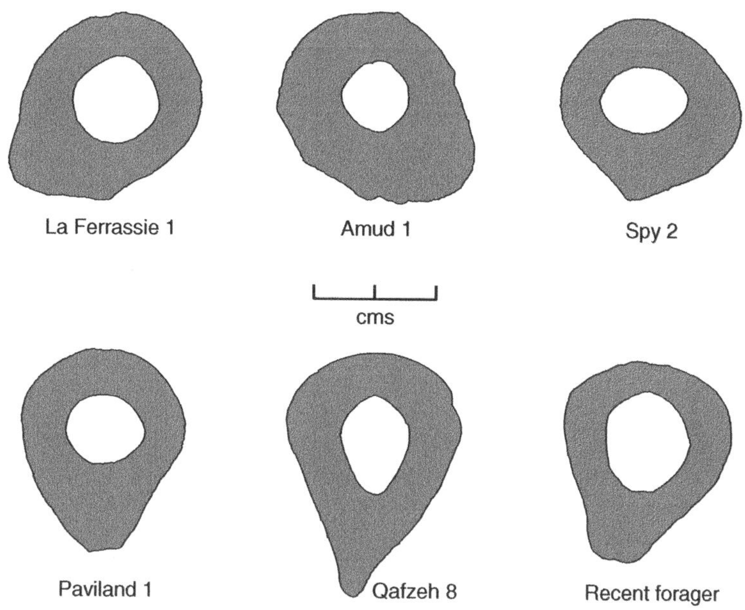

FIGURE 10.1

Representative femoral midshaft cross-sections of male Neandertals (top row) and modern humans (bottom row). Anterior is to the top, posterior to the bottom (note that the femoral section of the recent forager is only roughly to scale; all other sections are to scale with scalebar). La Ferrassie 1 and Spy 2 = European Neandertals (sections from Trinkaus 1997 and Trinkaus and Ruff 1989b), Amud 1 = Near Eastern Neandertal (section from Trinkaus and Ruff 1999); Paviland 1 = European early modern human (section from Trinkaus 1997), Qafzeh 8 = Near Eastern Mousterian-associated early modern human (section from Trinkaus and Ruff 1999), recent forager = Georgia Coast preagricultural native American (section from Ruff et al. 1984). Note the mediolateral expansion (and greater overall circularity) of the Neandertal sections, and the anteroposterior expansion of the modern human sections.

than 1: Table 10.5) indicate a midshaft femur that was, on average, somewhat stronger in the mediolateral than the anteroposterior direction (as opposed to the early modern human pattern of anteroposteriorly reinforced femoral shafts, which tend to have ratios greater than 1: Table 10.5). Among modern human foragers, femoral I_x/I_y ratios are positively correlated with terrestrial mobility (Stock 2006), which raises the possibility that Neandertals were actually *less* mobile than later, Upper Paleolithic modern humans. However, variation in the shape of the midshaft femur among modern humans mainly reflects variation in the anteroposterior second moment of inertia (I_x), which increases relative to that of the mediolateral plane (I_y) in individuals who walk or run greater distances (resulting in greater values of I_x/I_y). Neandertal femora tend to have midshaft I_x values that, when scaled to body size, are not significantly different from those of early Upper Paleolithic (Gravettian) modern humans, again suggesting parity in mobility levels between the groups (Trinkaus 1997). There is a reduction

TABLE 10.5 ■ Neandertal femoral midshaft I_x/I_y ratios

	I_x/I_y	Reference
Amud 1	0.95	Trinkaus and Ruff 1999
Fond-de-Forêt 1	0.98	Trinkaus and Ruff 1989b
Neandertal 1	1.07	Beauval et al. 2005
Rochers-de-Villenuve 1	1.10	Beauval et al. 2005
Shanidar 4	1.02	Trinkaus and Ruff 1999
Shanidar 5	1.01	Trinkaus and Ruff 1999
Shanidar 6	0.90	Trinkaus and Ruff 1999
Spy 2	0.89	Trinkaus and Ruff 1989b
St. Césaire 1	1.24	Beauval et al. 2005
Tabun 1	0.70	Trinkaus and Ruff 1999
Tabun 3	0.85	Trinkaus and Ruff 1999
Mean (SD, n) values:		
Neandertals	0.97 (0.14, 11)	
Gravettian modern humans	1.48 (0.21, 21)	Beauval et al. 2005
Late Upper Paleolithic humans	1.31 (0.08, 21)	Shackelford 2007

in I_x values, and thus I_x/I_y ratios, from Gravettian to Late Upper Paleolithic samples (Table 10.5), which *does* appear to represent a reduction in mobility between these modern human groups (Holt 1999, 2003; Shackelford 2007), but this does not seem to be the case with the Neandertal/Gravettian modern human comparison. Neandertal femoral sections have low I_x/I_y ratios not because they have low bending strength in the anteroposterior plane, but because they have greater reinforcement in the mediolateral plane.

A number of behavioral explanations have been posited to account for the enhanced mediolateral bending strength of Neandertal lower limb bones (Trinkaus and Ruff 1989a; Trinkaus and Hilton 1996), including (1) the possibility that Neandertal movement on the landscape was more irregular and less directed than those of modern humans (perhaps involving greater movement over rugged terrain: see Pearson et al. 2006), or (2) that Neandertals experienced high levels of lateral upper body movement (as in forcefully thrusting a spear) that required greater effort with the legs to stabilize the body (see also Black 1999), or (3) that the more lateral body build (i.e., wider trunks) of Neandertals created greater mediolateral bending moments in their lower limbs (see also Weaver 2003), or some combination of two or more of these explanations. Mediolateral bending stresses in the lower limb, caused by a wide trunk, would be expected to be inversely proportional to the distance from the hip, such that more distal areas (like the tibial shaft) should be less affected by body shape and thus be better indicators of mobility (Ruff et al. 2006b; Stock 2006; but see Pearson et al. 2006). Interestingly, a shift towards the modern human pattern (of greater I_x/I_y ratios) is seen in the relatively late MIS 3 Neandertals from Feldhofer, Les Rochers-de-Villeneuve, and St. Césaire (Beauval et al. 2005: see also Table 10.5). Since these Neandertals were likely as wide-bodied as their predecessors, the change in femoral shaft architecture probably reflects a change in mobility patterns (to more directed movement on the landscape, or less climbing over rugged terrain) or less need for trunk stabilization in the later part of the Mousterian.

Unfortunately, other developmentally-plastic aspects of lower limb morphology do not really resolve the question of Neandertal mobility levels. Neandertals and modern humans appear to differ in femoral shaft curvature, femoral neck-shaft angle, and in the morphology of at least one lower limb muscular enthesis, but the significance of these differences is not clear. With respect to femoral curvature, it has long been noted that Neandertals have femora with a greater degree of anterior curvature than are commonly seen in modern humans (Trinkaus and Shipman 1993; Churchill 1998; Weaver 2003). Femoral curvature varies among modern human groups, but does not covary with measures of diaphyseal robusticity, and has been suggested to perhaps be a better indicator of activity levels than are diaphyseal properties (De Groote 2008). While some studies failed to detect a difference in femoral curvature between Neandertals and early modern humans (Shackelford and Trinkaus 2002), size-standardized measures of curvature have been shown to differ between these groups (De Groote 2011). While this suggests a greater amount of mobility in the Neandertals, it is presently not clear if the differences reflect genetic differences between groups, differences in childhood levels of mobility, or differences in adult activity levels (De Groote 2011). Neandertals also tend to have low femoral neck-shaft angles, and the Levantine Neandertals contrast with their modern human sympatriots, which may suggest that differences in mobility levels existed between these groups (Trinkaus 1993). The orientation of the femoral neck is very responsive to loading during childhood (Tardieu and Trinkaus 1994), and the lower neck-shaft angle seen in Neandertals has been taken to imply differing levels of childhood mobility in the two Levantine Mousterian groups (Trinkaus 1993). This difference has been suggested to reflect differences in social organization between them, with Neandertal children having traveled with foraging parties (thus engaging in long distance mobility from an early age) while modern human children stayed at the residential site without accompanying foraging parties (consistent with the provision of alloparental care by "grandmothers": Hawkes et al. 1997, 1998). However, Weaver (2003) suggested that the low neck-shaft angles of Neandertals were a consequence of their cold-adpated body proportions, in which wide trunks produced different loading patterns on the developing femoral neck than those incurred by more narrow-bodied modern human children. This hypothesis is supported by a study of global variation in recent human neck-shaft angles (Gilligan 2010a), which finds that the angle of the femoral neck is highly variable within and between groups, but that about 20% of this variation can be explained by variation in mean annual temperature. While mobility levels during childhood may affect femoral neck-shaft angle, it remains to be determined how much of the difference in mean angle between Neandertals and modern humans is attributable to differences in activity level versus average body proportions. Finally, a review of the surface morphology of the insertion sites of several lower limb muscles in the Krapina (Croatia) Neandertals found no significant differences with modern humans in most of the sites examined (Mariotti and Belcastro 2011). The singular exception was the insertion of the *M. gluteus maximus*, which fell outside of the modern human range of variation in both its dimensions and surface morphology. The same morphology was seen in both adult and immature individuals at Krapina, which suggests either an early development of a developmentally-plastic trait or a genetic basis to the morphology, and the surface morphology suggests that Neandertals may have had a fibrocartilagenous enthesis at this location (as opposed to the fleshy insertion seen in modern humans) (Mariotti and Belcastro 2011). The functional significance of the morphology seen in Neandertals is not clear, as it may reflect activity-level differences, differences in the mechanical role of *M. gluteus maximus* given the wider body of Neandertals, or genetic fixation of a functionally-neutral trait (Mariotti and Belcastro 2011).

As a final paleontological indicator of mobility, it is worth noting that strontium isotope analysis of a Neandertal tooth from Lakonis, Greece, indicates that its owner grew up at a distance of at least 20 km (and possibly considerably more) from the fossil's final resting spot (Richards et al. 2008a). In terms of hunter-gatherer mobility, 20 km is not far (as Clive Finlayson was quoted in the media coverage of the research, "we're talking about humans, not trees"), and thus while this case does document at the least a minimum level of residential mobility, it does not tell us much about the overall mobility of the Neandertals. At present, the most we can say is that lower limb diaphyseal morphology in Neandertals, along with the evidence from femoral curvature and the femoral neck, is consistent with inferences of a high degree of mobility in the Middle Paleolithic.

10.4 THE ENERGETIC COST OF MOBILITY

The evidence reviewed above suggests that Neandertals exploited moderate-sized territories, that under most circumstances they were probably using high residential mobility to deal with biodepletion and as a positioning strategy, and that their activity levels (as reflected in lower limb morphology) were at least comparable to those of modern human hunter-gatherers, and were possibly higher. Beyond these broad observations, is it possible to say more about the energetic cost of locomotion among the Neandertals? Obviously, the cost of locomotion is a function of a host of variables that are unknown for our Mousterian foragers: average daily travel distances; proportion of travel time spent burdened (as when transporting carcass parts, or moving bedding and shelter material during a camp move) versus unburdened, and the average weight of the loads carried; the nature of the terrain traversed, and the average changes in elevation experienced during travel; the average depth of winter snow and the distances traveled in it; and the distances covered at running speeds during hunting, predator avoidance, and logistical forays. Direct estimates of mobility costs thus appear to be out of reach, but fortunately our goal is simply to evaluate the reasonableness of the PAL multipliers that have been suggested for Neandertals. In that regard, we might be able to produce some reasonable estimates of the caloric cost of mobility (again, likely the largest item in the activity energy budget) in the Middle Paleolithic based on the behavior of recent foragers.

Because of the many unknowns about Neandertal mobility, the most reasonable approach might be to use average foraging distances as observed in modern hunter-gatherers (while these data are not abundant, there is a little bit to go on). As a simplifying assumption we will consider subsistence locomotion (that is, travel during foraging and residential moves) as constituting the vast majority of Neandertal travel, and we'll assume that lithic raw materials were collected during subsistence travel (that is, that raw material acquisition was embedded in subsistence: see above). These assumptions are conservative, in that Neandertals probably traveled between groups for social purposes (mate acquisition, visiting, etc.) from time to time, and so by focusing only on subsistence effort we are more likely to underestimate true travel amounts. Likewise, if Neandertals were scheduling travel for raw material acquisition separately from their subsistence rounds, their actual locomotor costs would again be higher than estimated by considering subsistence travel alone. Binford (2001) provides data on round-trip foraging distances for females from five groups (Nunamiut, Alyawara, Pitjandjara, Anbarra, and Pume), which produce an average day trip of 10.6 ± 4.4 km. Data on three male groups (Nunamiut, Pume, and !Kung) produce an average foraging round trip distance of 16.8 ± 6.8 km. The male mean trip distance is greatly influenced by the large travel distances of the arctic Nunamiut – without them, the mean round trip distance for the tropical/subtropical Pume/!Kung drops to 12.9 km (the female mean drops only about a half a kilometer if the Nunamiut are excluded). Home range sizes are highly variable both within and between these groups, ranging from a low of 56 km^2 among the Anbarra to between 5200–20,500 km^2 for the Nunamiut (Kelly 1995), and the site settlement systems employed vary across the groups as well. Nonetheless, the variance in average female round trip distances is relatively low, which suggests that variation in productivity is adjusted for more via residential moves than by adjusting female foraging distances. As for males, the Nunamiut have logistical round trip distances roughly twice that of males in the other two groups, which is a reflection of the radiating mobility strategy they employed, combined with a heavy dependence on transportation technology. For Neandertals that were likely employing settlement strategies more dependent on residential mobility, and who seemingly lacked transportation technology, the round trip distances for the Pume and !Kung might be more appropriate (and certainly more conservative with respect to modeling energy budgets). Thus, as working values we might use 10.6 km per foraging trip for Neandertal females, and 12.9 km per trip for males. As with other variables, the number of trips in any given amount of time is difficult to estimate, and was likely seasonally variable. Binford (2001) suggests that producers work on average about 170 days per year, which accords reasonably well with observations in some groups that males and females alike take rest days every other or every third day (Lee 1968; Sahlins 1968). For simplicity, we'll assume that Neandertals were alternating foraging days with rest days, which produces weekly foraging distances of about 37 km for females and 45 km for males. Hunter-gatherers who are heavily dependent on terrestrial game average 18.5 ± 13.8 residential moves per year of an average distance of 32.2 ± 10.8 km per move (Binford 2001), which we might also take as being representative of Neandertals (for an average weekly residential move distance of 11.5 km for both sexes). Finally, we might assume that residential moves involved carrying burdens, but that only about one-quarter to one-half of foraging mobility was done while carrying loads. This provides us with some rough weekly mobility estimates with which to work (Table 10.6).

TABLE 10.6 ■ Estimated weekly travel distances for female and male Neandertals

	Foraging	Residential move	Burdened travel	Unburdened travel
Females	37	11.5	20.8–30.0	18.5–27.8
Males	45	11.5	22.8–34.0	22.5–33.8

All values in km.

Given some idea of amounts of mobility, we need to be able to translate that to average energetic cost of transport. The energetic expense of locomotion ($\dot{E}$, in W) is determined by body mass (*BM*, in kg), external loads carried (*L*, in kg), velocity (*V*, in m s^{-1}), grade (*G*, %) and the nature of the terrain (Pimental and Pandolf 1979). This last variable is expressed as a terrain factor (μ) ranging from 1.0 for firm, even surfaces to 2.2 for packed sand to 3.8 for "post-holing" through unpacked snow (Kramer 2004b). The cost of transport can thus be estimated (Pimental and Pandolf 1979) as:

$$\dot{E} = 1.5BM + 2(BM + L)\left(\frac{L}{BM}\right)^2 + \mu(BM + L)(1.5V^2 + 0.35VG)$$

If we set $\mu = 1.8$ and $V = 1.36$ m s^{-1} (following Kramer 2004b), we can estimate $\dot{E}$ for a male ($BM = 78$ kg) and female ($BM = 66$ kg) Neandertal walking on level ground ($G = 0\%$) in both the unburdened ($L = 0$) and burdened ($L = 0.35BM$) conditions (setting load mass = 35% body mass following Kramer 2004b). For the female Neandertal, we assume she is nongravid and is not carrying any offspring. The equation above is useful because it allows for the effects of variation in load, compliance of the substrate (terrain factor), velocity, and slope on the estimation of energy. Unfortunately, it does not take into consideration the effects of lower limb length on cost of transport (Steudel-Numbers and Tilkens 2004; Weaver and Steudel-Numbers 2005; Kramer and Sarton-Miller 2008; Kramer and Sylvester 2009). The cost of transport (VO_2, in ml O_2 m^{-1}) in modern humans of varying mass (*BM*, in kg) and leg length (*LL*, in cm), walking unburdened on flat, noncompliant substrates, was determined empirically by Steudel-Numbers and Tilkens (2004) as:

$$VO_2 = 8.301 + 0.234BM - 0.169LL$$

When representative Neandertal mass and leg length values are plugged into this equation, the results indicate that the cost to travel a given distance was about 30% higher for Neandertals than for the average modern human (Steudel-Numbers and Tilkens 2004), of which about 14% was attributable to leg length. Since our equation for estimating energy expenditure was based on modern human males (Pimental and Pandolf 1979), an upward adjustment of 14% is needed to more closely approximate the actual cost of transport in Neandertals. However, the shorter limbs of Neandertals, and especially their low crural indices, were likely advantageous on sloped terrains (Higgins and Ruff 2011), such that any energetic disadvantage (relative to modern humans) was diminished as grade increased. Table 10.7 provides estimated values of cost of transport for average-sized adult male and female Neandertals under a variety of conditions.

The values in Table 10.7 reflect costs of transport that are high relative to most modern humans. Combining these values with the weekly mobility estimates in Table 10.6 gives us some rough (very rough!) estimates of the weekly cost of movement for Neandertals. Neandertals were likely routinely ascending and descending hilly terrain during foraging and residential moves (see Higgins and Ruff 2011), and acsent of even moderate slopes can greatly increase the energetic cost of transport (Minetti et al. 2002). The weekly estimates were derived by assuming that Neandertal movement involved roughly equal amounts of walking on level ground, hill ascending, and hill

TABLE 10.7 ■ Representative cost of transport (kcal km^{-1}) values for adult female and male Neandertals under varying terrain and load conditions

	Unburdened	Burdened[a]
Female		
no snow, flat	86	113
snow cover, flat	159	212
no snow, 10% grade	199	266
no snow, 20% grade	293	393
no snow, 30% grade	373	502
snow cover, 30% grade	769	1035
Male		
no snow, flat	101	134
snow cover, flat	188	251
no snow, 10% grade	235	315
no snow, 20% grade	346	465
no snow, 30% grade	441	593
snow cover, 30% grade	908	1224

Based on average adult female (66 kg) and male (78 kg) masses, velocity = 1.36 m sec^{-1}, and terrain factors of 1.8 (no snow) and 3.8 (snow cover). To compensate for the interplay of leg length and terrain (see text), estimates were adjusted upwards by 14% for walking on level ground and 10% grade; and by 7% for walking up 20% grade.

[a] Burden set equal to 35% of body mass.

descending. Descending hills is more energetically costly than walking on level ground, but only moderately so (at a steep 45% decline, cost of transport is roughly two times greater than with level ground bipedalism: Minetti et al. 2002): accordingly, the cost of descending hill slopes was set equal to that of level ground walking in the model. A 0% grade was thus used to represent 66.6% of Neandertal locomotion; for the remaining 33.4%, hill climbing was assumed to involve equal proportions of ascent of grades of 10%, 20% and 30%. For simplicity's sake, locomotion in snow is not yet considered. This exercise produces estimates in the range of 8500–12,400 kcal wk^{-1} for females, and 14,500–17,000 kcal wk^{-1} for males. These equate with average daily mobility costs of about 1200–1800 kcal for females and 2100–2400 kcal for males.

These estimates are, of course, very speculative, but they do suggest that mobility costs were likely to have been high for Neandertals, and that limiting those costs was probably an important component of Neandertal subsistence strategies. Regardless of the accuracy of our estimates of distance, burden carrying, substrate compliance, or nature of the terrain (slope grade) over which Neandertals were walking, their relatively great mass combined with their short limbs resulted in high costs of transport relative to modern foragers. Again, mobility costs are arguably the highest priced item in forager activity budgets, and when paying more per unit distance traveled (as were the Neandertals), curtailing these costs as much as possible would have been advantageous. A high cost of transport also directly affects the optimal foraging radius around a residential camp (Verpoorte 2006; Macdonald et al. 2009). Following the logic of the patch choice model from optimal foraging theory (Charnov 1976; Winterhalder and Smith 1981), it is in the best interests of foragers to abandon a resource patch when the net return rate in the patch (which for most types of patches steadily decreases as the resource is harvested) drops below the mean return rate for the environment overall. When residential moves are used as a positioning strategy, the local environment around the camp can be considered the resource patch, and we would expect a group to abandon the patch (that is, move camp) when biodepletion around camp reduces the net return rate below the overall average for the environment. Mobility is costly, and thus the net return for any item decreases with its distance from camp, which places a constraint on the foraging radius. Furthermore, as the cost of transport increases the net return on all items decreases, which leads to earlier abandonment of the patch. Thus we would predict that a high cost of transport in Neandertals would result in a small foraging radius and would favor frequent residential moves (Verpoorte 2006; Macdonald et al. 2009), both of which appear to have been the norm based on lithic raw material collection patterns and evidence of ephemeral site occupation (see Sections 10.1 and 10.2).

High mobility costs may have also placed a premium on *not* moving when the average return rate of the environment became very low, as no doubt happened during the late winter and early spring. Given the lack of plant production (although starches stored by geophytes may still have provided a source of calories, if they could be located) and the poor nutritive quality of animal prey, it may have paid to reduce foraging activity (and even curtail it altogether) to avoid negative net return rates. This "hunkering down" strategy may have been especially important in light of the effects of snow cover on both mobility costs and potential foraging returns. While cold winters may have increased natural mortality in prey species (thus increasing the abundance of winter-scavengable carcasses: see Section 10.1), winter snow would have been doubly damning to overwintering Neandertals. First, as noted above, snow cover increases the energetic cost of transport. Expenditures increase as a linear function of snow depth up to about 15 cm, after which they continue to rise linearly but with a steeper slope (Heinonen et al. 1959). Bipedal locomotion through 40 cm of snow is almost six times more costly than walking through an accumulation of 10 cm (Heinonen et al. 1959). Not only does this greatly increase the energetic cost of transport, it likely made movement over certain types of terrain exorbitantly expensive: using the same estimation technique we used above, we can predict that a male Neandertal carrying a burden up a 30% grade in snow deep enough to require "post-holing" (lifting the leg up and out of the snow with each step, rather than pushing it through the snow) would have consumed a whopping 1224 kcal km^{-1} (Table 10.7)! Thus, snow (and ice) may have been a major impediment to Neandertal mobility, and along these lines Monnier (2006) has suggested that the high frequency of scrapers found in Mousterian assemblages from cold intervals reflects reduced raw material procurement because of reduced mobility (thus demanding greater retouch intensity on existing tools, which in turn results in assemblages typologically rich in scrapers: see Chapter 3.2).

The second problem produced by snow cover concerns prey quality during the winter months. As outlined in Chapter 4.4, snow reduces herbivore mobility and buries some portion of the vegetation on which they depend during the winter, with smaller species being more heavily affected by both of these considerations than larger species. As a result, wintertime loss of condition in herbivores is positively related to snow depth. Among North American wolves, for example, deep snow can lead to surplus killing of white-tailed deer (in which many carcasses are abandoned only partially eaten because they were of poor nutritional quality: DelGiudice 1998) or prey-switching to larger species, such as elk (red deer), which are less affected by snow depth (Huggard 1993; Kunkel et al. 2004). The net effect of reduced prey quality is a potentially dramatic decrease in net caloric return rates from hunting, which again may favor a strategy of minimal winter mobility. Climatic simulations suggest that snow cover was a greater problem for the more continental (central and eastern) parts of Europe, where even during a warm phase of MIS 3 they might have experienced snow covers up to 50 cm for as much 150 days per year (Barron et al. 2003). Further west, such as in the south of France, much shallower snows (5–20 cm, with depths up to 50 cm on the massif centrale) may have persisted for 30–120 days. Of course, during glacial episodes snow cover in these areas would have been greater, perhaps more so in duration than in depth. Climatic models for the last glacial maximum indicate depths of 50 cm or more for 180 to 210 days y^{-1} across most of central and eastern Europe, and 5–20 cm for 60–150 days y^{-1} in the south of France.

The overall picture that emerges from our consideration of the energetics of locomotion is that mobility was likely a costly component of their adaptive strategies (and relatively more so than among modern foragers, given differences in mass and limb proportions), that high mobility costs were likely to have affected Middle Paleolithic site settlement dynamics, and that there may have been substantial pressure to reduce the cost of mobility as much as possible (perhaps extending so far as to curtail foraging during the late winter and early spring). What do the estimated costs imply about the PAL multipliers that have been suggested for the Neandertals? If mobility were the only cost beyond basal metabolism (an impossible situation, but a heuritically-useful starting point), our estimated daily mobility costs (coupled with our BMR estimates for average-sized adult males and females) would equate with PALs of 1.6 to 2.3 for Neandertal females, and 1.9 to 2.3 for the males. What might domestic activities around camp have required of their energy budgets?

10.5 THE ENERGETIC COST OF DOMESTIC ACTIVITIES

When not moving about the landscape on subsistence forays or raw material gathering trips or social visits, activity costs would have been limited to things done around camp. It is important to note that large-bodied carnivores tend to have very tight energy budgets (Gittleman and Oftedal 1987; Oftedal and Gittleman 1989; Carbone et al. 1999, 2007), such that inactivity is as important to their behavioral ecology as is foraging activity. Lions, for example, spend more than 90% of their time at rest (Schaller 1972). Polar bear males, who do not normally engage in long-duration hibernation even though they appear to be physiologically capable of it (Nelson et al. 1983), do use short hibernation bouts as an energy saving mechanism (Stirling and Oritsland 1995). A similar energetic significance of rest likely also applied to heavily meat-dependent Neandertals, both in terms of rest as a part of daily life (that is, resting whenever possible during active days combined with regular rest days of minimal activity in camp) and seasonal periods (late winter and early spring) of greatly reduced activity (what I have referred to above as "hunkering down," or what might be better called pseudohibernation: see Section 10.4). Still, energy had to be expended on the domestic tasks necessary to survival. These tasks no doubt included behaviors associated with the production and maintenance of tools (flint knapping and retouching, retooling of hafted tools, working of wood and other organic raw materials, and collection of suitable wood for tool production); preparation of animal skins for clothing, bedding, and shelters (scraping, smoothing, and preserving of hides, possibly simple sewing); food preparation (butchery and marrow extraction from whole or partial carcasses, processing of vegetal foods, cooking); preparation and maintenance of bedding areas (collection of vegetal matter for use as bedding [Henry et al. 2004; Cabanes et al. 2010], erecting and maintaining shelter walls or tents [Vallverdú et al. 2010]); hearth tending (collection of firewood [see Fernández-Laso et al. 2011], construction of simple hearths [Speth 2006; Vallverdú et al. 2012]); housekeeping (at some sites with longer occupation, Neandertals sometimes cleared refuse from the living areas towards the periphery of the site [Speth 2006]); and of course resting, self-maintenance (personal grooming), socializing (including grooming others and interacting with children), and eating.

The time spent at these various activities, and thus their energetic costs, are of course not known for Middle Paleolithic foragers. Once again, recent human foragers serve as our only guide. Unfortunately good-quality hunter-gatherer time allocation data are scarce, and what data do exist come exclusively from tropical and neotropical foragers and forager-horticulturalists. Certainly the time and energy allocated to specific domestic activities vary across groups and are determined by environmental circumstances (for example, hide preparation for clothing is likely to comprise a greater proportion of the energy budgets of foragers living in cold temperate and arctic conditions: Gilligan 2010b; Shaw et al. 2012), and also varies seasonally and by sex, age, and individual ability within groups (Panter-Brick 2002; Gurven and Kaplan 2006). In our consideration of Neandertal mobility patterns above (Section 10.4) we assumed a sexual division of foraging effort and modeled the movement of males and females separately using travel distance averages from recent hunter-gatherers. However, it has yet to be established if Neandertals had a sexual division of labor similar to that seen among recent foragers, and arguments that this kind of labor organization postdates the Neandertals have been made (Kuhn and Stiner 2006; Balme and Bowdler 2006; Stiner and Kuhn 2009; but see Mellars 1996; Hayden 2012). Accordingly, in our consideration of Neandertal domestic activity budgets, and in estimating overall activity levels (Section 10.6), we should model activity both assuming a sexual division of labor and assuming no such division.

Time allocation data for a few foragers and forager-horticulturalist groups are provided in Table 10.8. Following Hawkes and colleagues (Hawkes et al. 1997), nonforaging activities are divided into one of three categories: food processing, household maintenance, and manufacture and repair of technology. Again, the data derived entirely from tropical/neotropical groups, and in most cases were collected in a single field season (and thus represent activity averages for a single season rather than averaged over the year). One additional problem is that ethnographers vary in the ways that they categorize activity, making it difficult to compare across studies (for example, Lee [1979] lumps food processing and household maintanence activities together as "housework," whereas Gurven and Kaplan [2006] combine household maintanence and manufacture/repair activities together as "other work"). Still, the data give us some idea of the time spent in domestic chores among foraging groups. Since the time devoted to domestic activities varies depending on whether a forager is in camp during a rest day or out actively foraging, and since we are interested in the total cost of

TABLE 10.8 ■ Time allocation (hrs wk^{-1}) of foraging (F) and foraging-horticultural (FH) groups

	Adult males				Adult females			
	FP	HM	M/R	Total	FP	HM	M/R	Total
Hadza (F)	4.2	3.2	8.0	15.4	8.2	2.9	6.2	17.3
!Kung (F)	15.4[a]		7.5	22.9	22.4[a]		5.1	27.5
Ache (F)	2.1	1.8	3.5	7.4	3.7	4.9	8.3	16.9
Machiguenga (FH)	1.3	5.4[b]		6.7	6.4	9.5[b]		15.9
Piro (FH)	1.2	19.3[b]		20.5	4.8	29.6		34.4
Yapú (FH)	–	–	–	–	1.4	0.6	–	–
Mean				**14.6 ± 7.4**				**22.4 ± 8.2**

Time spent in food processing (FP), household maintenance (HM), and manufacture and repair of technology (M/R). Data from Hawkes et al. 1997; Lee 1979; Hill et al. 1985; Hurtado et al. 1985; Gurven and Kaplan 2006; Dufour 1984.

[a] Lee (1979) appears to lump cooking (and presumably other aspects of food processing) with "housework."

[b] Gurven and Kaplan (2006) appear to lump household chores (HM) with M/R as "other work."

domestic activity (not just that performed in camp), the values in Table 10.8 are inclusive of both active foraging and stay-in-camp days. The values in the table reflect considerable variance in time spent on domestic chores, from a low of about 7.4 hours per week among Ache males to a high of about 34.4 among Piro females. Looking across groups, males spend an average of about 14.6 h wk^{-1} and females about 22.4 h wk^{-1} on domestic work (for a combined sex average of 18.5 h wk^{-1}).

If we assume the Neandertals were spending a similar amount of time in nonforaging related work, what were the associated energy costs? Again, we have little basis for knowing the actual time that Neandertals devoted to various tasks. In the colder parts of their range and during climatic downturns they may have devoted more effort to hide preparation than do warm-climate foragers (see Shaw et al. 2012), but given a simpler subsistence toolkit this may have been balanced out by a reduction in tool maintanence costs. Without some means of refining our time allocation estimates, the recent forager data appear to be our only avenue for estimating domestic activity costs among the Neandertals. Some domestic chores may have taken moderately high amounts of metabolic energy: experimental work suggests that flintknapping is about as energetically costly as walking (Heffter et al. 2013), and butchering an animal consumes about six times the energy expended at rest (Ainsworth et al. 2000). Among industrialized western peoples, most domestic activities are done at an energetic cost of between 1.5 and 3.5 times the resting metabolic rate (Montoye et al. 1996; Ainsworth et al. 2000). Work intensity is often expressed as a metabolic equivalent (MET), which is the ratio of work metabolic rate to resting metabolic rate (Jette et al. 1990). The measurement of resting metabolic rate (RMR) is not as stringent as is that of BMR, but the two measures are fairly equivalent (RMR tends to be slightly higher than BMR for any given subject, because RMR is not always measured with the subject in a fasting state, and thus metabolic increases having to do with the digestion and assimilation of food [the specific dynamic action of food] may add to the subject's metabolic rate). For

TABLE 10.9 ■ Estimated physical activity levels (PAL) for Neandertal males and females

	Division of labor	No division of labor
Males		
Mobility costs	2100–2400	2100–2400
Domestic activity	391–500	496–633
Total activity costs (AC)	2491–2900	2596–3033
TEE[a]	4327–5412	4383–5512
BMR_T	1800–2300	1800–2300
PAL[a]	2.1–2.7	2.1–2.8
Females		
Mobility costs	1200–1800	1405–2040
Domestic activity	467–650	386–537
Total activity costs (AC)	1667–2450	1791–2577
TEE[a]	2962–4545	3126–4703
BMR_T	1400–1950	1400–1950
PAL[a]	1.8–2.8	1.9–2.9

Total activity costs (AC) = mobility costs + domestic activity. All values except PAL in kcal d^{-1}.
[a] Minimum and maximum – see text for calculation method.

our purposes, it is conservative to use average Neandertal BMRs in place of RMR. In general practice, the MET is expressed relative to a standard RMR (Jette et al. 1990), although experimental work shows this practice to be unsound (Byrne et al. 2005), and accordingly we use here the average BMR for male and female adult Neandertals to standardize the MET values. If we use an overall MET of 2.5 for domestic work, combined with our range of BMR values for an average adult male and female Neandertal, we end up with domestic activity costs of about 2740–3500 kcal wk^{-1} (391–500 kcal d^{-1}) for males and 3270–4550 kcal wk^{-1} (467–650 kcal d^{-1}) for females. If domestic costs were shared equally between the sexes (no sexual division of labor), the male values increase to ca. 3470–4430 kcal wk^{-1} (496–633 kcal d^{-1}) and the female values drop to about 2700–3760 (386–537 kcal d^{-1}). If we also assume that most other camp activities, such as relaxing, attending to personal needs, and socializing, all occur at METs around 1, then the estimates above more-or-less round out (when combined with mobility costs) the overall activity budget of Neandertals.

10.6 NEANDERTAL PHYSICAL ACTIVITY LEVELS

The mobility and domestic activity costs can be combined to get an idea of overall physical activity (Table 10.9). Because the energetic cost of movement and other activities are estimated from empirical measures of actual energy consumption (via measurements of oxygen consumption), and since a portion of the metabolic expeditures of the subjects in the underlying studies was due to basal and other metabolic functions (that is, what is measured in the laboratory is total metabolic expenditure, not just the energy expended on the activity), it is important to subtract off the basal metabolic contribution for the time spent walking (calculated from velocity and distance) and doing domestic activities (based on the time allocation data). For purposes of calculating

PAL, we can thus think of the total basal metabolic rate (BMR_T) as the sum of the basal metabolic rate at rest (BMR_R) and while active (BMR_A). A second adjustment is also needed to produce the most reasonable estimates of physical activity costs. The product of BMR and PAL should equal the total energy expenditure (TEE) of an individual per 24 h period. We have so far accounted for the major metabolic expenses – BMR and physical activity – but have not factored in the energetic cost of digesting and assimilating food (the specific dynamic action [SDA] or "thermal effect" of food: see Chapter 4.3). Based on the SDA of protein, fats, and carbohydrates observed for humans (Tappy 1996), virtually any dietary mix of these three nutrients results in overall SDAs of about 9–11%. Thus an SDA of 10% was factored into the calculations of TEE as well. TEE can thus be calculated from BMR_T (in kcal d^{-1}), activity costs (AC, also in kcal d^{-1}), and SDA (here set to 0.10) as:

$$TEE = (BMR_T + AC - BMR_A)(1 + SDA)$$

The PALs (=TEE/BMR_T) presented in Table 10.9 represent the minimum (based on the highest BMR and lowest activity cost) and maximum (based on the lowest BMR and highest activity cost) for each sex. A number of interesting observations can be drawn from consideration of the values presented in Table 10.9.

First, all of the PAL values fall in the range of what the World Health Organization considers heavy activity (FAO/WHO/UNU 1985). For both sexes, the values exceed or are at the upper end of the range of the values determined empirically for recent foragers (Table 10.1). Leonard (2008) reports average PALs of 1.98 for males and 1.82 for females, based on a meta-analysis of the literature on subsistence-level populations, and our estimates for Neandertals generally exceed those values. Elsewhere (Churchill 2006), I used PALs for Neandertals of 2.0–2.5 based on the direct measurement of calorie consumption in Inuit that still practice a traditional foraging lifestyle, which indicated TEEs on the order of 2.0–2.5 times BMR (see Shephard and Rode 1996; Steegmann et al. 2002). The estimates derived here would suggest that those PALs are in the ballpark. Sorensen and Leonard (2001) argued for PALs of 2.0–3.0 for Neandertals, and again these results would be consistent with those values. They are also consistent with PALs employed by Snodgrass and Leonard (2009), who used values of 1.82–2.5 for females and 1.98–2.5 for males (note, however, that in this study PAL was exclusive of the SDA of food, so these values are not directly comparable to those of other studies). An alternative approach (Froehle and Churchill 2009) is to adjust PALs by sex and climate using data from recent foragers, on the assumption that activity levels vary by sex and that more productive environments require less energy to exploit. That exercise resulted in PALs of 1.7 (female) to 1.93 (male) for Neandertals living in temperate conditions, and 2.2 (female) to 2.5 (male) for Neandertals in cold environments – estimates that are low relative to the PALs determined here. Although concordant with the PAL values that have been argued to fit the Neandertals in earlier studies, the estimates produced in Table 10.9 are broad, ranging from 2.1–2.8 for males and 1.8–2.9 for females. It has also been suggested that estimates of caloric demand in Neandertals (TEE) in some of these studies may be too high (Hockett 2012), and certainly the maximum TEEs in Table 10.9 are well above those estimated in other studies (Figure 10.2). We might ask, then, just how reasonable are these estimates? Mammals have an average PAL of 2.65 (Karasov 1992), and consideration of energetic data from human groups living and working under a variety of conditions suggests

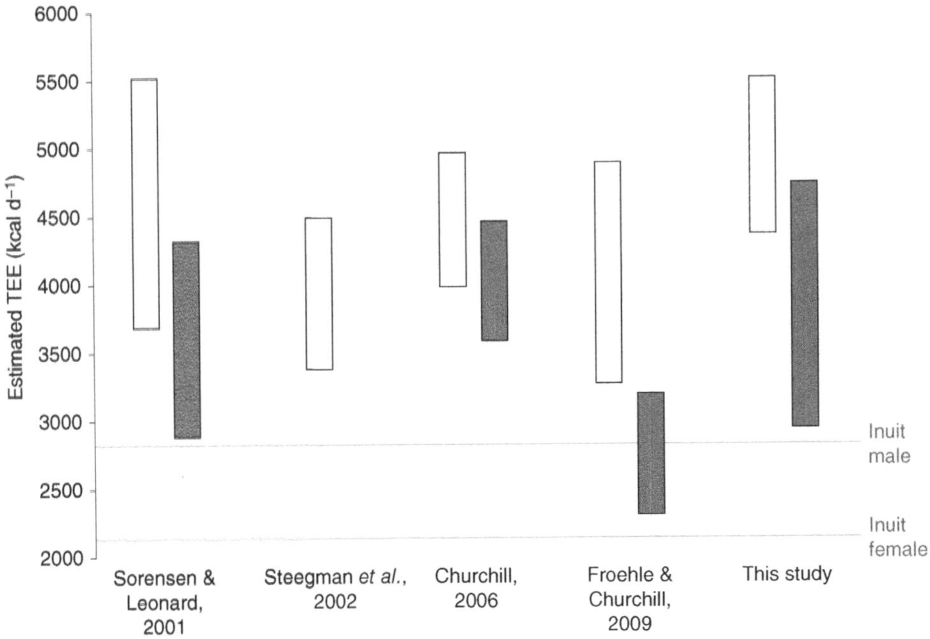

FIGURE 10.2

Estimated total daily energy expenditure (TEE, in kcal d^{-1}) in Neandertal males (□) and females (■) from several recent studies. Grey lines represent average TEEs for adult male and female Inuit, from Keene 1985.

that a PAL of 2.5 is likely to be the maximum for a sustainable lifestyle among human populations (Shetty 2005). Thus the higher estimates in Table 10.9 are likely to be exaggerated, and it seems reasonable to place Neandertal PALs in the range of 2.1–2.5 for males and 1.8–2.5 for females – still high relative to recent human foragers, but not remarkably so.

Second, the TEE estimates in Table 10.9 are quite high, relative even to most other studies of Neandertal energetics (Figure 10.2). Note that these estimates are not the product of BMR and PAL (which would produce a broader range of TEE estimates, up to 15% lower than the lowest estimate and up to 7% higher than the highest), but rather were determined by summing the lowest BMR and AC (and adjusting for SDA) to obtain the minimum TEE for each sex, and summing the highest BMR and AC (and adjusting for SDA) to obtain the maximum for each sex. While these estimates are high relative to other studies, they are quite similar to those of Sorensen and Leonard (2001), which is surprising given that that study used PAL values up to 3.0. However, average Neandertal body mass appears to be markedly underestimated in that study (65 kg for males; 55 kg for females: Sorensen and Leonard 2001), which results in very low estimates of average BMR (1674 kcal d^{-1} for males and 1305 kcal d^{-1} for females). Nonetheless, the estimates derived here support the contention that the energetic demands on Neandertals were great. Even the lowest of the TEE estimates exceed those observed in modern foragers (Leonard and Robertson 1992), including those living in the arctic (Steegmann et al. 2002). What is important to understand in this exercise is that the estimates derived here do not rely on the assumption of abnormally high activity levels among the Neandertals. On the contrary, these estimates are based on the assumption that Neandertal activity levels – both in terms of mobility and domestic chores – were similar to those of recent hunter-gatherers (and primarily from groups living in more productive tropical environments). The higher TEE values of the Neandertals derive entirely from their greater average mass (resulting in higher BMR and contributing to a greater cost of transport) and shorter lower limbs (which also contribute to a greater cost of transport). As we will explore in greater detail in Chapter 11, the energetic differential between Neandertals and the early modern humans with whom they may have competed are a function of morphological, not behavioral, differences between the groups. This means that, all other things being equal, the Neandertals would have needed to be more efficient foragers than were early modern Eurasians in order to maintain any type of competitive advantage. Even if the latest Neandertals invented or adopted (via cultural exchanges with modern humans) new technologies that improved their foraging efficiency, they were still saddled with relatively massive bodies that were costly to maintain and move about the landscape. A second interesting observation, in light of the fact that Neandertal-like PALs may be about average for mammals in general, is not that Neandertal values are high, but rather that recent foragers manage with relatively low levels of physical activity. This is no doubt a function of a heavy dependence on extractive technology that greatly reduces the energetic costs of foraging, and is certainly a part of the great adaptive success of modern humans from the Late Pleistocene onwards.

The third observation that derives from Table 10.9 is that energetic costs were likely to be higher for both sexes if the Neandertals did not have a modern forager-like sexual division of labor. Costs would be higher for males because their contribution to domestic work would have likely been higher (while their mobility levels remain the same), while for females any gains they enjoyed from a reduced domestic workload would be offset by increased mobility costs. Thus if Neandertals did not have a well-developed sexual division of labor, as has been suggested (Kuhn and Stiner 2006; Balme and Bowdler 2006; Stiner and Kuhn 2009), it likely contributed to what were already high energy demands. Seen another way, the division of labor that characterizes modern foragers may have evolved, in part at least, as a way of increasing foraging efficiency and reducing overall foraging costs (as well as reducing foraging risk: see Wrangham et al. 1999).

Finally, the TEE values in Table 10.9 may have important implications for understanding Neandertal respiratory anatomy, specifically with respect to their relatively large thoraces and their prognathic midfaces with wide nasal apertures. With respect to their thoracic morphology, comparisons of the size and shape of Neandertal ribs and mesosterna relative to those of recent humans suggest that the Neandertals had large chests for their stature (Churchill 1994a; Franciscus and Churchill 2002; Sawyer and Maley 2005; Weinstein 2008; Gómez-Olivencia et al. 2009, 2012), which appears to represent the primitive condition for archaic humans generally (Gómez-Olivencia et al. 2009, 2010). It is less clear if Neandertals had thoraces that were large for their mass, and which might then suggest ventilatory capacities and aerobic demands greater than that seen in modern humans. Attempts to evaluate the respiratory capacity of Neandertals and modern humans relative to body mass are hampered by small fossil sample

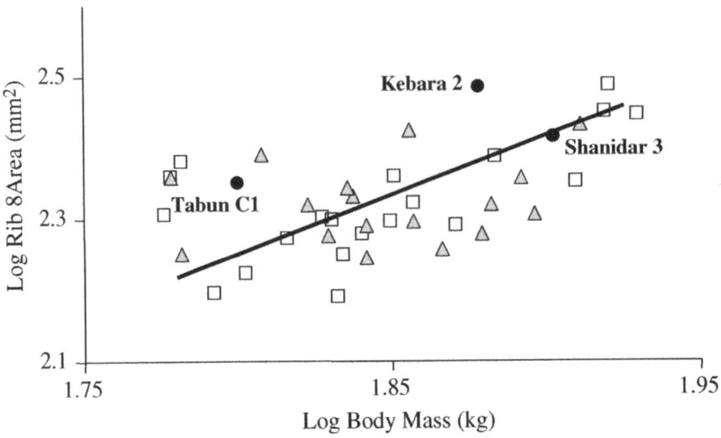

FIGURE 10.3

Log respiratory area of rib 8 (the area enclosed by the rib: see Franciscus and Churchill 2002) versus log body mass in Neandertals and subarctic (Aleutian Islanders: ▲) and temperate (Euro-Americans: □) recent humans. Ordinary least squares regression line ($y = 1.6152x - 0.661$; $r = 0.4823$) based on recent humans only. Values for Kebara 2 and Shanidar 3 measured by SEC; value for Tabun C1 based on data reported in Weinstein 2008.

sizes and by a great amount of variation within modern samples. The area enclosed by the eighth rib (following Franciscus and Churchill 2002) provides a crude measure of the respiratory area of the lower thorax. Regrettably, this measure does not take into account variation in the size of the costal cartilages, or in other features that contribute to ventilatory volume, such as the size of the diaphragm and the size and shape of the other ribs (see Bellemare et al. 2001), but it at least provides a rough idea of overall thoracic size. Only three Neandertals preserve eighth ribs sufficiently complete to allow calculation of their enclosed areas (Figure 10.3), and unfortunately all three specimens derive from the Near East (and thus may not represent the potentially more extreme morphology of Neandertals adapted to colder conditions: see Franciscus 1989; Gómez-Olivencia et al. 2012). Of these three specimens, the male Shanidar 3 has a respiratory area proportional to mass (as judged by modern human comparators), while the male Kebara 2 and female Tabun C1 have relatively large respiratory areas (Figure 10.3). The relative size of the eighth rib is also highly variable within samples: Aleutian Islanders – who are expected to have large chests for both thermoregulatory and activity reasons (see Churchill 1994a) – do not differ from European-Americans in mean eighth rib respiratory area relative to body size (Figure 10.3). While thoracic dimensions do in some cases reflect ventilatory capacity (as in high altitude populations: see references in Churchill 1994a), variation in the size of soft tissues may play a more important role in determining ventilatory capacities. In a radiographic study by Bellemare et al. (2001), the anteroposterior and mediolateral diameters of the thorax (taken at the level of the sixth rib) only accounted for 6.9% and 26.1%, respectively, of the variation in plethysmographically-measured total lung capacity, with the height of the diaphragm accounting for the remaining 67% of the variation (in males: similar values obtained for females). At present, we can only say that the overall constellation of Neandertal rib morphology suggests they had chests that were large for their body mass, and that this may reflect ventilatory capacities that were relatively great.

Might the high energy requirements of Neandertals explain their relatively large chest size? A liter of oxygen is needed for every 4.8 kcal burned, such that a Neandertal male would have required somewhere between 900 and 1150 l O_2 d^{-1} (based on the TEE values in Table 10.9). A male Neandertal at rest (MET = 1) would consume between 0.26–0.33 l O_2 min^{-1}, and given a constant O_2 capture rate of 3.1% (Schmidt-Nielsen 1984), this equates with average alveolar ventilation rates ($\dot{V}_E$) of about 8–11 l of air per minute. The average adult modern human moves a half liter of air with each breath, and breathes 12–18 times per minute at rest, for $\dot{V}_E$ of 6–9 l min^{-1}. Of course, as activity increases so also does ventilatory demand. At moderate activity levels such as walking or doing domestic tasks (METs = 2–4: see Ainsworth et al. 2000), adult male Neandertal $\dot{V}_E$ would increase to ca. 17–43 l min^{-1}. At high activity levels, such as burden carrying up hill, hill climbing through snow, or chasing prey during hunting, METs may have reached 10–12 (Ainsworth et al. 2000), with associated $\dot{V}_E$ in the range of 83–128 l min^{-1}, which approach the maxima of modern human athletes (123–155 l min^{-1}: Hermansen and Andersen 1965; Volkov et al. 1975). Thus a combination of greater ventilatory demand at rest, periodic episodes of extremely high ventilatory demand, and relatively short stature (and thus a shorter SI dimension to the thorax) may have necessitated in Neandertals chests that were large relative to both their stature and mass. Again, greater relative chest size in Neandertals does not necessarily indicate activity levels greater than what have obtained in early modern humans or recent foragers, but rather a greater energetic cost to activity thanks to their relative great mass and short stature.

Greater ventilatory demands may also be reflected in the craniofacial antomy of Neandertals. If archaic humans needed greater $\dot{V}_E$ to satisfy their energetic demands, they would necessarily require greater volume flow in the upper respiratory tract – the nasal passages and pharynx. As reviewed in Chapter 5.3, Neandertals had wide nasal apertures and wide internal passageways (Franciscus 1995; Yokley 2006), which is unexpected for humans living in cold environments. All other things being equal, narrow upper respiratory tracts generate high airflow resistance and greatly increase the work of breathing (resistance is inversely proportional to the fourth power of the average radius of the passageway: Strauss 1979). The work of breathing (also known as the oxygen cost of breathing) varies from about 1% of metabolic energy at rest to about 10% during heavy exercise (Strauss 1979), and thus represents a potentially important energetic target of selection in hominins with high ventilatory rates. Breathing through the mouth (technically called oronasal breathing, since some air continues to flow through the nasal passageways) affords a greater diameter passageway and reduced resistance (Hall 2005), and has an average onset at $\dot{V}_E$ of 35.3 l min^{-1} in adult humans (Niinimma et al. 1980). The $\dot{V}_E$ at which individuals switch from nasal to oronasal breathing is inversely related to nasal resistance (Niinimma et al. 1980). The larger nasal passageways in Neandertals would have both reduced the energetic cost of quiet breathing and also would have allowed them to maintain nasal breathing (with enhanced conditioning of inspired and expired air relative to mouth breathing: see Chapter 5.3) longer as ventilatory rate increased during activity. Furthermore, the ontogenetic development of nasal breadth appears to be tied to growth at the premaxillary suture (Maureille and Bar 1999; Maureille and Braga 2002), in that prolonged suture patency is associated with prognathic faces and broader noses (the condition seen in Neandertals: Holton and Franciscus 2008). Thus, high ventilatory demands and selection pressures favoring broader noses in Neandertals may have pliotropically contributed to their characteristic midfacial prognathism (Froehle et al. 2013).

The arguments presented in this chapter suggest that the energetic costs associated with metabolically-expensive Neandertal bodies had consequences ranging from their mobility patterns and site settlement systems to their craniofacial and thoracic morphology. In the next chapter, we'll consider the impact their energy budgets had on their life history and demography. Before moving on, however, it is important to address one last issue about energy budgets. Some readers will have noticed a seeming contradiction between the Neandertals having had very high energy demands (which must mean they were highly effective foragers to meet those demands) and the argument that they were at a competitive disadvantage relative to other carnivores (they were not the socially-dominant carnivores in their guild, but rather ranked somewhere in the middle). Sorensen and Leonard (2001) were the first to point out that Neandertals must have been very effective foragers to support their high energy demands. Elsewhere (Churchill 2006), I have estimated that, to meet their caloric demand, Neandertals must have had prey capture rates between 3.8–7.6 kg hunter^{-1} d^{-1}, which is comparable to that observed for wolves in the northern parts of Europe and North America (5.4–5.6 kg hunter^{-1} d^{-1}). Wolves are very efficient hunters, and thus Neandertals must have been as well. Wolves are also considered "apex predators" that, unlike smaller and less socially-dominant mesopredators, are relatively unbothered by interactions with other carnivores and exert considerable top-down control over the structure of the herbivore populations on which they prey (Ritchie and Johnson 2009; Roemer et al. 2009).

How can it be, then, that the Neandertals were highly efficient hunters yet not apex predators? The solution to this seeming paradox can be seen with an analogy to the concepts of daily sales rate and market share in business economics. Imagine two businesses that serve a single market in a city, one being a big box national retailer and the other a locally-owned Mom & Pop small business. Through aggressive advertising and marketing, the big box store controls 90% of the market share. One may find that the daily sales rate is equivalent in the two types of stores, but if that's the case, we would expect there to be nine big box stores for every Mom & Pop store. Expansion (more stores) can only be had for the Mom & Pop business if they manage to capture more of the market share, otherwise their new shops will simply compete with their existing shops for the 10% share that they control (and the daily sales rates will go down in their stores). This same kind of relationship holds in carnivore ecology. Cheetahs have kill rates (10 kg hunter^{-1} d^{-1}: Schaller 1968) that are roughly comparable to those of female lions (8.7–14.1 kg hunter^{-1} d^{-1}: Stander 1992), yet lions tend to live at considerably higher biomass (Table 9.3), and the biomass of cheetahs tends to be inversely proportional to that of lions across their range of sympatry (Kelly et al. 1998; Linnell and Strand 2000). Lions control more of the available prey calories (see Karanth et al. 2004), and thus cheetah population growth is limited, despite their being every bit as effective at capturing prey as lions. It is probably also worth noting that wolves of the Pleistocene were probably every bit as efficient at capturing prey as are modern wolves yet, unlike modern wolves who exist in depauperate guilds, Pleistocene wolves were most likely not the apex predators in their ecosystems (see Chapter 9).

NOTES

1. Among living humans, total daily energy expenditures (TEE) can be measured directly by a variety of techniques (O_2 consumption, doubly labeled water, or flex heart rate monitoring). These techniques are difficult to employ with living foragers, and of course cannot be applied to prehistoric ones, and thus many researchers resort to estimating TEE either by using a physical activity level (PAL) multiplier, or by employing time allocation studies (where the amount of time spent engaged in various activities and the estimates of the caloric cost of those activities are used to estimate total TEE). As with all other researchers who have tried to estimate TEE in Neandertals, I take a PAL approach. However, unlike other studies, here we will examine as best as possible the inferences that can be made about actual activity costs related to mobility, subsistence, and domestic activities (including the production and maintenance of technology) to determine if the PAL values commonly applied to Neandertals are reasonable. This effort to evaluate the choice of PAL multiplier uses the same logic behind time allocation studies. Time allocation studies tend to underestimate actual energy expenditure, as measured by flex heart rate monitors (Leonard 2003) or indirect calorimetry (O_2) (Katzmarzyk et al. 1996). It has also been argued that PAL values applied to Neandertals are likely to also underestimate their actual energy expenditures (Snodgrass and Leonard 2009). It is important to note, then, that if one of the central arguments of this book is that Neandertals had high energy expenses and very tight energy budgets, the use of techniques that tend to underestimate TEE are conservative with respect to the argument.
2. PAL = TEE/BMR. Some studies (e.g., Leonard and Robertson 1992) use resting metabolic rate rather than basal metabolic rate, which underestimates PAL slightly relative to using BMR (Ulijaszek 1995; Jenike 2001).
3. These values are for modern humans: costs would have been higher for Neandertals (see Chapter 10.4).
4. Throughout the book the term *forager* is used synonymously with *hunter-gatherer*, and not in the sense defined by Binford (1980).
5. Resource patches in the arctic (and presumably in cold steppic and tundra environments of the Pleistocene) tend to be mobile (for example, herds of reindeer or horses) and variable with respect to the times at which they can be harvested (reindeer migration, for example, might begin any time from late August to mid-October). Thus hunter-gatherers employing logistical collecting systems to exploit these patches tend to invest heavily in gathering information about the resources on which they depend, such that properly equipped labor parties can be in the right place at the right time to bulk harvest resources. While the gathering of information requires substantial logistical movement on the landscape, and is thus energetically quite expensive, it is critical to survival in these environments.
6. In the arctic, winter residential camps are sometimes tethered to wooded valleys where firewood can be procured throughout the winter. Among tropical foragers, such as the San, residential camps are tethered to water sources during the dry season. Neither wood nor water were likely to be limiting resources in Pleistocene Europe.
7. While it would be instructive to compare mean Neandertal femoral robusticity measures with those of Holocene foragers (who likely are more gracile than both Neandertals and early modern humans of the Pleistocene), studies of lower limb cross-sectional geometry vary so widely in methods of body size standardization that robusticity measures cannot be pooled across studies.

CHAPTER ELEVEN

Neandertal Social Life, Life History, and Demography

As an approach to understanding Neandertal reproductive ecology and its effects on demography, we will briefly explore the inferences that can be made about Neandertal socioecology (the relationship of social structure and organization to the environment), and then once again pick up the thread of Neandertal energetics. We thus begin this chapter by asking what can be said about the size of Neandertal social groups – both in terms of local groups (the economic groups in which individuals foraged together and interacted with one another on a daily basis) and in terms of their extended social and mating networks – and about their social organization (and how it might relate to subsistence labor demands). We then return to the issue of energy budgets. As has been well-articulated in the field of dynamic energy budgets (Nisbet et al. 2000), there exist tight linkages between an organism's energy budgets and its life history and demography. In the previous chapters, we have explored the flow of energy through the Neandertal world, in terms of the energetic resources available to them (the relative abundance of humanly-consumable calories in the various environments in which they

Thin on the Ground: Neandertal Biology, Archeology, and Ecology, First Edition. Steven Emilio Churchill.

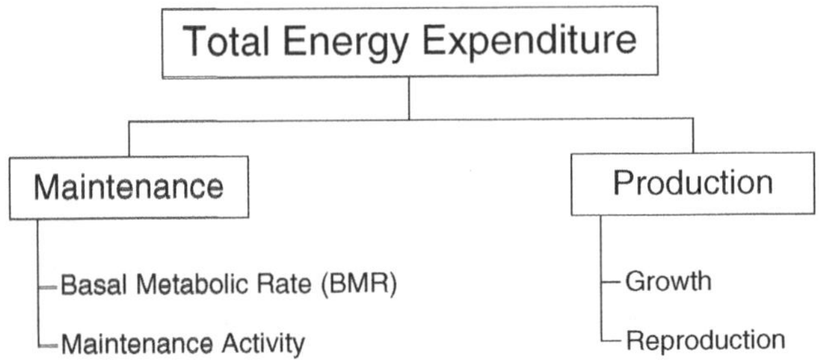

FIGURE 11.1

Simplified energy budget, adapted from WR Leonard et al. 2007.

found themselves, and which of the potential foodstuffs in their environments they were actually consuming), in terms of additional ecological constraints (competition with carnivores) that may have limited their potential energy capture, and in terms of the ways in which they must have been budgeting energy for their metabolic needs and activities (basal metabolism, thermoregulation, mobility, and domestic activity). That exercise suggested that Neandertal energy budgets were fairly tight – the energetic costs of somatic maintenance and activity were high, and competition for calories was likely intense. In this chapter, we examine the final component of Neandertal energy budgets: the amount of energy that was left over (after basic metabolic and activity costs had been paid) for reproduction.

Modeling the interaction between energy and reproduction relies upon three key concepts (see Ulijaszek 1995; McNab 2002): *energy budgets*, *energy balance*, and *trade-offs*. The concept of energy budgets (Figure 11.1) provides a useful heuristic for thinking about how organisms allocate gross energy expenditure to the important tasks of maintenance and production.[1] Maintenance includes energy for basic physiological functions and tissue turnover (BMR), as well as physical activity, while production includes the energy needed to grow to maturity, storage of energy in new tissues (such as fat), and the direct costs of reproduction (i.e., the energy costs of producing and provisioning offspring). Another useful concept is that of energy balance, which concerns an individual's ability to meet or exceed maintenance requirements through resource acquisition. Acquired energy in excess of maintenance demands (a positive energy balance) can then fund reproduction directly by promoting fertilization and offspring nutrition (Ellison 2003), or indirectly either through growth to adult body size and attainment of sexual maturity (Key and Ross 1999), or through investment in fat stores that can be used directly for reproduction (for females) or that can guard against future calorie shortfalls (for both sexes). Negative energy balance occurs when the organism fails to meet maintenance needs, which begin to deplete somatic energy stores (of fat and muscle) and results over time in a loss of condition. Negative energy balances can hamper growth and fecundity (Ellison and O'Rourke 2000; Ellison 2003; Bogin et al. 2007) and thus limit reproduction. For most animals, energy balance fluctuates between positive and negative throughout the course of a year, and the evidence (Chapter 4.3) strongly suggests that it did for Neandertals as well. While short-term changes in energy balance may influence ovarian functioning in human females (see Vitzthum 2009), what likely matters more with respect to completed fertility (the total number of children produced by the end of an individual's reproductive lifespan, as a measure of fitness) is the ability to attain a net positive energy balance over the course of a lifetime (to attain an adult body size sufficient to compete for mates and to materially invest in offspring, and for energy reserves to devote to production of offspring). A final useful concept is that of trade-offs, involving shifts in energy allocation between different subcategories of the budget. Research on energetic trade-offs in human evolution tend to focus on long-term, species-specific changes in life history and socioecology (such as trade-offs between greater brain size and patterns of maternal versus nonmaternal investment in offspring in early *Homo*: Aiello and Key 2002). However, trade-offs also occur at the individual level, and can be important short-term solutions to energy balance problems (as, for example, when a nutritionally-stressed pregnant female limits her physical activity or reduces her BMR and fat storage so as to be able to invest in the production of fetal tissues).

For any species, the ability to produce offspring depends upon a positive net energy balance, but human offspring are particularly energetically expensive. Human neonates are born with brains that are 50% larger relative to body size than are those of apes, and they also maintain a high fetal rate of brain growth for a full year postnatally, while other primates reduce brain growth rates at birth (Martin 1983, 2007). Chimpanzees achieve nutritional independence shortly after weaning but human children do not (Robson et al. 2006): humans have a unique childhood period (Bogin and Smith 1996; Bogin 1997, 1999) consisting of several years between weaning and the adolescent period when the child is able to obtain most of their own food. In the Hadza, for example, children are usually weaned by age 3 (Marlowe 2002) and are capable of obtaining 50% of their own calories by age 5 (Blurton-Jones et al. 1989, 1994; Marlowe 2003). Nonetheless, Hadza children, and those of other hunter-gatherers, only acquire the capacity for full nutritional independence around age 18 (Hrdy 2005). Although children do contribute to their own care and to the care of other children in a variety of ways (e.g., by holding infants while mothers forage or process food: Crittenden and Marlowe 2008), adults still provide a considerable portion of offspring calories before puberty. Thus, human children have energetically costly brains and a long dependency period (during which they are filling those brains with information that will allow them to be effective actors in their natural and social worlds), but they do alleviate somewhat the energetic demands upon those who provision them by having relatively slow somatic growth (with an adolescent growth spurt to adult size coming at a time when the child is able to provide for a sizable portion of their own energetic needs: Bogin 1999).

When her own BMR and activity costs are added to the cost of producing a baby, a modern human female expends roughly 2.5 million kcal between conception and weaning (Aiello and Key 2002). This does not include the cost of child carrying, which for foragers adds considerably to the energetic cost of transport (Kramer 1998; Wall-Scheffler et al. 2007; Watson et al. 2008). In addition, the uniquely-human long childhood period requires roughly 9–13 million kcal (Hrdy 2005; Gurven and Walker 2006), at least a third of which must come from provisioning (using rough estimates of age-specific growth rates from Walker et al. 2006, and the self-provisioning benchmarks for the Hadza cited above). Among nonhuman primates, nutritional independence coincides with weaning, such that once a female weans her offspring she can then begin devoting energy reserves to improving her condition in advance of the next pregnancy, and to producing her next offspring. Given the long period of nutritional dependency of human children, human mothers eventually find themselves caring for multiple, nutritionally-dependent offspring. The compound energy needs of all children at a woman's reproductive peak are enormous, requiring an additional 3000–6500 kcal d^{-1} (based on Ache and Ju'/hoansi foragers, from supplemental data in Gurven and Walker 2006) above and beyond her own maintenance requirements. Given the enormous energetic expense of producing human children, it is remarkable that females in foraging economies are able to produce them so rapidly (interbirth intervals range from about 2.5 to 3.75 years in hunter-gatherers, compared to about 5 years in wild chimpanzees: Hill 1993); all other things being equal, more expensive items are generally acquired at slower rates. Humans accomplish this through cooperative breeding – the non-maternal provisioning of dependents (or of reproducing families whose caloric demands outstrip the foraging returns of the family's producers) (Hrdy 2005; Kramer 2005; Hill and Hurtado 2009). Reasonable arguments can be made that modern human-like patterns of cooperative breeding first emerged in early *Homo* (most likely with *H. erectus*: Key 2000; Aiello and Key 2002), which means that Neandertals were also probably dependent on non-maternal provising as a central component of their reproductive strategies. We thus begin in Section 11.1 our consideration of Neandertal reproduction and demography by exploring the (scant) evidence pertaining to Neandertal social group size and structure (that is, what was the likely ratio of producers to consumers) by way of trying to evaluate the type of energetic assistance a Neandertal female might have been afforded by her group mates.

11.1 SUBSISTENCE LABOR DEMANDS, GROUP SIZE, AND SOCIAL STRUCTURE

In what ways was Neandertal social organization like that of modern humans, and in what ways might it have differed? Of course, social behavior doesn't fossilize, and as with most things about the Neandertals in which we're interested, we are compelled to make what inferences we can from indirect evidence. And as with everything having to do with the Neandertals, diametrically-opposed conclusions have been reached by different researchers based on differing interpretations of that evidence. Aspects of social organization have been inferred from the size and intrasite spatial organization of Middle Paleolithic sites, from evidence (or absence thereof) of symbolic behavior, from the practice of burial bestowed upon some Neandertals, and from patterns of raw material transport. In this constellation of archeological features some see evidence of very little social organization (perhaps not much different than might be seen among other higher primates) or social systems that differed in fundamental ways from those of modern human foragers (Binford in Fischman 1992; Gamble 1999, 2011; Pettitt 1997, 2000), while others see nothing to suggest that Neandertal social behavior differed in any important way from that of modern humans (Davies and Underdown 2006; Hayden 2012). This issue relates heavily to the question of potential cognitive differences between Neandertals and modern humans, in part because differences in cognitive constructs relating the self to others and to material culture may have had organizational consequences as they played out across larger social scales (see, for example, Gamble 2011), and also because of differences in interpretation of the *general* (but not total) lack of symbolic behavior among Neandertals (which to some may indicate less ability for symbolic thought among the Neandertals, while others may see it as reflecting a social world in which symbolic expression did not play an important function). Leaving the issue of cognition aside (but see Chapter 1), what can we reasonably infer about the Neandertals' social world? The following observations seem, to me, to be reasonable working hypotheses, based on current evidence, about Neandertal social organization:

1. As large-brained hominids, the Neandertals were undoubtedly social animals, and most likely lived in groups. Group living would have afforded important benefits in terms of cooperative breeding (see above), defense from predators (Majolo et al. 2008), and opportunities for cooperative hunting (Stiner and Kuhn 2006; Kuhn and Stiner 2006). While Neandertals may have foraged or traveled alone at times, and while deposits in some sites might represent accumulated debris from repeated site use by solitary actors (Pettitt 1997), their living arrangements were likely to be social. Furthermore, these groups appear to have been multi-male/multi-female, as is common in modern humans, chimpanzees, and bonobos (contra the suggestion of Binford [in Fischman 1992] that Neandertal males and females may have led largely separate social lives). This inference derives support from the site of El Sidrón in the north of Spain, which has produced the remains of 12 individuals (Rosas et al. 2006b; Lalueza-Fox et al. 2011). On the basis of taphonomic and genetic evidence, the El Sidrón Neandertals appear to represent a single social group (Lalueza-Fox et al. 2011), comprised of three adult males, three adult females, and six juveniles. While the sample at El Sidrón may not represent the entire social group, it clearly shows (if in fact all the skeletons represent social contemporaries) adult males and females together as group mates.
2. Like modern humans, the mating system in Neandertals was probably polygynous. Although modern humans do engage in a degree of pair-bonding, our mating systems are polygynous as defined biologically (greater variance in male than female reproductive success: Hammer et al. 2008). In terms of mating behavior, humans fall somewhere between pair-bonded gibbons and non-pair-bonded chimpanzees (Nelson et al. 2011). The average ratio of the length of the second to fourth finger (2D:4D) in Neandertals is signficantly lower than the mean for modern humans, suggesting greater exposure to prenatal androgens in Neandertals (Nelson et al. 2011). 2D:4D ratios across hominoids are correlated with mating systems, and the mean value for five Neandertals (0.928) falls almost at the lower confidence limit of a large sample of modern humans (mean of 320 individuals = 0.957) and is closer to the gorilla mean (0.919, n = 112) (Nelson et al. 2011). These results suggest a greater degree of male/male competition in Neandertals, and possibly a reduced incidence of pair-bonding relative to that seen in modern humans.
3. Again as with modern human foragers, Neandertals most likely lived in anucleated (open) social groups that were characterized by exogamy, free movement of people between groups, and exchange of resources between groups (Yellen and Harpending 1972). As with most primates (Lehmann et al. 2007), Neandertals likely employed a dynamic system of group fission and fusion to adjust social group size to local ecological conditions and labor demands, to promote

exogamous mating, to periodically reduce social tension, and to distribute foraging risk across a geographically-extended social network (Gamble 2011; Grove et al. 2012). Grove and colleagues (2012) argue that Neandertals, like modern human foragers, employed an hierarchically-inclusive system of nested social groups, going from nuclear and composite families, to consanguineal kin groups, to clans, to communities. When seen in this light, the very large (10,000 km^2 or more: see Chapter 10.2) lithic catchment areas that have been documented at some sites, such as Champ Grand where the raw material derives from sites spread over a 54,000 km^2 area of central France (Slimak and Giraud 2007), may reflect the areal extent of a Neandertal breeding population (a community), and movement of individuals (and with them, exotic flint) between social groups (or alternatively, reflecting occasional aggregations of smaller social groups: Hayden 2012). The (relatively) long-distance movement of raw material in the Middle Paleolithic is unlikely to reflect trade of flint nodules across social space, because exotic flints tend to appear as finished products (the end of the *chaine opératoire*) without evidence of having been reduced *in situ* (Gamble 2011). However, despite being probable participants in extended social networks, it is uncertain if Neandertals were able to maintain networks as large as or as geographically extensive as those of modern humans (see below, and Gamble 2011; Pearce et al. 2013).

4. Neandertals either practiced a system of female dispersal (patrilocality) or, like modern human foragers, had variable residence systems (patrilocal, matrilocal, and mixed: Kelly 1995). A patrilocal system has been inferred for the Neandertals from El Sidrón, where the three adult males in the social group all belonged to the same mtDNA lineage (Lalueza-Fox et al. 2011). While one of the adult females (for whom sex determination was based on morphological features, but could not be confirmed on genetic grounds) belonged to the same maternal lineage as the adult males, the two other adult females derived from two separate mtDNA lineages. Thus the Neandertals at El Sidrón appear to have had a patrilocal residence system, although the extent to which this characterized other Neandertal social groups is not known (although minimal phylogeographic structure in Neandertal mtDNA, as has been reported by Briggs et al. 2009, would be consistent with widespread practice of patrilocality: Lalueza-Fox et al. 2011).
5. Neandertal bands (*sensu* Layton and O'Hara 2010) were probably fairly small. The best evidence we have for social group size currently consists of the occupational surface areas of Middle Paleolithic sites. These surface areas are somewhat variable, and while occupation surfaces as large as 100 m^2 have been reported (e.g., at Figuier Cave, France: Daujeard and Moncel 2010), most sites have living areas on the order of about 30-40 m^2 (see Table 1 in Daujeard and Moncel 2010, Table 1 in Hayden 2012, and page 129 in Hoffecker 2002). Among modern foragers the areal extent of occupations varies seasonally, with living areas tending to contract during the cold seasons. Hayden (2012) suggests, based on ethnographic analogies, that winter habitations tend to occupy 2.5–3.0 m^2 per person. When applied to Middle Paleolithic site surfaces that likely represent winter occupations, estimated occupancies range from 12–18 individuals (Table 1 in Hayden 2012; note that Hayden's estimate of 23–28 individuals at Abric Romaní is likely to be an overestimate: see below). Applying the same occupant density to a broader range of sites (Table 1

in Daujeard and Moncel 2010) produces estimates primarily in the 11–16 individual range, with a few smaller (5–8 individuals at Pêcheurs Cave and 8–10 individuals at Baume Flandin, both in France) and a few larger (23–28 individuals at Saint-Marcel Cave and 33–40 individuals at Figuier Cave, both in France). A major problem in interpreting these living floors concerns the number of occupations represented: the larger surface areas are likely to represent palimpsests accumulated over multiple occupations (Farizy 1994), and even the smaller sites could potentially represent the accumulated debris of numerous occupations by solitary visitors to the site (Pettitt 1997). Consideration of the size of sleeping areas, however, leads to estimates that accord well with the usual size of occupation floors in the Middle Paleolithic. At the Abric Romaní, the size of the sleeping area and the number of warming hearths along the rear wall of the shelter imply a group of about 8–10 individuals (Vallverdú et al. 2010). It has been suggested that the tent structure at the MIS 6 Acheulean site of Grotte du Lazaret (France), presumably demarking a sleeping area (de Lumley 1969), may have accommodated only three individuals (Pettitt 1997). A possible tent at the open-air site of Buhlen, Germany, encloses an area of less than 20 m^2 and, like the Middle Paleolithic mammoth bone shelters at Molodova I in the Ukraine, would have only sheltered a small number of occupants (Pettitt 1997). A final line of evidence comes from the 12 Neandertal individuals recovered from the Spanish site of El Sidrón, which have been argued on taphonomic and genetic grounds to all be members of the same social group (Lalueza-Fox et al. 2011). This number of individuals is consistent with the inference of generally small social group size (and with the idea that bands may have been structured as composite families organized along patrilineal lines: see above), although the possibility exists that the El Sidrón Neandertals only sample a portion of a larger social group. Overall, the evidence seems to suggest that Neandertal bands tended to be relatively small (on the order of 8–16 individuals), but at times reached numbers of 25 or more, which is consistent with the group sizes observed in modern foragers living in relatively unproductive environments (Hayden 2012). While small groups may have been the norm for Neandertals and may also be found among some modern foragers, the modal group size across a global sample of recent foraging groups is 25 (Kelly 1995), and it has been argued on theoretical grounds that below this size the reproductive viability of groups is threatened by short-term fluctuations in fertility, mortality, and sex ratio.

6. Neandertal social groups probably frequently experienced high dependency ratios. Among modern foragers, observed dependency ratios (the number of consumers in a social group divided by the number of producers) vary between 1.25 and 3.00, with a mean value across foraging groups from varied environments of 1.75 ($\pm$ 0.14, n = 18 groups) (Binford 2001). Mortality profiles (Trinkaus 1995a) and ratios of old to young individuals (Caspari and Lee 2004, 2006; Trinkaus 2011b) indicate high young adult mortality in Neandertals. In a large sample (206 individuals) of Neandertals, fully 38.8% of the individuals represented died as young adults, with another 18.0% having died as adolescents (Trinkaus 1995a). A similar high adolescent and young adult mortality appears to have also characterized the preNeandertal population represented at the Sima de los Huesos in Spain (Bermúdez de Castro and Nicolás 1997; Bermúdez de Castro et al. 2004). It has been suggested that the skeletal sample at the Sima de los Huesos (representing a minimum number of individuals [MNI] of 32),

as well as that of the Neandertals from Krapina, Croatia (MNI = 24), represent mortality from a catastrophic event, and are thus representative of the age structure of preNeandertal (*H. heidelbergensis*) and Neandertal populations (Bocquet-Appel and Arsuaga 1999). Both of these samples exhibit a marked underrepresentation of adults older than about 25 years of age (Bocquet-Appel and Arsuaga 1999), supporting the inference of high young adult mortality drawn from analysis of the total Neandertal sample. Inferring mortality patterns from skeletal assemblages is made difficult by taphonomic and human behavioral factors that may bias the representation of certain age classes in skeletal assemblages, but even if adjustments are made to weight the importance of certain age groups or certain large sites, the picture of high young adult mortality remains unchanged (Trinkaus 1995a). The causes of adult Neandertal mortality are unknown, although complications during childbirth were likely a significant source of mortality for adult females, and cold winter temperatures and starvation were likely to cause deaths among both sexes (Sørensen 2011). The high frequency of trauma evident in Neandertal skeletons (Berger and Trinkaus 1995) lends itself to the suggestion that close-range hunting of large ungulates might also have been a significant source of mortal trauma (Trinkaus 1995a; Kennedy 2003; Sørensen 2011). It is also highly likely that aggressive encounters with other carnivores (see Chapter 9) in the context of intraguild interference competition were a significant source of mortality for the Neandertals. Of 393 large carnivore (lion, leopard, and hyena) attacks on humans in Uganda since 1923, 62.9% were fatal (Treves and Naughton-Treves 1999). And despite being largely eradicated from most of the country, lions attacked more than 1000 people in Tanzania between 1990 and 2007 (Kushnir et al. 2010), with fatality rates (based on data as of 2005) of 76.8% (Packer et al. 2005). Along these lines it is interesting to note that the Near Eastern Neandertals appear to have fared considerably better than their European counterparts (Caspari and Lee 2006). In a small sample (n = 18) of adult individuals from the Near East, fully half of them had lived to old age, whereas among the Europen sample (n = 139) only 26% had survived to old age (Caspari and Lee 2006). While the small size of the Near Eastern sample undermines the confidence we can place in this finding, it does suggest that young adult mortality was lower in western Asia, perhaps as a result of a greater availability of plant foods and a correspondingly reduced hunting intensity, which would have simultaneously reduced hunting-related trauma and the intensity of the competition experienced with other carnivores. At any rate, adolescent and young adult mortality appears to have been high among the Neandertals, which would have led to relatively high dependency ratios. Since relatively few Neandertals lived to old age, it is unlikely that Neandertal social groups were heavily burdened with dependent older individuals whose productivity had dropped off due to age (but who could otherwise contribute to childcare and domestic work: Hawkes et al. 1998), although some older or otherwise less productive (such as the heavily traumatized individual Shanidar 1 from Iraq: Trinkaus and Zimmerman 1982) might be expected. Still, it is likely that many Neandertal groups found themselves with a fairly high number of infants and children relative to adults. The dependency ratio of 2.00 suggested by the possible social group from El Sidrón, Spain (Lalueza-Fox et al. 2011) may have been fairly typical, or towards the low end, for Neandertals. Dependency ratios are the primary intrinsic variable that determines foraging effort among modern hunter-gatherers (Binford 2001): among two Australian aboriginals groups, for example, a 0.5 difference in dependency ratio equated with an 80% difference in the average work effort of the producers. Thus, high adult mortality among Neandertals would have had a negative impact on their adult activity levels and, ultimately, their fitness (see below).

7. The extended social network size of Neandertals may have been smaller than that generally seen in modern human foragers. Among recent hunter-gatherers, extended social networks (communities, *sensu* Grove et al. 2012) range in size from about 175 to 1500 (Kelly 1995). Wobst (1974) has argued on theoretical grounds that the minimum equilibrium size of breeding populations of hunter-gatherers falls between about 175–475 people, depending on population fertility rates, mortality rates, sex ratio, and cultural rules pertaining to mating. Neandertals would have needed to have regular-enough contact with several hundred other individuals for purposes of reproduction, and an extended social network would also serve to reduce foraging risk related to variance in local productivity (foragers tend to "store" calories in social bonds, such that when times become bad in their area they may disperse to go visiting social relations who live in distant camps where, presumably, conditions might be better). However, a low population density and high cost of transport may have constrained the geographic radius of the Neandertal social world, and thus the number of individuals within it. The relative size of the neocortex of the brain has been used as an index of social complexity in primates (Dunbar 1992), and thus to predict the size of the extended social network within which individuals of a given species normally operate. When Neandertal neocortex size is adjusted for their presumably larger (based on orbital area) occipital lobes, the results suggest median social networks that were smaller (ca. 115 people) than that of modern humans (ca. 144 people) (Pearce et al. 2013). These values do not reflect the total size of a breeding population, but rather the average number of known individuals within that population with whom the actor interacts (Gamble 2011). Still, the smaller number in Neandertals is consistent with the suggestion, on population density and energetic grounds, that they may have maintained smaller social networks. High adolescent and young adult mortality increases the minimal equilibrium size needed for a population to remain viable (Wobst 1974), and thus if Neandertal mobility costs limited the social network, regional populations may have faced a high probability of going locally extinct. This could be offset to a degree by reducing as much as possible the variance in fertility (through child spacing and possibly infanticide) and relaxing the cultural rules concerning reproduction (such as proscriptions against polygyny, or mating with certain classes of relatives) (Wobst 1974).

What do these observations about group size and social organization imply about Neandertal behavioral ecology? High adult mortality and high dependency ratios no doubt had major effects on Neandertal socioecology. First, given that relatively few Neandertals survived into older age, it is unlikely that they were able to employ the kind of three-generation cooperative breeding pattern ("grandmothering," in which post-reproductive females contribute material support to the reproductive efforts of their daughters: Hawkes et al. 1997) that is common among modern foragers. In groups with high dependency ratios, this would have produced pressure for juveniles to begin contributing to foraging returns as early as they possibly could. It would have also increased the required work effort of the active producers in these groups, thus exacerbating what were likely already tight energy budgets (see Chapter 10.6) and negatively impacting fitness (by reducing the ability of producers to invest calories in mating effort, and offspring production and nurturing). There may have also been a selective premium on early weaning (see Section 11.2) as a solution to high young adult mortality (since weaned young can then be provisioned by others, relaxing some of the burden on the mother: Kennedy 2003). Second, at small group sizes Neandertals may indeed have found it difficult to maintain a division of labor along gender lines (Kuhn and Stiner 2006; Stiner and Kuhn 2009; see also Chapter 10.6). The procurement of moderate-to-large terrestrial game with close-range weapon systems most likely entailed a significant degree of cooperative hunting. Modern human foragers often hunt individually, but their ability to do so is supported by a rich kit of extractive technology, including long-range projectile weapon systems. As noted above (Chapter 3.4) Neandertal subsistence toolkits tend to be simple, and do not vary in complexity across different environments, leading Kuhn and Stiner (2006; Stiner 1994; Kuhn 1995, 2011) to suggest that Neandertals were modulating environmental variability socially rather than technologically; that is, they were adjusting the labor characteristics of their foraging parties dynamically to compensate for the exigencies of particular environments. Social carnivores use coordinated hunting to improve net return rates and to increase the range of prey sizes that they can capture (Earle 1987), as well as to derive some measure of protection from aggressive encounters with other carnivores (Packer et al. 1990). If Neandertal bands included only a few adult males, hunting return rates may have significantly improved with the coordinated help of other members of the social group, even if they simply operated as drivers of game towards concealed hunters (Kuhn and Stiner 2006). Furthermore, small foraging parties – whether male or female – may have been vulnerable to harassment by carnivores. This might be especially true of a small group of adult females foraging with children in tow. All of these factors may have operated to curtail a division of foraging labor along gender lines. Third, Neandertals may have had a diminished capacity to use the broader social community to alleviate resource stress. One analysis of trauma frequency in Neandertals and modern foragers concluded that the high frequency of injury seen in Neandertals cannot be explained by reference to their subsistence strategies or foraging technology (Underdown 2004), leading to the suggestion that small social groups of Neandertals – without the safety net of an extended social network – may have had to resort to increased hunting intensity, or to hunting more dangerous prey, during times of crisis (Davies and Underdown 2006).

11.2 NEANDERTAL LIFE HISTORY

Life history refers to the timing of key events such as the ages of sexual maturity and senescence in the reproductive life of individuals. Life history theory holds that the schedule of these events is shaped by natural selection to maximize fitness given the physical and social environment in which a species operates (Harvey and Clutton-Brock 1985; Harvey et al. 1989; Charnov 1991). The evolution of the human life history, and its distinctiveness relative to other primates, is an area of active research (Hill 1993;

Hawkes et al. 1998; Key 2000; Robson et al. 2006). Given the tight relationships that exist between energetic ecology and life history across a wide range of organisms (Nisbet et al. 2000; see also Key 2000), and given the tight energy budgets in Neandertals, the challenges they faced in maintaining positive energy balances, and the possibility that energetic trade-offs that affect the scheduling of reproductive events may have been critical to Neandertal fertility under these conditions, we might wonder in what ways their life history might have differed from ours. Ideally, we would wish to know a whole host of life history characteristics of the Neandertals, such as gestation length, age at weaning, age at first reproduction, interbirth interval, and mean and maximum life span, but these characteristics are generally not known for any extinct hominins (Robson and Wood 2008). Given a large sample of individuals that can reasonably be assumed to represent a single population, these and other life history variables can be estimated probabilistically from mortality data, as has been done with the *H. heidelbergensis* sample from the Sima de los Huesos in Spain (Durand et al. 2000): no such opportunity exists in the Neandertal fossil record. We can, however, make some reasonable inferences about gestation length, age at weaning (and what that implies about interbirth intervals), and growth rates (and what they imply about the length of the childhood period and the energetic costs of raising offspring) in Neandertals.

Present evidence suggests that gestation length in Neandertals was effectively no different than that of modern humans (260–280 days, with medians of 274 and 269 days for first time and multiparous mothers, respectively: Mittendorf et al. 1990; Crews and Gerber 2003). Human gestation length is not markedly different from that of chimapnzees (240–260 days) or gorillas (260–280 days) (Crews and Gerber 2003), but is remarkably short (by 3–5 months) for mammals of our brain size (Sacher and Staffeldt 1974), such that humans produce neonates that are altricial relative to those of our closest living relatives (Dunsworth et al. 2012). In light of the large adult brain size of Neandertals, their elongated superior pubic rami (once thought to signal a larger birth canal), and likely socioecological constraints on their ability to manage with highly altricial young, Trinkaus (1984) argued that gestation in Neandertals may have taken 11 to 12 months, and that secondary altriciality was a derived feature of modern humans. The discovery of a partial male Neandertal pelvis at Kebara Cave (Israel: Rak and Arensburg 1987), and the virtual reassembly of the fragmentary pelvis of the female from Tabun (Weaver and Hublin 2009), reveal that the elongated pubic rami of Neandertals are related to differences in the shape of the pelvic aperture, not its relative size. Virtual reconstruction of the cranial vaults of a Neandertal neonate from Mezmaiskaya Cave (Russia) and two infants from Dederiyeh Cave (Syria) suggest that neonatal brain size was similar in Neandertals and modern humans (Ponce de León et al. 2008). Given similarity in neonatal brain size and pelvic aperture size in Neandertals and modern humans, and given the generally conservative nature of gestation length in the African hominids, there is nothing at present to suggest any real difference in Neandertal gestation lengths relative to that of modern humans.

Birth represents a major transition in the way that a mother invests energy in the production of her offspring: weaning represents another. Of course, weaning isn't an abrupt transition like birth, but rather begins with the introduction of solid food (usually during the first year of life in recent foragers: Skinner 1997) and a gradual shift over several years in the proportions of breast milk and solid foods (Kennedy 2005). Thus, detecting the biological consequences of weaning, such as indicators of systemic stress that may occur in the developing dentition or growing skeleton as the baby transitions off milk, or the onset of dental wear as the baby begins to chew weanling foods, may be difficult (Smith 2013). Furthermore, there is not necessarily a direct relationship between weaning, resumption of the mother's ovarian cycling, and interbirth interval (IBI). In general, suckling frequency and intensity downregulates the production of gonadal hormones that induce ovulation (Taylor et al. 1999), but the contraceptive effectiveness of breast-feeding remains uncertain (Ellison 1990). In some traditional, natural fertility populations, it is not uncommon for conception of the next child to occur while the mother is still nursing the previous child, or for a woman to nurse infants of different ages (Kennedy 2005). Still, there is a general relationship between the average age of weaning and the interbirth interval, and humans appear to have increased their fertility rates (relative to other apes) by weaning infants at earlier ages (Kennedy 2005; Gurven and Walker 2006). Foragers wean on average at 2.65 years ($\pm$ 0.84, n = 40) and have an average IBI of 3.29 years ($\pm$ 0.74, n = 16) (Marlowe 2005).

Several lines of evidence can be explored to infer something about weaning ages in Neandertals, and in one case to even say something about interbirth intervals. One such line of evidence is dental attrition to juvenile teeth, the onset of which likely signals supplementation of breast milk with solid foods. Examination of wear to the teeth of infant, juvenile, and young adult Neandertals and early modern Europeans relative to their inferred age at death suggests that significant supplementation with solid foods began during the fourth year of life for Neandertals, and a year earlier for Upper Paleolithic modern humans (Skinner 1997). However, potential differences in rates of dental development between these groups (see below) may produce a systematic overestimate of the ages of Neandertal children by a half year, and the comparison does not take into consideration differences between groups in enamel thickness (which affects rate of wear) and dietary consistency (Smith 2013). A second line of evidence concerns growth disturbances that are recorded in dental enamel as linear enamel hypoplasias (LEH). It has been suggested that the transition from protein-rich breast milk to protein-poor weanling foods represents a systemic shock to the infant, resulting in temporary cessation of growth, including an interruption of ameliogenesis which result in LEH. This weaning–LEH association, however, is not well supported in recent human populations (Smith 2013), and may be even more questionable for Eurasian Pleistocene groups which may have provided weanlings with relatively protein- and fat-rich foods. Nonetheless, a peak in LEH frequency between 2–5 (Ogilvie et al. 1989) or 3–5 (Skinner 1996) years of age in Neandertals has been suggested to reflect weaning during that interval. Again, potential differences in rates of dental development may mean the Neandertal ages need to be adjusted downwards by a half year (Smith 2013), and the extent to which this patterning in LEH incidence reflects weaning remains an open question. Bone and dental biochemistry provides a final line of evidence with which to explore weaning ages. An elevated $\delta^{15}N$ signal (12.6‰) in the 5–6-year-old Neandertal Engis 2 (Belgium), indicating feeding at a high trophic level (nursing infants are feeding at one trophic level above their mother), would be consistent with weaning around age 4 (Bocherens et al. 2001). Trace element analysis of barium in dental enamel has also recently been used to examine weaning in one Neandertal juvenile. Breast milk is heavily enriched in barium, and thus ratios of barium (Ba) to calcium (Ca) in developing dental enamel increase markedly at birth, decline during supplementation, and drop off after full weaning (Austin et al. 2013). Ba/Ca ratios in the dental enamel and dentine of a juvenile Neandertal from Scladina Cave (Belgium) indicate supplementation with solid foods beginning at seven months of age, followed by an abrupt cessation of nursing at 1.2 years of age (Austin et al. 2013). Since this has only been applied to a single Neandertal, it is not clear whether this might be an extraordinary case (e.g., the death of the mother and premature cessation of nursing), but it does suggest that supplementation may have begun at an early age. Again, although this is an isolated case, the timing of two nursing events (supplementation and weaning) is interesting from an energetic perspective. The upper limit of milk production among modern humans appears to be on the order of about 1.2 l d^{-1} (Kennedy 2005). Based on estimates of infant and toddler energy demands in Table 3.6, the needs of Neandertal babies may have started to exceed what their mothers could produce near the end of the first year of life, necessitating early supplementation. The overall nursing evidence is thus mixed, suggesting supplementation with solid foods before one year of age, with full weaning probably by 3.5–4.5 years of age (adjusting for bias in estimating age at death of Neandertal children). If this is the case (additional Ba/Ca analyses will contribute greatly to answering this question), Neandertals may have nursed their young on average a year or two longer than do most modern human foragers. Early supplementation would have eased the energetic burden on the mothers somewhat, although longer nursing periods would still equate to some degree with longer IBI than is typical of modern human foragers. Interestingly, the possible Neandertal social group recovered at El Sidrón (Spain) may provide some direct evidence for Neandertal interbirth intervals. Among the individuals at the site are a 5–6-year-old juvenile and a 2–3-year-old infant, both of which share the same mtDNA haplotype as one of the adult females, and which thus appear to be her offspring. This single case suggests an IBI of about 3 years (Lalueza-Fox et al. 2011), which is short relative to the weaning ages estimated from dental attrition and LEHs, and not markedly different from the mean reported for modern foragers (bearing in mind, however, that weaning ages and IBIs are highly variable between individuals within a social group, and between children of a single female). One final note about weaning: milk production occurs at about 80% energy efficiency (Dufour and Sauther 2002), and in a situation where calories are a valuable commodity (as in the Neandertal world), it would pay to supplement and wean children as quickly as possible. I strongly suspect that weaning behavior in Neandertals was on a schedule similar to that seen in modern foragers (or perhaps even earlier), and that the peaks in LEH in Neandertal 2–5-year-olds are reflecting something other than weaning.

Among the more intractable recent problems in Neandertal research is the question of whether they had "accelerated life histories," that is, if they grew rapidly and reached sexual maturity earlier than do modern humans. High adult mortality favors rapid development (such that individuals can begin reproducing at younger ages to maximize their fertility). For example, adult mortality among Hiwi foragers in Venezuela is about twice that of comparable foraging groups (Ache, Hadza, and !Kung), and the Hiwi grow faster and reach reproductive maturity earlier than other foragers (Hill et al. 2007). Accordingly, we might expect Neandertals to have had shorter childhood periods and earlier ages at first reproduction than do modern humans. On the other hand, rapid growth requires higher rates of energy acquisition and a higher burden on those provisioning dependent young, which might have over-stressed the already tight energy budgets of Neandertals and thus favored slower growth in them. Along these lines, the high metabolic needs of cold-adapted Evenki reindeer herders of Siberia have been argued to account for their relatively poor growth (Leonard et al. 1994), and Bogin and Rios (2002) have suggested that the short adult stature and short limbs of Neandertals may be the result of inadequate childhood nutrition. However, some of the energetic cost of more rapid attainment of adult body size may be offset by earlier entry into subsistence activity, and thus potentially a proportionally greater ability to self-provision, earlier self-sufficency (which doesn't develop until about age 18 in modern foragers: Hrdy 2005), and earlier contributions to the provisioning of other dependents. Since growth rates to maturity affect age at first reproduction, and age at first reproduction affects completed fertility and is a major determinant of the average fitness of a population, it would be useful to know something about Neandertal growth rates. Juvenile Neandertals are fairly well represented in the fossil record, but unfortunately there is little consensus on how best to interpret the developmental evidence (debate stems both from contradictions among studies and from unresolved issues about how best to interpret the data: Guatelli-Steinberg 2009). Patterns of tooth crown and enamel development suggest to some that Neandertals developed more rapidly than both recent and Middle Paleolithic modern humans (Thompson and Nelson 2000; Ramírez Rozzi and Bermúdez de Castro 2004; Smith et al. 2007b, 2010), and that their developmental schedules may have been accelerated by as much as 2–3 years (Smith et al. 2007a). *H. heidelbergensis* appears also to have been on an accelerated growth schedule relative to modern humans (Bermúdez de Castro et al. 2003, 2010; Ramírez Rozzi and Bermúdez de Castro 2004), although potentially one falling between that of Neandertals and modern humans (and possibly closer to modern humans: Bermúdez de Castro et al. 1999b). This suggests that Neandertals followed an evolutionary trajectory of accelerated growth (as might be expected given their high adult mortality) relative to their last common ancestor with modern humans. Furthermore, Wolpoff (1979) noted that age at death estimates from molar occlusal wear gave higher estimates for Neandertal third molars compared to the first and second molars. From this he inferred an age of M3 eruption of 15 (Wolpoff 1979), suggesting a faster rate and earlier cessation of growth in Neandertals. The dental evidence for rapid growth in Neandertals is consistent with evidence of faster growth rates (rate hypermorphosis) of the brain (Dean et al. 1986) and cranium (Manzi et al. 1996; Ponce de León and Zollikofer 2001) and an early attainment of high levels of postcranial robusticity (Arsuaga et al. 2007; Volpato et al. 2007; but see Cowgill et al. 2007). The available data suggest to others, however, that Neandertal developmental rates and schedules did not differ from those of modern humans. In terms of dental developmental patterns, Neandertals do not appear to differ from modern human samples (Tompkins 1996), and disagreement exists as to the evidence for faster rates of tooth formation. Contra claims of rapid dental development in Neandertals, studies of enamel formation rates in both the anterior teeth (Guatelli-Steinberg et al. 2005, 2007) and postcanine teeth (Macchiarelli et al. 2006) have been interpreted as showing parity in Neandertal and modern human dental growth rates. Added to this is evidence from the postcranial skeleton that suggests that rates of somatic growth in Neandertals appear to be retarded relative to dental development (Thompson and Nelson 2000), suggesting either that Neandertals had rapid dental development and modern human-like rates of somatic growth, or modern human-like rates of dental development and slow somatic growth.

The latter possibility is supported by consideration of stature attainment by age in Neandertal and modern human children (Martín-González et al. 2012), and again this may either be directly due to nutritional (and energetic) insufficiencies during childhood (Thompson and Nelson 2000; Nelson and Thompson 2002; Bogin and Rios 2002) or to genetic programs for slower somatic growth that were adaptive in the context of tight energy budgets. Regardless, the uncoupling of dental and somatic growth rates makes it all the more difficult to assess the life history implications of the available evidence for Neandertal gowth and development (Guatelli-Steinberg 2009).

Uncertainty about Neandertal growth rates and weaning behavior leaves us with two very different potential life history models. On the one hand, the data could be taken to infer that there was no substantial difference between Neandertal and modern human forager life history (O'Connell 2006). By this view, the evidence that LEHs begin to peak around two (Ogilvie et al. 1989) to three (Skinner 1996) years of age, and that supplementation began early (Austin et al. 2013) could be seen as consistent with average weaning ages and IBIs similar to those of modern foragers (2.65 and 3.29 years, respectively). The suggestion of a three-year IBI at El Sidrón (Lalueza-Fox et al. 2011) would also be consistent with this model. Furthermore, the somatic growth data could be taken to indicate an attainment of reproductive maturity on a schedule like that of modern foragers (18–20 years of age: O'Connell 2006; with a mean age of first reproduction of 19.7: Kaplan et al. 2000). Alternatively, the data could be construed as reflecting a life history very different from that of modern hunter-gatherers. The peak in LEHs at 3.5 and 4.5 years of age (Skinner 1996; adjusted following Smith 2013) might be seen as evidence of weaning occurring, on average, one to two years later than seen among modern foragers. Using the difference of ca. 8 months between mean weaning age and mean IBI in modern humans (Marlowe 2005), this would suggest average IBIs of 4.1–5.1 years in the Neandertals. The evidence for rapid dental growth, especially the potential evidence for early third molar eruption (Wolpoff 1979), might suggest attainment of adult body size and first reproduction as early as about 15 years of age. Resolving the issues that plague Neandertal life history research will clearly be critical to furthering our understanding of their socioecology; in the meantime, we can at least explore the implications of these different life histories for Neandertal demography.

11.3 NEANDERTAL DEMOGRAPHY

If Neandertals had high mortality rates (Trinkaus 1995a), they must have had high fertility rates in order to maintain viable populations (Wobst 1976). Demographic simulations suggest that Neandertals would have needed completed fertilities (average number of children born to a female cohort by the end of their reproductive careers) of 7.613 to maintain stable populations over a 200 Ka period (Sørensen 2011). This value is substantially higher than the mean fertility observed in modern foragers (5.35 ± 1.69, n = 47 groups), but is not outside the range (0.81–8.50) (Marlowe 2005). Completed fertility (F) is a function of the mean age of first reproduction (A_{FR}), mean age of last reproduction (A_{LR}) and interbirth interval (IBI), as (adapted from Kennedy 2005):

$$F = \frac{A_{LR} - A_{FR}}{IBI}$$

It is unlikely that most Neandertal females lived to postmenopausal age (Trinkaus and Tompkins 1990; Trinkaus 1995a), and thus we might take either the upper end of observed adult longevity or female life expectancy at the age of first reproduction as measures of the age at which females ended their reproductive lifespans. The overall age distribution of Neandertal fossils (assuming there is not some major behavioral bias that causes an under-representation of old adults in the fossil record relative to their actual abundance in Neandertal populations) suggests that only about 5–15% of Neandertals survived to age 40 or beyond (Trinkaus and Tompkins 1990; Trinkaus 1995a). The maximum longevity in the preNeandertal sample from the Sima de los Huesos appears to be about 40 years (Bermúdez de Castro and Nicolás 1997), and probabilistic modeling of life history parameters based on the mortality profiles at the site suggests that the reproductive period ended between 37 and 40 years of age (Durand et al. 2000). Mortality profiles also suggest that a Neandertal female who attained the age of 15 could be expected to live another 16 years, while one who attained the age of 20 could be expected to live an additional 14 years (Sørensen 2011). If Neandertal life history were like that of modern foragers, then A_{FR} = 19.7, A_{LR} = 34–40, and IBI = 3.29. This produces values of F = 4.347–6.170, which at the higher end comes close to the fertility values needed (based on demographic modeling) to sustain Neandertal population structure in the long term. Longer IBIs (4.1–5.1) produce comparable values of F (4.902–6.098) if the age at first reproduction is set to 15 and longevity is set to age 40. With either a modern human-like age of onset of the reproductive career or an earlier average age at death, levels of fertility (3.725–4.951) would have probably been too low to maintain viable population sizes. Only with a young age of first reproduction (15), maximum longevity (40), and a modern forager-like IBI (3.29) can we attain a value of F (7.599) that approaches that required for maintanence of a stable population. While there are clearly many unknowns about Neandertal life history, the high fertility that would be required to offset the high level of young adult mortality would have certainly created pressures to reach sexual maturity at an early age and to keep interbirth intervals low.

The Neandertals thus appear to have occupied a sociobiological position in which high adult mortality necessitated rapid growth, early sexual maturation, and short interbirth intervals, but in which production of offspring was costly (from gestation through growth to maturity), energy budgets were tight, and options for cooperative breeding were constrained (by small group sizes, high adult mortality which produced high dependency ratios, and insufficient longevity to allow for substantial grandmothering). The potential of Neandertal mothers to take advantage of energetic trade-offs was likely also limited. For example, gravid females may have been able to reduce their metabolic demands by limiting their activity (but which would reduce their foraging returns) and reducing their own late pregnancy storage of fat, and thus to devote more energy to the production of fetal tissues. However, both of these trade-offs result in poorer maternal condition post-partum, which in turn can negatively impact milk production (especially during annual periods of nutritional stress) and energy investment in children, and increase the risk of infant death. The combination of tight energy budgets, periodic caloric shortfalls, and cold winters was undoubtedly a major contributor to what may have been high infant mortality in the Neandertals (Trinkaus 1995a; Pettitt 2000; Sørensen 2011).[2] These considerations define the rather harsh realities of

Neandertal reproduction, and provide an understanding of why Neandertal population densities may have remained relatively low in most places at most times.

Opinions differ with respect to the possibility of contrasts in adult mortality and longevity between Neandertals and early modern humans (see Caspari and Lee 2004 vs. Trinkaus 2011b). If young adult mortality was higher in Neandertals than it was in early modern humans, the Neandertals would have needed correspondingly higher fertility in order to maintain stable populations. In this scenario, the ultimate demise of the Neandertals and the success of modern humans in Eurasia can be seen in the context of the differential effects of fertility in the two groups (modern human populations being able to grow at lower fertility rates, such that even if the groups had equal fertility the modern human population would grow faster: Flores 1998). Computer simulations suggest that as little as a 2% difference in mortality could have led to the rapid demise of the Neandertals once modern humans entered Europe (Zubrow 1989). Energetic differences in the costs of producing offspring (Snodgrass and Leonard 2009; Froehle et al. 2013), if they lowered Neandertal fertility, would have exacerbated the demographic disadvantage of the Neandertals. Alternatively, if adult mortality patterns did not differ between the two groups, the demise of the Neandertals and the modern human peopling of Eurasia may have largely been a function of differential fertility. Here the energetic differences in reproduction may be critical, as Neandertals may simply not have been able to sustain reproductive rates on a par with modern humans. A relatively simple model of reproductive energetics played out across the reproductive lifespan of Neandertal and modern human females (simple in that the only parameters considered to differ between the groups are adult BMR, cost of transport, and the age-specific energetic needs of offspring: all other life history, demographic, and energetic variables were held equal in the two groups) suggests that, by the peak of her reproductive career (when she would be provisioning multiple dependent offspring) a Neandertal mother would require upwards of 800 kcal d^{-1} more than an early modern human female to satisfy her own and her children's energy demands (Froehle et al. 2013). By this model, a Neandertal mother's caloric needs may have peaked at close to 10,000 kcal d^{-1}, which would require daily foraging returns that are likely to be well outside the realm of possibility for a single forager (Froehle et al. 2013), again suggesting that Neandertals must have relied on nonmaternal provisioning of offspring (cooperative breeding). Given infant and childhood mortality (which was not factored in to the model by Froehle et al. 2013), it is likely that Neandertal mothers rarely ever had more than two or possibly three dependent offspring at a time. Even so, a mother suckling one child and provisioning another would have needed between about 3500–7700 kcal d^{-1} for her and her children.[3] The combined high energetic costs of somatic maintenance (BMR + activity) and reproduction was, more than likely, a major constraint on Neandertal fertility (Snodgrass and Leonard 2009; Froehle et al. 2013), and may have made infanticide necessary even at fairly low population densities (see Harpending and Bertram 1975). It is thus likely that many local populations of Neandertals experienced inadequate fertility to maintain stable population sizes and were liable to extinction (Trinkaus 1995a). This situation may have produced a state of regular extinction of local populations (especially in less productive biomes to the north) and repopulation from other areas (such as regions of more constant productivity adjacent to the Mediterranean) (Dennell et al. 2011). Over much of the Neandertal geographic range, inadequate demographic growth may have produced population "sinks" (Pulliam 1988) that were only sustainable with inmigration from more stable source populations living in regions of greater productivity to the south; at times, all of Europe may have been a Neandertal population sink (Dennell et al. 2011: see also Chapter 2.5).

The abundance of Middle Paleolithic sites, the volume of artifacts within them, and the nature of the fauna contained therein – relative to those of the Upper Paleolithic – suggest that the metapopulation of Neandertals was small and that they lived at low population densities (Straus 1982; Stiner et al. 1999; Lahr and Foley 2003; Hovers and Belfer-Cohen 2006; Shea 2008; Hublin and Roebroeks 2009; Mellars and French 2011; Dennell et al. 2011). As with other aspects of Neandertal biology, determining the actual census size of Neandertals, and their population density, for any given time period is difficult. One often hears effective population size (N_e) estimates from genetic studies touted as evidence of small census sizes (N_c: sometimes referred to as actual population sizes, N_a) in Neandertals. However, there is no simple relationship between N_e and N_c, because N_e reflects the harmonic mean of population size across generations, and thus it is heavily influenced by population bottlenecks in the evolutionary history of the groups. In some respects, N_e is more a measure of genetic diversity than a measure of census size, and this is well illustrated with living humans: attempts to estimate the living human global effective population size produce N_e estimates in the tens of thousands of individuals (for example, Park 2011 estimates living human N_e to be between 622 and 10,437), while N_c is roughly seven billion. Another method to estimate Neandertal N_c has been to apply the population densities of modern foragers living in similar ecological and environmental conditions to the geographic area of Neandertal occupation. Such an approach produces fairly large metapopulation census sizes (250,000 overall [Biraben 2003]; and as high as 500,000–1,000,000 during interglacial periods [Dennell et al. 2011]), but are highly problematic in that they assume the same relationships between environment, carrying capacity, and population density in Neandertals and modern human foragers. Given an occupation area across Eurasia of ca. 24 * 10^6 km^2 (Flores 2011), this equates with population densities ranging from 0.01 to 0.04 people km^{-2} which, not surprisingly, are comparable to those of modern foragers living in colder environments (Kelly 1995). Modern humans living in cold temperate and arctic environments tend to employ complex extractive technology for acquiring terrestrial resources, and also tend to rely heavily on marine resources, and thus live at higher population densities than were likely to be sustainable for Neandertals. Estimates have also been derived from average territory sizes (based on raw material movement distances) on the assumption that social groups with a modal number of 25 individuals (but see Section 11.1) would have occupied each territory. This method produces estimates of between 478–746 individuals living in the 150,000 km^2 across which the Mousterian of Acheulean Tradition is distributed in France during MIS 3, and 1245–1946 individuals inhabiting the 391,000 km^2 covered by the central European Micoquian during MIS 3 (Richter 2008). These estimates by Richter (2008) produce population density estimates in the range of 0.003–0.005 individuals km^{-2}, which are quite low relative to modern forager densities (see Table 6-4 in Kelly 1995), but which are consistent with overall site density in the Middle Paleolithic. These estimates are, however, lower than the densities we estimated based on ecological grounds (assuming Neandertals ranked in the middle of the carnivore guild: Table 6.5). Dennel and colleagues (2011) have suggested that core populations on the order of 3000–5000 individuals during interstadials and 1500–2500 individuals during glacial periods were required to maintain *H. heidelbergensis* and early Neandertal populations during the Middle Pleistocene. These core populations may have been geographically restricted (possibly heavily concentrated in the Iberian, Apennine, and Balkan peninsulas), but if they were supporting (through migration) population sinks in other parts of Europe, the overall population density would have been quite low.

While a relatively small-sized metapopulation and low overall population density may have characterized most of the Neandertals' occupation of Europe and western Asia, there are indicators of population peaks and potentially high (relatively) population densities in parts of Europe during MIS 3, and we will explore this issue in greater detail in the next chapter. Before closing our consideration of population size, however, I would like to once more return to the issue of competitive interactions within the carnivore guild. One of the central tenets of this book has been that exploitation and (especially) interference competition with other carnivores was a major limiter of Neandertal population growth. Exploitation competition served to reduce the overall carrying capacity, negatively impacting the ability of Neandertals to capture calories and thus negatively impacting their fertility. Interference competition also likely reduced their net calorie capture (through loss of carcasses in aggressive encounters) and was more than likely a significant source of mortality, both of which also would have had negative demographic consequences. As covered in Chapter 9.4, this argument is predicated on the likelihood that Neandertals were not the behaviorally-dominant carniviores in the guild. There is little doubt that many of my colleagues will find this suggestion unpalatable, under the idea that packs of spear-wielding, muscular, and ill-tempered Neandertals would quickly find themselves the apex predators of any ecosystem. In response, I propose we examine the flip side of the population density issue – if the Neandertals were the socially-dominant, apex predators of Pleistocene Eurasia, *why were there seemingly so few of them?* In modern subSaharan African ecosystems in which lions are socially-dominant over hyenas, wild dogs, cheetahs, and leopards, lions constitute almost half of the biomass at their trophic level (based on body size times observed density in Table 6.3). If the Neandertals made up 50% of the secondary consumer biomass, we would expect (on the basis of biomass estimates in Chapter 9.5) population densities on the order of 0.03–0.10 individuals km^{-2} in steppic conditions and 0.04–0.35 individuals km^{-2} during interstadials. Even if the Neandertals only constituted 25% of the biomass at their trophic level, they would still have population densities one to two orders of magnitude larger than estimated on archeological grounds (Richter 2008) – densities that apparently were not obtained until the Upper Paleolithic (Bocquet-Appel et al. 2005; Mellars and French 2011). The fossil record of Middle and Upper Pleistocene Europe also suggests that cave lions existed at much higher biomass densities than did Neandertals. Rich cave lion assemblages have been recovered from sites in Spain, France, Italy, Germany, Poland, the Czech Republic, and Russia (Turner 1984; García et al. 1997; García and Arsuaga 1998; Mussi and Palombo 2001; Sotnikova and Nikolskiy 2006; Diedrich 2007, 2009a, 2009b, 2011a, 2011b, 2011c; Diedrich and Rathgeber 2012), and the odd one or two cave lions commonly appear on faunal lists from Middle Paleolithic sites across Europe. For example, Mussi (2001b) noted that 19 Middle Paleolithic sites in western and northeastern Italy had produced cave lion remains, whereas only eight had produced Neandertal remains. While it is unclear exactly how many individual lions are represented in the record, a rough estimate from available fossils suggests they may have been 4–5 more numerous than the Neandertals (Lars Werdelin, personal communication). Given that the body mass of adult cave lions was, on average, probably about three times that of an adult Neandertal, this suggests that they represented a much greater portion of the carnivore biomass than did the Neandertals.

NOTES

1. Much of the discussion in this chapter of energetics as it relates to reproduction is drawn from Froehle et al. 2013, and owes much to the intellectual work of my colleague Andy Froehle.
2. Infants (birth to one year of age) are underrepresented in the Neandertal fossil record (comprising only 4.9% of the total sample: Trinkaus 1995a), which is likely due to taphonomic factors. Attempts to correct taphonomic bias suggest a more appropriate figure of 22.5% infant mortality, which is consistent with values from recent foraging and horticultural groups (Trinkaus 1995a).
3. The minimum estimate derives from a BMR of 1400 kcal d^{-1}, PAL = 1.8, milk production costs (770 kcal d^{-1}) to satisfy the energetic needs of a neonate (see Table 4.8), and TEE for a child aged 3-10 (= 2200 kcal d^{-1}, based on estimated TEE in modern arctic children: Table 4.8). The maximum estimate assumes BMR = 1950 kcal d^{-1}, PAL = 1.8, milk production costs (1394 kcal d^{-1}) to satisfy the energetic needs of a 1-3 year old child (see Table 4.8), and TEE for a male child aged 10-15 (= 2800 kcal d^{-1}: Table 4.8).

CHAPTER TWELVE

From Thin to Thick: The African MSA

Fossils of African Middle Pleistocene archaic humans (variously attributed to *H. heidelbergensis* or *H. rhodesiensis*), the presumed source population from which modern humans arose, are not abundant (Rightmire 1998, 2007; McBrearty and Brooks 2000). Middle Pleistocene hominins are somewhat better represented in Europe (thanks in large part to the spectacular fossils that have been recovered from the Sierra de Atapuerca in Spain), which is surprising given the vast difference in land surface area between Africa (ca. 30 million km^2) and Europe (ca. 10 million km^2). This difference in relative abundance may be attributable to a number of factors, including differences in the intensity of site discovery, geological differences that favor preservation and discovery of European fossils, and possibly differences in treatment of the dead by hominins living in these two regions (McBrearty and Brooks 2000). Nonetheless, the relatively poor representation of African *H. heidelbergensis* suggests to me that, like their European counterparts, they also lived at relatively low population densities. This inference appears to be supported by the relative paucity of early MSA sites (McBrearty

Thin on the Ground: Neandertal Biology, Archeology, and Ecology, First Edition. Steven Emilio Churchill.

and Brooks 2000; McBrearty and Tryon 2006),[1] as well as by faunal indicators which suggest that pressure on slow-growing and presumably staple food items (such as tortoises and mollusks) was relatively light during the MSA (Klein and Cruz-Uribe 1983; Steele and Klein 2005). This raises the question: were the factors that conspired to keep Neandertals (and their ancestors) at low population densities also operating on contemporaneous hominin populations in other parts of the Old World?

Certainly the energetic ecology of these archaic humans, and the nature of their competitive interactions with their carnivore contemporaries, differed in important ways from those of their European counterparts. Still, factors similar to those that limited population density in Pleistocene Eurasian hominins may have also been at play in Africa. First, Middle Pleistocene archaic humans were large-bodied (Grine et al. 1995; Ruff et al. 1997; Kappelman 1997; Churchill et al. 2012), and thus growth, somatic maintenance, and reproduction were likely to have been relatively energetically costly for them. Middle Pleistocene fossils from Namibia (Berg Aukas: Grine et al. 1995) and Zambia (Kabwe: Pycraft 1928) suggest that archaic Africans were relatively wide-bodied and thus, like Neandertals, relatively massive for their stature. Head diameter in the proximal femur from Berg Aukas suggests a body mass on the order of about 93 kg (Grine et al. 1995), which is more than seven standard deviations above the grand mean of ten, non-pygmy, subSaharan African male samples (57.7 ± 4.8; data from Ruff 1994). These same African samples have a mean mass/stature ratio of 0.35 (± 0.02: Ruff 1994), such that if the individual from Berg Aukas had a similar body shape he (or she?) would have had to have been about 266 cm (almost 9 feet) tall! Estimates of the length of the reconstructed femur vary from 480 mm (Trinkaus et al. 1999b) to 518 mm (Grine et al. 1995), which would be consistent with a stature in the range of 173–188 cm (following the method of Feldesman and Fountain 1996). These stature estimates would in turn suggest mass/stature ratios on the order of 0.49–0.53, which are great relative even to Neandertal male values (which range from 0.41 to 0.52: Table 2.6). Postcranial remains from Kabwe suggest a similar, relatively massive physique for these African Middle Pleistocene humans (see Churchill et al. 2012). While the adaptive advantages of this kind of body form may have been different for European and African archaic humans (having less to do with thermoregulation and more to do with megafaunal community ecology for the African groups: Churchill et al. 2012), the energetic realities would have been similar, and tight energy budgets may likewise have been a constraint on fertility for African archaic humans. The competitive dynamics with sympatric carnivores was, on the other hand, likely to have been very different in Africa than it was in Europe. In tropical and subtropical environments, exploitation competition with carnivores is considerably less intense, since humans there derive the majority of their calories from plant foods (Kelly 1995). Accordingly, interference competition from carnivores was also likely much less problematic for African archaic humans (and may explain why a humanly-mediated extinction of large-bodied carnivores never occurred in Africa like it did in Eurasia). Still, carnivores were probably a significant source of mortality for African archaic humans (see Chapter 9.4), and predation probably served to increase the fertility rates that were required to maintain steady population sizes. Thus, while the situation in Middle Pleistocene Africa may not have been as severe as it was for the energy-strapped, relatively meat-dependent Neandertals, the combination of high energy demands and aggressive interactions with carnivores may have been limiting factors on population growth for both groups. By late Middle Pleistocene times modern humans had appeared in Africa, and by the Late Pleistocene they had clearly overcome whatever factors had been limiting earlier population growth. Neandertal populations in some places may have also, at times, reached appreciable densities. What conditions may have allowed these groups to grow to larger sizes?

12.1 TIPPING THE SCALES ON POPULATION GROWTH

Neandertal metapopulation size and population density appear to have been variable across the Late Pleistocene, as one would expect of a human population dealing with climatic and ecological oscillations as glacial episodes waxed and waned. While the abundance of archeological sites and the density of artifacts contained therein (van Andel et al. 2003; Mellars and French 2011), as well as an apparent lack of intensity of exploitation of food resources that are sensitive to over-harvesting (Stiner et al. 1999, 2000), indicate that Neandertal population densities never reached the levels seen in the European Upper Paleolithic, they do appear at times to have reached levels sufficient for a modicum of cumulative technological evolution, and to encourage symbolic expression (see Section 12.2). These apparent upturns in population density seem to be largely related to climate. Using the number of dated Micoquian and Mousterian levels per thousand years (based on 316 dated levels from 103 sites distributed across Europe), van Andel and colleagues (2003) found consistently low representation of the Middle Paleolithic (and thus presumably the Neandertals) from MIS 5c (the upper age of the sites in their sample) until the beginning of MIS 3. With the end of the glacial MIS 4 episode and the onset of ameliorated conditions in MIS 3 at ca. 60 Ka BP, the temporal density of sites increases from about 1–2 sites Ka^{-1} to a peak of about 7.5 sites Ka^{-1} at around 55 Ka BP (van Andel et al. 2003). The interval between 59 and 43 Ka BP sees both an increased abundance of sites in regions to the south that had remained occupied (at lower population densities) during MIS 4, and an expansion of sites into areas above 45° N latitude. This interval of population growth is punctuated, however, by a brief period of decline (to about four sites Ka^{-1}) during a short episode of cold–dry steppic conditions between 53–49 Ka BP (van Andel et al. 2003), before climbing again to roughly eight sites Ka^{-1} at 45 Ka BP. Across the remainder of MIS 3, Middle Paleolithic sites gradually decrease in abundance (ultimately reaching the low values seen prior to MIS 3) towards the last glacial maximum. This picture of increased population density during MIS 3 is consistent with faunal indicators (based on the intensity of exploitation of small game) which suggest that, in the eastern Mediterranean region, the first major population growth pulse occurred before the end of the Middle Paleolithic (Stiner et al. 2000). Using the same Middle Paleolithic site age data as van Andel et al. (2003), Stringer et al. (2003) found a similar pattern of site increase during MIS 3 (the peaks are slightly different as it appears they used 5000-year rather than 1000-year bins for the dates), and also report a slight increase in sites during MIS 5a. Stringer and colleagues argue that increases in relative site abundance occurred during periods of low climatic stress, which was partly a function of temperature but which was also greatly influenced by the degree of climatic stability (and thus the amount of evolutionary time that groups had to adapt to a certain set of environmental and ecological conditions). Lahr and Foley (2003) likewise detected some evidence to suggest that

Neandertal population sizes were responding more to climatic variability than climatic conditions per se. Climatic instability has similarly been argued to have been a major check on human population growth in Middle Pleistocene Europe (Dennell et al. 2011). The first part of MIS 3 (from about 60–40 Ka BP) appears to have been somewhat more stable than the end of the interval, and Mousterian site abundance peaks at around 40 Ka BP (Stringer et al. 2003; Lahr and Foley 2003). Neandertals living during this interval probably benefited from interpleniglacial conditions that had greater productivity than glacial episodes (but without the large-scale expansion of broadleaf forest that occurred during full interglacial episodes), which may have increased the plant food items available to them. The greater productivity overall would have increased the carrying capacity (relaxing slightly the energetic constraints on Neandertal population growth), while a greater availability of plant foods would have relaxed somewhat the exploitative and interference competition they faced from other carnivores, allowing for moderate gains in population density.

The role of climate as an agent of control over early modern human demography in the African MSA is more questionable. Deciphering the part played by climate in the expansion and contraction of early modern human populations is complicated by the fact that Africa is an enormous continent, and Pleistocene climate change was largely asynchronous in different areas (Blome et al. 2012). Sediment cores from lakes Malawi, Tanganyika, and Bosumtwi reveal long-lasting periods of intense aridity (megadroughts) in subSaharan Africa between about 135–75 Ka BP (Scholz et al. 2007; Burnett et al. 2011), while geomorphological evidence of desert expansion and dune formation in west and central Africa suggest possible arid advances at 250 Ka BP, 190–155 Ka BP, and 120–90 Ka BP (Nichol 1999). Arid-interval expansion of grasslands and deserts would promote ecosystems that were presumably not favorable to human occupation (Cohen et al. 2007; Blome et al. 2012), resulting in contraction of early modern human populations into wetter refugia (Basell 2008). A major Late Pleistocene human population bottleneck, by one estimate involving an effective population size as small as 600 (Fagundes et al. 2007), has been inferred from studies of modern human genetic variation (Haigh and Maynard Smith 1972; Rogers and Jorde 1995; Harpending et al. 1998; Fagundes et al. 2007: but see Hawks et al. 2000b; Li and Durbin 2011). This bottleneck might thus be attributable to the constriction of environments favorable to humans during the prolonged MIS 6–5 megadrought (Basell 2008), with the onset after 70 Ka BP of wetter conditions over most of Africa stimulating population expansion and migration (Scholz et al. 2007; Cohen et al. 2007). This scenario is consistent with mtDNA mismatch analysis (Harpending et al. 1993; Sherry et al. 1994) and studies of matrilineal trees (Watson et al. 1997; Forster 2004) that indicate a major population expansion in Africa between 80–60 Ka BP. Parallels in the pattern of inferred population growth pulses and migration events, as reflected in human genetic variation, with episodes of climate change seen in the paleoclimatic data have also been used to argue that climate change was a major component of the modern human diaspora and their colonization of Eurasia, Australasia, and the Americas (Parker 2009; Eriksson et al. 2012). However, Pleistocene African demography appears not to have been a simple matter of climatically-driven changes in carrying capacity: meso-scale climatic reconstructions from multiple terrestrial and oceanic proxies indicate that Africa was not uniformly dry in any part of MIS 6, 5, or 4 and, furthermore, MSA site abundance appears to have little to no relationship with aridity over this time period (Blome et al. 2012). It is also the case that the various mtDNA matrilines did not expand synchronously, and that some of these lineages may have subsequently contracted (Behar et al. 2008; Atkinson et al. 2009), which further argues against a simple post-drought climatic model of MSA demography. It also bears noting that the genetic evidence for rapid population growth in the Pleistocene (Harpending et al. 1993, 1998; Sherry et al. 1994; Rogers 1995; Watson et al. 1997; Relethford 1998; Reich and Goldstein 1998; Ingman et al. 2000; Forster 2004; Cox et al. 2009) is not unanimously accepted, since some genetic markers show patterns of variation that are more consistent with uniform population growth across the entire Pleistocene, and selective sweeps at various times may mimic the pattern expected with rapid growth from a small population (Hey 1997; Hawks et al. 2000b; Hawks and Wolpoff 2001a; Li and Durbin 2011). However, patterns of variation across multiple markers (mtDNA, microsatellites, autosomal DNA, X-chromosome SNPs) and across multiple autosomal loci make the selective sweep scenario unlikely (since simultaneous sweeps in hundreds of unlinked genetic regions would be required to produce the observed pattern: Weaver and Roseman 2005). The evidence thus seems to suggest that population densities in Africa were relatively low – perhaps on a par with those of Neandertals – and that hominin groups were distributed as small, isolated populations (Lahr and Foley 2003) through the early part of the MSA, and then population densities increased markedly sometime around MIS 4. While it has been suggested that unfavorable climatic and environmental conditions in north, west, and east Africa during MIS 5 impelled modern humans to migrate out of Africa (Carto et al. 2009), the emerging consensus seems to be that modern human expansion into Eurasia and beyond was demographically rather than ecologically driven (e.g., see Eswaran 2002 and comments therein). This view is certainly consistent with archeological and genetic evidence for greater population sizes and densities in the African LSA and Eurasian UP (Stiner et al. 1999; Cox et al. 2009; Steele and Klein 2008, 2009; Mellars and French 2011).

I suspect that the demographic expansion of modern humans during the later MSA was brought about by multiple, interrelated events and developments. Wetter conditions in parts of Africa during MIS 4 may have been part of the story, but appear to only be a part. The development of true long-range projectile weapon systems may also be an important part of the story. While the evidence seems to suggest that Neandertals were never heavily invested in projectile weapon technology (Chapter 3.4), there are indications that, by later MSA times, modern humans were. The size and shape of MSA and MP points suggest an origin and diffusion of projectile weapons between 100–50 Ka BP in Africa (McBrearty and Brooks 2000; Brooks et al. 2005; Shea 2009: Lombard and Phillipson 2010), with movement of this technology, in the hands of modern human colonizers, into the Near East by about 50–40 Ka BP (Shea 2006). There are a number of ways that the adoption of projectile weapon systems would be expected to positively impact demographic growth. Projectile weapons improve access to arboreal, volant, and aquatic prey, and can greatly reduce the handling costs of low-ranked prey items or make available prey that otherwise is out of reach. Projectiles also allowed hunting in new places (such as open country) with the use of new tactics, and thus lowered the search and handling costs of higher ranked prey (Churchill 1993; Marlowe 2005; O'Connell 2006). These weapons thus change the economics of prey choice (see Churchill 1993; Shott 1993; Kelly 1995; Churchill and Morris 1998) and positively alter the carrying capacity by facilitating diet breadth expansion (see Stiner et al.

2000; Stiner 2001). Projectile weapons likely also served to reduce mortality in early modern humans, by reducing starvation risk, by reducing the injury risks associated with hunting larger prey items and, perhaps most importantly, by altering the dynamics of interference competition within the carnivore guild (see Richerson et al. 2009). Once MSA humans were armed with spearthrower-propelled darts capable of being launched from 40–50 m, even large-bodied social carnivores may have become increasingly reluctant to prey upon hominin social groups or challenge them for carcasses, or may have become less reticent to abandon a carcass when confronted by a human foraging party. Projectile technology-mediated reductions in adult mortality might be part of the demographic shift that occurred between the Middle and Upper Paleolithic, in which older individuals become much more abundant in human fossil assemblages (Caspari and Lee 2004: but see Trinkaus 2011b). Greater adult survivorship in turn lowers the average dependency ratios within social groups and enhances opportunities for intergenerational provisioning and caretaking of dependent young (i.e., grandmothering: Hawkes et al. 1998), which would be expected to further enhance fertility in these groups. This single class of subsistence tools, then, very likely positively impacted both fertility – by improving the net caloric return efficiency, and by increasing the number of catchable prey species which increases the effective carrying capacity, both of which make more calories available for reproduction – and survivorship, and likely had a major impact on early modern human demography. Even small improvements in any of these parameters would, over time, have had a significant effect on population growth, and would have also been a factor in the competitive exclusion of Neandertals by early modern humans (Shea 2003b, 2005).

Before looking at other factors that might have served to enhance later MSA human population growth, it bears asking why, if projectile weapons gave early modern humans a competitive advantage, the Neandertals didn't just adopt this technology when confronted with competition from modern human colonists? Indeed, there's nothing to suggest that they didn't, and there's even limited evidence to suggest that they did. Châtelperronian points compare favorably in cross-sectional area to ethnohistorically-known dart points and to presumed projectile points from later Upper Paleolithic assemblages, and tend to exhibit impact damage characteristic of use as projectile armatures (Shea 2006). Again, however, it is not currently clear whether the Neandertals produced the Châtelperronian (see Chapter 3.1), and this issue needs to be resolved before we can address the possibility of dart-throwing Neandertals. Still, I suspect that even if the Neandertals added projectile weapons to their arsenal, it probably did not provide enough of an advantage to counter the demographic and energetic advantage possessed by modern humans (see below).

A second potentially demographically-important change in the later MSA was the development of extended social networks. Recent foragers are known to store calories in social bonds: when hard times strike they often rely on the generosity of neighboring groups for whom conditions are better (Cashdan 1985; Stein Mandryk 1993; Wiessner 2002; Whallon 2006). "Social storage" is clearly an important mechanism for managing risk in hunter-gatherer economic systems, but it is not the only benefit that accrues from extended social networks. Such networks no doubt facilitated exogamy, and probably also provided the potential to form coalitions to manage local carnivores (see Chapter 9.4). All of these benefits would have either increased fertility or reduced mortality, and fostered population growth. The initiation of extended networks may have revolved around the first establishment of stable pairbonds, which allows individuals to recognize a greater range of kin (through their fathers and his relatives) (Chapais 2010) and allows the establishment of a system of bilateral descent, which in turn facilitates multilocal residence and maximizes the number of kin ties across camps (Marlowe 2005). With these social developments would have emerged the incipient notion of a social environment that extended beyond the immediate band, that is, the nascent concept of the tribe (Chapais 2010). It is not clear when in human evolutionary history this sort of social organization first emerged, but there are indicators that the social world was becoming more complex in the later part of the MSA. Items of personal adornment, in the form of pierced shell beads, appear sporadically in MSA deposits ranging from MIS 5a–3 (McBrearty and Brooks 2000; Bouzouggar et al. 2007; Henshilwood et al. 2004; d'Errico et al. 2005), and are a common component of LSA and UP artifact assemblages (R White 1992; Nowell 2006). Pigments such as ochre (iron oxide) and manganese dioxide, which appear even in early MSA contexts (McBrearty and Brooks 2000), may also have been applied to skin or clothing items as a form of personal adornment, although these pigments also have utilitarian purposes (e.g., for hide preservation or as a loading agent in the production of hafting glues: Mellars 1996; Wadley 2005). Engraved artifacts also begin to appear towards the end of the MSA (Henshilwood et al. 2002, 2009; Texier et al. 2010). Beads and other items of personal adornment are social signals that can impart information about group identity, social status, or individual identity (R White 1992). Among foragers that live in relatively small bands, this information is generally known to the members of one's own social group, and thus the advent of personal adornment may signal an increased frequency of contact with out-group members in a more socially complex world, leading Kuhn and Stiner (2007: 40) to argue that their appearance in the Pleistocene reflects "expanding scales of social interaction and more complex social landscapes resulting from unprecedentedly large and internally differentiated human populations." Items of personal adornment, as well as engraved items and other forms of mobilary art, may also function as trade items that serve to strengthen and maintain social bonds across groups (as, for example, such items function in the *hxaro* exchange network of southern African foragers: Wiessner 2002). The Upper Paleolithic populations of Eurasia, in contrast to those of the Middle Paleolithic, appear to have been denser and more interconnected (Stiner and Kuhn 2006), which is reflected in long-distance exchange networks (Bar-Yosef 2002) and shared participation in systems of representational art across vast expanses of space (venus figurines, although diverse in their form and style [Nelson 1990], are found in Gravettian assemblages from the Mediterranean to Siberia [Beck, 2000 #19351}]). These extended social networks, which allow for the rapid diffusion of technological innovations, the distribution of economic risk, and the formation of coalitions to deal with ecological (e.g., carnivore competitors) or social (i.e., other tribes) problems, were likely of major adaptive significance to early modern humans expanding into Eurasia, and may have made the difference in their competitive encounters with the Neandertals (Horan et al. 2005).

Change in body size represents another factor that might have had important demographic consequences for MSA populations. As noted above, archaic humans (*H. heidelbergensis*) associated with the early MSA were large-bodied: the MIS 7–6 emergence of *H. sapiens* appears to have been associated with a substantial reduction in body size (Churchill et al. 2012). While Middle Pleistocene archaic humans appear to have been considerably larger than living, non-pygmy subSaharan Africans (see above), MSA-associated early modern humans do not appear to have differed significantly in body size from living comparators (Churchill et al. 2012). The evolutionary causes of this size reduction are unclear. It has been claimed that the modern human life history attribute of slow growth appeared first in MSA early modern humans (Smith et al. 2007a: though not everyone would agree; see Chapter 11.2), which is consistent with patterns of growth rate variation across small-scale human societies (in which growth is slower in groups of smaller average body size: Walker et al. 2006). This in turn suggests that mortality risk was reduced in early modern humans relative to their ancestors, since the opportunity costs of delaying reproduction are not offset if the probability of surviving to reproductive age is low or the reproductive lifespan is short (Charnov 1990, 1991). Regardless of the evolutionary causes of body size reduction, there is no doubt that it would have greatly reduced the energetic demands of somatic maintenance, activity, and reproduction in early *H. sapiens*. Differences in the energetics of growing, maintaining, and moving larger versus smaller bodies were no doubt a significant factor in population growth rate differences between archaic and modern humans (Froehle et al. 2013).

12.2 CULTURE CHANGE IN THE LATE MSA AND MOUSTERIAN

The later part of the MSA was marked by the sporadic emergence, and rapid evaporation, of innovations in technology or symbolic expression. Marine shells with evidence of having been strung, for example, appear at 92 Ka BP in the Near East (Bar-Yosef Mayer et al. 2009), again at 82 Ka BP in north Africa (Bouzouggar et al. 2007), and again at ca. 72–71 Ka BP in southern Africa (Henshilwood et al. 2004; d'Errico et al. 2008: date from Jacobs et al. 2008). With the possible exception of a few beads from potentially 100–135 Ka BP sediments at Skhul (d'Errico and Vanhaeren 2009), these limited occurrences represent the totality of the evidence for beads used as items of personal adornment prior to 40 Ka BP, at which time beads become relatively common in LSA and UP sites in Africa, the Near East, and Europe (d'Errico and Vanhaeren 2009). Likewise, finely made, bifacially-worked lanceolate points made a brief appearance (along with some of the shell beads mentioned above, as well as bone tools and engraved ochre: Henshilwood et al. 2001, 2004, 2009) in the Still Bay industry of southern Africa, but only for about 1000 years (ca. 72–71 Ka BP: Jacobs et al. 2008). Leptolithic and microlithic technology also made a brief debut in southern Africa at ca. 65 Ka BP with the Howieson's Poort industry, which disappeared some 5000 years later (Jacobs et al. 2008). As discussed above, these and other short-lived occurrences of technological innovation and symbolic behavior probably correspond with temporary increases in population density, in which groups became both larger and more interconnected and CTE took off (Shennan 2001; Powell et al. 2009). However, whatever behaviorally-adaptive strides were made during these episodes, they were apparently insufficient to prevent population losses as local environmental conditions worsened, and with population crashes came cultural loss (see Henrich 2004). A parallel process of population density-mediated increase in cultural complexity appears to have characterized the end of the European Middle Paleolithic.

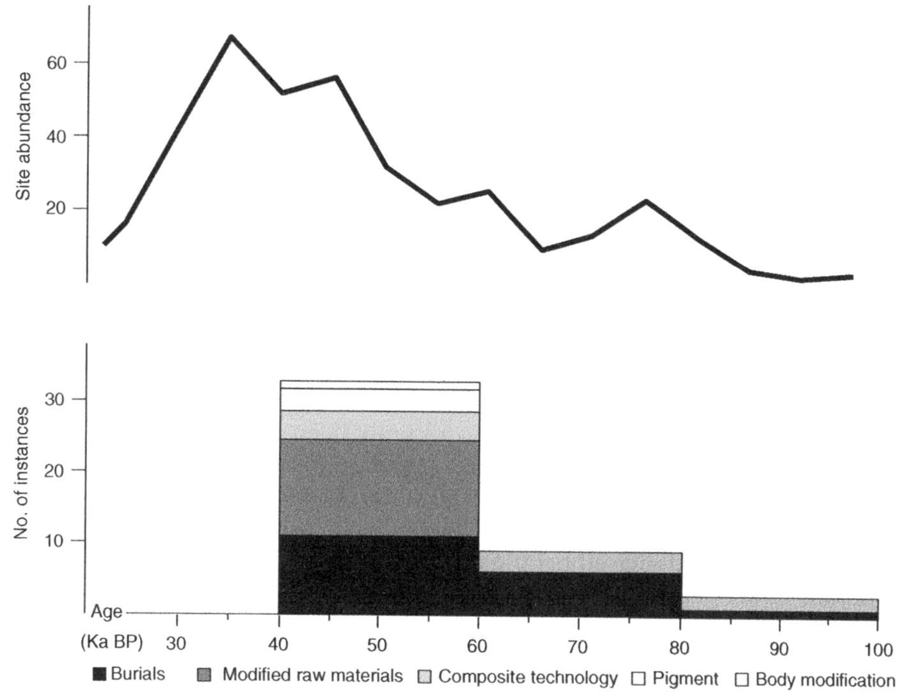

FIGURE 12.1

Neandertal population density and CTE: (top) number of Mousterian sites and Neandertal remains per 5 Ka from late MIS 5 to MIS 3; (bottom) number of instances of cultural complexity and symbolic behavior associated with Neandertals, by 20 Ka intervals, between 100–40 Ka BP. Site abundance data based on Figure 13.5 in Stringer et al. 2003; instances of cultural complexity and symbolism from Figure 2 in Langley et al. 2008.

Although increased cultural complexity in the terminal Mousterian was not as marked as that of the late MSA to LSA/UP transition, there does appear to have been a noticeable increase in innovation and symbolic behavior by the Neandertals in MIS 3, coincident with an increase in site abundance (Figure 12.1). There are a few instances of potentially symbolic behavior that antedate MIS 3, but they are relatively few (assuming one is careful to discount claims that are impossible to verify on current evidence, or that have been shown after initial publication to be the product of non-human taphonomic factors: Langley et al. 2008). These include the burial of the adult female Tabun C1 (Israel), which may date to MIS 6 or 5e (although the stratigraphic position of this specimen is contentious, and it may date to MIS 3 times: see Chapters 2.6 and 3.1) and multiple burials from the sites of La Ferrassie, La Quina and Roc de Marsal (France), Shanidar (Iraq), Kiik-Koba (Crimea), and Teshik Tash (Uzbekistan) that date to late MIS 5 and MIS 4 (Langley et al. 2008). It must be noted, however, that the act of burial does not necessarily imply ritual or symbolic behavior (see Hayden 1993), and the intentional nature of Neandertal burials has itself been challenged (Gargett 1989,

1999). Despite it being a fairly ancient practice for Neandertals, the frequency of burial does seem to pick up during MIS 3, at which time it tends to be localized in the areas of highest site density (Aquitaine and the Levant), suggesting a connection between this behavior and population size (Pettitt 2010). Otherwise, good cases that might signal increased behavioral complexity are pretty rare before MIS 3, and (following Langley et al. 2008) appear to be limited to an inscribed fossil nummulite from the site of Tata (Hungary: Bednarik 1992), inscribed bone artifacts from Taubach (Germany: Gaudzinski 2004), a piece of notched bone from La Ferrassie (France: Bednarik 1992), perforated bone and wolf tooth from Repolusthöhle (Austria: Bednarik 1992), and cut-marked raptor claws from Combe Grenal (France: Morin and Laroulandie 2012). Pigment use thus far appears to be limited to MIS 3 times, but seems to have been practiced quite widely by the Neandertals of the Interpleniglacial (Mellars 1996; Cârciumaru et al. 2002; d'Errico 2003; Soressi and d'Errico 2007). Evidence for personal adornment rivaling that seen in the later MSA has also recently been found in MIS 3 Mousterian contexts, including perforated and pigment-stained marine shells from Iberia (Zilhão et al. 2010) and the processing of non-food birds for their claws and feathers in France and Italy (Peresani et al. 2011; Morin and Laroulandie 2012). Simple mobilary art (in the form of a pecked and incised pebble in Spain [García-Diez et al. 2013] and a stone with multiple subparallel incisions from Bulgaria [Bednarik 1992]) likewise makes an appearance in MIS 3. Innovations in technology are also evident at this time, including the use of bone points (Gaudzinski 1999) and *lissoirs* (Soressi et al. 2013). Given the preponderance of evidence, it is hard to escape the conclusion that the Neandertals were fully capable, cognitively, of behaving in symbolic ways (d'Errico 2003), but that the conditions (demographic and social) necessary for the expression of this behavior were not in place until the end of the Mousterian.

An interesting question emerges from this perspective on culture change and demography towards the end of the Mousterian, concerning the metapopulation size and local population densities of the Neandertals at the time of contact with early modern human immigrants. The central premise of this book has been that the Neandertals' hold on Europe was always rather tenuous – that is, that low population density was the norm – and thus that they were vulnerable to extinction long before modern humans arrived in their lands. Yet numbers of Mousterian sites, along with the indirect evidence of CTE, strongly suggest that the Neandertals were not only doing well in some parts of Europe during MIS 3, but had attained population densities sufficient to promote the kinds of behavior changes that were elsewhere driving the MSA to LSA/UP transition. While the issue of who made the Initial Upper Paleolithic industries of Europe (the Châtelperronian, Ulluzian, Olschewian, etc: see Chapter 3.1) is currently unresolved, there is a possibility that, in places, the Neandertals had independently developed an Upper Paleolithic way of life before they ever encountered modern humans (d'Errico et al. 1998; d'Errico 2003; Zilhão 2006). Regardless of whether the Neandertals had autonomously invented the Upper Paleolithic, the site abundance data and evidence for accelerated cultural change during MIS 3 would suggest that when modern humans entered Europe they would have encountered what was likely the historically largest-ever Neandertal population, and one that was best positioned to adaptively respond to the competitive challenges that modern humans brought to them.

This question (and the related question of Neandertal–IUP associations) deserves attention. I suspect that the MIS 3 Mousterian was much like the MSA during MIS 5a and 4, in that local increases in population density (sufficient to stimulate CTE) were not sustained, and population declines and culture loss followed episodes of higher density. When examined in 1000 year intervals, Mousterian site abundance is highly variable across MIS 3, with marked declines during six intervals, the longest being between 54–49 Ka BP (Lahr and Foley 2003). There is an overall trend for decline in site numbers after 37 Ka BP, which does not appear to be explained by the presence of modern humans (as judged by abundance of Upper Paleolithic sites) (Lahr and Foley 2003). This suggests to me that, despite intervals of relatively high population densities, the Neandertals had not yet really tipped the balance on population growth, and that their metapopulation size was likely small at the time of first contact with the Cro-magnon.

At the time of Neandertal–modern human contact, modern human metapopulation sizes were growing, and cultural evolution was accelerating. Cultural ratcheting is an emergent property of population density: with greater density comes an absolutely greater number of innovators, along with conditions that favor dissemination of cultural traits within and between groups (Shennan 2001; Powell et al. 2009; Richerson et al. 2009). But increased population density in turn creates resource stress, which places a premium on innovations in extractive technology that reduce the handling costs of low ranked food items and facilitate the expansion of diet breadth (Stiner et al. 2000), leading to a population-technology-population positive feedback loop (Richerson et al. 2009). Thus it is not surprising that most of the technological innovations that emerged in the LSA and UP involved subsistence tools: spear throwers and darts, bows and arrows, fish gorges, leisters, weirs, harpoons, nets, traps, and grindstones (Julien 1982; Tyldesley and Bahn 1983; Straus 1990, 1993; Bergman 1993; Holliday and Churchill 2006). The population-technology-population feedback system appears to have been engaged by the end of the MSA, and was driving the expansion of modern humans out of Africa and towards their encounter with the Neandertals. Given a high degree of interconnectedness between social groups, and a subsistence toolkit that included projectile weapons, modern humans were able to effectively deal with the resource competitors they encountered as they grew into new lands. With higher fertility rates and lower mortality than the Neandertals, their eventual annexation of Eurasia was probably inevitable. The Neandertals, in turn, were largely victims of their own biology: saddled with relatively massive, energetically costly bodies, it is unlikely that they could have managed the fertility rates necessary to avoid being swamped by modern humans even if they had managed to attain parity in all other aspects of their ecology. While there is still much active debate about why the Neandertals went extinct, I have to believe that modern humans were a large part of the story and, therefore, that the Neandertal story is part of our story.

NOTE

1. As with the human fossil record from this time, the intensity of discovery across the large African landmass may be biasing the representation of early MSA sites. As noted by McBrearty and Brooks (2000: 487), "Africa is vast, researchers are few, and research history is short."

References

Abbott S, Trinkaus E, Burr DB (1996) Dynamic bone remodeling in later Pleistocene fossil hominids. Am. J. Phys. Anthropol. 99:585–601.

Abuabara A (2006) A review of facial injuries due to dog bites. Med. Oral Patol. Oral Cir. Bucal 11:348–350.

Adler DS, Tushabramishvili N (2004) Middle Palaeolithic patterns of settlement and subsistence in the southern Caucasus. In: Conard N (ed) Settlement Dynamics of the Middle Paleolithic and Middle Stone Age, Volume II. Kerns Verlag, Tübingen, pp 91–132.

Adler DS, Bar-Oz G, Belfer-Cohen A, Bar-Yosef O (2006a) Ahead of the game: Middle and Upper Palaeolithic hunting behaviors in the southern Caucasus. Curr. Anthropol. 47:89–118.

Adler DS, Belfer-Cohen A, Bar-Yosef O (2006b) Between a rock and a hard place: Neanderthal-modern human interactions in the southern Caucasus. In: Conard NJ (ed) Neanderthals and Modern Humans Met. Kerns Verlag, Tübingen, pp 165–187.

Agustí J, Antón M (2002) Mammoths, Sabertooths, and Hominids: 65 Million Years of Mammalian Evolution in Europe. Columbia University Press, New York.

Aiello LC, Dean MC (1990) An Introduction to Human Evolutionary Anatomy. Academic Press, London.

Thin on the Ground: Neandertal Biology, Archeology, and Ecology, First Edition. Steven Emilio Churchill.

Aiello LC, Key C (2002) Energetic consequences of being a *Homo erectus* female. Am. J. Hum. Biol. 14:551–565.

Aiello LC, Wheeler P (1995) The expensive-tissue hypothesis. Curr. Anthropol. 36:199–221.

Aiello LC, Wheeler P (2003) Neanderthal thermoregulation and the glacial climate. In: van Andel T, Davies W (eds) Neanderthals and Modern Humans in the European Landscape During the Last Glaciation. McDonald Institute for Archaeological Research, Cambridge, pp 147–166.

Ainsworth BE, Haskell WL, Whitt MC, Irwin ML, Swartz AM, Strath SJ, O'Brien WL, Bassett DRJ, Schmitz KH, Emplaincourt PO, Jacobs DRJ, Leon AS (2000) Compendium of physical activities: an update of activity codes and MET intensities. Med. Sci. Sports Exerc. 32:S498-S516.

Aitken MJ, Stringer CB, Mellars PA (eds) (1992) The Origin of Modern Humans and the Impact of Chronometric Dating. Royal Society, London.

Akazawa T, Muhesen S, Dodo Y, Kondo O, Mizoguchi Y, Abe Y, Nishiaki Y, Ohta S, Oguchi T, Haydal J (1995) Neanderthal infant burial from the Dederiyeh cave in Syria. Paléorient 21:77–86.

Albert RM, Weiner S, Bar-Yosef O, Meignen L (2000) Phytoliths in the Middle Palaeolithic deposits of Kebara Cave, Mt Carmel, Israel: study of the plant materials used for fuel and other purposes. J. Archaeol. Sci. 27:931–947.

Alexander RM (1985a) The maximum forces exerted by animals. J. Exp. Biol. 115:231–238.

Alexander RM (1985b) Mechanics of posture and gait of some large dinosaurs. Zool. J. Linn. Soc. 83:1–25.

Alexander RM, Jayes AS, Maloiy GMO, Wathuta EM (1979) Allometry of limb bones of mammals from shrews (*Sorex*) to elephant (*Loxodonta*). J. Zool. 189:305–314.

Allen JA (1877) The influence of physical conditions in the genesis of species. Rad. Rev. 1:108–140.

Álvarez-Lao D, García-García N (2006) A new site from the Spanish Middle Pleistocene with cold-resistant faunal elements: La Parte (Asturias, Spain). Quatern. Int. 142–143:107–118.

Amati D, Shallice T (2007) On the emergence of modern humans. Cognition 103:358–385.

Ambrose SH (1998a) Late Pleistocene human population bottlenecks, volcanic winter, and differentiation of modern humans. J. Hum. Evol. 34:623–651.

Ambrose SH (1998b) Prospects for stable isotopic analysis of later Pleistocene hominid diets in west Asia and Europe. In: Akazawa T, Aoki K, Bar-Yosef O (eds) Neandertals and Modern Humans in Western Asia. Plenum, New York, pp 277–289.

Ambrose SH (2001) Paleolithic technology and human evolution. Science 291:1748–1753.

Ambrose SH (2010) Coevolution of composite-tool technology, constructive memory, and language: implications for the evolution of modern human behavior. Curr. Anthropol. 51 Suppl. 1:S135–147.

An K-N, Linscheid RL, Brand PW (1991) Correlation of physiological cross-sectional areas of muscle and tendon. J. Hand Surg. 16B:66–67.

Anderson-Gerfaud P (ed) (1990) Aspects of Behaviour in the Middle Paleolithic: Functional Analysis of Stone Tools from Southwest France. Edinburgh University Press, Edinburgh.

Anderson-Gerfaud P, Helmer D (1987) L'emmanchement au Moustérien In: Stordeur D (ed) La Main et l'Outil. Manches et Emmanchements Préhistoriques. Travaux de la Maison de l'Orient, Lyon, pp 37–54.

Anderson CM (1989) Neandertal pelves and gestation length: hypotheses and holism in paleoanthropology. Am. Anthropol. 19:327–340.

Andrews JT (1983) Short ice age 230,000 years ago? Nature 303:21–22.

Andrews P, Fernández Jalvo Y (1997) Surface modifications of the Sima de los Huesos fossil humans. J. Hum. Evol. 33:191–217.

Andrews P, Stringer C (1989) Human Evolution: An Illustrated Guide. Cambridge University Press, Cambridge.

Antón M, Galobart A, Turner A (2005) Co-existence of scimitar-toothed cats, lions and hominins in the European Pleistocene. Implications of the post-cranial anatomy of *Homotherium latidens* (Owen) for comparative palaeoecology. Quat. Sci. Rev. 24:1287–1301.

Antón SC (1990) Neandertals and the anterior dental loading hypothesis: a biomechanical evaluation of bite force production. Kroeber Anthropol. Soc. Papers 71–72:67–76.

Antón SC (1994) Mechanical and other perspectives on Neandertal craniofacial morphology. In: Corrucini RS, Ciochon RL (eds) Integrative Paths to the Past. Palaeonthropological Advances in Honor of F. Clark Howell. Prentice Hall, Englewood Cliffs, pp 677–695.

Antón SC (1996) Tendon-associated bone features of the masticatory system in Neandertals. J. Hum. Evol. 31:391–408.

Anyonge W (1993) Body mass in large extant and extinct carnivores. J. Zool. 231:339–350.

Anyonge W (1996) Locomotor behaviour in Plio-Pleistocene sabre-toothed cats: a biomechanical analysis. J. Zool. 238:395–413.

Anyonge W, Roman C (2006) New body mass estimates for *Canis dirus*, the extinct Pleistocene dire wolf. J. Vert. Paleontol. 26:209–212.

Arcadi AC (2006) Species resilience in Pleistocene hominids that traveled far and ate widely: an analogy to the wolf-like canids. J. Hum. Evol. 51:383–394.

Arensburg B (2002) Human remains from Geula Cave, Haifa. Bull. Mem. Soc. Anthropol. Paris 14:141–148.

Arensburg B, Belfer-Cohen A (1998) Sapiens and Neandertals: rethinking the Levantine Middle Paleolithic hominids. In: Akazawa T, Aoki K, Bar-Yosef O (eds) Neandertals and Modern Humans in Western Asia. Plenum, New York, pp 311–322.

Arensburg B, Nathan H (1980) A Mousterian third cervical vertebra from Hayonim Cave, Israel. J. Hum. Evol. 9:193–194.

Arensburg B, Bar-Yosef O, Belfer-Cohen A, Rak Y (1990) Mousterian and Aurignacian human remains from Hayonim Cave, Israel. Paléorient 16:107–109.

Armstrong LE, DeLuca JP, Christensen EL, Hubbard RW (1990) Mass-to-surface area index in a large cohort. Am. J. Phys. Anthropol. 83:321–329.

Arribas A, Palmqvist P (1999) On the ecological connection between sabre-tooths and hominids: faunal dispersal events in the lower Pleistocene and a review of the evidence for the first human arrival in Europe. J. Archaeol. Sci. 26:571–585.

Arsuaga J-L, Martinez I, Gracia A, Carretero J-M, Carbonell E (1993) Three new human skulls from the Sima de los Huesos Middle Pleistocene site in the Sierra de Atapuerca, Spain. Nature 362:534–537.

Arsuaga J-L, Martínez I, Gracia A, Lorenzo C (1997a) The Sima de los Huesos crania (Sierra de Atapuerca, Spain). A comparative study. J. Hum. Evol. 33:219–281.

Arsuaga J-L, Carretero JM, Lorenzo C, Gracia A, Martínez I, Bermúdez de Castro M, Carbonell E (1997b) Size variation in Middle Pleistocene humans. Science 277:1086–1088.

Arsuaga J-L, Lorenzo C, Carretero J-M, Gracia A, Martínez I, García N, Bermúdez de Castro J-M, Carbonell E (1999a) A complete human pelvis from the Middle Pleistocene of Spain. Nature 399:255–258.

Arsuaga J-L, Martínez I, Lorenzo C, Gracia A, Muñoz A, Alonso O, Gallego J (1999b) The human cranial remains from Gran Dolina Lower Pleistocene site (Sierra de Atapuerca, Spain). J. Hum. Evol. 37:431–457.

Arsuaga J-L, Villaverde V, Quam R, Martínez I, Carretero JM, Lorenzo C, Gracia A (2007) New Neandertal remains from Cova Negra (Valencia, Spain). J. Hum. Evol. 52:31–58.

Arsuaga J-L, Baquedano E, Pérez-González A, Sala N, Quam RM, Rodríguez L, García R, García N, Álvarez-Lao DJ, Laplana C, Huguet R, Sevilla P, Maldonado E, Blain H-A, Ruiz-Zapata MB, Sala P, Gil-García MJ, Uzquiano P, Pantoja A, Márquez B (2010) Understanding the ancient habitats of the last-interglacial (late MIS 5) Neanderthals of central Iberia: paleoenvironmental and taphonomic evidence from the Cueva del Camino (Spain) site. Quatern. Int. 275: 55–75.

Ascenzi A, Biddittu I, Cassoli PF, Segre AG, Segre-Naldini E (1996) A calvarium of late *Homo erectus* from Ceprano, Italy. J. Hum. Evol. 31:409–423.

Ascenzi A, Mallegni F, Manzi G, Segre AG, Segre Naldini E (2000) A re-appraisal of Ceprano calvaria affinities with *Homo erectus*, after the new reconstruction. J. Hum. Evol. 39:443–450.

Asfaw B, Gilbert WH, Beyene Y, Hart WK, Renne PR, WoldeGabriel G, Vrba ES, White TD (2002) Remains of *Homo erectus* from Bouri, Middle Awash, Ethiopia. Nature 416:317–320.

Ashton KG, Tracy MC, de Queiroz A (2000) Is Bergmann's rule valid for mammals? Am. Nat. 156:390–415.

Ashton NM (2002) Absence of humans in Britain during the last interglacial (Oxygen Isotope Stage 5e). In: Tuffreau A, Roebroeks W (eds) Le Dernier Interglaciaire et les Occupations Humaines du Paléolithique Moyen. Publications du CERP, Lille, pp 93–103.

Atkinson SN, Ramsay MA (1995) The effects of prolonged fasting on the body composition and reproductive success of female polar bears (*Ursus maritimus*). Func. Ecol. 9:559–567.

Atkinson QD, Gray RD, Drummond AJ (2009) Bayesian coalescent inference of major human mitochondrial DNA haplogroup expansions in Africa. Proc. R. Soc. B 276:367–373.

Auerbach BM, Ruff CB (2004) Human body mass estimation: a comparison of "morphometric" and "mechanical" methods. Am. J. Phys. Anthropol. 125:331–342.

Auger F, Jamison PL, Balslev-Jorgensen J, Lewin T, DePena JF, Skrobak-Kaczynski J (1980) Anthropometry of circumpolar populations. In: Milan FA (ed) The Human Biology of Circumpolar Populations. Cambridge University Press, Cambridge, pp 213–255.

Austin C, Smith TM, Bradman A, Hinde K, Joannes-Boyau R, Bishop D, Hare DJ, Doble P, Eskenazi B, Arora M (2013) Barium distributions in teeth reveal early-life dietary transitions in primates. Nature 498:216–220.

Avis SP (1999) Dog pack attack: hunting humans. Am. J. For. Med. Pathol. 20:243–246.

Backwell L, d'Errico F, Wadley L (2008) Middle Stone Age bone tools from the Howiesons Poort layers, Sibudu Cave, South Africa. J. Archaeol. Sci. 35:1566–1580.

Backwell L, Pickering R, Brothwell D, Berger L, Witcomb M, Martill D, Penkman K, Wilson A (2009) Probable human hair found in a fossil hyaena coprolite from Gladysvale cave, South Africa. J. Archaeol. Sci. 36:1269–1276.

Bailey BJR, Briars GL (1996) Estimating the surface area of the human body. Stat. Med. 15:1325–1332.

Bailey GN (1975) The role of molluscs in coastal economies: the results of midden analysis in Australia. J. Archaeol. Sci. 2:45–62.

Bailey H (1960) A method of determining warmth and temperateness of climate. Geo. Ann. 43:1–16.

Bailey RC, Head G, Jenike M, Owen B, Rechtman R, Zechenter E (1989) Hunting and gathering in tropical rain forest: is it possible? Am. Anthropol. 91:59–82.

Bailey SE (2002) A closer look at Neanderthal postcanine dental morphology: the mandibular dentition. Anat. Rec. 269:148–156.

Bailey SE (2004) A morphometric analysis of maxillary molar crowns of Middle-Late Pleistocene hominins. J. Hum. Evol. 47:183–198.

Bailey SE, Hublin JJ (2006a) Dental remains from the Grotte du Renne at Arcy-sur-Cure (Yonne). J. Hum. Evol. 50:485–508.

Bailey SE, Hublin JJ (2006b) Did Neanderthals make the Châtelperronian assemblage from La Grotte du Renne (Arcy-sur-Cure, France)? In: Harvati K, Harrison T (eds) Neanderthals Revisited: New Approaches and Perspectives. Springer, Dordrecht, pp 191–210.

Bailey SE, Lynch JM (2005) Diagnostic differences in mandibular P4 shape between Neandertals and anatomically modern humans. Am. J. Phys. Anthropol. 126:268–277.

Bailey SE, Glantz M, Weaver TD, Viola B (2008) The affinity of the dental remains from Obi-Rakhmat Grotto, Uzbekistan. J. Hum. Evol. 55:238–248.

Bailey TN (1993) The African Leopard: Ecology and Behavior of a Solitary Felid. Columbia University Press, New York.

Baldeon A (1993) El yacimiento de Lezetxiki (Gipuzkoa, País Vasco). Los niveles musterienses. Munibe Antropologia-Arkeologia 45:3–97.

Balme J, Bowdler S (2006) Spear and digging stick: the origin of gender and its implications for the colonization of new continents. J. Soc. Arch. 6:379–401.

Balter V, Simon L (2006) Diet and behavior of the Saint Césaire Neanderthal inferred from biogeochemical data inversion. J. Hum. Evol. 51:329–338.
Balter V, Bocherens H, Person A, Labourdette N, Renard M, Vandermeersch B (2002) Ecological and physiological variability of Sr/Ca and Ba/Ca in mammals of West European mid-Würmian food webs. Palaeogeogr. Palaeoclimatol. Palaeoecol. 186:127–143.
Balzeau A, Rougier H (2010) Is the suprainiac fossa a Neandertal autapomorphy? A complementary external and internal investigation. J. Hum. Evol. 58:1–22.
Balzeau A, Holloway RL, Grimaud-Hervé D (2012) Variations and asymmetries in regional brain surface in the genus *Homo*. J. Hum. Evol. 62:696–706.
Banks WE, d'Errico F, Peterson AT, Kageyama M, Sima A, Sánchez-Goñi M-F (2008) Neanderthal extinction by competitive exclusion. PLoS One 3:e3972.
Bannikov AG (1967) Some remarks concerning determination of biomass of wild ungulates in natural geographical zones of the USSR. In: Petrusewicz K (ed) Secondary Productivity of Terrestrial Ecosystems (Principles and Methods), Vol. 1. Panstwowe Wydawnictwo Naukowe, Warsaw, pp 255–260.
Bar-Yosef Mayer DE, Vandermeersch B, Bar-Yosef O (2009) Shells and ochre in Middle Paleolithic Qafzeh Cave, Israel: indications for modern behavior. J. Hum. Evol. 56:307–314.
Bar-Yosef O (1989) Upper Pleistocene cultural stratigraphy in Southwest Asia. In: Trinkaus E (ed) The Emergence of Modern Humans. Cambridge University Press, Cambridge, pp 154–180.
Bar-Yosef O (1992) Middle Paleolithic human adaptations in the Mediterranean Levant. In: Akazawa T, Aoki K, Kimura T (eds) The Evolution and Dispersal of Modern Humans in Asia. Hokusen-Sha, Tokyo, pp 189–215.
Bar-Yosef O (1998) The chronology of the Middle Paleolithic of the Levant. In: Akazawa T, Aoki K, Bar-Yosef O (eds) Neanderals and Modern Humans in Western Asia. Plenum, New York, pp 39–56.
Bar-Yosef O (2002) The Upper Paleolithic revolution. A. Rev. Anthropol. 31:363–393.
Bar-Yosef O (2006) Neanderthals and modern humans: a different interpretation. In: Conard NJ (ed) When Neanderthals and Modern Humans Met. Kerns Verlag, Tübingen, pp 467–482.
Bar-Yosef O, Bordes JG (2010) Who were the makers of the Châtelperronian culture? J. Hum. Evol. 59:586–593.
Bar-Yosef O, Callander J (1999) The woman from Tabun: Garrod's doubts in historical perspective. J. Hum. Evol. 37:879–885.
Bar-Yosef O, Kuhn SL (1999) The big deal about blades: laminar technologies and human evolution. Am. Anthropol. 101:322–338.
Bar-Yosef O, Pilbeam D (1993) Dating hominid remains. Nature 366:415.
Bar-Yosef O, Vandermeersch B (eds) (1991) Le Squelette Moustérien de Kébara 2. Editions du C.N.R.S., Paris.
Bar-Yosef O, Vandermeersch B, Arensburg B, Belfer-Cohen A, Goldberg P, Laville H, Meignen L, Rak Y, Speth JD, Tchernov E, Tillier A-M, Weiner S (1992) The excavations in Kebara Cave, Mt. Carmel. Curr. Anthropol. 33:497–550.
Barkai R, Gopher A, Lauritzen SE, Frumkin A (2003) Uranium series dates from Qesem Cave, Israel, and the end of the Lower Palaeolithic. Nature 423:977–979.
Barkai R, Lemorini C, Shimelmitz R, Lev Z, Stiner MC, Gopher A (2009) A blade for all seasons? Making and using Amudian blades at Qesem Cave, Israel. Hum Evol 24:57–75.
Barrett L, Gaynor D, Rendall D, Mitchell D, Henzi SP (2003) Habitual cave use and thermoregulation in chacma baboons (*Papio hamadryas ursinus*). J. Hum. Evol. 46:215–222.
Barron E, van Andel TH, Pollard D (2003) Glacial environments II: reconstructing the climate of Europe in the last glaciation. In: van Andel TH, Davies W (eds) Neanderthals and Modern Humans in the European Landscape During the Last Glaciation. McDonald Institute for Archaeological Research, Cambridge, pp 57–78.
Barroso Ruiz C, Hublin JJ (1994) The late Neanderthal site of Zafarraya (Andalucía, España). AEQUA Monografias 2:61–70.

Barton CM (1997) Stone tools, style, and social identity: an evolutionary perspective on the archaeological record. In: Barton CM, Schifer MB (eds) Rediscovering Darwin: Evolutionary Theory in Archeological Explanation. Archeological Papers of the American Anthropological Association, pp 141–156.

Barton RNE, Currant AP, Fernandez-Jalvo Y, Finlayson JC, Goldberg P, Macphail R, Pettitt PB, Stringer CB (1999) Gibraltar Neanderthals and results of recent excavations in Gorham's, Vanguard and Ibex Caves. Antiquity 73:13–23.

Baryshnikov GF, Hoffecker JF, Burgess RL (1996) Paleontology and zooarchaeology of Mezmaiskaya Cave (northwestern Caucasus, Russia). J. Archaeol. Sci. 23:313–335.

Basell LS (2008) Middle Stone Age (MSA) site distributions in eastern Africa and their relationship to Quaternary environmental change, refugia and the evolution of *Homo sapiens*. Quat. Sci. Rev. 27:2484–2498.

Basmajian JV (1979) Muscles Alive: Their Functions Revealed by Electromyography, 4th Edition. Williams & Wilkins, Baltimore.

Bassinot FC, Labeyrie LD, Vincent E, Quidelleur X, Shackleton NJ, Lancelot Y (1994) The astronomical theory of climate and the age of the Brunhes-Matuyama magnetic reversal. Earth Planet. Sci. Lett. 126:91–108.

Bastien GJ, Scheoens B, Willems PA, Heglund NC (2005) Energetics of load carrying in Nepalese porters. Science 308:1755.

Bastir M, Rosas A, Gunz P, Peña-Melian A, Manzi G, Harvati K, Kruszynski R, Stringer C, Hublin JJ (2011) Evolution of the base of the brain in highly encephalized human species. Nat. Commun. 2:588.

Bateman JE (1978) The Shoulder and Neck, 2nd Edition. Saunders, Philadelphia.

Beall CM, Steegmann ATJ (2000) Human adaptation to climate: temperature, ultraviolet, radiation, and altitude. In: Stinson S, Bogin B, Huss-Ashmore R, O'Rourke D (eds) Human Biology: An Evolutionary and Biocultural Perspective. Wiley-Liss, New York, pp 163–224.

Beals KL (1974) Head form and climatic stress. Am. J. Phys. Anthropol. 37:85–92.

Beals KL, Smith CL, Dodd SM (1983) Climate and the evolution of brachycephalization. Am. J. Phys. Anthropol. 62:425–437.

Beals KL, Smith CL, Dodd SM (1984) Brain size, cranial morphology, climate and time machines. Curr. Anthropol. 25:301–330.

Beaupré GS, Orr TE, Carter DR (1990a) An approach for time-dependent bone modeling and remodeling – application: a preliminary remodeling situation. J. Orthop. Res. 8:662–670.

Beaupré GS, Orr TE, Carter DR (1990b) An approach for time-dependent bone modeling and remodeling – theoretical development. J. Orthop. Res. 8:651–661.

Beauval C, Bourguignon L, Costamagno S, Couchoud I, Marquet J-C, Maureille B, Meignen L, Mann AE, Texier J-P, Vandermeersch B (2002) Recent discoveries of Neandertal remains from Les Pradelles (Marillac-le-Franc, Charente, France). J. Hum. Evol. 42:A5-A6 (abs).

Beauval C, Maureille B, Lacrampe-Cuyaubère F, Serre D, Peressinotto D, Bordes JG, Cochard D, Couchoud I, Dubrasquet D, Laroulandie V, Lenoble A, Mallye JB, Pasty S, Primault J, Rohland N, Pääbo S, Trinkaus E (2005) A late Neandertal femur from Les Rochers-de-Villeneuve, France. Proc. Nat. Acad. Sci. 102:7085–7090.

Beauval C, Lacrampe-Cuyaubère F, Maureille B, Trinkaus E (2006) Direct radiocarbon dating and stable isotopes of the neandertal femur from Les Rochers-de-Villeneuve (Lussac-les-Châteaux, Vienne). Bull. Mem. Soc. Anthropol. Paris 18:35–42.

Bednarik RG (1992) Paleoart and archaeological myths. Cambridge Archaeol. J. 2:27–57.

Been E, Gómez-Olivencia A, Kramer PA (2011) Lumbar lordosis of extinct hominins. Am. J. Phys. Anthropol. 147:64–77.

Begon M, Harper JL, Townsend CR (1996) Ecology: Individuals, Populations and Communities, 3rd Edition. Blackwell Science, Oxford.

Behar DM, Villems R, Soodyall H, Blue-Smith J, Pereira L, Metspalu E, Scozzari R, Makkan H, Tzur S, Comas D, Bertranpetit J, Quintana-Murci L, Tyler-Smith C, Wells RS, Rosset S, The

Genographic Consortium (2008) The dawn of human matrilineal diversity. Am. J. Hum. Genet. 82:1130–1140

Behre KE, van der Plicht J (1992) Towards an absolute chronology for the last glacial period in Europe: radiocarbon dates from Oerel, northern Germany. Veget. Hist. Archaeobot. 1: 111–117.

Bellemare J-F, Cordeau M-P, Leblanc P, Bellemare F (2001) Thoracic dimensions at maximum lung inflation in normal subjects and in patients with obstructive and restrictive lung diseases. Chest 119:376–386.

Ben-David M, McColl CJ, Boonstra R, Karels TJ (1999) ^{15}N signatures do not reflect body condition in arctic ground squirrels. Can. J. Zool. 77:1373–1378.

Benazzi S, Douka K, Fornai C, Bauer CC, Kullmer O, Svoboda J, Pap I, Mallegni F, Bayle P, Coquerelle M, Condemi S, Ronchitelli A, Harvati K, Weber GW (2011) Early dispersal of modern humans in Europe and implications for Neanderthal behavior. Nature 479:525–528.

Bentley GR (1985) Hunter-gatherer energetics and fertility: a reassessment of the !Kung San. Hum. Ecol. 13:79–109.

Berger AL (1978) Long-term variations of daily insolation and Quaternary climatic changes. J. Atmosph. Sci. 16:390–403.

Berger J (1999) Anthropogenic extinction of top carnivores and interspecific animal behaviour: implications of the rapid decoupling of a web involving wolves, bears, moose and ravens. Proc. R. Soc. Lond. B 266:2261–2267.

Berger J, Cunningham C (1994) Active intervention and conservation: Africa's pachyderm problem. Science 263:1241–1242.

Berger LR, Parkington JE (1995) A new Pleistocene hominid-bearing locality at Hoedjiespunt, South Africa. Am. J. Phys. Anthropol. 98:601–609.

Berger TD, Trinkaus E (1995) Patterns of trauma among Neandertals. J. Archaeol. Sci. 22:841–852.

Bergmann C (1847) Über die Verhältnisse der Wärmeökonomie der Thiere zu ihrer Grösse. Göttinger Studien 3:595–708.

Bergman CA (1993) The development of the bow in western Europe: a technological and functional perspective. In: Peterkin GL, Bricker HM, Mellars P (eds) Hunting and Animal Exploitation in the Later Palaeolithic and Mesolithic of Europe. Archeological Papers of the American Anthropological Association No. 4, pp 95–105.

Bergman CA, McEwen E, Miller R (1988) Experimental archery: projectiles' velocities and comparison of bow performances. Antiquity 62:658–670.

Bermúdez de Castro JM, Nicolás ME (1997) Palaeodemography of the Atapuerca-SH Middle Pleistocene hominid sample. J. Hum. Evol. 33:333–355.

Bermúdez de Castro JM, Arsuaga JL, Carbonell E, Rosas A, Martínez I, Mosquera M (1997) A hominid from the Lower Pleistocene of Atapuerca, Spain: possible ancestor to Neanderthals and modern humans. Science 276:1392–1395.

Bermúdez de Castro JM, Carbonell E, Cáceres I, Díez JC, Fernández-Jalvo Y, Mosquera M, Ollé A, Rodríguez J, Rodríguez XP, Rosas A, Rosell J, Sala R, Vergés JM, van der Made J (1999a) The TD6 (Aurora stratum) hominid site. Final remarks and new questions. J. Hum. Evol. 37:695–700.

Bermúdez de Castro JM, Rosas A, Carbonell E, Nicolás ME, Rodríguez J, Arsuaga JL (1999b) A modern human pattern of dental development in Lower Pleistocene hominids from Atapuerca-TD6 (Spain). Proc. Nat. Acad. Sci. 96:4210–4213.

Bermúdez de Castro JM, Ramírez Rozzi F, Martinón-Torres M, Sarmiento Perez S, Rosas A (2003) Patterns of dental development in Lower and Middle Pleistocene hominins from Atapuerca (Spain). In: Thompson JL, Krovitz GE, Nelson AJ (eds) Patterns of Growth and Development in the Genus *Homo*. Cambridge University Press, Cambridge, pp 246–270.

Bermúdez de Castro JM, Martinón-Torres M, Lozano M, Sarmiento S, Muela A (2004) Paleodemography of the Atapuerca-Sima de los Huesos hominin sample: a revision and new approaches to the paleodemography of the European Middle Pleistocene population. J. Anthropol. Res. 60:5–26.

Bermúdez de Castro JM, Martinón-Torres M, Prado L, Gómez-Robles A, Rosell J, López-Polín L, Arsuaga JL, Carbonell E (2010) New immature hominin fossil from European Lower Pleistocene shows the earliest evidence of a modern human dental development pattern. Proc. Nat. Acad. Sci. 107:11739–11744.

Beyries S (1987a) Quelques exemples de stigmates d'emmanchements observés sur des outils du Paléolithique moyen. In: Stordeur D (ed) La Main et l'Outil. Manches et Emmanchements Préhistoriques. Travaux de la Maison de l'Orient, Lyon, pp 55–62.

Beyries S (1987b) Variabilité de l'Industrie Lithique au Moustérien: Approche Fonctionelle sur Quelques Gisements Français. British Archaeological Reports IS 328, Oxford.

Beyries S (1988) Functional variability of lithic sets in the Middle Paleolithic. In: Dibble H, Montet-White A (eds) Upper Pleistocene Prehistory of Western Eurasia. University of Pennsylvania Museum Press, Philadelphia, pp 213–224.

Beyries S (1990) Problems of interpreting the functional results for ancient periods. Aun 14:71–76.

Bianco AC, Sheng XY, Silva JE (1988) Triiodothyronine amplifies norepinephrine stimulation of uncoupling protein gene transcription by a mechanism not requiring protein synthesis. J. Biol. Chem. 263:18168–18175.

Bicho N, Haws J (2008) At the land's end: marine resources and the importance of fluctuations in the coastline in the prehistoric hunter-gatherer economy of Portugal. Quat. Sci. Rev. 27:2166–2175.

Binford LR (1976) Forty-seven trips: a case study in the character of archaeological formation process. In: Wright RVS (ed) Stone Tools as Cultural Markers. Australian Institute of Aboriginal Studies, Canberra, pp 24–36.

Binford LR (1978) Nunamiut Ethnoarchaeology. Academic Press, New York.

Binford LR (1979) Organization and formation processes: looking at curated technologies. J. Anthropol. Res. 35:255–273.

Binford LR (1980) Willow smoke and dog's tails: hunter-gatherer settlement systems and archaeological site formation. Am. Antiquity 45:4–20.

Binford LR (1982a) The archaeology of place. J. Anthropol. Archaeol. 1:5–31.

Binford LR (1982b) Comment on R. White, Rethinking the Middle/Upper Paleolithic transition. Curr. Anthropol. 23:177–181.

Binford LR (1983) Working at Archaeology. Academic Press, New York.

Binford LR (1984a) An Alyawara day: flour, spinifex gum, and shifting perspectives. J. Anthropol. Res. 40:157–182.

Binford LR (1984b) Faunal Remains from Klasies River Mouth. Academic Press, New York & London.

Binford LR (1985) Human ancestors: changing views of their behavior. J. Anthropol. Archaeol. 4:292–327.

Binford LR (1987) Were there elephant hunters at Torralba? In: Nitecki MH, Nitecki DV (eds) The Evolution of Human Hunting. Plenum, New York, pp 47–105.

Binford LR (1989) Isolating the transition to cultural adaptations: an organizational approach. In: Trinkaus E (ed) The Emergence of Modern Humans: Biocultural Adaptations in the Later Pleistocene. Cambridge University Press, Cambridge, pp 18–41.

Binford LR (1991) Review of P. Mellars & C. Stringer (eds.) The Human Revolution. J. Field Archaeol. 18:111–115.

Binford LR (2001) Constructing Frames of Reference: An Analytical Method for Archaeological Theory Building Using Hunter-Gatherer and Environmental Data Sets. University of California Press, Berkeley.

Binns CW (1998) Infant feeding and growth. In: Ulijaszek SJ, Johnston FE, Preece MA (eds) The Cambridge Encyclopedia of Human Growth and Development. Cambridge University Press, Cambridge, pp 320–323.

Biraben JN (2003) L'évolution du nombre des hommes. Popul. Soc. 394:1–4.
Birket-Smith K (1935) The Eskimos. E.P.Dutton, New York.
Bischoff JL, Shamp DD, Aramburu A, Arsuaga JL, Carbonell E, de Castro JMB (2003) The Sima de los Huesos hominids date to beyond U/Th equilibrium (>350 kyr) and perhaps to 400–500 kyr: new radiometric dates. J. Archaeol. Sci. 30:275–280.
Bischoff JL, Williams RW, Rosenbauer RJ, Aramburu A, Arsuaga JL, García N, Cuenca-Bescós G (2007) High-resolution U-series dates from the Sima de los Huesos hominids yields 600 + ∞/ - 66 kyrs: implications for the evolution of the early Neanderthal lineage. J. Archaeol. Sci. 34:763–770.
Bisson MS (2001) Interview with a Neanderthal: an experimental approach for reconstructing scraper production rules, and their implications for imposed form in Middle Palaeolithic tools. Cambridge Archaeol. J. 11:165–184.
Black MT (1999) The "trunk torsion hypothesis" and Neandertal superior pubic ramal morphology. J. Hum. Evol. 36:A2.
Blackburn TM, Hawkins BA (2004) Bergmann's rule and the mammal fauna of northern North America. Ecology 27:715–724.
Blackburn TM, Gaston KJ, Loder N (1999) Geographic gradients in body size: a clarification of Bergmann's rule. Divers. Dist. 5:165–174.
Blanc AC (1937) Fauna a ippopotamo ed industrie paleolitiche nel riempimento delle grotte litoranee del monte Circeo; I, La grotta delle Capre; II, La grotta del Fossellone. Atti della Accademia Nazionale dei Lincei, Rendiconti, Classe di Scienze Fisiche, Matematiche e Naturali 25:88–93.
Blasco MF (1997) In the pursuit of game: the Mousterian cave site of Gabasa 1 in the Spanish Pyrenees. J. Anthropol. Res. 53:177–217.
Blasco R, Fernándes Peris J (2009) Middle Pleistocene bird consumption at level XI of Bolomor cave (Valencia, Spain). J. Archaeol. Sci. 36:2213–2223.
Blaxter K (1989) Energy Metabolism in Animals and Man. Cambridge University Press, Cambridge.
Bleed P (1986) The optimal design of hunting weapons. Am. Antiquity 51:737–747.
Blehr O (1990) Communal hunting as a prerequisite for caribou (wild reindeer) as a human resource. In: Davis LB, Reeves BOK (eds) Hunters of the Recent Past. Unwin Hyman Ltd, London, pp 304–326.
Bliss LC, Richards JH (1982) Present-day arctic vegetation and ecosystems as a predictive tool for the arctic-steppe mammoth biome. In: Hopkins DM, Matthews JV, Schweger CE, Young SB (eds) Paleoecology of Beringia. Academic Press, New York, pp 241–257.
Blix AS, Steen JB (1979) Temperature regulation in newborn polar homeotherms. Physiol. Rev. 59:285–304.
Blome MW, Cohen AS, Tryon CA, Brooks AS, Russell J (2012) The environmental context for the origins of modern human diversity: a synthesis of regional variability in African climate 150,000-30,000 years ago. J. Hum. Evol. 62:563–592.
Blumenfeld J (2002) Neandertal facial morphology and cold adaptation: a comparative approach. Am. J. Phys. Anthropol. Suppl 34:45.
Blumenschine RJ, Madrigal TC (1993) Variability in long bone marrow yields of east African ungulates and its zooarchaeological implications. J. Archaeol. Sci. 20:555–587.
Blurton-Jones NG, Hawkes K, O'Connell JFO (1989) Modelling and measuring costs of children in two foraging societies. In: Standen V, Foley RA (eds) Comparative Socioecology. Blackwell Scientific Publications, Oxford, pp 367–390.
Blurton-Jones NG, Hawkes K, Draper P (1994) Differences between Hadza and !Kung children's work: affluence and practical reason? In: Burch ESJ, Ellanna LJ (eds) Key Issues in Hunter-Gatherer Research. Berg, Oxford, pp 189–215.
Boaz NT, Ciochon RL, Xu Q, Liu J (2000) Large mammalian carnivores as a taphonomic factor in the bone accumulation at Zhoukoudian. Acta Anthropol. Sinica 19 Suppl.:224–234.

Bocherens H (2011) Diet and ecology of Neanderthals: implications from C and N isotopes: insights from bone and tooth biogeochemistry. In: Conard NJ, Richter J (eds) Neanderthal Lifeways, Subsistence and Technology: One Hundred and Fifty Years of Neanderthal Study. Springer, Dordrecht, pp 73–85.

Bocherens H, Billiou D, Mariotti A, Patou-Mathis M, Otte M, Bonjean D, Toussaint M (1999) Palaeoenvironmental and palaeodietary implications of isotopic biogeochemistry of last interglacial Neanderthal and mammal bones in Scladina Cave (Belgium). J. Archaeol. Sci. 26:599–607.

Bocherens H, Billiou D, Mariotti A, Toussaint M, Patou-Mathis M, Bonjean D, Otte M (2001) New isotopic evidence for dietary habits of Neandertals from Belgium. J. Hum. Evol. 40:497–505.

Bocherens H, Argant A, Argant J, Billiou D, Crégut-Bonnoure E, Donat-Ayache B, Philippe M, Thinon M (2004) Diet reconstruction of ancient brown bears (*Ursus arctos*) from Mont Ventoux (France) using bone collagen stable isotope biogeochemistry (^{13}C, ^{15}N). Can. J. Zool. 82:576–586.

Bocherens H, Drucker DG, Billiou D, Patou-Mathis M, Vandermeersch B (2005) Isotopic evidence for diet and subsistence pattern of the Saint-Césaire I Neanderthal: review and use of a multi-source mixing model. J. Hum. Evol. 49:71–87.

Bocquet-Appel J-P, Arsuaga J-L (1999) Age distributions of hominid samples at Atapuerca (SH) and Krapina could indicate accumulation by catastrophe. J. Archaeol. Sci. 26:327–338.

Bocquet-Appel J-P, Tuffreau A (2009) Technological responses of Neanderthals to macroclimatic variations (240,000 - 40,000 BP). Hum. Biol. 81:287–307.

Bocquet-Appel J-P, Demars P-Y, Noiret L, Dobrowsky D (2005) Estimates of Upper Paleolithic meta-population size in Europe from archaeological data. J. Archaeol. Sci. 32:1656–1668.

Boëda E (1988) Le concept Levallois et évaluation de son champ d'application. In: Otte M (ed) L'Homme de Néandertal: La Technique. Etudes et Recherches Archéologiques de l'Université de Liège (E.R.A.U.L.), Liège, pp 13–26.

Boëda E (1993) Le Concept Levallois: Variabilité des Méthodes. Editions du CNRS, Paris.

Boëda E, Connan J, Dessort D, Muhesen S, Mercier N, Valladas H, Tisnérat N (1996) Bitumen as a hafting material on Middle Palaeolithic artefacts. Nature 380:336–338.

Boëda E, Connan J, Muhesen S (1998) Bitumen as hafting material on Middle Paleolithic artifacts from the El Kowm basin, Syria. In: Akazawa T, Aoki K, Bar-Yosef O (eds) Neandertals and Modern Humans in Western Asia. Plenum, New York, pp 181–204.

Boëda E, Geneste JM, Griggo C, Mercier N, Muhesen S, Reyss JL, Taha A, Valladas H (1999) A Levallois point embedded in the vertebra of a wild ass (*Equus africanus*): hafting, projectiles and Mousterian hunting weapons. Antiquity 73:394–402.

Boëda E, Bonilauri S, Connan J, Jarvie D, Mercier N, Tobey M, Valladas H, al Sakhe H (2008) New evidence for significant use of bitumen in Middle Palaeolithic technical systems at Umm el Tlel (Syria) around 70,000 BP. Paléorient 34:67–83.

Bogin B (1997) Evolutionary hypotheses for human childhood. Yearb. Phys. Anthropol. 40:63–89.

Bogin B (1999) Evolutionary perspective on human growth. A. Rev. Anthropol. 28:109–153.

Bogin B, Rios L (2002) Rapid morphological change in living humans: implications for modern human origins. Comp. Biochem. Physiol. Part A 136:71–84.

Bogin B, Smith BH (1996) Evolution of the human life cycle. Am. J. Hum. Biol. 8:703–716.

Bogin B, Silva MIV, Rios L (2007) Life history trade-offs in human growth: adaptation or pathology? Am. J. Hum. Biol. 19:631–642.

Boismier WA (2006) A Middle Paleolithic site at Lynford Quarry, Mundford, Norfolk: interim statement. Proc. Prehist. Soc. 69:315–324.

Bond GC, Lotti R (1995) Iceberg discharges into the North Atlantic on millennial time scales during the last glaciation. Science 267:1005–1010.

Bond GC, Broecker WD, Johnsen S, McManus J, Labeyrie L, Jouzel J, Bonani G (1993) Correlations between climate records from North Atlantic sediments and Greenland ice. Nature 365:143–147.

Bordes F (1953) Essai de classification des industries 'mousteriennes'. Bull. Soc. Préhist. Française 50:457–466.

Bordes F (1961a) Mousterian cultures in France. Science 134:803–810.

Bordes F (1961b) Typologie du Paleolithique Ancien et Moyen. Imprimeries Delmas, Bordeaux.

Bordes F (1972) A Tale of Two Caves. Harper & Row, New York.

Bosinski G (1967) Die mittelpaläolithischen Funde im westlichen Mitteleuropa. Fundamenta A/4, Köln.

Bosinski G (1992) Die ersten Menschen in Eurasien. Jahrbuch des Römisch-Germanischen Zentralmuseums Mainz 39:131–181.

Botkin DB, Melillo JM, Wu LS (1981) How ecosystem processes are linked to large mammal population dynamics. In: Fowler CW, Smith T (eds) Population Dynamics of Large Mammals. John Wiley and Sons, New York, pp 373–387.

Bourguignon L, Sellami F, Deloze V, Sellier-Segard N, Beyries S, Emery-Barbier E (2002) L'habitat moustérien de "La Folie" (Poitiers, Vienne): synthèse des premiers résultats. Paléo 14:29–48.

Bouzouggar A, Barton N, Vanhaeren M, d'Errico F, Collcutt S, Higham T, Hodge E, Parfitt S, Rhodes E, Schwenninger J-L, Stringer C, Turner E, Ward S, Moutmir A, Stambouli A (2007) 82,000-year-old shell beads from North Africa and implications for the origins of modern human behavior. Proc. Nat. Acad. Sci. 104:9964–9969.

Boyd E, Scammon RE, Lawrence D (1927) The determination of surface area of living children. Anat. Rec. 35:445–449.

Boyle KV (2000) Reconstructing Middle Palaeolithic subsistence strategies in the south of France. In: Burke A (ed) Hunting in the Middle Paleolithic. Int. J. Osteoarchaeol. 10:336–356.

Boyle KW (1996) Mousterian settlement patterning in southern France. In: Conard NJ, Wendorf F (eds) Middle Palaeolithic and Middle Stone Age Settlement Systems. A.B.A.C.O. Edizioni, Forlí, pp 241–245.

Bradtmöller M, Pastoors A, Weninger B, Weniger G-C (2012) The repeated replacement model – rapid climate change and population dynamics in Late Pleistocene Europe. Quatern. Int. 247:38–49.

Braidwood RJ, Reed C (1957) The achievement and early consequences of food-production: a consideration of the archaeological and natural-historical evidence. Cold Spring Harb. Symp. Quant. Biol. 22:19–32.

Brajkovic D, Ducharme MB, Frim J (2001) Relationship between body heat content and finger temperature during cold exposure. J. Appl. Physiol. 90:2445–2452.

Brantingham PJ (2006) Measuring forager mobility. Curr. Anthropol. 47:435–459.

Brantingham PJ, Kuhn SL (2001) Constraints on Levallois core technology: a mathematical model. J. Archaeol. Sci. 28:747–761.

Bratton SP, Stromberg LL, Harmon ME (1982) Firewood-gathering impacts in backcountry campsites in Great Smoky Mountains National Park. Environ. Manage. 6:63–71.

Bräuer G (1984) The 'Afro-European sapiens hypothesis' and hominid evolution in East Asia during the late Middle and Upper Pleistocene. Courier Forschungsinstitut Senckenberg 69:145–165.

Brennan MU (1991) Health and Disease in the Middle and Upper Paleolithic of Southwestern France: A Bioarcheological Study. Ph.D. Dissertation, New York University.

Breuil H (1912) Les subdivisions du Paléolithique supérieur et leur signification. Comptes Rendus de la XIVeme Session, Cong. Intern. Anthrop. Archéol. Préhist., Genève:5–78.

Bridges PS (1985) Changes in Long Bone Structure with the Transition to Agriculture: Implications for Prehistoric Activities. Ph.D. Dissertation, University of Michigan.

Bridges PS (1989) Changes in activites with the shift to agriculture in the southeastern United States. Curr. Anthropol. 30:385–394.

Briggs AW, Good JM, Green RE, Krause J, Maricic T, Stenzel U, Lalueza-Fox C, Rudan P, Brajković D, Kućan Ž, Gušić I, Schmitz R, Doronichev VB, Golovanova LV, de la Rasilla M, Fortea J, Rosas A, Pääbo S (2009) Targeted retrieval and analysis of five Neandertal mtDNA genomes. Science 325:318–321.

Briggs JM, Knapp AK (2004) Determinants of C3 for growth and production in a C4 dominated grassland. Plant Ecol. 152:93–100.

Brink JS (1993) Postcranial evidence for the evolution of the black wildebeest, *Connochaetes gnou*: an exploratory study. Palaeont. Afr. 30:61–69.

Brink JW (1997) Fat content in leg bones of *Bison bison*, and applications to archaeology. J. Archaeol. Sci. 24:259–274.

Britton K, Grimes V, Niven L, Steele TE, McPherron S, Soressi M, Kelly TE, Jaubert J, Hublin J-J, Richards MP (2011) Strontium isotope evidence for migration in late Pleistocene *Rangifer*: implications for Neanderthal hunting strategies at the Middle Palaeolithic site of Jonzac, France. J. Hum. Evol. 61:176–185.

Broadhurst CL, Wang YQ, Crawford MA, Cunnane SC, Parkington JE, Schmidt WF (2002) Brain-specific lipids from marine, lacustrine, or terrestrial food resources: potential impact on early African *Homo sapiens*. Comp. Biochem. Phys. B 131:653–673.

Brody S (1945) Bioenergetics and Growth. Reinhold, Baltimore.

Broglio A, De Stefani M, Guriolo F, Pallecchi P, Giachi G, Higham T, Brock F (2009) L'art aurignacien dans la décoration de la Grotte de Fumane. L'Anthropologie 113:753–761.

Brook BW, Bowman DMJS (2004) The uncertain blitzkrieg of Pleistocene megafauna. J. Biogeogr. 31:517–523.

Brooks AS, Helgren DM, Cramer JS, Franklin A, Hornyak, Keating JM, Klein RG, Rink WJ, Schwarcz H, Leith Smith JN, Stewart K, Todd NE, Verniers J, Yellen JE (1995) Dating and context of three Middle Stone Age sites with bone points in the Upper Semliki Valley, Zaire. Science 268:548–553.

Brooks AS, Yellen JE, Nevell L, Hartman G (2005) Projectile technologies of the African MSA: implications for modern human origins. In: Hovers E, Kuhn S (eds) Transitions Before the Transition: Evolution and Stability in the Middle Paleolithic and Middle Stone Age. Kluwer, New York, pp 233–255.

Brown JH, Maurer BA (1986) Body size, ecological dominance and Cope's rule. Nature 324:248–250.

Brown KS, Marean CW, Herries AIR, Jacobs Z, Tribolo C, Braun D, Roberts DL, Meyer MC, Bernatchez J (2009) Fire as an engineering tool of early modern humans. Science 325:859–862.

Bruner E (2010) Morphological differences in the parietal lobes within the human genus: a neurofunctional perspective. Curr. Anthropol. 51:577–588.

Bruner E, Holloway RL (2010) A bivariate approach to the widening of the frontal lobes in the genus *Homo*. J. Hum. Evol. 58:138–146.

Bruner E, Manzi G (2006) Saccopastore 1: the earliest Neanderthal? A new look at an old cranium. In: Harvati K, Harrison T (eds) Neanderthals Revisited: New Approaches and Perspectives. Springer, Dordrecht, pp 23–36.

Bryant HN, Russell AP, Laroiya R, Powell GL (1996) Claw retraction and protraction in the Carnivora: skeletal microvariation in the phalanges of the Felidae. J. Morphol. 229:289–308.

Buck LT, Stringer CB (2014) Having the stomach for it: a contribution to Neanderthal diet? Quat. Sci. Rev. (in press).

Büdel J (1949) Die räumliche und zeitliche Gliederung des Eiszeitklimas. Naturwissenschaften 36:105–112, 133–139.

Burke A (ed) (2000a) Hunting in the Middle Paleolithic. Int. J. Osteoarchaeol. 10.

Burke A (2000b) The view from Starosele: faunal exploitation at a Middle Palaeolithic site in western Crimea. In: Burke A (ed) Hunting in the Middle Paleolithic. Int. J. Osteoarchaeol. 10:325–335.

Burke A (ed) (2004) The Ecology of Neanderthals. Int. J. Osteoarchaeol. 14.

Burke A (2006) Neanderthal settlement patterns in Crimea: a landscape approach. J. Anthropol. Archaeol. 25:510–523.

Burke A (2012) Spatial abilities, cognition and the pattern of Neanderthal and modern human dispersals. Quatern. Int. 247:230–235.

Burke A, Ebert D, Cardille J, Dauth D (2008) Paleoethology as a tool for the development of archaeological models of land-use: the Crimean Middle Palaeolithic. J. Archaeol. Sci. 35:894–904.

Burness GP, Diamond J, Flannery T (2001) Dinosaurs, dragons, and dwarfs: the evolution of maximal body size. Proc. Nat. Acad. Sci. 98:14518–14523.

Burnett AP, Soreghan MJ, Scholz CA, Brown ET (2011) Tropical east African climate change and its relation to global climate: a record from Lake Tanganyika, tropical east Africa, over the past 90+ kyr. Palaeogeogr. Palaeoclimatol. Palaeoecol. 303:155–167.

Burt WH, Grossenheider RP (1976) A Field Guide to the Mammals, 3rd Edition. Houghton Mifflin Company, Boston.

Burton AC, Edholm OG (1955) Man in a Cold Environment: Physiological and Pathological Effects of Exposure to Low Temperatures. Edward Arnold, London.

Butzer KW (1971) Environment and Archeology: An Ecological Approach to Prehistory. Aldine-Atherton, Inc., Chicago.

Butzer KW (1975) Pleistocene littoral-sedimentary cycles of the Mediterranean basin: a Mallorquin view. In: Butzer KW, Isaac GL (eds) After the Australopithecines. Mouton, The Hague, pp 25–71.

Byrne NM, Hills AP, Hunter GR, Weinsier RL, Schutz Y (2005) Metabolic equivalent: one size does not fit all. J. Appl. Physiol. 99:1112–1119.

Cabanac M, Caputa M (1979) Natural selective cooling of the human brain: evidence of its occurrence and magnitude. J. Physiol. (Lond). 286:255–264.

Cabanes D, Mallol C, Expósito I, Baena J (2010) Phytolith evidence for hearths and beds in the late Mousterian occupations of Esquilleu cave (Cantabria, Spain). J. Archaeol. Sci. 37:2947–2957.

Cachel S (1997) Dietary shifts and the European Upper Paleolithic transition. Curr. Anthropol. 38:579–603.

Calder WA (1996) Size, Function, and Life History. Dover Publications, Mineola, NY.

Calhoun KH, House W, Hokanson JA, Quinn FB (1990) Normal nasal airway resistance in noses of different sizes and shapes. Otolaryngol. Head Neck Surg. 103:605–609.

Callow P, Walton D, Shell CA (1986) The use of fire at La Cotte de St. Brelade. In: Callow P, Cornford JM (eds) La Cotte de St. Brelade 1961–1978. Excavations by C.B.M. McBurney. Geo Books, Norwich, UK, pp 193–195.

Campbell B (1995) Human Ecology, 2nd Edition. Aldine de Gruyter, New York.

Caramelli D, Lalueza-Fox C, Condemi S, Longo L, Milani L, Manfredini A, de Saint Pierre M, Adoni F, Lari M, Giunti P, Ricci S, Casoli A, Calafell F, Mallegni F, Bertranpetit J, Stanyon R, Bertorelle G, Barbujani G (2006) A highly divergent mtDNA sequence in a Neandertal individual from Italy. Curr. Biol. 16:R630-R632.

Carbone C, Gittleman JL (2002) A common rule for the scaling of carnivore density. Science 295:2273–2276.

Carbone C, Du Toit JT, Gordon IJ (1997) Feeding success in African wild dogs: does kleptoparasitism by spotted hyenas influence hunting group size? J. Animal Ecol. 66:318–326.

Carbone C, Mace GM, Roberts SC, Macdonald DW (1999) Energetic constraints on the diet of terrestrial carnivores. Nature 402:286–288.

Carbone C, Cowlishaw G, Isaac NJB, Rowcliffe JM (2005a) How far do animals go? Determinants of day range in mammals. Am. Nat. 165:290–297.

Carbone C, Frame L, Frame G, Malcolm J, Fanshawe J, FitzGibbon C, Schaller G, Gordon IJ, Rowcliffe JM, Du Toit JT (2005b) Feeding success of African wild dogs (*Lycaon pictus*) in the Serengeti: the effects of group size and kleptoparasitism. J. Zool. 266:153–161.

Carbone C, Teacher A, Rowcliffe JM (2007) The costs of carnivory. PLoS Biology 5:1–6.

Carbonell E, Castro-Curel Z (1992) Palaeolithic wooden artefacts from the Abric Romani (Capellades, Barcelona, Spain). J. Archaeol. Sci. 19:707–719.

Carbonell E, Bermúdez de Castro JM, Parés JM, Pérez-González A, Cuenca-Bescós G, Ollé A, Mosquera M, Huguet R, van der Made J, Rosas A, Sala R, Vallverdú J, García N, Granger DE,

Martinón-Torres M, Rodríguez XP, Stock GM, Vergès JM, Allué E, Burjachs F, Cáceres I, Canals A, Benito A, Díez C, Lozano M, Mateos A, Navazo M, Rodríguez J, Rosell J, Arsuaga JL (2008) The first hominin of Europe. Nature 452:465–469.
Carbonell E, Cebrià A, Allué E, Cáceres I, Castro Z, Díaz R, Esteban M, Ollé A, Pastó I, Rodríguez XP, Rosell J, Sala R, Vallverdú J, Vaquero M, Vergés JM (1996a) Behavioural and organizational complexity in the Middle Palaeolithic from the Abric Romaní. In: Carbonell E, Vaquero M (eds) The Last Neandertals, the First Anatomically Modern Humans. A Tale about the Human Diversity. Cultural Change and Human Evolution: the Crisis at 40 KA BP. Universitat Rovira I Virgili, Tarragona, pp 385–434.
Carbonell E, Mosquera M, Rodriguez XP, Sala R (1996b) The first human settlement of Europe. J. Anthropol. Res. 51:107–114.
Cârciumaru M (1989) Contexte stratigraphique, paléoclimatique et géochronologique des civilisations du Paléolithique Moyen et Supérieur en Roumanie. L'Anthropologie 93:99–122.
Cârciumaru M, Moncel MH, Anghelinu M, Cârciumaru R (2002) The Cioarei-Borosteni Cave (Carpathian Mountains, Romania): Middle Palaeolithic finds and technological analysis of the lithic assemblages. Antiquity 76:681–690.
Cârciumaru M, Ion R-M, Niţu E-C, Ştefănescu R (2012) New evidence of adhesive as hafting material on Middle and Upper Palaeolithic artefacts from Gura Cheii-Râşnov Cave (Romania). J. Archaeol. Sci. 39:1942–1950.
Carey JW, Steegmann ATJ (1981) Human nasal protrusion, latitude, and climate. Am. J. Phys. Anthropol. 56:313–319.
Carlson KJ, Grine FE, Pearson OM (2007) Robusticity and sexual dimorphism in the postcranium of modern hunter-gatherers from Australia. Am. J. Phys. Anthropol. 134:9–23.
Caro TM, Graham CM, Stoner CJ, Vargas JK (2004) Adaptive significance of antipredator behaviour in artiodactyls. Anim. Behav. 67:205–228.
Caron F, d'Errico F, Del Moral P, Santos F, Zilhao J (2011) The reality of Neandertal symbolic behavior at the Grotte du Renne, Arcy-sur-Cure, France. PLoS ONE 6:e21545.
Carrier DR (1984) The energetic paradox of human running and hominid evolution. Curr. Anthropol. 25:483–495.
Carrión JS, Yll EI, Walker MJ, Legaz AJ, Chaín C, López A (2003) Glacial refugia of temperate, Mediterranean and Ibero-North African flora in south-eastern Spain: new evidence from cave pollen at two Neanderthal man sites. Global Ecol. Biogeogr. 12:119–129.
Carto SL, Weaver AJ, Hetherington R, Lam Y, Wiebe EC (2009) Out of Africa and into an ice age: on the role of global climate change in the late Pleistocene migration of early modern humans out of Africa. J. Hum. Evol. 56:139–151.
Cashdan EA (1985) Coping with risk: reciprocity among the Basarwa of northern Botswana. Man 20:454–474.
Caspari R, Lee S-H (2004) Older age becomes common late in human evolution. Proc. Nat. Acad. Sci. 101:10895–10900.
Caspari R, Lee S-H (2006) Is human longevity a consequence of cultural change or modern biology? Am. J. Phys. Anthropol. 129:512–517.
Caspari R, Radovčić J (2006) New reconstruction of Krapina 5, a male Neandertal cranial vault from Krapina, Croatia. Am. J. Phys. Anthropol. 130:294–307.
Castro-Curel Z, Carbonell E (1995) Wood pseudomorphs from level I at Abric Romani, Barcelona, Spain. J. Field Archaeol. 22:376–384.
Cattelain P (1989) Un crochet de propulseur solutréen de la grotte de Combe- Saunière 1 (Dordogne). Bull. Soc. Préhist. Française 86:213–216.
Cena K, Clark JA (1978) Thermal insulation of animal coats and human clothing. Phys. Med. Biol. 23:565–591.
Çep B, Waiblinger J (2001) The use of cave and open-air sites in southern Germany. In: Conard N (ed) Settlement Dynamics of the Middle Paleolithic and Middle Stone Age. Kerns Verlag, Tübingen, pp 251–263.
Chaney MS, Ross ML (1972) Nutrition, 8th Edition. Houghton Mifflin, Boston.

Chang ML, Meyer MR (2007) Phylogenetic evaluation of adaptive explanations for Neandertal nasal morphology. Am. J. Phys. Anthropol. Suppl. 44:86.
Chapais B (2010) The deep structure of human society: primate origins and evolution. In: Kappeler PM, Silk JB (eds) Mind the Gap: Tracing the Origins of Human Universals. Springer-Verlag, Berlin, pp 19–51.
Charnov EL (1976) Optimal foraging, the marginal value theorem. Theor. Pop. Biol. 9:129–136.
Charnov EL (1990) On evolution of age at maturity and the adult lifespan. J. Evol. Biol. 3:139–144.
Charnov EL (1991) Evolution of life-history variation among female mammals. Proc. Nat. Acad. Sci. 88:1134–1137.
Chase PG (1986a) The Hunters of Combe Grenal: Approaches to Middle Paleolithic Subsistence in Europe. British Archaeological Reports IS 286, Oxford.
Chase PG (1986b) Relationships between Mousterian lithic and faunal assemblages at Combe Grenal. Curr. Anthropol. 27:69–71.
Chase PG (1988) Scavenging and hunting in the Middle Paleolithic: the evidence from Europe. In: Dibble H, Monet-White A (eds) The Upper Pleistocene Prehistory of Western Eurasia. University of Pennsylvania Press, Philadelphia, pp 225–232.
Chase PG (1989) How different was Middle Palaeolithic subsistence? A zooarchaeological perspective on the Middle to Upper Palaeolithic transition. In: Mellars P, Stringer CB (eds) The Human Revolution: Behavioural and Biological Perspectives on the Origins of Modern Humans. Princeton University Press, Princeton, NJ, pp 321–337.
Chase PG (1990) Tool-making tools and Middle Paleolithic behavior. Curr. Anthropol. 31:443–447.
Chase PG (2006) The Emergence of Culture: The Evolution of a Uniquely Human Way of Life. Springer, New York.
Chiarelli B (2004) Spongiform encephalopathy, cannibalism and Neanderthals extinction. Hum. Evol. 19:81–92.
Chiswick ML (1985) Intrauterine growth retardation. Br. Med. J. 291:845–848.
Christiansen P (2004) Body size in proboscideans, with notes on elephant metabolism. Zool. J. Linn. Soc. 140:523–549.
Churchill SE (1993) Weapon technology, prey size selection, and hunting methods in modern hunter-gatherers: implications for hunting in the Palaeolithic and Mesolithic. In: Peterkin GL, Bricker HM, Mellars P (eds) Hunting and Animal Exploitation in the Later Palaeolithic and Mesolithic of Europe. Archeological Papers of the American Anthropological Association No. 4, pp 11–24.
Churchill SE (1994a) Human Upper Body Evolution in the Eurasian Later Pleistocene. Ph.D. Dissertation, University of New Mexico.
Churchill SE (1994b) Medial clavicular length and upper thoracic shape in Neandertals and European early modern humans. Am. J. Phys. Anthropol. Suppl 18:67–68 (abs).
Churchill SE (1998) Cold adaptation, heterochrony and the Neandertals. Evol. Anthropol. 7:46–61.
Churchill SE (2002) Of assegais and bayonets: reconstructing prehistoric spear use. Evol. Anthropol. 11:185–186.
Churchill SE (2006) Bioenergetic perspectives on Neandertal thermoregulatory and activity budgets. In: Havarti K, Harrison T (eds) Neanderthals Revisited: New Approaches and Perspectives. Springer, Dordrecht, pp 113–134.
Churchill SE, Morris AG (1998) Muscle marking morphology and labor intensity in prehistoric Khoisan foragers. Int. J. Osteoarchaeol. 8:390–411.
Churchill SE, Rhodes JA (2006) How strong were the Neandertals? Leverage and muscularity at the shoulder and elbow in Mousterian foragers. Periodicum Biologorum 108:457–470.
Churchill SE, Rhodes JA (2009) The evolution of the human capacity for "killing at a distance": the human fossil evidence for the evolution of projectile weaponry. In: Hublin JJ, Richards M (eds) The Evolution of Hominid Diets: Integrating Approaches to the Study of Palaeolithic Subsistence. Springer, Dordrecht, pp 201–210.

Churchill SE, Schmitt D (2002) Biomechanics in palaeoanthropology: engineering and experimental approaches to the study of behavioural evolution in the genus *Homo*. In: Harcourt CS, Sherwood BR (eds) New Perspectives in Primate Evolution and Behaviour. Westbury, West Yorkshire, pp 71–102.

Churchill SE, Smith FH (2000a) Makers of the early Aurignacian of Europe. Yearb. Phys. Anthropol. 43:61–115.

Churchill SE, Smith FH (2000b) A modern human humerus from the early Aurignacian of Vogelherdhöhle (Stetten, Germany). Am. J. Phys. Anthropol. 112:251–273.

Churchill SE, Trinkaus E (1990) Neandertal scapular glenoid morphology. Am. J. Phys. Anthrop. 83:147–160.

Churchill SE, Pearson OM, Grine FE, Trinkaus E, Holliday TW (1996a) Morphological affinities of the proximal ulna from Klasies River Main Site: archaic or modern? J. Hum. Evol. 31:213–237.

Churchill SE, Weaver AH, Niewoehner WA (1996b) Late Pleistocene human technological and subsistence behavior: functional interpretations of upper limb morphology. In: Bietti A, Grimaldi S (eds) Reduction Processes ("Chaines Opératoires") in the European Mousterian. Quaternaria Nova 6:18–51.

Churchill SE, Brink JS, Gruss LT (2000) Middle Pleistocene to Holocene postcranial gracilization in black wildebeest, *Connochaetes gnou*, and its implications for understanding diachronic changes in robusticity in the genus *Homo*. Am. J. Phys. Anthropol. Suppl. 30:124–125 (abs).

Churchill SE, Shackelford LL, Georgi JN, Black MT (2004) Morphological variation and airflow dynamics in the human nose. Am. J. Hum. Biol. 16:625–638.

Churchill SE, Franciscus RG, McKean-Peraza HA, Daniel JA, Warren BR (2009) The Shanidar 3 Neandertal rib puncture wound and Paleolithic weaponry. J. Hum. Evol. 57:163–178.

Churchill SE, Berger LR, Hartstone-Rose A, Zondo BH (2012) Body size in African Middle Pleistocene *Homo*. In: Reynolds SC, Gallagher A (eds) African Genesis: Perspectives on Hominid Evolution. Cambridge University Press, Cambridge, pp 319–346.

Clarke D (1976) Mesolithic Europe: the economic basis. In: Sieveking GdG, Longworth IH, Wilson KE (eds) Problems in Economic and Social Archaeology. Duckworth & Co., London, pp 449–481.

Clark JD (1992) African and Asian perspectives on the origins of modern humans. Phil. Trans. R. Soc. B 337:148–178.

Clark JGD (1968) World Prehistory: A New Outline, 2nd Edition. Cambridge University Press, Cambridge.

Clarke RJ (2000) A corrected reconstruction and interpretation of the *Homo erectus* calvaria from Ceprano, Italy. J. Hum. Evol. 39:433–442.

Clottes J (ed) (2003) Return to Chauvet Cave. Thames & Hudson, London.

Cochard D, Brugal J-P, Morin E, Meignen L (2012) Evidence of small fast game exploitation in the Middle Paleolithic of Les Canalettes Aveyron, France. Quatern. Int. 264:32–51.

Coe MD, Cummings DH, Phillipson J (1976) Biomass and production of large herbivores in relation to rainfall and primary production. Oecologia 22:341–354.

Cohen AS, Stone JR, Beuning KRM, Park LE, Reinthal PN, Dettman D, Scholz CA, Johnson TC, King JW, Talbot MR, Brown ET, Ivory SJ (2007) Ecological consequences of early Late Pleistocene megadroughts in tropical Africa. Proc. Nat. Acad. Sci. 104:16422–16427.

Cohen JA (1978) *Cuon alpinus*. Mammal Species 100:1–3.

Cohen JA, Fox MW, Johnsingh AJT, Barnett BD (1978) Food habits of the dhole in south India. J. Wildl. Manage. 42:933–936.

Cohen ML (1977) Measurement of the thermal properties of human skin. A review. J. Invest. Dermatol. 69:333–338.

Colbert EH (1962) The weights of dinosaurs. Am. Mus. Novit. 2076:1–16.

Cole P (1982a) Modification of inspired air. In: Proctor DF, Andersen I (eds) The Nose: Upper Airway Physiology and the Atmospheric Environment. Elsevier Biomedical Press, Amsterdam, pp 351–375.

Cole P (1982b) Upper respiratory airflow. In: Proctor DF, Andersen I (eds) The Nose: Upper Airway Physiology and the Atmospheric Environment. Elsevier Biomedical Press, Amsterdam, pp 163–185.

Cole TJ (1995) Conditional reference charts to assess weight gain in British infants. Arch. Disease Childhood 73:8–16.

Colinvaux P (1978) Why Big Fierce Animals are Rare: an Ecologist's Perspective. Princeton University Press, Princeton, NJ.

Conard NJ (ed) (2001) Settlement Dynamics of the Middle Paleolithic and Middle Stone Age. Kerns Verlag, Tübingen.

Conard NJ (2003) Palaeolithic ivory sculptures from southwestern Germany and the origins of figurative art. Nature 426:830–832.

Conard NJ (ed) (2004) Settlement Dynamics of the Middle Paleolithic and Middle Stone Age, Volume II. Kerns Verlag, Tübingen.

Conard NJ (2009) A female figurine from the basal Aurignacian of Hohle Fels Cave in southwestern Germany. Nature 459:248–252.

Conard NJ, Bolus M (2003) Radiocarbon dating the appearance of modern humans and timing of cultural innovations in Europe: new results and new challenges. J. Hum. Evol. 44:331–371.

Conard NJ, Delagnes A (eds) (2010) Settlement Dynamics of the Middle Paleolithic and Middle Stone Age, Volume III. Kerns Verlag, Tübingen.

Conard NJ, Fischer B (2000) Are there recognizable cultural entities in the German Middle Paleolithic. In: Ronen A, Weinstein-Evron M (eds) Toward Modern Humans: Yabrudian and Micoquian, 400-50 ka Years Ago. BAR International Series 850, pp 7–24.

Conard NJ, Prindiville TJ (2000) Middle Palaeolithic hunting economies in the Rhineland. In: Burke A (ed) Hunting in the Middle Paleolithic. Int. J. Osteoarchaeol. 10:286–309.

Conard NJ, Wendorf F (eds) (1996) Middle Palaeolithic and Middle Stone Age Settlement Systems. A.B.A.C.O. Edizioni, Forlí.

Conard NJ, Malina M, Münzel SC (2009) New flutes document the earliest musical tradition in southwestern Germany. Nature 460:737–740.

Condemi S (1988) Caracteres plesiomorphes et apomorphes de l'os temporal des Neanderthaliens Europeens Würmiens. In: Trinkaus E (ed) L'Homme de Néandertal: L'Antomie. Etudes et Recherches Archéologiques de l'Université de Liège (E.R.A.U.L.), Liège, pp 49–52.

Condemi S (1992) Les Hommes Fossiles de Saccopastore et leurs Relations Phylogénétiques. CNRS Éditions, Paris.

Condemi S (2001) Les Néandertaliens de La Chaise. Comité des Travaux Historiques et Scientifiques, Paris.

Condemi S, Mounier A, Giunti P, Lari M, Caramelli D, Longo L (2013) Possible interbreeding in late Italian Neanderthals? New data from the Mezzena jaw (Monti Lessini, Verona, Italy). PLoS ONE 8:e59781.

Coon CS (1955) Some problems of human variability and natural selection in climate and culture. Am. Nat. 89:257–279.

Coon CS (1962) The Origin of Races. Knopf, New York.

Coope GR (2002) Changes in the thermal climate in Northwestern Europe during marine oxygen isotope stage 3, estimated from fossil insect assemblages. Quatern. Res. 57:401–408.

Copeland L (1975) The Middle and Upper Palaeolithic of Lebanon and Syria in the light of recent research. In: Wendorf E, Marks AE (eds) Problems in Prehistory: North Africa and the Levant. SMU Press, Dallas, pp 317–350.

Coppa A, Grün R, Stringer C, Eggins S, Vargiu R (2005) Newly recognized Pleistocene human teeth from Tabun Cave, Israel. J. Hum. Evol. 49:301–315.

Coqueugniot H, Hublin JJ, Veillon F, Houët F, Jacob T (2004) Early brain growth in *Homo erectus* and implications for cognitive ability. Nature 431:299–302.

Cordain L, Watkins BA, Florant GL, Kelher M, Rogers L, Li Y (2002) Fatty acid analysis of wild ruminant tissues: evolutionary implications for reducing diet-related chronic disease. Eur. J. Clin. Nutr. 56:181–191.

Cordy J-M (1988) Apport de la paléozoologie à la paléoécologie et à la chronostratigraphie en Europe du nord-occidental. In: Laville H (ed) L'Homme de Néandertal: L'Environment. Etudes et Recherches Archéologiques de l'Université de Liège (E.R.A.U.L.), Liège, pp 55–64.

Cordy JM, Bastin B (1992) Synthèse des études paléontologiques réalisées dans les dépôts de la grotte Scladina (Sclayn, Province de Namur). In: Otte M (ed) Recherches aux Grottes de Sclayn, vol. 1, Le Contexte. Etudes et Recherches Archéologiques de l'Université de Liège (E.R.A.U.L.), Liège, pp 153–156.

Cosner AJ (1956) Fire hardening of wood. Am. Antiquity 22:179–180.

Costamagno S (2013) Bone grease rendering in Mousterian contexts: the case of Noisetier Cave (Fréchet-Aure, Hautes-Pyrénées, France). In: Clark JL, Speth JD (eds) Zooarchaeology and Modern Human Origins: Human Hunting Behavior during the Later Pleistocene. Springer, Dordrecht, pp 209–225.

Costamagno S, Beauval C, Vandermeersch B, Mann A, Maureille B (2003) Middle Paleolithic human teeth digested by carnivores (Les Pradelles, Marillac-le-Franc, France). 2003 Paleoanthropology Society Meetings (abstract: www.paleoanthro.org/abst2003.htm).

Costamagno S, Meignen L, Beauval C, Vandermeersch B, Maureille B (2006) Les Pradelles (Marillac-le-Franc, France): a Mousterian reindeer hunting camp? J Anthropol. Archaeol. 25:466–484

Cottle MH (1955) The structure and function of the nasal vestibule. Arch. Otolaryngol. 62:173–181.

Coulson T, Albon S, Guinness F, Pemberton J, Clutton-Brock T (1997) Population substructure, local density, and calf winter survival in red deer (*Cervus elaphus*). Ecology 78:852–864.

Courtiss EH, Goldwyn RM (1983) The effects of nasal surgery on airflow. Plastic Reconstr. Surg. 72:9–19.

Cowgill L, Trinkaus E, Zeder M (2007) Shanidar 10: a Middle Paleolithic immature distal lower limb from Shanidar Cave, Iraqi Kurdistan. J. Hum. Evol. 53:213–223.

Cox MP, Morales DA, Woerner AE, Sozanski J, Wall JD, Hammer MF (2009) Autosomal resequence data reveal Late Stone Age signals of population expansion in Sub-Saharan African foraging and farming populations. PLOS One 4:e6366.

Creel S (2001) Four factors modifying the effect of competition on carnivore population dynamics as illustrated by African wild dogs. Consv. Biol. 15:271–274.

Crête M (1999) The distribution of deer biomass in North America supports the hypothesis of exploitation ecosystems. Ecol. Lett. 2:223–227.

Crews DE, Gerber LM (2003) Reconstructing life history of hominids and humans. Coll. Anthropol. 27:7–22.

Crittenden AN, Marlowe FW (2008) Allomaternal care among the Hadza of Tanzania. Hum. Nature 19:249–262.

Crockford SJ (2002) Thyroid hormone in Neandertal evolution: a natural or a pathological role? Geogr. Rev. 92:73–88.

Crockford SJ (2003) Thyroid rhythm phenotypes and hominid evolution: a new paradigm implicates pulsatile hormone secretion in speciation and adaptation changes. Comp. Biochem. Physiol. A 135:105–129.

Crummett T (1995) The three dimensions of shovel-shaping. In: Moggi-Cecchi J (ed) Aspects of Dental Biology: Palaeontology, Anthropology and Evolution. International Institute for the Study of Man, Florence, pp 305–313.

Crummett TL (1994) The Evolution of Shovel Shaping: Regional and Temporal Variations in Human Incisor Morphology. Ph.D. Dissertation, University of Michigan.

Culley EV, Popescu G, Clark GA (2013) An analysis of the compositional integrity of the Levantine Mousterian facies. Quatern. Int. 25:213–233.

Cunningham DJ (1908) The evolution of the eyebrow region of the forehead, with special reference to the excessive supraorbital development in the Neanderthal race. Trans. R. Soc. Edinburg 46:283–311.

Currat M, Excoffier L (2004) Modern humans did not admix with Neanderthals during their range expansion into Europe. PLoS Biol. 2:e421.

Currat M, Excoffier L (2011) Strong reproductive isolation between humans and Neanderthals inferred from observed patterns of introgression. Proc. Nat. Acad. Sci. 108:15129–15134.

Currey J (1984) The Mechanical Adaptation of Bones. Princeton University Press, Princeton.

Cyrek K, Cyrek M (2009) Composite tools in the oldest levels of the Biśnik Cave, Poland. In: Burdukiewicz JM, Cyrek K, Dyczek P, Szymczak K (eds) Understanding the Past. Papers Offered to Stefan Karol Kozłowski. University of Warsaw, Warsaw, pp 91–96.

Czarnetzki A (1995) Morphological evidence of adaptive characters in the genus *Homo*. In: Ullrich H (ed) Man and Environment in the Palaeolithic. Etudes et Recherches Archéologiques de l'Université de Liège (E.R.A.U.L.), Liège, pp 97–110.

d'Errico F (2003) The invisible frontier: a multiple species model for the origin of behavioral modernity. Evol. Anthropol. 12:188–202.

d'Errico F, Henshilwood CS (2007) Additional evidence for bone technology in the southern African Middle Stone Age. J. Hum. Evol. 52:142–163

d'Errico F, Sánchez Goñi MF (2003) Neandertal extinction and the millennial scale climatic variability of OIS 3. Quat. Sci. Rev. 23:769–788.

d'Errico F, Sánchez Goñi MF (2004) "A Garden of Eden for the Gibraltar Neandertals? A reply to Finlayson et al." Quat. Sci. Rev. 23:1210–1216.

d'Errico F, Stringer CB (2011) Evolution, revolution or saltation scenario for the emergence of modern cultures? Phil. Trans. R. Soc. B 366:1060–1069.

d'Errico F, Vanhaeren M (2009) Earliest personal ornaments and their significance for the origin of language debate. In: Botha R, Knight C (eds) The Cradle of Language. Oxford University Press, New York, pp 16–40.

d'Errico F, Zilhão J, Julien M, Baffier D, Pelegrin J (1998) Neanderthal acculturation in western Europe? A critical review of the evidence and its interpretation. Curr. Anthropol. 39 Suppl.:S1-S44.

d'Errico F, Henshilwood C, Vanhaeren M, van Niekerk K (2005) *Nassarius kraussianus* shell beads from Blombos Cave: evidence for symbolic behaviour in the Middle Stone Age. J. Hum. Evol. 48:3–24.

d'Errico F, Vanhaeren M, Wadley L (2008) Possible shell beads from the Middle Stone Age layers of Sibudu Cave, South Africa. J. Archaeol. Sci. 35:2675–2685.

d'Errico F, Vanhaeren M, Barton N, Bouzouggar A, Mienis H, Richter D, Hublin J-J, McPherron SP, Lozouet P (2009) Additional evidence on the use of personal ornaments in the Middle Paleolithic of North Africa. Proc. Nat. Acad. Sci. 106:16051–16056.

Dalén L, Orlando L, Shapiro B, Brandström-Durling M, Quam R, Gilbert MTP, Díez Fernández-Lomana CJ, Willerslev E, Arsuaga JL, Götherström A (2012) Partial genetic turnover in Neandertals: continuity in the east and population replacement in the west. Mol. Biol. Evol. 29:1893–1897.

Dalke PD (1953) Yields of seeds and mast in second growth hardwood forest, south central Missouri. J. Wildl. Manage. 17:378–380.

Damuth J (1981) Population density and body size in mammals. Nature 290:699–700.

Damuth J (1993) Cope's rule, the island rule and the scaling of mammalian population density. Nature 365:748–750.

Daniau A-L, d'Errico F, Sánchez Goñi MF (2010) Testing the hypothesis of fire use for ecosystem management by Neanderthal and Upper Palaeolithic modern human populations. PLoS ONE 5:e9157.

Dansgaard W, Johnsen SJ, Clausen HB, Dahl-Jensen D, Gundestrup NS, Hammer CU, Hvidberg CS, Steffensen JP, Sveinbjörnsdottir AE, Jouzel J, Bond G (1993) Evidence for general instability of past climate from a 250-kyr ice-core record. Nature 364:218–220.

Danzl DF, Pozos RS (1994) Accidental hypothermia. New Eng. J. Med. 331:1756–1760.

Daujeard C, Moncel M-H (2010) On Neanderthal subsistence strategies and land use: a regional focus on the Rhone Valley area in southeastern France. J. Anthropol. Archaeol. 29:368–391.

Davenport J (1992) Animal Life at Low Temperature. Springer, London.
Davidson I (1976) Les Mallaetes and Mondúver: the economy of a human group in prehistoric Spain. In: Sieveking GdG, Longworth IH, Wilson KE (eds) Problems in Economic and Social Archaeology. Duckworth, London, pp 483–499.
Davidson I, Noble W (1992) Why the first colonisation of the Australian region is the earliest evidence of modern human behaviour. Archaeol. Oceania 27:113–119.
Davidson S, Passmore R, Eastwood MA (1986) Human Nutrition and Dietetics, 8th Edition. Churchill Livingston, Edinburgh.
Davies A (1932) A resurvey of the morphology of the nose in relation to climate. J. R. Anthropol. Inst. 62:337–360.
Davies R, Underdown S (2006) The Neanderthals: a social synthesis. Cambridge Archaeol. J. 16:145–164.
Davies W, Gollop P (2003) The human presence in Europe during the last glacial period II: climate tolerance and climate preferences of Mid- and Late Glacial hominids. In: van Andel TH, Davies W (eds) Neanderthals and Modern Humans in the European Landscape during the Last Glaciation. McDonald Institute for Archaeological Research, Cambridge, pp 131–146.
Davis SJM (1981) The effects of temperature change and domestication on the body size of late Pleistocene to Holocene mammals of Israel. Paleobiology 7:101–114.
Dayan T, Simberloff D (2005) Ecological and community-wide character displacement: the next generation. Ecol. Lett. 8:875–894.
Dayan T, Simberloff D, Tchernov E, Yom-Tov Y (1990) Feline canines: community-wide character displacement among the small cats of Israel. Am.Nat. 136:39–60.
Dayan T, Simberloff D, Tchernov E, Yom-Tov Y (1991) Calibrating the paleothermometer: climate, communities, and the evolution of size. Paleobiology 17:189–199.
de Beaulieu J-L, Reille M (1992) Long Pleistocene pollen sequences from the Velay Plateau (Massif Central, France), 1. Ribans Maar. Veget. Hist. Archaeobot. 1:223–242.
de Beaune SA (1993) Nonflint stone tools of the Early Upper Paleolithic. In: Knecht H, Pike-Tay A, White R (eds) Before Lascaux: The Complex Record of the Early Upper Paleolithic. CRC Press, Boca Raton, pp 163–191.
de Beaune SA (2003) Du grain à moudre sur les Néandertaliens. La Recherche 360:56–59.
De Groote I (2008) A Comprehensive Analysis of Long Bone Curvature in Neanderthals and Modern Humans Using 3D Morphometrics. Ph.D. Dissertation, University College, London.
De Groote I (2011) Femoral curvature in Neanderthals and modern humans: a 3D geometric morphometric analysis. J. Hum. Evol. 60:540–548.
de Lange P, Lanni A, Beneduce L, Moreno M, Lombardi A, Silvestri E, Goglia F (2001) Uncoupling protein-3 is a molecular determinant for the regulation of resting metabolic rate by thyroid hormone. Endocrinology 142:3414–3420.
de Lumley H (1969) Une Cabane Acheuléenne dans la Grotte du Lazaret. Mémoires de la Société Préhistorique Française No. 7.
de Lumley H (1969–72) Le Paléolithique Inférieur et Moyen du Midi Méditerranéen dans son Cadre Géologique. CNRS Éditions, Paris.
de Lumley MA (1973) Anténéandertaliens et Néandertaliens du bassin méditerranéen occidental européen. CNRS Éditions, Paris.
Deacon HJ (1993) Planting an idea: an archaeology of Stone Age gatherers in South Africa. S. Afr. Archaeol. Bull. 48:86–93.
Dean D, Hublin J-J, Holloway R, Ziegler R (1998) On the phylogenetic position of the pre-Neandertal specimen from Reilingen, Germany. J. Hum. Evol. 34:485–508.
Dean MC (1988) Another look at the nose and the functional significance of the face and nasal mucous membrane for cooling the brain in fossil hominids. J. Hum. Evol. 17:715–718.
Dean MC, Stringer CB, Bromage TG (1986) Age at death of the Neanderthal child from Devil's Tower, Gibraltar and the implications for studies of general growth and development in Neanderthals. Am. J. Phys. Anthropol. 70:301–309.
Debénath A, Dibble HL (1994) Handbook of Paleolithic Typology, Vol. 1: Lower and Middle Paleolithic of Europe. Cypher, Philadelphia.

Defleur A, White T, Valensi P, Slimak L, Crégut-Bonnoure É (1999) Neanderthal cannibalism at Moula-Guercy, Ardèche, France. Science 286:128–131.

Delagnes A, Rendu W (2011) Shifts in Neandertal mobility, technology and subsistence strategies in western France. J. Archaeol. Sci. 38:1771–1783.

Delagnes A, Jaubert J, Meignen L (2007) Les technocomplexes du paléolithique moyen en Europe occidentale dans leur cadre diachronique et géographique. In: Vandermeersch B, Maureille B (eds) Les Neandertaliens. Biologie et Cultures. Éditions du CTHS, Paris, pp 213–229.

DelGiudice GD (1998) Surplus killing of white-tailed deer by wolves in northcentral Minnesota. J. Mammol. 79:227–235.

Delson E, Harvati K (2006) Return of the last Neanderthal. Nature 443:762–763.

Demes B (1987) Another look at an old face: biomechanics of the neandertal facial skeleton reconsidered. J. Hum. Evol. 16:297–303.

Demes B, Stern JTJ, Hausman MR, Larson SG, Mcleod K, Rubin CT (1998) Patterns of strain in the macaque ulna during functional activity. Am. J. Phys. Anthropol. 106:87–100.

Demes B, Qin Y-X, Stern JTJ, Larson SG, Rubin CT (2001) Patterns of strain in the macaque tibia during functional activity. Am. J. Phys. Anthropol. 116:257–265.

Demment MW, Greenwood GB (1988) Forage ingestion: effects of sward characteristics and body size. J. Anim. Sci. 66:2380–2392.

Dennell RW, Martinón-Torres M, Bermúdez de Castro JM (2011) Hominin variability, climatic instability and population demography in Middle Pleistocene Europe. Quat. Sci. Rev. 30:1511–1524.

Derevianko AP, Postnov AV, Rybin EP, Kuzmin YV, Keates SG (2005) The Pleistocene peopling of Siberia: a review of environmental and behavioural aspects. Indo-Pacific Prehist. Assoc. Bull. 25:57–68.

Desse G, Desse J (1976) La pêche. In: de Lumley H (ed) La Préhistoire Française. Tome I: Les Civilisations Paléolithiques et Mésolithiques de la France. CNRS, Paris, pp 697–702.

Dewailly E, Ayotte P, Bruneau S, Gingras S, Belles-Isles M, Roy R (2000) Susceptibility to infections and immune status in Inuit infants exposed to organochlorines. Env. Health Pers. 108:205–211.

Dewey KG, Heinig MJ, Nommsen LA, Peerson JM, Lönnerdal B (1992) Growth of breast-fed and formula-fed infants from 0 to 18 months: the DARLING study. Pediatrics 89:1035–1041

di Vincenzo F, Churchill SE, Manzi G (2012) The Vindija Neanderthal scapular glenoid fossa: comparative shape analysis suggests evo-devo changes among Neanderthals. J. Hum. Evol. 62:274–285.

Diamond J (1997) Guns, Germs, and Steel: The Fates of Human Societies. W.W. Norton & Company, New York.

Dibble HL (1984a) Interpreting typological variation of Middle Paleolithic scrapers: function, style, or sequence of reduction? J. Field Archaeol. 11:431–436.

Dibble HL (1984b) The Mousterian industry from Bisitun Cave (Iran). Paléorient 10:23–34.

Dibble HL (1987) The interpretation of Middle Paleolithic scraper morphology. Am. Antiquity 52:109–117.

Dibble HL (1991) Mousterian assemblage variability on an interregional scale. J. Anthropol. Res. 47:239–257.

Dibble HL, Holdaway SJ (1993) The Middle Paleolithic industries of Warwasi. In: Olszewski D, Dibble HL (eds) The Paleolithic Prehistory of the Zagros-Taurus. University of Pennsylvania Museum Press, Philadelphia, pp 75–99.

Dibble HL, Rolland N (1992) On assemblage variability in the Middle Paleolithic of western Europe. In: Dibble HL, Mellars P (eds) The Middle Paleolithic: Adaptation, Behavior, and Variability. University of Pennsylvania Museum Press, Philadelphia, pp 1–28.

Diedrich CG (2007) Upper Pleistocene *Panthera leo spelaea* (Goldfuss, 1810) skeleton remains from Praha-Podbaba and other lion finds from loess and river terrace sites in Central Bohemia (Czech Republic). Bull. Geosci. 82:99–117.

Diedrich CG (2009a) Steppe lion remains imported by Ice Age spotted hyenas into the Late Pleistocene Perick Caves hyena den in northern Germany. Quatern. Res. 71:361–374.

Diedrich CG (2009b) Upper Pleistocene *Panthera leo spelaea* (Goldfuss, 1810) remains from the Bilstein Caves (Sauerland Karst) and contribution to the steppe lion taphonomy, palaeobiology and sexual dimorphism. Ann. Paléontol. 95:117–138.

Diedrich CG (2011a) The largest European lion *Panthera leo spelaea* (Goldfuss 1810) population from the Zoolithen Cave, Germany: specialised cave bear predators of Europe. Historical. Biol. 23:271–311.

Diedrich CG (2011b) Late Pleistocene steppe lion *Panthera leo spelaea* (Goldfuss, 1810) footprints and bone records from open air sites in northern Germany – evidence of hyena-lion antagonism and scavenging in Europe. Quat. Sci. Rev. 30:1883–1906.

Diedrich CG (2011c) Pleistocene *Panthera leo spelaea* (Goldfuss, 1810) remains from the Balve Cave (nw Germany): a cave bear, hyena den and Middle Palaeolithic human cave - and review of the Sauerland Karst lion cave sites. Quaternaire 22:105–127.

Diedrich CG, Rathgeber T (2012) Late Pleistocene steppe lion *Panthera leo spelaea* (Goldfuss, 1810) skeleton remains of the Upper Rhine Valley (SW Germany) and contributions to their sexual dimorphism, taphonomy and habitus. Historical. Biol. 24:1–28.

Diedrich CG, Žák K (2006) Prey deposits and den sites of the Upper Pleistocene hyena *Crocuta crocuta spelaea* (Goldfuss, 1823) in horizontal and vertical caves of the Bohemian Karst (Czech Republic). Bull. Geosci. 81:237–276.

Dobson SD, Trinkaus E (2002) Cross-sectional geometry and morphology of the mandibular symphysis in Middle and Late Pleistocene *Homo*. J. Hum. Evol. 43:67–87.

Domanski M, Webb JA (1992) Effect of heat treatment on siliceous rocks used in prehistoric lithic technology. J. Archaeol. Sci. 19:601–614.

Donadio E, Buskirk SW (2006) Diet, morphology, and interspecific killing in carnivora. Am. Nat. 167:524–536.

Dorst J, Dandelot P (1970) A Field Guide to the Larger Mammals of Africa. Collins, London.

Downey B (2006) Wolf-grizzly interaction in Yellowstone National Park. Int. Wolf 16:22–25.

Draper HH (1977) The aboriginal Eskimo diet in modern perspective. Am. Anthropol. 79:309–316.

Driver HE, Massey WC (1957) Comparative studies of North American Indians. Trans. Am. Phil. Soc. 47:165–456.

Drucker DG, Bocherens H (2004) Carbon and nitrogen stable isotopes as tracers of change in diet breadth during Middle and Upper Palaeolithic in Europe. Int. J. Osteoarchaeol. 14:162–177.

Duarte C, Maurício J, Pettitt P, Souto P, Trinkaus E, van der Plicht H, Zilhão J (1999) The early Upper Paleolithic human skeleton from the Abrigo do Lagar Velho (Portugal) and modern human emergence in Iberia. Proc. Nat. Acad. Sci. 96:7604–7609.

DuBois D, DuBois EF (1916) A formula to estimate the approximate surface area if height and weight are known. Arch. Int. Med. 17:863–871.

Dufour DL (1984) The time and energy expenditure of indigenous women horticulturalists in the northwest Amazon. Am. J. Phys. Anthropol. 65:37–46.

Dufour DL, Sauther ML (2002) Comparative and evolutionary dimensions of the energetics of human pregnancy and lactation. Am. J. Hum. Biol. 14:584–602.

Dunbar RIM (1992) Neocortex size as a constraint on group size in primates. J. Hum. Evol. 20:469–493.

Dunbar RIM (2003) The social brain: mind, language, and society in evolutionary perspective. A. Rev. Anthropol. 32:163–181.

Dunbrack RL, Ramsay MA (1993) The allometry of mammalian adaptdons to seasonal environments: a critique of the fasting endurance hypothesis. Oikos 66:336–342.

Dunsworth HJ, Warrener AG, Deacon T, Ellison PT, Pontzer H (2012) Metabolic hypothesis for human altriciality. Proc. Nat. Acad. Sci. 109:15212–15216.

Durand AI, Ipiña SL, Bermúdez de Castro JM (2000) A probabilistic approach to the assessment of some life history pattern parameters in a Middle Pleistocene human population. Math. Biosci. 165:147–162.

Durant SM (1998) Competition refuges and coexistence: an example from Serengeti carnivores. J. Animal Ecol. 87:370–386.

Durnin JVGA, McKillop FM, Grant S, Fitzgerald G (1985) Is nutritional status endangered by virtually no extra intake during pregnancy? Lancet ii:823–825.

Dusseldorp GL (2011) Studying Pleistocene Neanderthal and cave hyena dietary habits: combining isotopic and archaeozoological analyses. J. Archaeol. Method Theory 18:224–255.

Dusseldorp GL (2013) Neanderthals and cave hyenas: co-existence, competition or conflict? In: Clark JL, Speth JD (eds) Zooarchaeology and Modern Human Origins: Human Hunting Behavior during the Later Pleistocene. Springer, Dordrecht, pp 191–208.

Earle M (1987) A flexible body mass in social carnivores. Am. Nat. 129:755–760.

Eaton RL (1979) Evolution of sociality in the Felidae. Carnivore 2 Suppl.:82–89.

Ecker M, Bocherens H, Julien M-A, Rivals F, Raynal J-P, Moncel M-H (2013) Middle Pleistocene ecology and Neanderthal subsistence: insights from stable isotope analyses in Payre (Ardèche, southeastern France). J. Hum. Evol. 65:363–373.

Eckert R (1988) Animal Physiology: Mechanisms and Adaptations, 3rd Edition. W.H. Freeman and Company, New York.

Economos AC (1980) Taxonomic differences in the mammalian life span-body weight relationship and the problem of brain weight. Gerontology 26:90–98.

Eisenberg JF (1990) The behavioral/ecological significance of body size in the Mammalia. In: Damuth J, MacFadden BJ (eds) Body Size in Mammalian Paleobiology: Estimation and Biological Implications. Cambridge University Press, Cambridge, pp 25–37.

Ekblaw WE (1928) The material response of the Polar Eskimo to their far arctic environment (continued). Ann. Assoc. Am. Geogr. 18:1–24.

El Zaatari S (2010) Occlusal microwear texture analysis and the diets of historical/prehistoric hunter-gatherers. Int. J. Osteoarchaeol. 20:67–87.

El Zaatari S, Grine FE, Ungar PS, Hublin JJ (2011) Ecogeographic variation in Neandertal dietary habits: evidence from occlusal molar microwear texture analysis. J. Hum. Evol. 61:411–424.

Elad D, Liebenthal R, Wenig BL, Einav S (1993) Analysis of air flow patterns in the human nose. Med. Biol. Eng. Comp. 31:585–592.

Ellison PT (1990) Human ovarian function and reproductive ecology: new hypotheses. Am. Anthropol. 92:933–952.

Ellison PT (2003) Energetics and reproductive effort. Am. J. Hum. Biol. 15:342–351.

Ellison PT, O'Rourke MT (2000) Population growth and fertility regulation. In: Stinson S, Bogin B, Huss-Ashmore R, O'Rourke D (eds) Human Biology: An Evolutionary and Biocultural Perspective. Wiley-Liss, New York, pp 553–586.

Eloff FC (1980) Cub mortality in the Kalahari lion *Panthera leo vernayi* (Roberts, 1948). Koedoe 23:163–179.

Emami P, Kaiser TM, Regelsberger J, Goebell E, Fiehler J, Westphal M, Heese O (2012) Case report: surviving a tiger attack. Neurosurg. Rev. 35:621–624.

Endicott P, Ho SYW, Stringer C (2010) Using genetic evidence to evaluate four palaeoanthropological hypotheses for the timing of Neanderthal and modern human origins. J. Hum. Evol. 59:87–95.

Endo B (1971) Some characteristics of the deltoid tuberosity of the humerus in west-Asian and the European "classic" Neandertals. J. Anthropol. Soc. Nippon 79:249–258.

Endo B, Kimura T (1970) Postcranial skeleton of the Amud Man. In: Suzuki H, Takai F (eds) The Amud Man and his Cave Site. University of Tokyo, Tokyo, pp 231–406.

Enloe JG, David F (1997) *Rangifer* herd behavior: seasonality of hunting in the Magdalenian of the Paris Basin. In: Jackson LJ, Thacker PT (eds) Caribou and Reindeer Hunters of the Northern Hemisphere. Avebury Press, Aldershot, pp 52–68.

Enloe JG, David F, Baryshnikov G (2000) Hyenas and hunters: zooarchaeological investigations at Prolom II Cave, Crimea. In: Burke A (ed) Hunting in the Middle Paleolithic. Int. J. Osteoarchaeol. 10:310–324.

Epstein HE, Lauenroth WK, Burke IC, Coffin DP (1997) Productivity patterns of C_3 and C_4 functional types in the U.S. Great Plains. Ecology 78:722–731

Ereifej KI (2004) Seed characteristics of wild legume (*Tetragonolobus palaestinus*) as compared with *Lens culinaris* and *Pisum sativum*. Int. J. Food Properties 7:639–646.

Eriksson A, Betti L, Friend AD, Lycett SJ, Singarayer JS, von Cramon-Taubadel N, Valdes P, Balloux F, Manica A (2012) Late Pleistocene climate change and the global expansion of anatomically modern humans. Proc. Nat. Acad. Sci. 109:16089–16094.

Estévez J (2004) Vanishing carnivores: what can the disappearance of large carnivores tell us about the Neanderthal world? Int. J. Osteoarchaeol. 14:190–200.

Eswaran V (2002) A diffusion wave out of Africa: the mechanism of the modern human revolution? Curr. Anthropol. 43:749–774.

Eveleth PB, Tanner JM (1976) Worldwide Variation in Human Growth. Cambridge University Press, Cambridge.

Excoffier L, Schneider S (1999) Why hunter-gatherer populations do not show signs of Pleistocene demographic expansions. Proc. Nat. Acad. Sci. 96:10597–10602.

Fa JE, Stewart JR, Lloveras L, Vargas JM (2013) Rabbits and hominin survival in Iberia. J. Hum. Evol. 64:233–241.

Fabre V, Condemi S, Degioanni A (2009) Genetic evidence of geographical groups among Neandertals. PLoS One 4:e5151.

Fagundes NJ, Ray N, Beaumont M, Neuenschwander S, Salzano FM, Bonatto SL, Excoffier L (2007) Statistical evaluation of alternative models of human evolution. Proc. Nat. Acad. Sci. 104:17614–17619.

Fahlström G (1981) The glenohumeral joint in man. Ossa 8:3–154.

Fancy SG, White RG (1985) Energy expenditures by caribou while cratering in snow. J. Wildl. Manage. 49:987–993.

FAO/WHO/UNU (1985) Energy and Protein Requirements. Technical Report Series World Health Organization, Geneva.

Farizy C (1994) Spatial patterning of Middle Paleolithic sites. J. Anthropol. Archaeol. 13:153–160.

Farizy C, David F (1989) Chasse et alimentation carnée au Paléolithique moyen, l'apport des gisements de plein air. In: Otte M (ed) L'Homme de Néandertal: La Subsistance. Etudes et Recherches Archéologiques de l'Université de Liège (E.R.A.U.L.), Liège, pp 59–62.

Farizy C, David F (1992) Subsistence and behavioral patterns of some Middle Paleolithic local groups. In: Dibble HL, Mellars P (eds) The Middle Paleolithic: Adaptation, Behavior, and Variability. University of Pennsylvania Museum Press, Philadelphia, pp 87–96.

Farlow JO, Pianka ER (2002) Body size overlap, habitat partitioning and living space requirements of terrestrial vertebrate predators: implications for the paleoecology of large Theropod dinosaurs. Historical Biol. 16:21–40.

Farlow JO, Smith MB, Robinson JM (1995) Body mass, bone "strength indicator," and cursorial potential of *Tyrannosaurus rex*. J. Vert. Paleontol. 15:713–725.

Féblot-Augustins J (1993) Mobility strategies in the late Middle Palaeolithic of central Europe and western Europe: elements of stability and variability. J. Anthropol. Archaeol. 12:211–265.

Féblot-Augustins J (1997) La Circulation des Matières Premières au Paléolithique. Etudes et Recherches Archéologiques de l'Université de Liège (E.R.A.U.L.), Liège.

Féblot-Augustins J (1999) Raw material transport patterns and settlement systems in the European Lower and Middle Palaeolithic: continuity, change and variability. In: Roebroeks W, Gamble C (eds) The Middle Palaeolithic Occupation of Europe. Leiden University, Leiden, pp 193–214.

Fedele FG, Giaccio B, Hajdas I (2008) Timescales and cultural processes at 40,000 BP in the light of the Campanian Ignimbrite eruption, western Eurasia. J. Hum. Evol. 55:834–857.

Fedriani JM, Fuller TK, Sauvajot RM, York EC (2000) Competition and intraguild predation among three sympatric carnivores. Oecologia 125:258–270.

Feldesman MR, Fountain RL (1996) "Race" specificity and the femur/stature ratio. Am. J. Phys. Anthropol. 100:207–224.

Feldesman MR, Kleckner JG, Lundy JK (1990) The femur/stature ratio and estimates of stature in mid- and late-Pleistocene fossil hominids. Am. J. Phys. Anthropol. 83:359–372.

Feranec R, García N, Díez JC, Arsuaga JL (2010) Understanding the ecology of mammalian carnivorans and herbivores from Valdegoba cave (Burgos, northern Spain) through stable isotope analysis. Palaeogeogr. Palaeoclimatol. Palaeoecol. 297:263–272.

Fernandes P, Raynal JP, Moncel MH (2008) Middle Palaeolithic raw material gathering territories and human mobility in the southern Massif Central, France: first results from a petro-archaeological study on flint. J. Archaeol. Sci. 35:2357–2370.

Fernández-Jalvo Y, Andrews P (2003) Experimental effects of water abrasion on bone fragments. J. Taphonomy 1:147–163.

Fernández-Jalvo Y, Díez JC, Cáceres I, Rosell J (1999) Human cannibalism in the Early Pleistocene of Europe (Gran Dolina, Sierra de Atapuerca, Burgos, Spain). J. Hum. Evol. 37:591–622.

Fernández-Laso MC, Chacón Navarro MG, García-Antón MD, Rivals F (2011) Territorial mobility of Neanderthal groups: a case study from Level M of Abric Romaní (Capellades, Barcelona, Spain). In: Conard NJ, Richter J (eds) Neanderthal Lifeways, Subsistence and Technology: One Hundred and Fifty Years of Neanderthal Study. Springer, Dordrecht, pp 187–202.

Fernandez P, Legendre S (2003) Mortality curves for horses from the Middle Palaeolithic site of Bau de l'Aubesier (Vaucluse, France): methodological, palaeo-ethnological, and palaeo-ecological approaches. J. Archaeol. Sci. 30:1577–1598.

Finlayson C (2004) Neanderthals and Modern Humans: An Ecological and Evolutionary Perspective. Cambridge University Press, Cambridge.

Finlayson C (2005) Biogeography and evolution of the genus *Homo*. Trends Ecol. Evol. 20:457–463.

Finlayson C (2008) On the importance of coastal areas in the survival of Neanderthal populations during the Late Pleistocene. Quat. Sci. Rev. 27:2246–2252.

Finlayson C, Carrión JS (2007) Rapid ecological turnover and its impact on Neanderthal and other human populations. Trends Ecol. Evol. 22:213–222.

Finlayson C, Fa DA, Finlayson G, Giles Pacheco F, Rodríguez Vidal J (2004) Did the moderns kill off the Neanderthals? A reply to F. d'Errico and Sánchez Goñi. Quat. Sci. Rev. 23:1205–1209.

Finlayson C, Giles Pacheco F, Rodriguez-Vidal J, Fa DA, Gutierrez López JM, Santiago Pérez A, Finlayson G, Allue E, Baena Preysler J, Cáceres I, Carrión JS, Fernández Jalvo Y, Gleed-Owen CP, Jimenez Espejo FJ, López P, López Sáez JA, Riquelme Cantal JA, Sánchez Marco A, Giles Guzman F, Brown K, Fuentes N, Valarino CA, Villalpando A, Stringer CB, Martinez Ruiz F, Sakamoto T (2006) Late survival of Neanderthals at the southernmost extreme of Europe. Nature 443:850–853.

Finlayson C, Fa DA, Espejo FJ, Carrión JS, Finlayson G, Giles Pacheco F, Rodriguez-Vidal J, Stringer C, Martinez Ruiz F (2008) Gorham's Cave, Gibraltar – the persistence of a Neanderthal population. Quatern. Int. 181:64–71.

Fiore I, Gala M, Tagliacozzo A (2004) Ecology and subsistence strategies in the eastern Italian Alps during the Middle Palaeolithic. Int. J. Osteoarchaeol. 1:273–286.

Fiorenza L, Benazzi S, Tausch J, Kullmer O, Bromage TG, Schrenk F (2011) Molar macrowear reveals Neanderthal eco-geographic dietary variation. PLoS One 6:e14769.

Fischman J (1992) Hard Evidence. Discover 13:44–51.

Fizet M, Mariotti A, Bocherens H, Lange-Badré B, Vandermeersch B, Borel JP, Bellon G (1995) Effect of diet, physiology and climate on carbon and nitrogen stable isotopes of collagen in a Late Pleistocene anthropic palaeoecosystem: Marillac, Charente, France. J. Archaeol. Sci. 22:67–79.

Flinn MV, Geary DC, Ward CV (2005) Ecological dominance, social competition, and coalitionary arms races: why humans evolved extraordinary intelligence. Evol. Hum. Behav. 26:10–46.

Flores JC (1998) A mathematical model for Neanderthal extinction. J. Theor. Biol. 191:295–298.

Flores JC (2011) Diffusion coefficient of modern humans outcompeting Neanderthals. J. Theor. Biol. 280:189–190.

Fogel ML, Tuross N, Owsley DW (1989) Nitrogen isotope tracers of human lactation in modern and archeological populations. Annual Report of the Director of the Geophysical Laboratory, Carnegie Institution of Washington, 1988–1989. Carnegie Institution, Washington, DC, pp 111–117.

Foley RA (1999) Hominid behavioural evolution: missing links in comparative primate socioecology. In: Lee PC (ed) Comparative Primate Socioecology. Cambridge University Press, Cambridge, pp 363–386.

Foote DC (1965) Exploration and Resource Utilization in Northwestern Arctic Alaska Before 1855. McGill University, Montreal.

Forster P (2004) Ice Ages and the mitochondrial DNA chronology of human dispersals: a review. Phil. Trans. R. Soc. Lond. B 359:255–264.

Fox MW (1984) The Whistling Hunters: Fields Studies of the Asiatic Wild Dog (*Cuon alpinus*). State University of New York Albany, Albany, NY.

Franciscus RG (1989) Neandertal mesosterna and noses: implications for activity and biogeographical patterning. Am. J. Phys. Anthropol. 78:223 (abs).

Franciscus RG (1995) Later Pleistocene Nasofacial Variation in Western Eurasia and Africa and Modern Human Origins. Ph.D. Dissertation, University of New Mexico.

Franciscus RG (1999) Neandertal nasal structures and upper respiratory tract "specialization." Proc. Nat. Acad. Sci. 96:1805–1809.

Franciscus RG (2003) Internal nasal floor configuration in *Homo* with special reference to the evolution of Neandertal facial form. J. Hum. Evol. 44:701–729.

Franciscus RG, Churchill SE (2002) The costal skeleton of Shanidar 3 and a reappraisal of Neandertal thoracic morphology. J. Hum. Evol. 42:303–356.

Franciscus RG, Long JC (1991) Variation in human nasal height and breadth. Am. J. Phys. Anthropol. 85:419–427.

Franciscus RG, Trinkaus E (1988a) Nasal morphology and the emergence of *Homo erectus*. Am. J. Phys. Anthropol. 75:517–528.

Franciscus RG, Trinkaus E (1988b) The Neandertal nose. Am. J. Phys. Anthropol. 75:209–210.

Franciscus RG, Trinkaus E (1995) Determinants of retromolar space presence in Pleistocene *Homo* mandibles. J. Hum. Evol. 28:577–595.

Frank DA, McNaughton SJ, Tracy BF (1998) The ecology of the Earth's grazing ecosystems. Bioscience 48:513–521.

Freeman LG (1973) The significance of mammalian faunas from Paleolithic occupations in Cantabrian Spain. Am. Antiquity 38:3–44.

Freeman LG (1978) Mousterian worked bone from Cueva Morín (Santander, Spain): a preliminary description. In: Freeman LG (ed) Views of the Past: Essays in Old World Prehistory and Paleanthropology. Mouton, The Hague, pp 29–51.

Freeman LG (1983) More on the Mousterian: flaked bone from Cueva Morín. Curr. Anthropol. 24:366–377.

Frenzel B (1973) Climatic Fluctuations of the Ice Age. Case Western Reserve University, Cleveland.

Frenzel B, Pecsi M, Velichko AA (1992) Atlas of Paleoclimates and Paleoenvironments of the Northern Hemisphere, Late Pleistocene-Holocene. Gustav Fischer, Jena.

Fresia AE, Ruff CB, Larsen CS (1990) Temporal decline in bilateral asymmetry of the upper limb on the Georgia Coast. In: Larsen CS (ed) The Archaeology of Mission Santa Catalina de Guale: 2. Biocultural Interpretations of a Population in Transition. Anthropological Papers of the American Museum of Natural History, pp 121–132.

Freund G (1968) Mikrolithen aus dem Mittelpaläolithikum der Sesselfelsgrotte im unteren Altmühltal, Lkr. Kelheim. Quartär 19:133–154.

Friedman E, Goren-Inbar N, Rosenfeld A, Marder O, Burian F (1995) Hafting during Mousterian times: further indication. Mitekufat Haeven 26:8–31.

Friesen TM (2001) A zooarchaeological signature for meat storage: re-thinking the drying utility index. Am. Antiquity 66:315–331.

Friesen TM (2013) The impact of weapon technology on caribou drive system variability in the prehistoric Canadian Arctic. Quatern. Int. 297:13–23.

Frisancho AR (1993) Human Adaptation and Accommodation. University of Michigan Press, Ann Arbor.

Frison GC (1986) Mammoth hunting and butchering from a perspective of African elephant culling. In: Frison GC, Todd LC (eds) The Colby Mammoth Site. University of New Mexico Press, Albuquerque, pp 115–134.

Frison GC (1989) Experimental use of Clovis weaponry and tools on African elephants. Am. Antiquity 54:766–784.

Froehle AW (2008) Climate variables as predictors of basal metabolic rate: new equations. Am. J. Hum. Biol. 20:510–529.

Froehle AW, Churchill SE (2009) Energetic competition between Neandertals and anatomically modern humans. Paleoanthropology 2009:96–116.

Froehle AW, Yokley TR, Churchill SE (2013) Energetics and the origin of modern humans. In: Smith FH, Ahern JCM (eds) The Origins of Modern Humans: Biology Reconsidered. Wiley-Blackwell, Hoboken, NJ, pp 285–320.

Funston PJ, Mills MGL, Biggs HC (2001) Factors affecting the hunting success of male and female lions in the Kruger National Park. J. Zool., Lond. 253:419–431.

Gábori M (1979) Type of industry and ecology. Archaeologica Academiae Scientiarum Hungaricae 31:239–248.

Gabunia L, Vekua A, Lordipanidze D, Swisher III CC, Ferring R, Justus A, Nioradze M, Tvalchrelidze M, Antón SC, Bosinski G, Jöris O, de Lumley M-A, Majsuradze G, Mouskhelishvili A (2000) Earliest Pleistocene hominid cranial remains from Dmanisi, Republic of Georgia: taxonomy, geological setting, and age. Science 288:1019–1025.

Gaillard J-M, Delorme D, Boutin J-M, Van Laere G, Boisaubert B, Pradel R (1993) Roe deer survival patterns: a comparative analysis of contrasting populations. J. Animal Ecol. 62:778–791.

Gainor BJ, Piotrowski G, Puhl J, Allen WC, Hagen R (1980) The throw: biomechanics and acute injury. Amer. J. Sports Med. 8:114–118.

Galloway VA, Leonard WR, Ivakine E (2000) Basal metabolic adaptation of the Evenki reindeer herders of central Siberia. Am. J. Hum. Biol. 12:75–87.

Gambaryan PP (1974) How Animals Run: Anatomical Adaptations. Halstead Press, New York.

Gamble C (1986) The Paleolithic Settlement of Europe. Cambridge University Press, Cambridge.

Gamble C (1987) Man the shoveler: alternative models for Middle Pleistocene colonization and occupation of northern latitudes. In: Soffer O (ed) The Pleistocene Old World: Regional Perspectives. Plenum, New York, pp 81–98.

Gamble C (1992) Comment on "Dense forests, cold steppes, and the Palaeolithic settlement of northern Europe" by W. Roebroeks, N. J. Conard and T. van Kolfschoten. Curr. Anthropol. 33:569–571.

Gamble C (1999) The Palaeolithic Societies of Europe. Cambridge University Press, Cambridge.

Gamble C (2011) The social and material life of Neanderthals. In: Conard NJ, Richter J (eds) Neanderthal Lifeways, Subsistence and Technology: One Hundred and Fifty Years of Neanderthal Study. Springer, Dordrecht, pp 157–166.

García-Diez M, Ochoa Fraile B, Maestu IB (2013) Neanderthal graphic behavior: the pecked pebble from Axlor Rockshelter (northern Spain). J. Anthropol. Res. 69:397–410.

García N, Arsuaga JL (1998) The carnivore remains from the hominid-bearing Trinchera-Galería, Sierra de Atapuerca, Middle Pleistocene site (Spain). Geobios 31:659–674.

García N, Arsuaga JL, Torres T (1997) The carnivore remains from the Sima de los Huesos Middle Pleistocene site (Sierra de Atapuerca, Spain). J. Hum. Evol. 33:155–174.

Gardeisen A (1999) Middle Palaeolithic subsistence in the west cave of "Le Portel" (Pyrénées, France). J. Archaeol. Sci. 26:1145–1158.

Gardner JC, Smith FH (2006) The paleopathology of the Krapina Neandertals. Periodicum Biologorum 108:471–484.

Gargett RH (1989) Grave shortcomings: the evidence for Neanderthal burial. Curr. Anthropol. 30:157–190.
Gargett RH (1999) Middle Palaeolithic burial is not a dead issue: the view from Qafzeh, Saint-Césaire, Kebara, Amud, and Dederiyeh. J. Hum. Evol. 37:27–90.
Garrigan D, Hammer MF (2006) Reconstructing human origins in the genomic era. Nat. Rev. Genetics 7:669–680.
Garrod D, Bate D (1937) The Stone Age of Mount Carmel, Vol. 1. Oxford University Press, Oxford.
Garth WP, Leberte MA, Cool TA (1988) Recurrent fractures of the humerus in a baseball pitcher: a case report. J. Bone Jt. Surg. 70:305–306.
Gat A (1999) Social organization, group conflict and the demise of Neanderthals. Mankind Quart. 39:437–454.
Gaudzinski S (1995) Wallertheim revisited: a re-analysis of the fauna from the Middle Paleolithic site of Wallertheim (Rheinhessen/ Germany). J. Archaeol. Sci. 22:51–66.
Gaudzinski S (1999) Middle Palaeolithic bone tools from the open-air site Salzgitter-Lebenstedt (Germany). J. Archaeol. Sci. 26:125–141.
Gaudzinski S (2004) A matter of high resolution? The Eemian interglacial (OIS 5e) in north-central Europe and Middle Palaeolithic subsistence. Int. J. Osteoarchaeol. 14:201–211.
Gaudzinski S, Roebroeks W (2000) Adults only. Reindeer hunting at the Middle Paleolithic site Salzgitter Lebenstedt, Northern Germany. J. Hum. Evol. 38:497–521.
Gaudzinski-Windheuser S, Kindler L (2012) Research perspectives for the study of Neandertal subsistence strategies based on the analysis of archaeozoological assemblages. Quatern. Int. 247:59–68.
Gaudzinski-Windheuser S, Roebroeks W (2011) On Neanderthal subsistence in Last Interglacial forested environments in northern Europe. In: Conard NJ, Richter J (eds) Neanderthal Lifeways, Subsistence and Technology: One Hundred and Fifty Years of Neanderthal Study. Springer, Dordrecht, pp 61–71.
Gause GF (1934) The Struggle for Existence. Williams & Wilkins, Baltimore, MD.
Gautier A (1989) Preliminary notes on the fauna of the Middle Palaeolithic site at Zwolen (Poland). In: Patou M, Freeman LG (eds) L'Homme de Néandertal: La Subsistance. Etudes et Recherches Archéologiques de l'Université de Liège (E.R.A.U.L.), Liège, pp 69–73.
Gehan EA, George SL (1970) Estimation of human body surface area from height and weight. Can. Chemo. Rep. 54:225–235.
Geist V (1981) Neanderthal the hunter. Nat. Hist. 90:26–36.
Geist V (1987) Bergmann's rule is invalid. Can. J. Zool. 65:1035–1038.
Geneste J-M (1985) Analyse Lithique d'Industries Moustériennes du Périgord: Une Approche Technologique du Comportement des Groupes Humains au Paléolithique Moyen. Ph.D. Dissertation, Université de Bordeaux I.
Geneste JM (1988a) Les industries de la grotte Vaufrey: technologie du débitage, économie et circulation de la matière première. In: Rigaud JP (ed) La Grotte Vaufrey: Paléoenvironnement, Chronologie, Activités Humaines. Société Préhistorique Française, Paris, pp 441–517.
Geneste JM (1988b) Systèmes d'approvisionnement en matières premières au Paléolithique moyen et au Paléolithique supérieur en Aquitaine. In: Kozlowski JK (ed) L'Homme de Néandertal: La Mutation. Etudes et Recherches Archéologiques de l'Université de Liège (E.R.A.U.L.), Liège, pp 61–70.
Gerhart KL, White RG, Cameron RD, Russell DE (1996) Body composition and nutrient reserves of arctic caribou. Can. J. Zool. 74:136–146.
Gilligan I (2007) Neanderthal extinction and modern human behaviour: the role of climate change and clothing. World Archaeol. 39:499–514.
Gilligan I (2010a) Femoral neck-shaft angle and climate: preliminary report on a global study. In: Weiss KE (ed) Trends in Physical Anthropology. Nova Science Publishers, Inc., Hauppauge NY, pp 123–153.
Gilligan I (2010b) The prehistoric development of clothing: archaeological implications of a thermal model. J. Archaeol. Method Theory 17:15–80.

Girard TL, Bork EW, Nielsen SE, Alexander MJ (2013) Seasonal variation in habitat selection by free-ranging feral horses within Alberta's Forest Reserve. Rangeland Ecol. Manage. 66:428–437.

Girardin M, Bilgen E, Arbour P (1983) Experimental study of velocity fields in a human nasal fossa by laser anemometry. Ann. Otol. Rhinol. Laryngol. 92:231–236.

Gittleman JL (1985) Carnivore body size: ecological and taxonomic correlates. Oecologia 67:540–554.

Gittleman JL, Harvey PH (1982) Carnivore home-range size, metabolic needs and ecology. Behav. Ecol. Sociobiol. 10:57–63.

Gittleman JL, Oftedal OT (1987) Comparative growth and lactation energetics in carnivores. Symp. Zool. Soc. Lond. 57:41–77.

Godin G, Shephard RJ (1973) Activity patterns of the Canadian Eskimo. In: Edholm OG, Gunderson EKE (eds) Polar Human Biology. Heinemann Medical Books, London, pp 193–215.

Golovanova LV, Doronichev VB, Cleghorn NE, Koulkova MA, Sapelko TV, Shackley MS (2010) Significance of ecological factors in the Middle to Upper Paleolithic transition. Curr. Anthropol. 51:655–691.

Gómez-Olivencia A, Eaves-Johnson KL, Franciscus RG, Carretero JM, Arsuaga JL (2009) Kebara 2: new insights regarding the most complete Neandertal thorax. J. Hum. Evol. 57:75–90.

Gómez-Olivencia A, Carretero JM, Lorenzo C, Arsuaga JL, Bermúdez de Castro JM, Carbonell E (2010) The costal skeleton of *Homo antecessor*. Preliminary results. J. Hum. Evol. 59:620–640.

Gómez-Olivencia A, Franciscus RG, Couture-Veschambre C, Maureille B, Arsuaga JL (2012) The mesosternum of the Regourdou 1 Neandertal revisited. J. Hum. Evol. 62:511–519.

Goodale JC (1971) Tiwi Wives: A Study of the Women of Melville Island, North Australia. University of Washington Press, Seattle.

Goodship AE, Lanyon LE, McFie H (1979) Functional adaptation of bone to increased stress. J. Bone Joint Surg. 61A:539–546.

Goren-Inbar N, Sharon G, Melamed Y, Kislev ME (2002) Nuts, nut cracking, and pitted stones at Gesher Benot Ya'aqov, Israel. Proc. Nat. Acad. Sci. 99:2455–2460.

Gorjanovic-Kramberger D (1909) Pracovjek iz Krapine kanibal. Glasnik hrvatskog prirodoslovnog drustva 21:62–67.

Goudie A (1983) The chronology and nature of the Pleistocene. In: Goudie A (ed) Environmental Change, 2nd Edition. Clarendon Press, Oxford, pp 29–70.

Graham RW, Group FW (1996) Spatial response of mammals to late Quaternary environmental fluctuations. Science 272:1601–1606.

Grand TI (1997) How muscle mass is part of the fabric of behavioral ecology in east African bovids (*Madoqua, Gazella, Damaliscus, Hippotragus*). Anat. Embryol. 195:375–386.

Grant J, Hopcraft C, Sinclair ARE, Packer C (2005) Planning for success: Serengeti lions seek prey accessibility rather than abundance. J. Animal Ecol. 74:559–566.

Grant JWA, Chapman CA, Richardson KS (1992) Defended versus undefended home range size of carnivores, ungulates and primates. Behav. Ecol. Sociobiol. 31:149–161.

Grant PR (1972) Convergent and divergent character displacement. Biol. J. Linn. Soc. 4:39–68.

Grant PR (1994) Ecological character displacement. Science 266:746–747.

Grayson DK, Delpech F (2001) The Upper Paleolithic at Grotte XVI (Dordogne, France): richness, evenness, and cave bears. In: Hays MA, Thacker P (eds) Questioning the Answers: Resolving Fundamental Problems of the Early Upper Paleolithic. British Archaeological Reports IS 1005, Oxford, pp 187–197.

Grayson DK, Delpech F (2002) Specialized Early Upper Palaeolithic hunters in southwestern France? J. Archaeol. Sci. 29:1439–1449.

Greaves RD (1997) Hunting and multifunctional use of bows and arrows: ethnoarchaeology of technological organization among Pumé hunters of Venezuela. In: Knecht H (ed) Projectile Technology. Plenum, New York, pp 287–320.

Green RE, Malaspinas AS, Krause J, Briggs AW, Johnson PL, Uhler C, Meyer M, Good JM, Maricic T, Stenzel U, Prüfer K, Siebauer M, Burbano HA, Ronan M, Rothberg JM, Egholm M, Rudan P,

Brajković D, Kućan Z, Gusić I, Wikström M, Laakkonen L, Kelso J, Slatkin M, Pääbo S (2008) A complete Neanderthal mitochondrial genome sequence determined by high-throughput sequencing. Cell 134:416–426.

Green RE, Krause J, Briggs AW, Maricic T, Stenzel U, Kircher M, Patterson N, Li H, Zhai W, Fritz MH-Y, Hansen NF, Durand EY, Malaspinas A-S, Jensen JD, Marques-Bonet T, Alkan C, Prüfer K, Meyer M, Burbano HA, Good JM, Schultz R, Aximu-Petri A, Butthof A, Höber B, Höffner B, Siegemund M, Weihmann A, Nusbaum C, Lander ES, Russ C, Novod N, Affourtit J, Egholm M, Verna C, Rudan P, Brajkovic D, Kucan Z, Guic I, Doronichev VB, Golovanova LV, Lalueza-Fox C, de la Rasilla M, Fortea J, Rosas A, Schmitz RW, Johnson PLF, Eichler EE, Falush D, Birney E, Mullikin JC, Slatkin M, Nielsen R, Kelso J, Lachmann M, Reich D, Pääbo S (2010) A draft sequence of the Neandertal genome. Science 328:710–722.

Greene Ch (1986) Patterns of prey selection: implications of predator foraging tactics. Am. Nat. 128:824–839.

Grenier P (1979) New regulations on moose hunting in Quebec. Can. Wildl. Admin. 4:28–32.

Grine FE, Jungers WL, Tobias PV, Pearson OM (1995) Fossil *Homo* femur from Berg Aukas, Northern Namibia. Am. J. Phys. Anthropol. 97:151–185.

GRIP (1993) Climate instability during the last interglacial period recorded in the GRIP ice core. Nature 364:203–207.

Grootes PM, Stuiver M, White JWC, Johnsen SJ, Jouzel J (1993) Comparison of oxygen isotope records from the GISP2 and GRIP Greenland ice cores. Nature 366:552–554.

Grove M, Pearce E, Dunbar RIM (2012) Fission-fusion and the evolution of hominin social systems. J. Hum. Evol. 62:191–200.

Grün R, Stringer CB (1991) Electron spin resonance dating and the evolution of modern humans. Archaeometry 33:153–199.

Grün R, Stringer C (2000) Tabun revisited: revised ESR chronology and new ESR and U-series analyses of dental material from Tabun C1. J. Hum. Evol. 39:601–612.

Grün R, Brink JS, Spooner NA, Taylor L, Stringer CB, Franciscus RG, Murray AS (1996) Direct dating of Florisbad hominid. Nature 382:500–501.

Guatelli-Steinberg D (2009) Recent studies of dental development in Neandertals: implications for Neandertal life histories. Evol. Anthropol. 18:9–20.

Guatelli-Steinberg D, Larsen CS, Hutchinson DL (2004) Prevalence and the duration of linear enamel hypoplasia: a comparative study of Neandertals and Inuit foragers. J. Hum. Evol. 47:65–84.

Guatelli-Steinberg D, Reid DJ, Bishop TA, Larsen CS (2005) Anterior tooth growth periods in Neandertals were comparable to those of modern humans. Proc. Nat. Acad. Sci. 102:14197–14202.

Guatelli-Steinberg D, Reid DJ, Bishop TA (2007) Did the lateral enamel of Neandertal anterior teeth grow differently from that of modern humans. J. Hum. Evol. 52:72–84.

Guglielmino-Matessi CR, Gluckman P, Cavalli-Sforza LL (1979) Climate and the evolution of skull metrics in man. Am. J. Phys. Anthropol. 50:549–564.

Guiot J, Pons A, de Beaulieu JL, Reille M (1989) A 140,000-year continental climate reconstruction from two European pollen records. Nature 338:309–313.

Guipert G, de Lumley MA, Tuffreau A, Mafart B (2011) A late Middle Pleistocene hominid: Biache-Saint-Vaast 2, north France. C. R. Palevol 10:21–33.

Gunther KA, Smith DW (2004) Interactions between wolves and female grizzly bears with cubs in Yellowstone National Park. Ursus 15:232–238.

Gunz P, Neubauer S, Maureille B, Hublin JJ (2011) Virtual reconstruction of the Le Moustier 2 newborn skull. Implications for Neandertal Ontogeny. Paléo 22:155–172.

Gunz P, Neubauer S, Golovanova L, Doronichev V, Maureille B, Hublin JJ (2012) A uniquely modern human pattern of endocranial development. Insights from a new cranial reconstruction of the Neandertal newborn from Mezmaiskaya. J. Hum. Evol. 62:300–313.

Gurven M, Kaplan H (2006) Determinants of time allocation across the lifespan. Hum. Nature 17:1–49.

Gurven M, Walker R (2006) Energetic demand of multiple dependents and the evolution of slow human growth. Proc. Biol. Sci. 273:835–841.
Guthrie RD (1968) Paleoecology of the large-mammal community in interior Alaska during the Late Pleistocene. Am. Midl. Natur. 79:346–363.
Guthrie RD (1982) Mammals of the mammoth steppe as palaeoenvironmental indicators. In: Hopkins DM, Matthews JV, Schweger CE, Young SB (eds) Palaeoecology of Beringia. Academic Press, New York, pp 307–326.
Guthrie RD (1983) Osseous projectile points: biological considerations affecting raw material selection and design among Paleolithic and Paleoindian peoples. In: Clutton-Brock J, Grigson C (eds) Animals and Archaeology. Volume 1. Hunters and their Prey. British Archaeological Reports IS 163, Oxford, pp 273–294.
Guthrie RD (1990) Frozen Fauna of the Mammoth Steppe. University of Chicago Press, Chicago.
Guthrie RD (2003) Rapid body size decline in Alaskan Pleistocene horses before extinction. Nature 426:169–171.
Gysel LW (1957) Acorn production on good, medium and poor oak sites in southern Michigan. J. Forest. 55:570–574.
Hahn I, Scherer PW, Mozell MM (1993) Velocity profiles measured for airflow through a large-scale model of the human nasal cavity. J. Appl. Physiol. 75:2273–2287.
Haigh J, Maynard Smith J (1972) Population size and protein variation in man. Genet. Res. 19:73–89.
Hall RL (2005) Energetics of nose and mouth breathing, body size, body composition, and nose volume in young adult males and females. Am. J. Hum. Biol. 17:321–330.
Hambücken A (1993a) Révision des particularités de l'humérus des Néandertaliens européens. C R Acad Sci Paris 317, Série II:109–114.
Hambücken A (1993b) Variabilité Morphologique et Métrique de l'Humérus, du Radius et de l'Ulna des Néandertaliens. Comparaison avec l'Homme Moderne. Ph.D. Dissertation, Université de Bordeaux I.
Hammel HT (1955) Thermal properties of fur. Am. J. Physiol. 182:369–376.
Hammel HT (1956) Infrared emissivities of some arctic fauna. J. Mammol. 37:375–378.
Hammer MF, Mendez FL, Cox MP, Woerner AE, Wall JD, Petrov DA (2008) Sex-biased evolutionary forces shape genomic patterns of human diversity. PLoS Genetics 4:e1000202.
Hardin G (1960) The competitive exclusion principle. Science 131:1292–1297.
Hardy BL (1997) Evidence for plant exploitation at the Middle Paleolithic site of Starosele, Crimea. J. Hum. Evol. 32:A7-A8 (abs).
Hardy BL (2004) Neanderthal behaviour and stone tool function at the Middle Palaeolithic site of La Quina, France. Antiquity 78:547–565.
Hardy BL (2010) Climatic variability and plant food distribution in Pleistocene Europe: implications for Neanderthal diet and subsistence. Quat. Sci. Rev. 29:662–679.
Hardy BL, Moncel MH (2011) Neanderthal use of fish, mammals, birds, starchy plants and wood 125-250,000 years ago. PLoS ONE 6:e23768.
Hardy BL, Kay M, Marks AE, Monigal K (2001) Stone tool function at the paleolithic sites of Starosele and Buran Kaya III, Crimea: behavioral implications. Proc. Nat. Acad. Sci. 98:10972–10977.
Hardy BL, Moncel M-H, Daujeard C, Fernandes P, Béarez P, Desclaux E, Chacon Navarro MG, Puaud S, Gallotti R (2013) Impossible Neanderthals? Making string, throwing projectiles and catching small game during Marine Isotope Stage 4 (Abri du Maras, France). Quat. Sci. Rev. 82:23–40.
Hardy JD, DuBois EF (1938) Basal metabolism, radiation, convection and vaporization at temperatures of 22 to 35C. J. Nutr. 15:477–497.
Hardy K, Buckley S, Collins MJ, Estalrrich A, Brothwell D, Copeland L, García-Tabernero A, García-Vargas S, de la Rasilla M, Lalueza-Fox C, Huguet R, Bastir M, Santamaría D, Madella M, Wilson J, Fernández Cortés A, Rosas A (2012) Neanderthal medics? Evidence for food, cooking, and medicinal plants entrapped in dental calculus. Naturwissenschaften 99:617–626.

Harpending H, Bertram J (1975) Human population dynamics in archaeological time: some simple models. Am. Antiquity 40:82–91.
Harpending H, Sherry ST, Rogers AR, Stoneking M (1993) The genetic structure of ancient human populations. Curr. Anthropol. 34:483–496.
Harpending HC, Batzer MA, Gurven M, Jorde LB, Rogers AR, Sherry ST (1998) Genetic traces of ancient demography. Proc. Nat. Acad. Sci. 95:1961–1967.
Harrison GA, Tanner JM, Pilbeam DR, Baker PT (1988) Human Biology, 3rd Edition. Oxford University Press, Oxford.
Harrison SP, Kutzbach JE, Prentice IC, Behling PJ, Sykes MT (1995) The response of northern hemisphere extratropical climate and vegetation to orbitally induced changes in insolation during the last interglaciation. Quatern. Res. 43:174–184.
Hart TB, Hart JA (1986) The ecological basis of hunter-gatherer subsistence in African rain forests. Hum. Ecol. 14:29–55.
Harvati K (2007) Neanderthals and their contemporaries. In: Henke W, Tattersall I (eds) Handbook of Paleoanthropology. Springer-Verlag, Berlin, pp 1717–1748.
Harvati K, Darlas A, Bailey SE, Rein TR, El Zaatari S, Fiorenza L, Kullmer O, Psathi E (2013) New Neanderthal remains from Mani peninsula, Southern Greece: The Kalamakia Middle Paleolithic cave site. J. Hum. Evol. 64:486–499.
Harvey PH, Clutton-Brock TH (1985) Life history variation in primates. Evolution 39:559–581.
Harvey PH, Read AF, Promislow DEL (1989) Life history variation in placental mammals: unifying the data with theory. In: Harvey PH, Partridge L (eds) Oxford Surveys in Evolutionary Biology. Oxford University Press, Oxford, pp 13–31.
Hauser O (1916) La Micoque, die Kultur einer neuen Diluvialrasse. Verlag von Veit & Comp., Leipzig.
Hawkes EW (1916) Skeletal measurements and observations on the Point Barrow Eskimos with comparisons from other Eskimo groups. Am. Anthropol. 18:203–234.
Hawkes K, O'Connell JF, Blurton Jones NG (1997) Hadza women's time allocation, offspring provisioning, and the evolution of long postmenopausal life spans. Curr. Anthropol. 38:551–577.
Hawkes K, O'Connell JF, Blurton Jones NG, Alvarez H, Charnov EL (1998) Grandmothering, menopause, and the evolution of human life histories. Proc. Nat. Acad. Sci. 95:1336–1339.
Hawks J (2012) Dynamics of genetic and morphological variability within Neandertals. J Anthropol Sci 90:1–17.
Hawks J, Wolpoff MH (2001a) Paleoanthropology and the population genetics of ancient genes. Am. J. Phys. Anthropol. 114:269–272.
Hawks J, Wolpoff MH (2001b) The accretion model of Neandertal evolution. Evolution 55:1474–1485.
Hawks J, Dobson S, Wolpoff MH (2000a) Variation in nasal breadth as a test of genetic drift in European Neandertals. Bull. Mem. Soc. Anthropol. Paris 12:431–436.
Hawks J, Hunley K, Lee S-H, Wolpoff M (2000b) Population bottlenecks and Pleistocene human evolution. Mol. Biol. Evol. 17:2–22.
Hayden B (1993) The cultural capacities of Neandertals: a review and re-evaluation. J. Hum. Evol. 24:113–146.
Hayden B (2012) Neandertal social structure? Oxford J. Archaeol. 31:1–26.
Hayward MW (2006) Prey preferences of the spotted hyaena (*Crocuta crocuta*) and degree of dietary overlap with the lion (*Panthera leo*). J. Zool. 270:606–614.
Hayward MW, Kerley GIH (2005) Prey preferences of the lion (*Panthera leo*). J. Zool. 267:309–322.
Hayward MW, Henschel P, O'Brien J, Hofmeyr M, Balme G, Kerley GIH (2006) Prey preferences of the leopard (*Panthera pardus*). J. Zool. 270:298–313.
Hazzah LN (2006) Living among Lions (*Panthera leo*): Coexistence or Killing? Community Attitudes towards Cconservation Initiatives and the Motivations behind Lion Killing in Kenyan

Maasailand. Conservation Biology and Sustainable Development. M.S. Thesis, University of Wisconsin-Madison.
Hedges REM, Pettitt PB, Bronk Ramsey C, van Klinken GJ (1998) Radiocarbon dates from the Oxford AMS system: Archaeometry datelist 25. Archaeometry 40:227–239.
Heffter E, Raichlen D, Kuhn S (2013) Working hard or hardly working? A preliminary study of the metabolic costs of stone knapping. Paleoanthropology 2013:A17-A18 (abs).
Heglund NC, Cavagna GA (1985) Efficiency of vertebrate locomotory muscles. J. Exp. Biol. 115:283–292.
Heglund NC, Cavagna GA (1987) Mechanical work, oxygen-consumption, and efficiency in isolated frog and rat muscle. Am. J. Physiol. 253:C22-C29.
Heinonen AO, Karvonen MJ, Ruosteenoja R (1959) The energy expenditure of walking on snow at various depths. Ergonomics 2:389–394.
Henrich J (2004) Demography and cultural evolution: how adaptive cultural processes can produce maladaptive losses: the Tasmanian case. Am. Antiquity 69:197–214.
Henry-Gambier D, White R (2003) New chrono-cultural data on the Cro-Magnon and Combe-Capelle human remains (Dordogne, France): consequences for the biocultural origins of modern humans in Europe. 2003 Paleoanthropology Society Meetings (abstract: www.paleoanthro.org/abst2003.htm).
Henry AG, Brooks AS, Piperno DR (2011) Microfossils in calculus demonstrate consumption of plants and cooked foods in Neanderthal diets (Shanidar III, Iraq; Spy I and II, Belgium). Proc. Nat. Acad. Sci. 108:486–491.
Henry DO (1995) Prehistoric Cultural Ecology and Evolution: Insights from Southern Jordan. Plenum, New York.
Henry DO (1998) Intrasite spatial patterns and behavioral modernity: indications from the late Levantine Mousterian rockshelter of Tor Faraj, southern Jordan. In: Akazawa T, Aoki K, Bar-Yosef O (eds) Neandertals and Modern Humans in Western Asia. Plenum, New York, pp 127–142.
Henry DO, Hietala HJ, Rosen AM, Demidenko YE, Usik VI, Armagan TL (2004) Human behavioral organization in the Middle Paleolithic: were Neanderthals different? Am. Anthropol. 106:17–31.
Henschel JR, Tilson RL (1988) How much does a spotted hyaena eat? Perspective from the Namib Desert. Afr. J. Ecol. 4:199–230.
Henshilwood CS, Marean CW (2003) The origin of modern human behavior. Curr. Anthropol. 44:627–651.
Henshilwood CS, Sealy JC (1997) Bone artefacts from the Middle Stone Age at Blombos Cave, Southern Cape, South Africa. Curr. Anthropol. 38:890–895.
Henshilwood CS, d'Errico F, Marean CW, Milo RG, Yates R (2001) An early bone tool industry from the Middle Stone Age at Blombos Cave, South Africa: implications for the origins of modern human behaviour, symbolism and language. J. Hum. Evol. 41:631–678
Henshilwood CS, d'Errico F, Yates R, Jacobs Z, Tribolo C, Duller GAT, Mercier N, Sealy JC, Valladas H, Watts I, Wintle AG (2002) Emergence of modern human behaviour: Middle Stone Age engravings from South Africa. Science 295:1278–1280.
Henshilwood C, d'Errico F, Vanhaeren M, van Niekerk K, Jacobs Z (2004) Middle Stone Age shell beads from South Africa. Science 304:404.
Henshilwood CS, d'Errico F, Watts I (2009) Engraved ochres from the Middle Stone Age levels at Blombos Cave, South Africa. J. Hum. Evol. 57:27–47.
Hermansen L, Andersen KL (1965) Aerobic work capacity in young Norwegian men and women. J. Appl. Physiol. 20:425–431.
Hey J (1997) Mitochondrial and nuclear genes present conflicting portraits of human origins. Mol. Biol. Evol. 14:166–172.
Hibbert D (1982) History of the steppe-tundra concept. In: Hopkins DM, Matthews JV, Schweger CE, Young SB (eds) Paleoecology of Beringia. Academic Press, New York, pp 153–156.

Higgins RW, Ruff CB (2011) The effects of distal limb segment shortening on locomotor efficiency in sloped terrain: implications for Neandertal locomotor behavior. Am. J. Phys. Anthropol. 146:336–345.

Higham T, Bronk Ramsey C, Karavanić I, Smith FH, Trinkaus E (2006) Revised direct radiocarbon dating of the Vindija G1 Upper Paleolithic Neandertals. Proc. Nat. Acad. Sci. 103:553–557.

Higham T, Jacobi R, Julien M, David F, Basell L, Wood R, W D, Bronk Ramsey C (2010) Chronology of the Grotte du Renne (France) and implications for the context of ornaments and human remains within the Châtelperronian. Proc. Nat. Acad. Sci. 107:20234–20239.

Higham T, Compton T, Stringer C, Jacobi R, Shapiro B, Trinkaus E, Chandler B, Gröning F, Collins C, Hillson S, O'Higgins P, FitzGerald C, Fagan M (2011) The earliest evidence for anatomically modern humans in northwestern Europe. Nature 479:521–524.

Higham T, Basell L, Jacobi R, Wood R, Bronk Ramsey C, Conard N (2012) Testing models for the beginnings of the Aurignacian and the advent of figurative art and music: the radiocarbon chronology of Geißenklösterle. J. Hum. Evol. 62:664–676.

Hill AF (1952) Economic Botany, 2nd Edition. McGraw-Hill, New York.

Hill K (1993) Life history theory and evolutionary anthropology. Evol. Anthropol. 2:78–88.

Hill K, Hawkes K (1983) Neotropical hunting among the Aché of eastern Paraguay. In: Hames RB, Vickers WT (eds) Adaptive Responses of Native Amazonians. Academic Press, New York, pp 139–188.

Hill K, Hurtado AM (2009) Cooperative breeding in South American hunter-gatherers. Proc. R. Soc. B 276:3863–3870.

Hill K, Kaplan H, Hawkes K, Hurtado AM (1985) Men's time allocation to subsistence work among the Ache of Eastern Paraguay. Hum. Ecol. 13:29–47.

Hill K, Hurtado AM, Walker RS (2007) High adult mortality among Hiwi hunter-gatherers: implications for human evolution. J. Hum. Evol. 52:443–454.

Hilton CE, Greaves RD (2004) Age, sex, and resource transport in Venezuelan foragers. In: Meldrum DJ, Hilton CE (eds) From Biped to Strider: The Emergence of Modern Human Walking, Running, and Resource Transport. Kluwer Academic/Plenum Publishers, New York, pp 163–181.

Hinton RJ (1981) Form and patterning of anterior tooth wear among aboriginal human groups. Am. J. Phys. Anthropol. 54:555–564.

Hirst SM (1969) Predation as a limiting factor of large ungulate populations in a Transvaal lowveld nature reserve. Zool. Afr. 4:199–230.

Hitchcock R, Bleed P (1997) Each according to need and fashion: spear and arrow use among San hunters of the Kalahari. In: Knecht H (ed) Projectile Technology. Plenum, New York, pp 345–368.

Hobson KA, Alisaukas RT, Clark RG (1993) Stable-nitrogen isotope enrichment in avian tissues due to fasting and nutritional stress: implications for isotopic analyses of diet. Condor 95:388–394.

Hockett B (2012) The consequences of Middle Paleolithic diets on pregnant Neanderthal women. Quatern. Int. 264:78–82.

Hockett B, Haws JA (2003) Nutritional ecology and diachronic trends in Paleolithic diet and health. Evol. Anthropol. 12:211–216.

Hockett B, Haws JA (2005) Nutritional ecology and the human demography of Neandertal extinction. Quatern. Int. 137:21–34.

Hoffecker JF (1999) Neanderthals and modern humans in eastern Europe. Evol. Anthropol. 8:129–141.

Hoffecker JF (2002) Desolate Landscapes: Ice-Age Settlement in Eastern Europe. Rutgers University Press, New Brunswick, NJ.

Hoffecker JF (2005) Innovation and technological knowledge in the Upper Paleolithic of northern Eurasia. Evol. Anthropol. 14:186–198.

Hoffecker JF (2009) The spread of modern humans in Europe. Proc. Nat. Acad. Sci. 106:16040–16045.

Hoffecker JF, Cleghorn N (2000) Mousterian hunting patterns in the northwestern Caucasus and the ecology of the Neanderthals. In: Burke A (ed) Hunting in the Middle Paleolithic. Int. J. Osteoarchaeol. 10:368–378.
Hofreiter M, Serre D, Rohland N, Rabeder G, Nagel D, Conard N, Münzel S, Pääbo S (2004) Lack of phylogeography in European mammals before the last glaciation. Proc. Nat. Acad. Sci. 101:12963–12968.
Holdaway S (1989) Were there hafted projectile points in the Mousterian? J. Field Archaeol. 16:79–85.
Holdaway S (1990) Mousterian projectile points - reply to Shea. J. Field Archaeol. 17:114–115.
Holliday TW (1995) Body Size and Proportions in the Late Pleistocene Western Old World and the Origins of Modern Humans. Ph.D. Dissertation, University of New Mexico.
Holliday TW (1997a) Body proportions in Late Pleistocene Europe and modern human origins. J. Hum. Evol. 32:423–447.
Holliday TW (1997b) Postcranial evidence of cold adaptation in European Neandertals. Am. J. Phys. Anthropol. 104:245–258.
Holliday TW (1998) The ecological context of trapping among recent hunter-gatherers: implications for subsistence in terminal Pleistocene Europe. Curr. Anthropol. 39:711–719.
Holliday TW (1999) Brachial and crural indices of European Late Upper Paleolithic and Mesolithic humans. J. Hum. Evol. 36:549–566.
Holliday TW (2000) Evolution at the crossroads: modern human emergence in western Asia. Am. Anthropol. 102:54–68.
Holliday TW (2003) Species concepts, reticulation, and human evolution. Curr. Anthropol. 44:653–673.
Holliday TW (2006) Neanderthals and modern humans: an example of a mammalian syngameon? In: Harvati K, Harrison T (eds) Neanderthals Revisited: New Approaches and Perspectives. Springer, Dordrecht, pp 281–298.
Holliday TW, Churchill SE (2006) Mustelid hunting by recent foragers and the detection of trapping in the European Paleolithic. In: Haws J, Hockett B, Brugal J-P (eds) Paleolithic Zooarchaeology in Practice. British Archaeological Reports IS 1564, Oxford, pp 45–56.
Holliday TW, Hilton CE (2010) Body proportions of circumpolar peoples as evidenced from skeletal data: Ipiutak and Tigara (Point Hope) versus Kodiak Island Inuit. Am. J. Phys. Anthropol. 142:287–302.
Holliday TW, Ruff CB (1997) Ecogeographical patterning and stature prediction in fossil hominids: comment on M.R. Feldesman and R.L. Fountain, American Journal of Physical Anthropology (1996) 100:207–224. Am. J. Phys. Anthropol. 103:137–140.
Holliday TW, Ruff CB (2001) Relative variation in human proximal and distal limb segment lengths. Am. J. Phys. Anthropol. 116:26–33.
Holliday TW, Trinkaus E (1991) Limb/trunk proportions in Neandertals and early anatomically modern humans. Am. J. Phys. Anthropol. Suppl. 12:93–94 (abs).
Holloway RL (1976) Some problems of hominid brain endocast reconstruction, allometry, and neural reorganization. In: Tobias PV, Coppens Y (eds) Les Plus Anciens Hominidés. CNRS, Nice, pp 66–119.
Holloway RL (1978) The relevance of endocasts for studying primate brain evolution. In: Noback CR (ed) Sensory Systems in Primates. Plenum, New York, pp 181–200.
Holloway RL (1981) Volumetric and asymmetry determinations on recent hominid endocasts: Spy I and II, Djebel Irhoud 1, and the Salé *Homo erectus* specimens, with some notes on Neandertal brain size. Am. J. Phys. Anthropol. 55:385–393.
Holloway RL (1985) The poor brain of *Homo sapiens neanderthalensis*: see what you please… In: Delson E (ed) Ancestors: The Hard Evidence. Alan R. Liss, New York, pp 319–324.
Holt BM (1999) Biomechanical Evidence of Decreased Mobility in Upper Paleolithic and Mesolithic Europe. Ph.D. Dissertation, University of Missouri.
Holt BM (2003) Mobility in Upper Paleolithic and Mesolithic Europe: evidence from the lower limb. Am. J. Phys. Anthropol. 122:200–215.

Holton NE, Franciscus RG (2008) The paradox of a wide nasal aperture in cold-adapted Neandertals: a causal assessment. J. Hum. Evol. 56:942–951.
Holton NE, Yokley TR, Franciscus RG (2011) Climatic adaptation and Neandertal facial evolution: a comment on Rae et al. (2011). J. Hum. Evol. 61:624–627.
Höner OP, Wachter B, East ML, Runyoro VA, Hofer H (2005) The effect of prey abundance and foraging tactics on the population dynamics of a social, territorial carnivore, the spotted hyena. Oikos 108:544–554.
Hooijer D (1962) The sabre-toothed cat *Homotherium* found in the Netherlands. Lutra 4: 24–26.
Hopkins DM, Matthews JVJ, Schweger CE, Young SB (eds) (1982) Paleoecology of Beringia. Academic Press, New York.
Horan RD, Bulte E, Shogren JF (2005) How trade saved humanity from biological exclusion: an economic theory of Neanderthal extinction. J. Econ. Behav. Org. 58:1–29.
Hortolà P, Martínez-Navarro B (2013) The Quaternary megafaunal extinction and the fate of Neanderthals: an integrative working hypothesis. Quatern. Int. 296:69–72.
Hovers E (2001) Territorial behavior in the Middle Paleolithic of the southern Levant. In: Conard NJ (ed) Settlement Dynamics of the Middle Paleolithic and Middle Stone Age. Kerns Verlag, Tübingen, pp 123–152.
Hovers E (2006) Neandertals and modern humans in the Middle Paleolithic of the Levant: what kind of interaction? In: Conard NJ (ed) When Neanderthals and Modern Humans Met. Kerns Verlag, Tübingen, pp 65–85.
Hovers E (2009) The Lithic Assemblages of Qafzeh Cave. Oxford University Press, Oxford.
Hovers E, Belfer-Cohen A (2006) "Now you see it, now you don't" – modern human behavior in the Middle Paleolithic. In: Hovers E, Kuhn SL (eds) Transitions Before the Transition: Evolution and Stability in the Middle Paleolithic and Middle Stone Age. Springer, New York, pp 295–304.
Howell FC (1952) Pleistocene glacial ecology and the evolution of "classic Neandertal" man. Southwest J. Anthropol. 8:377–410.
Howell FC (1957) The evolutionary significance of variation and varieties of "Neandertal" man. Quart. Rev. Biol. 32:330–347.
Howell SM, Galinat BJ, Renzi AJ, Marone PJ (1988) Normal and abnormal mechanics of the glenohumeral joint in the horizontal plane. J. Bone Jt. Surg. 70-A(2):227–232.
Howes FN (1948) Nuts: Their Production and Everyday Uses. Faber & Faber, London.
Hrdlicka A (1930) Anthropological Survey in Alaska. 46th Annual Report of the Bureau of American Ethnology, 1928–1929, Washington, D.C., pp 19–374.
Hrdy SB (2005) Evolutionary context of human development. The cooperative breeding model. In: Carter CS, Ahnert L, Grossmann KE, Hrdy SB, Lamb ME, Porges SW, Sachser N (eds) Attachment and Bonding: A New Synthesis. The MIT Press, Cambridge, MA, pp 9–32.
Hublin JJ (1978) Quelques caractères apomorphes du crâne néandertalien et leur interprétation phylogénique. C. R. Acad. Sci. Paris 287:923–926.
Hublin JJ (1980) La Chaise Suard, Engis 2, et La Quina H 18: développement de la morphologie externe chez 1'enfant prénéandertalien et néandertalien. C. R. Acad. Sci. Paris 291:669–672.
Hublin JJ (1988a) Caracteres derives de la region occipito-mastoïdienne chez les Néandertaliens. In: Trinkaus E (ed) L'Homme de Néandertal: L'Anatomie. Etudes et Recherches Archéologiques de l'Université de Liège (E.R.A.U.L.), Liège, pp 67–73.
Hublin JJ (1988b) Les plus anciens representants de la lignee prenéandertalienne. In: Trinkaus E (ed) L'Homme de Néandertal: L'Anatomie. Etudes et Recherches Archéologiques de l'Université de Liège (E.R.A.U.L.), Liège, pp 81–94.
Hublin JJ (1996) The first Europeans. Archaeology 49:36–44.
Hublin JJ (1998) Climatic changes, paleogeography, and the evolution of the Neandertals. In: Akazawa T, Aoki K, Bar-Yosef O (eds) Neandertals and Modern Humans in Western Asia. Plenum, New York, pp 295–310.

Hublin JJ (2012) The earliest modern human colonization of Europe. Proc. Nat. Acad. Sci. 109:13471–13472.

Hublin JJ, Coqueugniot H (2006) Absolute or proportional brain size: that is the question. A reply to Leigh's (2006) comments. J. Hum. Evol. 50:109–113.

Hublin JJ, Roebroeks W (2009) Ebb and flow or regional extinctions? On the character of Neandertal occupation of northern environments. C. R. Palevol. 8:503–509.

Hublin JJ, Barroso Ruiz C, Medina Lara P, Fontugne M, Reyss J-L (1995) The Mousterian site of Zafarraya (Andalucia, Spain): dating and implications on the Paleolithic peopling process of Western Europe. C. R. Acad. Sci. Paris 321:931–937.

Hublin JJ, Spoor F, Braun M, Zonneveld F, Condemi S (1996) A late Neanderthal associated with Upper Palaeolithic artefacts. Nature 381:224–226.

Hublin JJ, Weston D, Gunz P, Richards M, Roebroeks W, Glimmerveen J, Anthonis L (2009) Out of the North Sea: the Zeeland Ridges Neandertal. J. Hum. Evol. 57:777–785.

Hublin JJ, Talamo S, Julien M, David F, Connet N, Bodu P, Vandermeersch B, Richards MP (2012) Radiocarbon dates from the Grotte du Renne and Saint-Césaire support a Neandertal origin for the Châtelperronian. Proc. Nat. Acad. Sci. 109:18743–18748.

Hudson G, Samuels DC, Chinnery PF (2008) Mutation meltdown of mitochondrial DNA and Neanderthal extinction. 58th Annual Meeting of the American Society of Human Genetics, Philadelphia, PA

Huggard DJ (1993) Effect of snow depth on predation and scavenging by gray wolves. J. Wildl. Manage. 57:382–388.

Huntley B, Allen JRM (2003) Glacial environments III: palaeo-vegetation patterns in Last Glacial Europe. In: van Andel T, Davies W (eds) Neanderthals and Modern Humans in the European Landscape During the Last Glaciation. McDonald Institute for Archaeological Research, Cambridge, pp 79–102.

Hurtado AM, Hill KR (1990) Seasonality in a foraging society: variation in diet, work effort, fertility, and sexual division of labor among the Hiwi of Venezuela. J. Anthropol. Res. 46:293–346.

Hurtado AM, Hawkes K, Hill K, Kaplan H (1985) Female subsistence strategies among Ache hunter-gatherers of Eastern Paraguay. Hum. Ecol. 13:1–28.

Hutchinson DL, Larsen CS, Choi I (1997) Stressed to the max? Physiological perturbation in the Krapina Neandertals. Curr. Anthropol. 38:904–914.

Hutchinson V (2000) Late Pleistocene circum-Mediterranean population interaction: evidence from the Tangier (Morocco) subadult fossil maxilla. Am. J. Phys. Anthropol. Suppl. 30:184–185 (abs).

Hylander WL (1972) The Adaptive Significance of Eskimo Craniofacial Morphology. Ph.D. Dissertation, University of Chicago.

Hylander WL (1977) The adaptive significance of Eskimo craniofacial morphology. In: Dahlberg AA, Graber TM (eds) Orofacial Growth and Development. Mouton Publishers, The Hague, pp 129–169.

Illius AW, Gordon IJ (1987) The allometry of food intake in grazing ruminants. J. Animal Ecol. 56:989–999.

Illius AW, Gordon IJ (1992) Modelling the nutritional ecology of ungulate herbivores: evolution of body size and competitive interactions. Oecologia 89:428–434.

INFORM (1998) International Organization Summary: Recommended intakes concerning n-6 to n-3 fatty acid ratios and long chain n-3 fatty acids. Washington DC.

Ingelstedt S (1970) Humidifying capacity of the nose. Ann. Otol. Rhinol. Laryngol. 79:475–480.

Ingman M, Gyllensten U (2007) A recent genetic link between Sami and the Volga-Ural region of Russia. Eur. J. Hum. Gen. 15:115–120.

Ingman M, Kaessmann H, Pääbo S, Gyllensten U (2000) Mitochondrial genome variation and the origin of modern humans. Nature 408:708–713.

Innis SM, Kuhnlein HV (1988) Long-chain *n*-3 fatty acids in breast milk of Inuit women consuming traditional foods. Early Hum. Dev. 18:185–189.

Institute of Medicine (2002) Dietary Reference Intake for Energy, Carbohydrates, Fiber, Fat, Fatty Acids, Cholesterol, Protein and Amino Acids. Food and Nutrition Board, National Academy Press, Washington, DC.

Ion PG, Kershaw GP (1989) The selection of snowpatches as relief habitat by woodland caribou (*Rangifer tarandus caribou*), Macmillan Pass, Selwyn/Mackenzie Mountains, N.W.T., Canada. Arctic Alpine Res. 21:203–211.

Jacobs Z, Roberts RG (2009) Human history written in stone and blood. Am. Sci. 97:2–9.

Jacobs Z, Roberts RG, Galbraith RF, Deacon HJ, Grün R, Mackay A, Mitchell P, Vogelsang R, Wadley L (2008) Ages for the Middle Stone Age of southern Africa: implications for human behavior and dispersal. Science 322:733–735.

James SR (1989) Hominid use of fire in the Lower and Middle Pleistocene. Curr. Anthropol. 30:1–26.

Janković I, Karavanić I, Ahern J, Brajković D, Leonardić J, Smith FH (2011) Archaeological, paleontological and genomic perspectives on late European Neandertals at Vindija cave, Croatia. In: Condemi S, Weniger G-C (eds) Continuity and Discontinuity in the Peopling of Europe. Springer, Dordrecht, pp 299–311.

Jędrzejewski W, Schmidt K, Miłkowski L, Jędrzejewska B, Okarma H (1993) Foraging by lynx and its role in ungulate mortality: the local (Białowieża Forest) and the Palaearctic viewpoints. Acta Theriol. (Warsz.) 38:385–403.

Jędrzejewski W, Schmidt K, Theuerkauf J, Jędrzejewska B, Kowalczyk R (2007) Territory size of wolves *Canis lupus*: linking local (Białowieża Primeval Forest, Poland) and Holarctic-scale patterns. Ecography 30:66–76.

Jelinek AJ (1981) The Middle Palaeolithic in the southern Levant from the perspective of the Tabun Cave. In: Cauvin J, Sanlaville P (eds) Préhistoire du Levant. CNRS Éditions, Paris, pp 265–280.

Jelinek AJ (1982a) The Middle Palaeolithic in the southern Levant with comments on the appearance of modern *Homo sapiens*. In: Ronen A (ed) The Transition from Lower to Middle Palaeolithic and the Origin of Modern Homo sapiens. British Archaeological Reports IS 151, Oxford, pp 57–104.

Jelinek AJ (1982b) The Tabun Cave and Palaeolithic man in the Levant. Science 216:1369–1375.

Jelinek AJ, Debénath A, Dibble HL (eds) (1989) A Preliminary Report on Evidence Related to the Interpretation of Economic and Social Activities of Neandertals at the Site of La Quina (Charente), France. Etudes et Recherches Archéologiques de l'Université de Liège (E.R.A.U.L.), Liège.

Jenike MR (2001) Nutritional ecology: diet, physical activity and body size. In: Panter-Brick C, Layton RH, Rowley-Conwy P (eds) Hunter-Gatherers: An Interdisciplinary Perspective. Cambridge University Press, Cambridge, pp 205–238.

Jenness D (1922) The Life of the Copper Eskimo. F.A. Acland, Ottawa.

Jenness D (1959) People of the Twilight. University of Chicago Press, Chicago.

Jette M, Sidney K, Blumchen G (1990) Metabolic equivalents (METS) in exercise testing, exercise prescription, and evaluation of functional capacity. Clin. Cardiol. 13:555–565.

Jiménez-Espejo F, Martínez-Ruiz F, Finlayson C, Paytan A, Sakamoto T, Ortega-Huertas M, Finlayson G, Iijima K, Gallego-Torres D, Fa D (2007) Climate forcing and Neanderthal extinction in Southern Iberia: insights from a multiproxy marine record. Quat. Sci. Rev. 26:836–852.

Jobin A, Molinari P, Breitenmoser U (2000) Prey spectrum, prey preference and consumption rates of Eurasian lynx in the Swiss Jura Mountains. Acta Theriol. (Warsz.) 45:243–252.

Johnson CN (2002) Determinants of loss of mammal species during the Late Quaternary 'megafauna' extinctions: life history and ecology, but not body size. Proc. R. Soc. Lond. B 269:2221–2227.

Jolicoeur P (1959) Multivariate geographical variation in the wolf *Canis lupus* L. Evolution 13:283–299.

Jolly CJ (2001) A proper study for mankind: analogies from the papionin monkeys and their implications for human evolution. Yearb. Phys. Anthropol. 44:177–204.

Jones M (1997) Character displacement in Australian dasyurid carnivores: size relationships and prey patterns. Ecology 78:2569–2587.
Jones RL, Keen DH (1993) Pleistocene Environments in the British Isles. Chapman & Hall, London.
Jorde LB, Bamshad MJ, Rogers AR (1998) Using mitochondrial and nuclear DNA markers to reconstruct human evolution. Bioessays 20:126–136.
Jørgensen JB (1953) The Eskimo skeleton: contributions to the physical anthropology of the aboriginal Greenlanders. Meddelelser om Grønland 146:1–154.
Jöris O (2006) Bifacially backed knives ("keilmesser") in the central European Middle Palaeolithic. In: Goren-Inbar N, Sharon G (eds) Axe Age: Acheulian Tool-making from Quarry to Discard. Equinox, London, pp 287–310.
Joubert D (2006) Hunting behaviour of lions (*Panthera leo*) on elephants (*Loxodonta africana*) in the Chobe National Park, Botswana. Afr. J. Ecol. 44:279–281.
Julien M (1982) Les Harpons Magdaléniens. 17e supplément a "Gallia prehistoire," Éditions du CNRS, Paris.
Jungers WL, Burr DB (1994) Body size, long bone geometry and locomotion in quadrupedal monkeys. Z. Morph. Anthrop. 80:89–97.
Jungers WL, Burr DB, Cole MS (1998) Body size and scaling of long bone geometry, bone strength and positional behavior in cercopithecoid primates. In: Strasser E, Fleagle J, Rosenberger A, McHenry H (eds) Primate Locomotion: Recent Advances. Plenum, New York, pp 309–330.
Kaagan LM (2000) The Horse in Late Pleistocene and Holocene Britain. Ph.D. Dissertation, University College London.
Kaiser K (1969) The climate of Europe during the Quaternary Ice Age. In: Wright HEJ (ed) Quaternary Geology and Climate. National Academy of Sciences, Washington, pp 10–37.
Kannus P, Haapasalo H, Sankelo M, Sievänen H, Pasanen M, Heinonen A, Oja P, Vuori I (1995) Effect of starting age and physical activity on bone mass in the dominant arm of tennis and squash players. Ann. Intern. Med. 123:27–31.
Kaplan H, Hill K, Lancaster J, Hurtado AM (2000) A theory of human life history evolution: diet, intelligence, and longevity. Evol. Anthropol. 9:156–185.
Kappelman J (1996) Evolution of body mass and relative brain size in fossil hominids. J. Hum. Evol. 30:243–276.
Kappelman J (1997) They might be giants. Nature 387:126–127.
Karanth KU, Sunquist ME (2000) Behavioural correlates of predation by tiger (*Panthera tigris*), leopard (*Panthera pardus*) and dhole (*Cuon alpinus*) in Nagarahole, India. J. Zool. 250:255–265.
Karanth KU, Nichols JD, Kumar NS, Link WA, Hines JE (2004) Tigers and their prey: predicting carnivore densities from prey abundance. Proc. Nat. Acad. Sci. 101:4854–4858.
Karasov WH (1992) Daily energy expenditure and the cost of activity in mammals. Amer. Zool. 32:238–248.
Karatsori E, Renault-Miskovsky J, de Lumley H, Lebreton V (2005) Environnement de l'homme de Néandertal en Ligurie au Pléistocène supérieur. Analyse pollinique de la Caverna delle Fate (Finale Ligure, Italie). C. R. Palevol 4:395–404.
Katch F, Michael ED, Horvath SM (1967) Estimation of body volume by underwater weighing: description of a simple method. J. Appl. Physiol. 23:811–813.
Katzenberg MA (1992) Advances in stable isotope analysis of prehistoric bones. In: Saunder SR, Katzenberg MA (eds) Skeletal Biology of Past Peoples: Research Methods. Wiley-Liss, New York, pp 105–119.
Katzmarzyk PT, Leonard WR (1998) Climatic influences on human body size and proportions: ecological adaptations and secular trends. Am. J. Phys. Anthropol. 106:483–503.
Katzmarzyk PT, Leonard WR, Crawford MH, Sukernik RI (1994) Resting metabolic rate and daily energy expenditure among two indigenous Siberian populations. Am. J. Hum. Biol. 6:719–730.

Katzmarzyk PT, Leonard WR, Stephen MA, Berti PR, Ross AGP (1996) Differences between observed and predicted energy costs at rest and during exercise in three subsistence-level populations. Am. J. Phys. Anthropol. 99:537–545.
Kaufman D (1999) Archaeological Perspectives on the Origins of Modern Humans. Bergin & Garvey, Westport, CT.
Keene AS (1985) Nutrition and economy: models for the study of prehistoric diet. In: Gilbert RIJ, Mielke JH (eds) The Analysis of Prehistoric Diets. Academic Press, Orlando, pp 155–190.
Kelly MJ, Laurenson MK, FitzGibbon CD, Collins DA, Durant SM, Frame GW, Bertram BCR, Caro TM (1998) Demography of the Serengeti cheetah (*Acinonyx jubatus*) population: the first 25 years. J. Zool., Lond. 244:473–488.
Kelly R (1983) Hunter-gatherer mobility strategies. J. Anthropol. Res. 37:277–306.
Kelly R (1995) The Foraging Spectrum. Smithsonian Institution, Washington DC.
Kennedy GE (2003) Palaeolithic grandmothers? Life history theory and early *Homo*. J. R. Anthropol. Inst. 9:549–572.
Kennedy GE (2005) From the ape's dilemma to the weanling's dilemma: early weaning and its evolutionary context. J. Hum. Evol. 48:123–145.
Kennedy KAR (1983) Morphological variations in ulnar supinator crests and fossae as identifying markers of occupational stress. J. For. Sci. 28:871–876.
Kennedy KAR (1989) Skeletal markers of occupational stress. In: İşcan, MY, Kennedy KAR (eds) Reconstruction of Life from the Skeleton. Alan R. Liss, Inc., New York, pp 129–160.
Kennedy KAR (2004) Slings and arrows of predaceous fortune: Asian evidence of prehistoric spear use. Evol. Anthropol. 13:127–131.
Kennett JP, Roark EB, Cannariato KG, Ingram BL, Tada R (2000) Latest Quaternary paleoclimatic and radiocarbon chronology, Hole 1017E, southern California margin. In: Lyle M, Koizumi I, Richter C, Moore TCJ (eds) Proceedings of the Ocean Drilling Program, Scientific Results. Integrated Ocean Drilling Program College Station, TX, pp 249–254.
Kerr D, Morton A, Dick I, Prince R (1996) Exercise effects on bone mass in postmenopausal women are site-specific and load-dependent. J. Bone Miner. Res. 11:218–225.
Key C, Ross C (1999) Sex differences in energy expenditure in non-human primates. Proc. Biol. Sci. 266:2479–2485.
Key CA (2000) The evolution of human life history. World Archaeol. 31:329–350.
Kimbel W (1991) Species, species concepts and hominid evolution. J. Hum. Evol. 20:355–371.
Kimbel WH, Martin LB (eds) (1993) Species, Species Concepts and Primate Evolution. John Wiley, New York.
Kind C-J (1992) Bemerkungen zur Diferenzierung der Süddeutschen Mittelpaläolithikums. Archäologisches Korrespondezblatt 22:151–159.
King AI, Evans FG (1967) Analysis of fatigue strength of human compact bone by the Weibull method. In: Jacobson B (ed) Digest of the 7th International Conference on Medical and Biological Engineering, Stockholm, p 514.
Kleiber M (1961) The Fire of Life: An Introduction to Animal Energetics. John Wiley, New York.
Klein DR (1970) Tundra ranges north of the boreal forest. J. Range Manage. 23:8–14.
Klein RG (1989a) The Human Career. University of Chicago Press, Chicago.
Klein RG (1989b) Why does skeletal part representation differ between smaller and larger bovids at Klasies River Mouth and other archeological sites? J. Archaeol. Sci. 6:363–381.
Klein RG (1992) The impact of early people on the environment: the case of large mammal extinctions. In: Jacobsen JE, Firor J (eds) Human Impact on the Environment: Ancient Roots, Current Challenges. Westview Press, Boulder, CO, pp 13–34.
Klein RG (1995) Anatomy, behavior, and modern human origins. J. World Prehist. 9:167–197.
Klein RG (1999) The Human Career: Human Biological and Cultural Origins, 2nd Edition. University of Chicago Press, Chicago.
Klein RG (2000) Archeology and the evolution of human behavior. Ev. Anthropol. 9:17–36.
Klein RG (2003) Whither the Neanderthals? Science 299:1525–1527.
Klein RG (2008) Out of Africa and the evolution of human behavior. Ev. Anthropol. 17:267–281.

Klein RG, Cruz-Uribe K (1983) Stone Age population numbers and average tortoise size at Byneskranskop Cave 1 and Die Kelders Cave 1, Southern Cape Province, South Africa. S. Afr. Arch. Bull. 38:26–30.

Klein RG, Cruz-Uribe K (1996) Size variation in the rock hyrax (*Procavia capensis*) and Late Quaternary climatic change in South Africa. Quatern. Res. 46:193–207.

Klein RG, Cruz-Uribe K (1998) Comment on "Mousterian large-mammal remains from Kobeh Cave: behavioral implications for Neanderthals and early modern humans" by C.W. Marean and S.Y. Kim. Curr. Anthropol. 39:S96-S97.

Klein RG, Scott K (1989) Glacial/interglacial size variation in fossil spotted hyenas (*Crocuta crocuta*) from Britain. Quatern. Res. 32:88–95.

Klein RG, Steele TE (2008) Gibraltar data are too sparse to inform on Neanderthal exploitation of coastal resources. Proc. Nat. Acad. Sci. 105:E115.

Kline MA, Boyd R (2010) Population size predicts technological complexity in Oceania. Proc. R. Soc. B 277:2559–2564.

Knecht H (1997) Projectile points of bone, antler, and stone: experimental explorations of manufacture and use. In: Knecht H (ed) Projectile Technology. Plenum, New York, pp 191–212.

Koertvelyessy T (1972) Relationships between the frontal sinus and climatic conditions: a skeletal approach to cold adaptation. Am. J. Phys. Anthropol. 37:161–172.

Kok OB, Nel JAJ (2004) Convergence and divergence in prey of sympatric canids and felids: opportunism or phylogenetic constraint? Biol. J. Linn. Soc. 83:527–538.

Kolen J (1999) Hominids without homes: on the nature of Middle Palaeolithic settlement in Europe. In: Roebroeks W, Gamble C (eds) The Middle Palaeolithic Occupation of Europe. Leiden University Press, Leiden, pp 139–175.

Koller J, Baumer U, Mania D (2001) High-tech in the Middle Paleolithic: Neandertal-manufactured pitch identified. Eur. J. Archaeol. 4:385–397.

Kontulainen S, Sievänen H, Kannus P, Pasanen M, Vuori I (2002) Effect of long-term impact-loading on mass, size, and estimated strength of humerus and radius of female racquet-sports players: a peripheral quantitative computed tomography study between young and old starters and controls. J. Bone Miner. Res. 17:2281–2289.

Kormondy EJ, Brown DE (1998) Fundamentals of Human Ecology. Prentice Hall, Upper Saddle River, NJ.

Kortlandt A (2002) Neanderthal anatomy and the use of spears. Evol. Anthropol. 11:183–184.

Kozlowski JK (2003) From bifaces to leaf points. In: Soressi M, Dibble HL (eds) Multiple Approaches to the Study of Bifacial Technologies. University of Pennsylvania, Philadelphia, pp 149–164.

Kramer KL (2005) Children's help and the pace of reproduction: cooperative breeding in humans. Evol. Anthropol. 14:224–237.

Kramer PA (1998) The costs of human locomotion: maternal investment in child transport. Am. J. Phys. Anthropol. 107:71–85.

Kramer PA (2004a) The behavioral ecology of locomotion. In: Meldrum DJ, Hilton CE (eds) From Biped to Strider: The Emergence of Modern Human Walking, Running, and Resource Transport. Kluwer Academic/Plenum Publishers, New York, pp 101–115.

Kramer PA (2004b) Burden transport: when, how and how much? Res. Econ. Anthropol. 23:247–267.

Kramer PA, Sarton-Miller I (2008) The energetics of human walking: is Froude number (*Fr*) useful for metabolic comparisons? Gait Posture 27:209–215.

Kramer PA, Sylvester AD (2009) Bipedal form and locomotor function: understanding the effects of size and shape on velocity and energetics. Paleoanthropology 2009:238–251.

Krause J, Lalueza-Fox C, Orlando L, Enard W, Green RE, Burbano HA, Hublin JJ, Hänni C, Fortea J, de la Rasilla M, Bertranpetit J, Rosas AP, Pablo S (2007a) The derived FOXP2 variant of modern humans was shared with Neandertals. Curr. Biol. 17:1908–1912.

Krause J, Orlando L, Serre D, Viola B, Prüfer K, Richards MP, Hublin JJ, Hänni C, Derevianko AP, Pääbo S (2007b) Neanderthals in central Asia and Siberia. Nature 449:902–904

Krause J, Fu Q, Good JM, Viola B, Shunkov MV, Derevianko AP, Pääbo S (2010) The complete mitochondrial DNA genome of an unknown hominin from southern Siberia. Nature 464:894–897.

Krings M, Capelli C, Tschentscher F, Geisert H, Meyer S, von Haeseler A, Grossschmidt K, Possnert G, Paunovic M, Pääbo S (2000) A view of Neandertal genetic diversity. Nat. Gen. 26:144–146.

Krings M, Stone A, Schmitz RW, Krainitzki H, Stoneking M, Pääbo S (1997) Neandertal DNA sequences and the origin of modern humans. Cell 90:19–30.

Krings M, Geisert H, Schmitz RW, Krainitzki H, Paabo S (1999) DNA sequence of the mitochondrial hypervariable region II from the Neandertal type specimen. Proc. Nat. Acad. Sci. 96:5581–5585.

Kruuk H (1972) The Spotted Hyena: A Study of Predation and Social Behavior. University of Chicago Press, Chicago.

Kruuk H (1976) Feeding and social behaviour of the striped hyaena (*Hyaena vulgaris* Desmarest). E. Afr. Wildl. J. 14:91–111.

Kuhn SL (1989) Projectile weapons and investment in extractive technology in the Middle Paleolithic of western Eurasia. Am. J. Phys. Anthropol. 78:257 (abs).

Kuhn SL (1991) "Unpacking" reduction: lithic raw material economy in the Mousterian of west-central Italy. J Anthropol. Archaeol. 10:76–106.

Kuhn SL (1993) Mousterian technology as adaptive response: a case study. In: Peterkin GL, Bricker HM, Mellars P (eds) Hunting and Animal Exploitation in the Later Palaeolithic and Mesolithic of Europe. Archeological Papers of the American Anthropological Association No. 4, pp 25–31.

Kuhn SL (1995) Mousterian Lithic Technology: An Ecological Perspective. Princeton University Press, Princeton.

Kuhn SL (2011) Neanderthal technoeconomics: an assessment and suggestions for future developments. In: Conard NJ, Richter J (eds) Neanderthal Lifeways, Subsistence and Technology: One Hundred and Fifty Years of Neanderthal Study. Springer, Dordrecht, pp 99–110.

Kuhn SL, Stiner MC (2006) What's a Mother to Do? Curr. Anthropol. 47:953–981.

Kuhn SL, Stiner MC (2007) Paleolithic ornaments: implications for cognition, demography and identity. Diogenes 54:40–48.

Kulemeyer J (1988) Die alt- und mittelpaläolithischen Funde von Kärlich. Ph.D. Dissertation, University of Köln.

Kunkel K, Pletscher DH (2001) Winter hunting patterns of wolves in and near Glacier National Park, Montana. J. Wildl. Manage. 65:520–530

Kunkel KE, Ruth TK, Pletscher DH, Hornocker MG (1999) Winter prey selection by wolves and cougars in and near Glacier National Park, Montana. J. Wildl. Manage. 63:901–910.

Kunkel KE, Pletscher DH, Boyd DK, Ream RR, Fairchild MW (2004) Factors correlated with foraging behavior of wolves in and near Glacier National Park, Montana. J. Wildl. Manage. 68:167–178.

Kurtén B (1965) The carnivora of the Palestine caves. Acta Zool. Fennica 107:1–74.

Kurtén B (1968) Pleistocene Mammals of Europe. Weidenfeld & Nicolson, London.

Kurtén B (1976) The Cave Bear Story. Columbia University Press, New York.

Kurtén B (1985) The Pleistocene lion of Beringia. Ann. Zool. Fennici 22:117–121.

Kushnir H, Leitner H, Ikanda D, Packer C (2010) Human and ecological risk factors for unprovoked lion attacks on humans in southeastern Tanzania. Hum. Dimen. Wildl. 15:315–331.

Kutzbach JE, Wright HE (1985) Simulation of the climate of 18,000 years BP. Results for the North American/North Atlantic/European sector and comparison with the geological record of North America. Quat. Sci. Rev. 4:147–187.

Lahr MM, Foley RA (2003) Demography, dispersal and human evolution in the Last Glacial period. In: van Andel T, Davies W (eds) Neanderthals and Modern Humans in the European Landscape During the Last Glaciation. McDonald Institute for Archaeological Research, Cambridge, pp 241–256.

Lai CSL, Fisher SE, Hurst JA, Vargha-Khadem F, Monaco AP (2001) A forkhead-domain gene is mutated in a severe speech and language disorder. Nature 413:519–523.

Lalueza-Fox C, Pérez-Pérez A (1993) The diet of the Neanderthal Child Gibraltar 2 (Devil's Tower) through the study of the vestibular striation pattern. J. Hum. Evol. 24:29–41.

Lalueza C, Pérez-Pérez A, Turbón D (1996) Dietary inferences through buccal microwear analysis of Middle and Upper Pleistocene human fossils. Am. J. Phys. Anthropol. 100:367–387.

Lalueza-Fox C, Sampietro ML, Caramelli D, Puder Y, Lari M, Calafell F, Martínez-Maza C, Bastir M, Fortea J, de la Rasilla M, Bertranpetit J, Rosas A (2005) Neandertal evolutionary genetics: mitochondrial DNA data from the Iberian peninsula. Mol. Biol. Evol. 22:1077–1081.

Lalueza-Fox C, Römpler H, Caramelli D, Stäubert C, Catalano G, Hughes D, Rohland N, Pilli E, Longo L, Condemi S, de la Rasilla M, Fortea J, Rosas A, Stoneking M, Schöneberg T, Bertranpetit J, Hofreiter M (2007) A melanocortin 1 receptor allele suggests varying pigmentation among Neanderthals. Science 318:1453–1455.

Lalueza-Fox C, Rosas A, Estalrrich A, Gigli I, Campos PF, García-Tabernero A, García-Vargas S, Sánchez-Quinto F, Ramírez O, Civit S, Bastir M, Huguet R, Santamaría D, Gilbert MTP, Willerslev E, de la Rasilla M (2011) Genetic evidence for patrilocal mating behavior among Neandertal groups. Proc. Nat. Acad. Sci. 108:250–253.

Landa A, Strand O, Swenson JE, Skogland T (1997) Wolverines and their prey in southern Norway. Can. J. Zool. 75:1292–1299.

Langley MC, Clarkson C, Ulm S (2008) Behavioural complexity in Eurasian Neanderthal populations: a chronological examination of the archaeological evidence. Cambridge Archaeol. J. 18:289–307.

Lanyon LE, Rubin CT (1984) Regulation of bone formation by applied dynamic loads. J. Bone Jt. Surg. 66A:397–402.

Lanyon LE, Magee PT, Baggott DG (1979) The relationship of functional stress and strain to the processes of bone remodelling. An experimental study of the sheep radius. J. Biomech. 12:593–600.

Larivière S, Crête M (1993) The size of eastern Coyotes (*Canis latrans*): a comment. J. Mammol. 74:1072–1074.

Larkin JL, Grimes RA, Cornicelli L, Cox JJ, Maehr DS (2001) Returning elk to Appalachia: foiling Murphy's Law. In: Maehr DS, Noss RF, Larkin JL (eds) Large Mammal Restoration. Island Press, Washington, pp 101–117.

Larsen CS (1997) Bioarchaeology: Interpreting Behavior from the Human Skeleton. Cambridge University Press, Cambridge.

Larsen E, Sejrup HP, Johnsen SJ, Knudsen KL (1995) Do Greenland ice cores reflect NW European interglacial climate variation? Quatern. Res. 43:125–132.

Lauder GV (1995) On the inference of function from structure. In: Thomasen JJ (ed) Functional Morphology in Vertebrate Paleontology. Cambridge University Press, Cambridge, pp 1–18.

Laughlin WS (1951) The Alaska gateway viewed from the Aleutian Islands. In: Laughlin WS (ed) Physical Anthropology of the American Indian. The Viking Fund, Inc., New York, pp 98–126.

Lawrence F, Whitehead RG (1984) Maintenance energy cost of pregnancy in rural Gambian women and influence of dietary status. Lancet ii:363–365.

Layton R, O'Hara S (2010) Human social evolution: a comparison of hunter-gatherer and chimpanzee social organization. Proc. Brit. Acad. 158:83–113.

Lazenby R, Smashnuk A (1999) Osteometric variation in the Inuit second metacarpal: a test of Allen's rule. Int. J. Osteoarchaeol. 9:182–188.

Le Gall O (1992) Poissons et pêches au Paléolithique: quelques données de l'Europe occidentale. L'Anthropologie 96:121–134.

Leakey LN, Milledge SAH, Leakey SM, Edung J, Haynes P, Kiptoo DK, McGeorge A (1999) Diet of striped hyaena in northern Kenya. Afr. J. Ecol. 37:314–326.

Lebel S, Trinkaus E, Faure M, Fernandez P, Guérin C, Richter D, Mercier N, Valladas H, Wagner GA (2001) Comparative morphology and paleobiology of Middle Pleistocene human remains from the Bau de l'Aubesier, Vaucluse, France. Proc. Nat. Acad. Sci. 98:11097–11102.

Lee RB (1968) What hunters do for a living, or, how to make out on scarce resouces. In: Lee RB, DeVore I (eds) Man the Hunter. Aldine, Chicago, pp 30–48.
Lee RB (1979) The !Kung San: Men, Women, and Work in a Foraging Society. Cambridge University Press, Cambridge.
Lee RB, DeVore I (eds) (1968) Man the Hunter. Aldine de Gruyter, Chicago.
Lehmann J, Korstjens AH, Dunbar RIM (2007) Fission–fusion social systems as a strategy for coping with ecological constraints: a primate case. Evol. Ecol. 21:613–634.
Lemorini C (2000) Reconnaître des Tactiques d'Exploitation du Milieu au Paléolithique Moyen: La Contribution de l'Analyse Fonctionelle: Etude Fonctionelle des Industries Lithiques de Grotta Breuil (Latium, Italie) et de La Combette (Bonnieux, Vaucluse, France). British Archaeological Reports IS 858, Oxford.
Leonard JA, Vilà C, Fox-Dobbs K, Koch PL, Wayne RK, Van Valkenburgh B (2007) Megafaunal extinctions and the disappearance of a specialized wolf ecomorph. Curr. Biol. 17:1146–1150.
Leonard WR (2003) Measuring human energy expenditure: what have we learned from the flex-heart rate method? Am. J. Hum. Biol. 15:479–489.
Leonard WR (2008) Lifestyle, diet, and disease: comparative perspectives on the determinants of chronic health risks. In: Stearns SC, Koella JC (eds) Evolution in Health and Disease, 2nd Edition. Oxford University Press, Oxford, pp 265–276.
Leonard WR, Robertson ML (1992) Nutritional requirements and human evolution: a bioenergetics model. Am. J. Hum. Biol. 4:179–195.
Leonard WR, Katzmarzyk PT, Comuzzie AG, Crawford MH, Sukernik RI (1994) Growth and nutritional status of the Evenki reindeer herders of Siberia. Am. J. Hum. Biol. 6:339–350.
Leonard WR, Katzmarzyk PT, Crawford MH (1996) Energetics and population ecology of Siberian herders. Am. J. Hum. Biol. 8:275–289.
Leonard WR, Sorensen MV, Galloway VA, Spencer GJ, Mosher MJ, Osipova L, Spitsyn VA (2002) Climatic influences on basal metabolic rates among circumpolar populations. Am. J. Hum. Biol. 14:609–620.
Leonard WR, Snodgrass JJ, Sorensen MV (2005) Metabolic adaptation in indigenous Siberian populations. A. Rev. Anthropol. 34:451–471.
Leonard WR, Robertson ML, Snodgrass JJ (2007) Energetic models of human nutritional evolution. In: Ungar P (ed) Evolution of the Human Diet: the Known, the Unknown, and the Unknowable. Oxford University Press, New York, pp 344–359.
Lev E, Kislev ME, Bar-Yosef O (2005) Mousterian vegetal food in Kebara Cave, Mt. Carmel. J. Archaeol. Sci. 32:475–484.
Lévêque F, Vandermeersch B (1980) Découverte de restes humains dans un niveau castelperronien à Saint-Césaire (Charente-Maritime). C. R. Acad. Sci. Paris 291:187–189.
Lewis-Williams D (2002) The Mind in the Cave. Thames & Hudson, London.
Lewis ME (1997) Carnivoran paleoguilds of Africa: implications for hominid food procurement strategies. J. Hum. Evol. 32:257–288.
Lewis ME, Werdelin L (2007) Patterns of change in the Plio-Pleistocene carnivorans of eastern Africa: implications for hominin evolution. In: Bobe R, Alemseged Z, Behrensmeyer AK (eds) Hominin Environments in the East African Pliocene: An Assessment of the Faunal Evidence. Springer, Dordrecht, pp 77–105.
Lewis T (1930) Observations upon the reactions of the vessels of the human skin to cold. Heart 15:177–208.
Leyton J, Drury PJ, Crawford MA (1987) Differential oxidation of saturated and unsaturated fatty acids in vivo in the rat. Br. J. Nutr. 57:383–393.
Li H, Durbin R (2011) Inference of human population history from individual whole-genome sequences. Nature 475:493–496.
Liapunova R (1975) Essays on the Ethnography of the Aleuts. University of Alaska Press, Fairbanks.

Lichtenfeld LL (2005) Our Shared Kingdom at Risk: Human-Lion Relationships in the 21st Century. Ph.D. Dissertation, Yale University.
Lieberman DE (1993) The rise and fall of seasonal mobility among hunter-gatherers: the case of the southern Levant. Curr. Anthropol. 34:599–614.
Lieberman DE (1998) Neandertal and early modern human mobility patterns. In: Akazawa T, Aoki K, Bar-Yosef O (eds) Neandertals and Modern Humans in Western Asia. Plenum, New York, pp 263–275.
Lieberman DE (2011) The Evolution of the Human Head. Harvard University Press, Cambridge, MA.
Lieberman DE, Shea JJ (1994) Behavioral differences between archaic and modern humans in the Levantine Mousterian. Am. Anthropol. 96:300–332.
Lieberman DE, Devlin MJ, Pearson OM (2001) Articular area responses to mechanical loading: effects of exercise, age, and skeletal location. Am. J. Phys. Anthropol. 116:266–277.
Lieberman DE, McBratney BM, Krovitz G (2002) The evolution and development of cranial form in *Homo sapiens*. Proc. Nat. Acad. Sci. 99:1134–1139.
Lieberman DE, Polk JD, Demes B (2003) Predicting long bone loading from cross-sectional geometry. Am. J. Phys. Anthropol. 123:156–171.
Lillie M, Richards MP (2000) New radiocarbon dates and palaeodietary evidence from the Ukrainian Mesolithic. J. Archaeol. Sci. 27:965–972.
Lindstedt SL, Boyce MS (1985) Seasonality, fasting endurance, and body size in mammals. Am. Nat. 125:873–878.
Linn RM, Kriegshauser LA (1991) Ball thrower's fracture of the humerus: a case report. Am. J. Sports Med. 19:194–197.
Linnell JDC, Strand O (2000) Interference interactions, co-existence and conservation of mammalian carnivores. Divers. Dist. 6:169–176.
Lioubine VP (1993) La chronostratigraphie du Paléolithique du Caucase. L'Anthropologie 97:291–298.
Lofroth EC, Krebs JA, Harrower WL, Lewis D (2007) Food habits of wolverine *Gulo gulo* in montane ecosystems of British Columbia, Canada. Wildl. Biol. 13 Suppl 2:31–37.
Loison A, Langvatn R, Solberg E (1999) Body mass and winter mortality in red deer calves: disentangling sex and climate effects. Ecogeography 22:20–30.
Lombard M, Phillipson L (2010) Indications of bow and stone-tipped arrow use 64 000 years ago in KwaZulu-Natal, South Africa. Antiquity 84:1–14.
Lovejoy CO, Trinkaus E (1980) Strength and robusticity of the Neandertal tibia. Am. J. Phys. Anthropol. 53:465–470.
Lowe J, Barton N, Blockley S, Ramsey CB, Cullen VL, Davies W, Gamble C, Grant K, Hardiman M, Housley R, Lane CS, Lee S, Lewis M, MacLeod A, Menzies M, Müller W, Pollard M, Price C, Roberts AP, Rohling EJ, Satow C, Smith VC, Stringer CB, Tomlinson EL, White D, Albert P, Arienzo I, Barker G, Boric D, Carandente A, Civetta L, Ferrier C, Guadelli JL, Karkanas P, Koumouzelis M, Müller UC, Orsi G, Pross J, Rosi M, Shalamanov-Korobar L, Sirakov N, Tzedakis PC (2012) Volcanic ash layers illuminate the resilience of Neanderthals and early modern humans to natural hazards. Proc. Nat. Acad. Sci. 109:13532–13537.
Loy TH, Hardy BL (1992) Blood residue analysis of 90,000-year-old stone tools from Tabun Cave, Israel. Antiquity 66:24–35.
Lucas G, Rigaud J-P, Simek JF, Soressi M (2007) The Châtelperronian of Grotte XVI, Cénac-et-Saint-Julien (Dordogne, France). In: Riel-Salvatore J, Clark GA (eds) New Approaches to the Study of Early Upper Paleolithic 'Transitional' Industries in Western Eurasia: Transitions Great and Small. Archaeopress, Oxford, pp 33–40.
Lusk G (1931) The Science of Nutrition. W.B. Saunders & Co., Philadelphia.
Macchiarelli R, Bondioli L, Debénath A, Mazurier A, Tournepiche JF, Birch W, Dean MC (2006) How Neanderthal molar teeth grew. Nature 444:748–751.
MacDonald DH, Hewlett BS (1999) Reproductive interests and forager mobility. Curr. Anthropol. 40:501–523.

Macdonald DW (1984) Encyclopedia of Mammals. Facts on File Inc., New York.
Macdonald K, Roebroeks W, Verpoorte A (2009) An energetics perspective on the Neandertal record. In: Hublin JJ, Richards M (eds) The Evolution of Hominid Diets: Integrating Approaches to the Study of Palaeolithic Subsistence. Springer, Dordrecht, pp 211–220.
MacDonald ML, Rogers QR, Morris JG (1984) Nutrition of the domestic cat, a mammalian carnivore. Annu. Rev. Nutr. 4:521–562.
Madella M, Jones M, Goldberg P, Goren Y, Hovers E (2002) The exploitation of plant resources by Neanderthals in Amud Cave, Israel: the evidence from phytolith studies. J. Archaeol. Sci. 29:703–719.
Madrigal TC, Holt JZ (2002) White-tailed deer meat and marrow return rates and their application to eastern woodlands archaeology. Am. Antiquity 67:745–759.
Majolo B, de Bortoli Vizioli A, Schino G (2008) Costs and benefits of group living in primates: group size effects on behaviour and demography. Anim. Behav. 76:1235–1247.
Maki JM (2012) Biomechanics of spear throwing, with implications for fossil hominins. Am. J. Phys. Anthropol. Suppl. 54:200–201 (abs).
Mallye J-B, Thiébaut C, Mourre V, Costamagno S, Claud É, Weisbecker P (2012) The Mousterian bone retouchers of Noisetier Cave: experimentation and identification of marks. J. Archaeol. Sci. 39:1131–1142.
Mania D, Toepfer V (1973) Königsaue: Gliederung, Oekologie und mittelpaläolithische Funde der Letzten Eiszeit. VEB Deutscher Verlag der Wissenschaften, Berlin.
Manning TH (1944) Hunting implements and methods of the present-day Eskimos of North-West Hudson Bay, Melville Peninsula, and South-West Baffin Island. Geographr. J. 103:137–152.
Manzi G, Vienna A, Hauser G (1996) Developmental stress and cranial hypostosis by epigenetic trait occurence and distribution: an exploratory study on the Italian Neandertals. J. Hum. Evol. 30:511–527.
Manzi G, Gracia A, Arsuaga J-L (2000) Cranial discrete traits in the Middle Pleistocene humans from the Sima de los Huesos (Sierra de Atapuerca, Spain). Does hypostosis represent any increase in "ontogenetic stress" along the Neandertal lineage? J. Hum. Evol. 38:425–446.
Manzi G, Bruner E, Caprasecca S, Gualdi G, Passarello P (2001) CT-scanning and virtual reproduction of the Saccopastore Neandertal crania. Riv. Antropol. 79:61–72.
Manzi G, Magri D, Milli S, Palombo MR, Margari V, Celiberti V, Barbieri M, Barbieri M, Melis RT, Rubini M, Ruffo M, Saracino B, Tzedakis PC, Zarattini A, Biddittu I (2010) The new chronology of the Ceprano calvarium (Italy). J. Hum. Evol. 59:580–585.
Marean CW (1998) A critique of the evidence for scavenging by Neandertals and early modern humans: new data from Kobeh Cave (Zagros Mountains, Iran) and Die Kelders Cave 1 Layer 10 (South Africa). J. Hum. Evol. 35:111–136.
Marean CW, Assefa Z (1999) Zooarcheological evidence for the faunal exploitation behavior of Neandertals and early modern humans. Evol. Anthropol. 8:22–37.
Marean CW, Ehrhardt CL (1995) Paleoanthropological and paleoecological implications of the taphonomy of a sabretooth's den. J. Hum. Evol. 29:515–547.
Marean CW, Kim SY (1998) Mousterian large-mammal remains from Kobeh Cave: behavioral implications for Neanderthals and early modern humans. Curr. Anthropol. 39:S79-S113.
Marean CW, Bar-Matthews M, Bernatchez J, Fisher E, Goldberg P, Herries AIR, Jacobs Z, Jerardino A, Karkanas P, Minichillo T, Nilssen PJ, Thompson E, Watts I, Williams HM (2007) Early human use of marine resources and pigment in South Africa during the Middle Pleistocene. Nature 449:905–908.
Mariotti V, Belcastro MG (2011) Lower limb entheseal morphology in the Neandertal Krapina population (Croatia, 130 000 BP). J. Hum. Evol. 60:694–702.
Marks A (1989) Early Mousterian settlement patterns in the central Negev, Israel: their social and economic implications. In: Patou M, Freeman LG (eds) L'Homme de Neandertal: La Subsistance. Etudes et Recherches Archéologiques de l'Université de Liège (E.R.A.U.L.), Liège, pp 115–126.

Marks AE (1992) Typological variability in the Levantine Middle Paleolithic. In: Dibble HL, Mellars PA (eds) The Middle Paleolithic: Adaptation, Behavior and Variability. University of Pennsylvania Museum Press, Philadelphia, pp 127–141.
Marks AE, Chabai VP (2001) Constructing Middle Paleolithic settlement systems in Crimea: potentials and limitations. In: Conard NJ (ed) Settlement Dynamics of the Middle Paleolithic and Middle Stone Age. Kerns Verlag, Tübingen, pp 179–204.
Marks AE, Chabai VP (2006) Stasis and change during the Crimean Middle Paleolithic. In: Hovers E, Kuhn SL (eds) Transitions Before the Transition: Evolution and Stability in the Middle Paleolithic and Middle Stone Age. Springer, New York, pp 121–135.
Marlowe FW (2002) Why the Hadza are still hunter-gatherers. In: Kent S (ed) Ethnicity, Hunter-Gatherers, and the "Other": Association or Assimilation in Africa. Smithsonian Institution Press, Washington, DC, pp 247–275.
Marlowe FW (2003) A critical period for provisioning by Hadza men. Implications for pair bonding. Evol. Hum. Behav. 24:217–229.
Marlowe FW (2005) Hunter-gatherers and human evolution. Evol. Anthropol. 14:54–67.
Márquez S, Gannon P, Lawson W, Reidenberg J, Laitman J (2002) Were Neanderthals full of "NO" gas? The relationship between paranasal sinus morphology and nitric oxide production. Am. J. Phys. Anthropol. Suppl. 34:107 (abs).
Márquez S, Gannon PJ, Mowbray K, Laitman JT, Copes L, Lawson W (2005) Neanderthal peripheral nasal apparatus reconstruction: a comparative approach to understand adaptive plasticity. Am. J. Phys. Anthropol. Suppl. 40:145 (abs).
Martín-González JA, Mateos A, Goikoetxea I, Leonard WR, Rodríguez J (2012) Differences between Neandertal and modern human infant and child growth models. J. Hum. Evol. 63:140–149.
Martin PS, Klein RG (eds) (1984) Quaternary Extinctions: A Prehistoric Revolution. University of Arizona Press, Tucson, AZ.
Martin RB, Burr DB (1989) Structure, Function, and Adaptation of Compact Bone. Raven Press, New York.
Martin RD (1983) Human Brain Evolution in an Ecological Context. American Museum of Natural History, New York.
Martin RD (2007) The evolution of human reproduction: a primatological perspective. Yearb. Phys. Anthropol. 50:59–84.
Martínez-Moreno J, Mora R, de la Torre I (2004) Methodological approach for understanding Middle Palaeolithic settlement dynamics at La Roca dels Bous (Noguera, Catalunya, northeast Spain). In: Conard N (ed) Settlement Dynamics of the Middle Paleolithic and Middle Stone Age, Volume II. Kerns Verlag, Tübingen, pp 393–414.
Martini FH (1998) Fundamentals of Anatomy and Physiology, 4th Edition. Prentice Hall, Upper Saddle River, NJ.
Martinón-Torres M, Bermúdez de Castro JM, Gómez-Robles A, Prado-Simón L, Arsuaga JL (2012) Morphological description and comparison of the dental remains from Atapuerca-Sima de los Huesos site (Spain). J. Hum. Evol. 62:7–58.
Masterton JM, Richardson FA (1979) Humidex, A Method of Quantifying Human Discomfort Due to Excessive Heat and Humidity. Environment Canada, Downsview ON.
Matsutani A (1987) Plant remains from the 1984 excavations at Douara Cave. In: Akazawa T, Sakaguchi Y (eds) Paleolithic Site of Doura Cave. Bulletin no. 21. Tokyo University, Tokyo, pp 117–122.
Maureille B, Bar D (1999) The premaxilla in Neandertal and early modern children: ontogeny and morphology. J. Hum. Evol. 37:137–152.
Maureille B, Braga J (2002) Between the incisive bone and premaxilla: from African apes to *Homo sapiens*. In: Minugh-Purvis N, McNamara KJ (eds) Human Evolution through Developmental Change. Johns Hopkins University Press, Baltimore, pp 464–478.
Maureille B, Houët F (1998) La variabilité morpho-métrique du nez: dérive génique dans la lignée néandertalienne? Biometr. Hum. Anthropol. 16:27–33.

Mazza PPA, Martini F, Sala B, Magi M, Colombini MP, Giachi G, Landucci F, Lemorini C, Modugno F, Ribechini E (2006) A new Palaeolithic discovery: tar-hafted stone tools in a European Mid-Pleistocene bone-bearing bed. J. Archaeol. Sci. 33:1310–1318.
McBrearty S, Brooks AS (2000) The revolution that wasn't: a new interpretation of the origin of modern human behavior. J. Hum. Evol. 39:453–563.
McBrearty S, Tryon C (2006) From Acheulean to Middle Stone Age in the Kapthurin Formation, Kenya. In: Hovers E, Kuhn SL (eds) Transitions Before the Transition: Evolution and Stability in the Middle Paleolithic and Middle Stone Age. Springer, New York, pp 257–277.
McCall S, Naples V, Martin L (2003) Assessing behavior in extinct animals: was *Smilodon* social? Brain Beh. Evol. 61:159–164.
McCown TD, Keith A (1939) The Stone Age of Mount Carmel II: The Fossil Human Remains from the Levalloiso-Mousterian. Clarendon Press, Oxford.
McDougall I, Brown FH, Fleagle JG (2005) Stratigraphic placement and age of modern humans from Kibish, Ethiopia. Nature 433:733–736.
McHenry HM (1992) Body size and proportions in early hominids. Am. J. Phys. Anthropol. 87:407–431.
McNab BK (1971) On the ecological significance of Bergmann's rule. Ecology 52:845–854.
McNab BK (1974) The energetics of endotherms. Ohio J. Sci. 74:370–380.
McNab BK (1983) Energetics, body size, and the limits to endothermy. J. Zool. 199:1–29.
McNab BK (1990) The physiological significance of body size. In: Damuth J, MacFadden BJ (eds) Body Size in Mammalian Paleobiology: Estimation and Biological Implications. Cambridge University Press, Cambridge, pp 11–23.
McNab BK (2002) The Physiological Ecology of Vertebrates: A View from Energetics. Cornell University Press, Ithaca, NY.
Meade GE (1961) The saber-toothed cat, *Dinobastis serus*. Tex. Meml. Mus. Bull. 2:23–60.
Mech DL (1970) The Wolf: The Ecology and Behavior of an Endangered Species. University of Minnesota Press, Minneapolis.
Mech LD (1994) Travel speeds of arctic wolves. J. Mammal. 75:741–742.
Mech LD, Boitani L (2003) Wolf social ecology. In: Mech LD, Boitani L (eds) Wolves: Behavior, Ecology, and Conservation. University of Chicago Press, Chicago, pp 1–34.
Mech LD, Korb M (1978) An unusually long pursuit of a deer by a wolf. J. Mammal. 59:860–861.
Meehan JPJ (1955) Individual and racial variations in a vascular response to a cold stimulus. Mil. Med. 116:330–334.
Meignen L, Bar-Yosef O (1988) Kebara et le Paléolithique moyen du Mont Carmel. Paléorient 14:123–130.
Meignen L, Bar-Yosef O (1991) Les outillages lithiques Moustériens de Kébara (fouilles 1982–1985). In: Bar-Yosef O, Vandermeersch B (eds) Le Squelette Moustérien de Kébara 2. Éditions CNRS, Paris, pp 49–75.
Meignen L, Bar-Yosef O, Goldberg P (1989) Les structures de combustion moustériennes de la grotte de Kébara (Mont Carmel, Israël). In: Olive M, Taborin Y (eds) Nature et Fonction des Foyers Préhistoriques. Mémoires du Musée de Préhistoire d'Ille de France 2, Nemours, pp 141–146.
Meignen L, Brugal J-P, Théry-Parisot I, Marquet JC (2001) Territorial exploitation, technical traditions and environment in a mid-altitude context: the Canalettes rockshelter (Grands Causses, France). In: Conard NJ (ed) Settlement Dynamics of the Middle Paleolithic and Middle Stone Age. Kerns Verlag, Tübingen, pp 463–483.
Meikle AW, Stringham JD, Woodward MG, Nelson JC (1988) Hereditary and environmental influences on the variation of thyroid hormones in normal male twins. J. Clin. Endocrinol. Metab. 66:588–592.
Meiri S, Dayan T (2003) On the validity of Bergmann's rule. J. Biogeogr. 30:331–351.
Meiri S, Dayan T, Simberloff D (2004) Carnivores, biases and Bergmann's rule. Biol. J. Linn. Soc. 81:579–588.

Meister K (2000) Injuries to the shoulder in the throwing athlete. Part one: Biomechanics/pathophysiology/classification of injury. Am. J. Sports Med. 28:265–275.

Mellars P (1986) A new chronology for the French Mousterian period. Nature 322:410–411.

Mellars P (1989a) Major issues in the emergence of modern humans. Curr. Anthropol. 30:349–385.

Mellars P (1989b) Technological changes across the Middle-Upper Palaeolithic transition: economic, social and cognitive perspectives. In: Mellars P, Stringer CB (eds) The Human Revolution: Behavioural and Biological Perspectives on the Origins of Modern Humans. Princeton University Press, Princeton, NJ, pp 338–365.

Mellars P (1996) The Neanderthal Legacy: An Archaeological Perspective from Western Europe. Princeton University Press, Princeton.

Mellars P (2005) The impossible coincidence. A single-species model for the origins of modern human behavior in Europe. Evol. Anthropol. 14:12–27.

Mellars P (2006a) A new radiocarbon revolution and the dispersal of modern humans in Eurasia. Nature 439:931–935.

Mellars P (2006b) Why did modern human populations disperse from Africa ca. 60,000 years ago? A new model. Proc. Nat. Acad. Sci. 103:9381–9386.

Mellars P (2007) Rethinking the human revolution: Eurasian and African perspectives. In: Mellars P, Bar-Yosef O, Stringer C, Boyle K (eds) Rethinking the Human Revolution. Cambridge University Press, Cambridge, pp 1–11.

Mellars P, French JC (2011) Tenfold population increase in western Europe at the Neandertal-to-modern human transition. Science 333:623–627.

Mellars PA (1965) Sequence and development of Mousterian traditions in south-western France. Nature 205:626–627.

Mellars PA (1969) The chronology of Mousterian industries in the Périgord region of southwest France. Proc. Prehist. Soc. 6:134–171.

Mercader J (2002) Forest people: the role of African rainforests in human evolution and dispersal. Evol. Anthropol. 11:117–124.

Mercier N, Valladas H (2003) Reassessment of TL age estimates of burnt flints from the Paleolithic site of Tabun Cave, Israel. J. Hum. Evol. 45:401–409.

Mercier N, Valladas H, Joron J-L, Reyss J-L, Lévêque F, Vandermeersch B (1991) Thermoluminescence dating of the late Neanderthal remains from Saint-Césaire. Nature 351:737–739.

Mercier N, Valladas H, Valladas G, Reyss JL, Jelinek A, Meignen L, Joron JL (1995) TL dates of burnt flints from Jelinek's excavations at Tabun and their implications. J. Archaeol. Sci. 22:495–509.

Mercier N, Vallidas H, Bar-Yosef O, Vandermeersch B, Stringer CB, Joron J-L (1993) Thermoluminescence date for the Mousterian site of Es Skhul, Mount Carmel. J. Archaeol. Sci. 20:169–174.

Meyer M, Kircher M, Gansauge MT, Li H, Racimo F, Mallick S, Schraiber JG, Jay F, Prufer K, de Filippo C, Sudmant PH, Alkan C, Fu Q, Do R, Rohland N, Tandon A, Siebauer M, Green RE, Bryc K, Briggs AW, Stenzel U, Dabney J, Shendure J, Kitzman J, Hammer MF, Shunkov MV, Derevianko AP, Patterson N, Andres AM, Eichler EE, Slatkin M, Reich D, Kelso J, Pääbo S (2012) A high-coverage genome sequence from an archaic Denisovan individual. Science 338:222–226.

Miles AEW (1997) The glenoid notch and the shape of the glenoid cavity of the scapula. J. Anat. 191:475.

Miller JE (1960) Javelin thrower's elbow. J. Bone Jt. Surg. 42B:788–792.

Milliken S (2007) Neanderthals, anatomically modern humans, and 'Modern human behaviour' in Italy. Oxford J. Archaeol. 26:331–358.

Mills MGL (1990) Kalahari Hyenas: Comparative Behavioral Ecology of Two Species. Unwin Hyman Ltd, London.

Mills MGL, Harvey M (2001) African Predators. Struik, Cape Town.

Mills MGL, Hofer H (eds) (1998) Hyaenas: Status Survey and Conservation Action Plan. World Conservation Union (IUCN), Gland, Switzerland.

Mills MGL, Shenk TM (1992) Predator-prey relationships: the impact of lion predation on wildebeest and zebra populations. J. Animal Ecol. 61:693–702.

Minagawa M, Wada E (1984) Stepwise enrichment of ^{15}N along food chains: further evidence and the relation between ^{15}N and animal age. Geochim. Cosmochim. Acta 48:1135–1140.

Minetti AE, Moia C, Roi GS, Susta D, Ferretti G (2002) Energy cost of walking and running at extreme uphill and downhill slopes. J. Appl. Physiol. 93:1039–1046.

Miracle P (2005) Late Mousterian subsistence and cave use in Dalmatia: the zooarchaeology of Mujina Pećina, Croatia. Int. J. Osteoarchaeol. 15:84–105.

Mishmar D, Ruiz-Pesini E, Golik P, Macaulay V, Clark AG, Hosseini S, Brandon M, Easley K, Chen E, Brown MD, Sukernik RI, Olckers A, Wallace DC (2002) Natural selection shaped regional mtDNA variation in humans. Proc. Nat. Acad. Sci. 100:171–176.

Mittendorf R, Williams MA, Berkey CS, Cotter PF (1990) The length of uncomplicated human gestation. Obstet. Gynecol. 75:929–932.

Moen AN (1973) Wildlife Ecology: an Analytical Approach. W.H. Freeman and Company, San Francisco.

Mohanty C, Prasad R, Reddy AS, Ghosh JK, Singh TB, Dasb BK (2005) Maternal anthropometry as predictors of low birth weight. J. Trop. Ped. 52:24–29.

Moncel MH (2001) Microlithic Middle Palaeolithic assemblages in central Europe and elephant remains. In: Cavarretta G, Giola P, Mussi M, Palombo MR (eds) La terra delgli Elefanti: The World of Elephants, Proceedings of the 1st International Congress. Consiglio nazionale delle ricerche, Rome, pp 314–317.

Moncel MH (2011) Technological behavior and mobility of human groups deduced from lithic assemblages in the late Middle and early Late Pleistocene of the Middle Rhône Valley (France). In: Conard NJ, Richter J (eds) Neanderthal Lifeways, Subsistence and Technology: One Hundred and Fifty Years of Neanderthal Study. Springer, Dordrecht, pp 261–287.

Moncel MH, Rivals F (2011) On the question of short-term Neanderthal site occupations: Payre, France (MIS 8-7) and Taubach/Weimar, Germany (MIS 5). J. Anthropol. Res. 67:47–75.

Moncel MH, Moigne AM, Combier J (2005) Pre-Neanderthal behaviour during isotope stage 9 and the beginning of stage 8. New data concerning fauna and lithics in the different occupation levels of Orgnac 3 (Ardèche, south-east France): occupation types. J. Archaeol. Sci. 32:1283–1301.

Moncel MH, Moigne AM, Sam Y, Combier J (2011) The emergence of Neanderthal technical behavior: new evidence from Orgnac 3 (Level 1, MIS 8), southeastern France. Curr. Anthropol. 52:37–75.

Monnier GF (2006) The Lower/Middle Paleolithic periodization in western Europe: an evaluation. Curr. Anthropol. 47:709–744.

Montet-White A (1996) Le Paléolithique en ancienne Yougoslavie. Jérôme Million, Grenoble.

Montoye HJ, Kemper HCG, Saris WHM, Washburn RA (1996) Measuring Physical Activity and Energy Expenditure. Human Kinetics, Champaign, IL.

Moote I (1955) The thermal insulation of caribou pelts. Tex. Res. J. 25:832–837.

Mora R, de la Torre I, Martínez-Moreno J (2004) Middle Palaeolithic mobility and land use in the Southwestern Pyrenees: the example of Level 10 in La Roca dels Bous (Noguera, Catalunya, northeast Spain). In: Conard NJ (ed) Settlement Dynamics of the Middle Paleolithic and Middle Stone Age. Volume II. Kerns Verlag, Tübingen, pp 415–435.

Moran EF (1981) Human adaptation to arctic zones. A. Rev. Anthropol. 1981:1–25.

Morin E, Laroulandie V (2012) Presumed symbolic use of diurnal raptors by Neanderthals. PLoS ONE 7:e32856.

Moss E (1987) Polish G and the question of hafting. In: Stordeur D (ed) La Main et l'Outil. Manches et Emmanchements Préhistoriques. Travaux de la Maison de l'Orient, Lyon, pp 97–102.

Mounier A, Condemi S, Manzi G (2011) The stem species of our species: a place for the archaic human cranium from Ceprano, Italy. PLoS ONE 6:e18821.

Mount LN (1979) Adaptations to a Thermal Environment. Arnold, London.

Mourre V, Villa P, Henshilwood CS (2010) Early use of pressure flaking on lithic artifacts at Blombos Cave, South Africa. Science 330:659–662.
Movius HL (1950) A wooden spear of Third Interglacial age from Lower Saxony. Southwest J. Anthropol. 6:139–142.
Movius HL (1953) Mousterian cave of Teshik-Tash, southeastern Uzbekistan, central Asia. Bull. Am. Sch. Prehist. Res. 17:11–71.
Müller UC, Pross J, Tzedakis PC, Gamble C, Kotthoff U, Schmiedl G, Wulf S, Christanis K (2011) The role of climate in the spread of modern humans into Europe. Quat. Sci. Rev. 30:273–279.
Munroe M, Shah CP, Badgley R, Bain HW (1984) Birth weight, length, head circumference and bilirubin level in Indian newborns in the Sioux Lookout Zone, northwestern Ontario. Can. Med. Assoc. J. 131:453–456.
Murdock G (1967) The Ethnographic Atlas: A summary. Ethnology 6.
Murphy NJ, Bulkow LR, Schraer CD, Lanier AP (1993) Prevalence of diabetes mellitus in pregnancy among Yup'ik Eskimos, 1987–1988. Diabetes Care 16:315–317.
Mussi M (2001a) Earliest Italy: An Overview of the Italian Paleolithic and Mesolithic. Kluwer Academic/Plenum Publishers, New York.
Mussi M (2001b) Humans, carnivores and caves in the Italian Middle Paleolithic record. In: Conard NJ (ed) Settlement Dynamics of the Middle Paleolithic and Middle Stone Age. Kerns Verlag, Tübingen, pp 507–517.
Mussi M, Palombo MR (2001) Human/carnivore interaction in the Middle Pleistocene of Latium (Central Italy): an open question. In: Cavarretta G, Gioia P, Mussi M, Palombo MR (eds) La terra delgli Elefanti: The World of Elephants. Consiglio Nazionale delle Ricerche Rome, pp 67–75.
Muttoni G, Scardia G, Kent DV, Swisher CC, Manzi G (2009) Pleistocene magnetochronology of early hominin sites at Ceprano and Fontana Ranuccio, Italy. Earth Planet Sci. Lett. 286:255–268.
Myhre R, Myrberget S (1975) Diet of wolverines (*Gulo gulo*) in Norway. J. Mammol. 56:752–757.
Myo-Thein, Lammert O, Garby L (1985) Effect of trunk load on the energy of walking and ergometer cycling. Eur. J. Appl. Physiol. 54:122–124.
Nagel D, Hilsberg S, Benesch A, Scholz J (2003) Functional morphology and fur patterns in recent and fossil *Panthera* species. Scripta Geol. 126:227–240.
Nelson AJ, Thompson JL (2002) Neanderthal adolescent postcranial growth. In: Minugh-Purvis N, McNamara K (eds) Human Evolution through Developmental Change. Johns Hopkins University Press, Baltimore, pp 442–463.
Nelson BK, DeNiro MJ, Schoeninger MJ, DePaolo DJ (1986) Effects of diagenesis on strontium, carbon, nitrogen and oxygen concentration and isotopic composition of bone. Geochim. Cosmochim. Acta 50:1941–1949.
Nelson E, Rolian C, Cashmore L, Shultz S (2011) Digit ratios predict polygyny in early apes, *Ardipithecus*, Neanderthals and early modern humans but not in *Australopithecus*. Proc. R. Soc. B 278:1556–1563.
Nelson RA, Folk GEJ, Pfeiffer EW, Craighead JJ, Jonkel CJ, Steiger DL (1983) Behavior, biochemistry, and hibernation in black, grizzly, and polar bears. Bears: Their Biol. Manage. 5:284–290.
Nelson SM (1990) Diversity of the Upper Paleolithic "venus" figurines and archeological mythology. In: Kehoe AB, Nelson SM (eds) Powers of Observation: Alternative Views in Archeology. American Anthropological Association Archaeological Paper, pp 11–22.
Newman MT (1953) The application of ecological rules to the racial anthropology of the New World. Am. Anthropol. 55:311–327.
Newman MT (1956) Adaptation of man to cold climates. Evolution 10:101–105.
Nichol JE (1999) Geomorphological evidence and Pleistocene refugia in Africa. Geogr. J. 165:79–89.
Niewoehner WA, Weaver AH, Trinkaus E (1997) Neandertal capitate-metacarpal articular morphology. Am. J. Phys. Anthropol. 103:219–233.

Niinimma V, Cole P, Mintz S, Shephard RJ (1980) The switching point from nasal to oronasal breathing. Resp. Phys. 42:61–71.

Nisbet RM, Muller EB, Lika K, Kooijman SALM (2000) From molecules to ecosystems through dynamic energy budget models. J. Animal Ecol. 69:913–926.

Niven L, Steele TE, Rendu W, Mallye J-B, McPherron SP, Soressi M, Jaubert J, Hublin J-J (2012) Neandertal mobility and large-game hunting: the exploitation of reindeer during the Quina Mousterian at Chez-Pinaud Jonzac (Charente-Maritime, France). J. Hum. Evol. 63:624–635.

Noakes TD (2000) Exercise and the cold. Ergonomics 43:1461–1479.

Nobel PS (1999) Physiochemical and Environmental Plant Physiology, 2nd Edition. Academic Press, San Diego.

Noonan J, Coop G, Kudaravalli S, Smith D, Krause J, Alessi J, Chen F, Platt D, Pääbo S, Pritchard J, Rubin E (2006) Sequencing and analysis of Neandertal genomic DNA. Science 314:1113–1118.

Nordburg M (1998) On the probability of Neanderthal ancestry. Am. J. Hum. Gen. 63:1237–1240.

Nowell A (2006) From a Paleolithic art to Pleistocene visual cultures (introduction to two special issues on 'advances in the study of Pleistocene imagery and symbol use'). J. Archaeol. Method Theory 13:239–249.

Nowell A (2010) Defining behavioral modernity in the context of Neandertal and anatomically modern human populations. A. Rev.Anthropol 39:437–452.

O'Connell JF (2006) How did modern humans displace Neanderthals? Insights from hunter-gatherer ethnography and archaeology. In: Conard NJ (ed) When Neanderthals and Modern Humans Met. Kerns Verlag, Tübingen, pp 43–64.

O'Connell JF, Hawkes K, Blurton Jones N (1988) Hadza scavenging: implications for Plio-Pleistocene hominid subsistence. Curr. Anthropol. 29:356–363.

O'Connor CF, Franciscus RG, Holton NE (2005) Bite force production capability and efficiency in Neandertals and modern humans. Am. J. Phys. Anthropol. 127:129–151.

O'Regan HJ, Turner A, Wilkinson DM (2002) European quaternary refugia: a factor in large carnivore extinction? J. Quat. Sci. 17:789–795.

Oakley KP (1956) Fire as Palaeolithic tool and weapon. Proc. Prehist. Soc. 21:36–46.

Oakley KP, Andrews P, Keeley LH, Clark JD (1977) A reappraisal of the Clacton spearpoint. Proc. Prehist. Soc. 43:13–30.

Odell GH, Cowan F (1986) Experiments with spears and arrows on animal targets. J. Field Archaeol. 13:195–212.

Odum EP (1975) Ecology, 2nd Edition. Holt, Rinehart and Winston, New York.

Oftedal OT, Gittleman JL (1989) Patterns of energy output during reproduction in carnivores. In: Gittleman JL (ed) Carnivore Behavior, Ecology, and Evolution. Chapman and Hall, London, pp 355–378.

Ogilvie M, Curran BK, Trinkaus E (1989) Incidence and patterning of dental enamel hypoplasia among the Neandertals. Am. J. Phys. Anthropol. 79:25–42.

Ohki M, Naito K, Cole P (1991) Dimensions and resistances of the human nose: racial differences. Laryngoscope 101:276–278.

Oksanen L, Oksanen T (2000) The logic and realism of the hypothesis of exploitation ecosystems. Am. Nat. 155:703–723.

Olariu A, Skog G, Hellborg R, Stenström K, Faarinen M, Persson P, Alexandrescu E (2003) Dating of two Paleolithic human fossil bones from Romania by accelerator mass spectrometry. arXiv:physics/0309110v1 [physics.soc-ph]

Olive M, Taborin Y (eds) (1989) Nature et Fonction des Foyers Préhistoriques. Mémoires du Musée de Préhistoire d'Ille de France 2, Nemours.

Olsen SL (1989) Solutré: a theoretical approach to the reconsruction of Upper Palaeolithic hunting strategies. J. Hum. Evol. 18:295–327.

Olszewski DI, Dibble HL (1994) The Zagros Aurignacian. Curr. Anthropol. 35:68–75.

Oppert JM, Dussault JH, Tremblay A, Despres JP, Theriault G, Bouchard C (1994) Thyroid hormones and thyrotropin variations during long term overfeeding in identical twins. J. Clin. Endocrinol. Metab. 79:547–553.

Orlando L, Darlu P, Toussaint M, Bonjean D, Otte M, Hänni C (2006) Revisiting Neandertal diversity with a 100,000 year old mtDNA sequence. Curr. Biol. 16:R400-R402.
Osborn A (1980) Comment on "Maritime hunter-gatherers: ecology and prehistory" by D.R. Yesner. Curr. Anthropol. 21:740–741.
Oswalt WH (1976) An Anthropological Analysis of Food-Getting Technology. John Wiley & Sons, New York.
Outram A, Rowley-Conwy P (1998) Meat and utility indices for horse (*Equus*). J. Archaeol. Sci. 25:839–849.
Ovchinnikov IV, Götherström A, Romanova GP, Kharitonov VM, Lidén K, Goodwin W (2000) Molecular analysis of Neanderthal DNA from the northern Caucasus. Nature 404:490–493.
Ovington JD (1965) Organic production, turnover and mineral cycling in woodlands. Biol. Rev. 40:295–336.
Owen-Smith N (1987) Pleistocene extinctions: the pivotal role of megaherbivores. Paleobiology 13:351–362.
Packer C, Scheel D, Pusey AE (1990) Why lions form groups: food is not enough. Am. Nat. 136:1–19.
Packer C, Ikanda D, Kissui B, Kushnir H (2005) Conservation biology: lion attacks on humans in Tanzania. Nature 436:927–928.
Palma di Cesnola A, Messeri P (1967) Quatre dents humaines paléolithiques trouvées dans des cavernes de l'Italie méridionale. L'Anthropologie 71:249–262.
Palmqvist P, Arribas A, Martínez-Navarro B (1999) Ecomorphological study of large canids from the lower Pleistocene of southeastern Spain. Lethaia 32:75–88.
Palmqvist P, Mendoza M, Arribas A, Gröcke DR (2002) Estimating the body mass of Pleistocene canids: discussion of some methodological problems and a new 'taxon free' approach. Lethaia 35:358–360.
Palomares F, Caro TM (1999) Interspecific killing among mammalian carnivores. Am. Nat. 153:492–508.
Panter-Brick C (2002) Sexual division of labor: energetic and evolutionary scenarios. Am. J. Hum. Biol. 14:627–640.
Pappas AM, Zawacki RM, Sullivan TJ (1985) Biomechanics of baseball pitching: a preliminary report. Amer. J. Sports Med. 13:216–222.
Park LY (2011) Effective population size of current human population. Gen. Res. 93:105–114.
Park R, Epstein S (1960) Carbon isotope fractionation during photosynthesis. Geochim. Cosmochim. Acta 21:110–126.
Parker AC (1910) Iroquois Uses of Maize and Other Food Plants. New York State Mus. Bull. 144.
Parker AG (2009) Pleistocene climate change in Arabia: developing a framework for hominin dispersal over the last 350 ka. In: Petraglia MD, Rose JJ (eds) The Evolution of Human Populations in Arabia. Springer, Dordrecht, pp 39–49.
Partington MW, Roberts N (1969) The heights and weights of Indian and Eskimo school children on James Bay and Hudson Bay. Can. Med. Assoc. J. 100:502–509.
Patou M (1989) Subsistance et approvisionnement au Paléolithique moyen. In: Otte M (ed) L'Homme de Néandertal: La Subsistance. Etudes et Recherches Archéologiques de l'Université de Liège (E.R.A.U.L.), Liège, pp 11–18.
Patou-Mathis M (1996) Stress biologiques et comportements de subsistance au Paléolithique moyen et supérieur en Europe. In: Otte M (ed) Nature et Culture. Etudes et Recherches Archéologiques de l'Université de Liège (E.R.A.U.L.), Liège, pp 443–453.
Patou-Mathis M (2000) Neanderthal subsistence behaviors in Europe. In: Burke A (ed) Hunting in the Middle Paleolithic. Int. J. Osteoarchaeol. 10:379–395.
Patou-Mathis M (2004) Subsistence behaviours in a Middle Palaeolithic site in Poland: the Raj Cave. Int. J. Osteoarchaeol. 14:244–255.
Paul GS (1988) Predatory Dinosaurs of the World. Simon and Schuster, New York.
Paunovic M, Smith FH (2002) Taphonomy of lower vertebrates from Vindija cave (Croatia): delicacy on the Neandertal table or animal prey? J. Hum. Evol. 42:A27 (abs).

Pavlov P, Svendsen JI, Indrelid S (2001) Human presence in the European arctic nearly 40,000 years ago. Nature 413:64–67.

Pavlov P, Roebroeks W, Svendsen JI (2004) The Pleistocene colonization of northeastern Europe: a report on recent research. J. Hum. Evol. 47:3–17.

Pawlik AF, Thissen JP (2011) Hafted armatures and multi-component tool design at the Micoquian site of Inden-Altdorf, Germany. J. Archaeol. Sci. 38:1699–1708.

Pearce E, Stringer C, Dunbar RIM (2013) New insights into differences in brain organization between Neanderthals and anatomically modern humans. Proc. R. Soc. B 280:20130168.

Pearson O, Jungers W, Grine F, Mowbray K (2001) The reliability of estimates of hominin body mass derived from bi-iliac breadth and stature. Am. J. Phys. Anthropol. Suppl. 32:118 (abs).

Pearson OM (2000) Postcranial differences between Neandertals and cold-adapted recent humans. Am. J. Phys. Anthropol. Suppl. 30:247 (abs).

Pearson OM, Lieberman DE (2004) The aging of Wolff's "law": ontogeny and responses to mechanical loading in cortical bone. Yearb. Phys. Anthropol. 47:63–99.

Pearson OM, Cordero RM, Busby AM (2006) How different were Neanderthals' habitual activities? A comparative analysis with diverse groups of recent humans. In: Harvati K, Harrison T (eds) Neanderthals Revisited: New Approaches and Perspectives. Springer, New York, pp 135–156.

Pennisi E (2013) More genomes from Denisova Cave show mixing of early human groups. Science 340:799.

Peresani M, Fiore I, Gala M, Romandini M, Tagliacozzo A (2011) Late Neandertals and the intentional removal of feathers as evidenced from bird bone taphonomy at Fumane Cave 44 ky B.P., Italy. Proc. Nat. Acad. Sci. 108:3888–3893.

Pérez-Pérez A, Lalueza C, Turbón D (1994) Intraindividual and intragroup variability of buccal tooth striation pattern. Am. J. Phys. Anthropol. 94:175–187.

Pérez-Pérez A, Espurz V, Bermúdez de Castro JM, de Lumley MA, Turbón D (2003) Non-occlusal dental microwear variability in a sample of Middle and Late Pleistocene human populations from Europe and the Near East. J. Hum. Evol. 44:497–513.

Perlès C (1976) Le feu. In: de Lumley H (ed) La Préhistoire Française, Vol. 1. CNRS, Paris, pp 679–683.

Perrot J (1968) La Préhistoire Palestinienne. Letouzey & Ané, Paris.

Perry J, Glousman RE (1995) Biomechanics of throwing. In: Nicholas JA, Hershman EB, Posner MA (eds) The Upper Extremity in Sports Medicine, 2nd Edition. Mosby, St. Louis, pp 99–117.

Peterkin GL, Bricker HM, Mellars P (eds)(1993) Hunting and Animal Exploitation in the Later Palaeolithic and Mesolithic of Eurasia. Archeological Papers of the American Anthropological Association No.4.

Peters J, Gautier A, Brink JS, Haenen W (1994) Late Quaternary extinction of ungulates in sub-Saharan Africa: a reductionist's approach. J. Archaeol. Sci. 21:17–28.

Peterson J (1998) The Natufian hunting conundrum: spears, atlatls, or bows? Musculoskeletal and armature evidence. Int. J. Osteoarchaeol. 8:378–389.

Pettitt PB (1997) High resolution Neanderthals? Interpreting Middle Palaeolithic intrasite spatial data. World Archaeol. 29:208–224.

Pettitt PB (1998) Neanderthal extinction: radiocarbon chronology, problems, prospects and interpretation of the existing data. Revue d'Archéometrie Suppl. (proceedings of the 3rd international radiocarbon in archeology conference, Lyon):165–177.

Pettitt PB (2000) Neandertal lifecycles: developmental and social phases in the lives of the last archaics. World Archaeol. 31:351–366.

Pettitt PB (2003) The Mousterian in action: chronology, mobility, and Middle Palaeolithic variability. In: Moloney N, Shott MJ (eds) Lithic Analysis at the Millennium. Institute of Archaeology, University College London, London, pp 29–44.

Pettitt PB (2010) The Palaeolithic Origins of Human Burial. Routledge, London.

Pettitt PB, Pike AWG (2001) Blind in a cloud of data: problems with the chronology of Neanderthal extinction and anatomically modern human expansion. Antiquity 75:415–420.

Pettitt PB, Davies W, Gamble CS, Richards MB (2003) Palaeolithic radiocarbon chronology: quantifying our confidence beyond two half-lives. J. Archaeol. Sci. 30:1685–1693.

Peyrony D (1930) Le Moustier: ses gisements, ses industries, ses couches géologiques. Rev. Anthropol. 40:48–76, 155–176.

Pianka ER (1994) Evolutionary Ecology, 5th Edition. Harper Collins College Publishers, New York.

Pienaar UdeV (1969) Predator-prey relationships amongst the larger mammals of the Kruger National Park. Koedoe 12:108–176.

Pierce BM, Bleich VC, Bowyer T (2000) Selection of mule deer by mountain lions and coyotes: effects of hunting styles, body size, and reproductive status. J. Mammal. 81:462–472.

Pimental N, Pandolf K (1979) Energy expenditure while standing or walking slowly uphill or downhill with loads. Ergonomics 22:963–973.

Pinhasi R, Higham TFG, Golovanova LV, Doronichev VB (2011) Revised age of late Neanderthal occupation and the end of the Middle Paleolithic in the northern Caucasus. Proc. Nat. Acad. Sci. 108:8611–8616.

Pinhasi R, Nioradze M, Tushabramishvili N, Lordkipanidze D, Pleurdeau D, Moncel MH, Adler DS, Stringer C, Higham TF (2012) New chronology for the Middle Palaeolithic of the southern Caucasus suggests early demise of Neanderthals in this region. J. Hum. Evol. 63:770–780.

Ponce de León MS, Zollikofer CPE (1999) New evidence from Le Moustier 1: computer-assisted reconstruction and morphometry of the skull. Anat. Rec. 254:474–489.

Ponce de León MS, Zollikofer CPE (2001) Neanderthal cranial ontogeny and its implications for late hominid diversity. Nature 412:534–538.

Ponce de León MS, Golovanova L, Doronichev V, Romanova G, Akazawa T, Kondo O, Ishida H, Zollikofer CPE (2008) Neanderthal brain size at birth provides insights into the evolution of human life history. Proc. Nat. Acad. Sci. 105:13764–13768.

Porat N, Chazan M, Schwarcz H, Horwitz LK (2002) Timing of the Lower to Middle Paleolithic boundary: new dates from the Levant. J. Hum. Evol. 43:107–122.

Porter AMW (1999) Modern human, early modern human and Neanderthal limb proportions. Int. J. Osteoarchaeol. 9:54–67.

Powell A, Shennan S, Thomas MG (2009) Late Pleistocene demography and the appearance of modern human behavior. Science 324:1298–1301.

Pratt RM, Putman RJ, Ekins JR, Edwards PJ (1986) Use of habitat by free-ranging cattle and ponies in the New Forest, southern England. J. Appl. Ecol. 23:539–557.

Proctor DF (1977) The upper airways: I. Nasal physiology and defense of the lungs. Am. Rev. Resp. Disease 115:97–129.

Proctor DF (1982) The upper airway. In: Proctor DF, Andersen I (eds) The Nose, Upper Airway Physiology and the Atmospheric Environment. Elsevier Biomedical Press, New York, pp 23–41.

Pulliam HR (1988) Sources, sinks, and population regulation. Am. Nat. 132:652–661.

Pushkina D, Raia P (2008) Human influence on distribution and extinctions of the late Pleistocene Eurasian megafauna. J. Hum. Evol. 54:769–782.

Puymerail L, Volpato V, Debénath A, Mazurier A, Tournepiche JF, Macchiarelli R (2012) A Neanderthal partial femoral diaphysis from the "grotte de la Tour," La Chaise-de-Vouthon (Charente, France): outer morphology and endostructural organization. C. R. Palevol 11:581–593.

Pycraft WP (1928) Rhodesian Man: description of the skull and other human remains from Broken Hill. In: Pycraft WP, Elliot Smith G, Yearsley M, Carter JT, Smith RA, Hopwood AT, Bate DMA, Swinton WE (eds) Rhodesian Man and Associated Remains. British Museum, London, pp 1–51.

Quam RM, Smith FH (1998) A reassessment of the Tabun C2 mandible. In: Akazawa T, Aoki K, Bar-Yosef O (eds) Neandertals and Modern Humans in Western Asia. Plenum, New York, pp 405–421.

Quam RM, Arsuaga J-L, Bermúdez de Castro JM, Díez CJ, Lorenzo C, Carretero M, García N, Ortega AI (2001) Human remains from Valdegoba Cave (Huérmeces, Burgos, Spain). J. Hum. Evol. 41:385–435.

Queiroz MS, Shao Y, Ismail-Beigi F (2004) Effect of thyroid hormone on uncoupling protein-3 mRNA expression in rat heart and skeletal muscle. Thyroid 14:177–185.

Rabinovich R, Hovers E (2004) Faunal analysis from Amud Cave: preliminary results and interpretations. In: Burke A (ed) The Ecology of Neanderthals. Int. J. Osteoarchaeol. 14:287–306.

Rabinowitz AR, Walker SR (1991) The carnivore community in a dry tropical forest mosaic in Huai Kha Khaeng Wildlife Sanctuary, Thailand. J. Trop. Ecol. 7:37–47.

Rae T, Hill R, Hamada Y, Koppe T (2003) Clinal variation of maxillary sinus volume in Japanese macaques (*Macaca fuscata*). Am. J. Primatol. 59:153–158.

Rae TC, Strand Viðarsdóttir U, Jeffery N, Steegmann ATJ (2006) Developmental response to cold stress in cranial morphology of *Rattus*: implications for the interpretation of climatic adaptation in fossil hominins. Proc. R. Soc. B 273:2605–2610.

Rae TC, Koppe T, Stringer CB (2011) The Neanderthal face is not cold adapted. J. Hum. Evol. 60:234–239.

Raichlen DA, Armstrong H, Lieberman DE (2011) Calcaneus length determines running economy: implications for endurance running performance in modern humans and Neandertals. J. Hum. Evol. 60:299–308.

Rak Y (1986) The Neanderthal: a new look at an old face. J. Hum. Evol. 15:151–164.

Rak Y (1998) Does any Mousterian cave present evidence of two hominid species? In: Akazawa T, Aoki K, Bar-Yosef O (eds) Neandertals and Modern Humans in Western Asia. Plenum, New York, pp 353–366.

Rak Y, Arensburg B (1987) Kebara 2 Neanderthal pelvis: first look at a complete inlet. Am. J. Phys. Anthropol. 81:323–332.

Rak Y, Kimbel WH, Hovers E (1996) On Neandertal autapomorphies discernible in Neandertal infants: a response to Creed-Miles *et al.* J. Hum. Evol. 30:155–158.

Ramírez Rozzi FV, Bermúdez de Castro JM (2004) Surprisingly rapid growth in Neanderthals. Nature 428:936–939.

Ramírez Rozzi FV, d'Errico F, Vanhaeren M, Grootes PM, Kerautret B, Dujardin V (2009) Cutmarked human remains bearing Neandertal features and modern human remains associated with the Aurignacian at Les Rois. J. Anthropol. Sci. 87:153–185.

Ransom JE (1946) Aleut natural-food economy. Am. Anthropol. 48:607–623.

Rasmussen K (1931) The Netsilik Eskimos. Report of the 5th Thule Expedition. Gyldendal, Copenhagen.

Rawn-Schatzinger V (1992) The scimitar cat *Homotherium serum* Cope: osteology, functional morphology, and predatory behavior. Ill. St. Mus. Rep. Invest. 47:1–80.

Ready E (2013) Neandertal foraging during the late Mousterian in the Pyrenees: new insights based on faunal remains from Gatzarria Cave. J. Archaeol. Sci. 40:1568–1578.

Recheis W, Macchiarelli R, Seidler H, Weaver DS, Schäfer K, Bondioli L, Weber GW, zur Nedden D (1999) Re-evaluation of the endocranial volume of the Guattari 1 Neandertal specimen (Monte Circeo). Coll. Antropol. 23:397–405.

Redmann RE (1982) Production and diversity in contemporary grasslands. In: Hopkins DM, Matthews JV, Schweger CE, Young SB (eds) Paleoecology of Beringia. Academic Press, New York, pp 223–239.

Reich D, Green RE, Kircher M, Krause J, Patterson N, Durand EY, Viola B, Briggs AW, Stenzel U, Johnson PLF, Maricic T, Good JM, Marques-Bonet T, Alkan C, Fu Q, Mallick S, Li H, Meyer M, Eichler EE, Stoneking M, Richards M, Talamo S, Shunkov MV, Derevianko AP, Hublin JJ, Kelso J, Slatkin M, Pääbo S (2010) Genetic history of an archaic hominin group from Denisova Cave in Siberia. Nature 468:1053–1060.

Reich DE, Goldstein DB (1998) Genetic evidence for a Paleolithic human population expansion in Africa. Proc. Nat. Acad. Sci. 95:8119–8123.

Reille M, de Beaulieu JL (1990) Pollen analysis of a long upper Pleistocene continental sequence in a Velay maar (Massif Central, France). Palaeogeogr. Palaeoclimatol. Palaeoecol. 80:35–48.

Relethford JH (1998) Mitochondrial DNA and ancient population growth. Am. J. Phys. Anthropol. 105:1–7.

Relethford JH (2001) Genetics and the Search for Modern Human Origins. Wiley-Liss, New York.
Rendu W (2010) Hunting behavior and Neanderthal adaptability in the Late Pleistocene site of Pech-de-l'Azé I. J. Archaeol. Sci. 37:1798–1810.
Rendu W, Costamagno S, Meignen L, Soulier M-C (2012) Monospecific faunal spectra in Mousterian contexts: implications for social behavior. Quatern. Int. 247:50–58.
Rennie DW, Covino BG, Blair MR, Rodahl K (1962) Physical regulation of temperature in Eskimos. J. Appl. Physiol. 17:326–332.
Rensch B (1938) Some problems of geographical variation and species-formation. Proc. Linn. Soc. Lond. 150:275–285.
Reumer JWF, Rook L, van der Borg K, Post K, Mol D, De Vos J (2003) Late Pleistocene survival of the saber-toothed cat *Homotherium* in northwestern Europe. J. Vert. Paleontol. 23:260–262.
Reynolds SC (2007) Mammalian body size changes and Plio-Pleistocene environmental shifts: implications for understanding hominin evolution in eastern and southern Africa. J. Hum. Evol. 53:528–548.
Rhoads JG, Trinkaus E (1977) Morphometrics of the Neandertal talus. Am. J. Phys. Anthropol. 46:29–44.
Rhodes JA, Churchill SE (2009) Throwing in the Middle and Upper Paleolithic: inferences from an analysis of humeral retroversion. J. Hum. Evol. 56:1–10.
Richards MP, Schmitz RW (2008) Isotope evidence for the diet of the Neanderthal type specimen. Antiquity 82:553–557.
Richards MP, Trinkaus E (2009) Isotopic evidence for the diets of European Neanderthals and early modern humans. Proc. Nat. Acad. Sci. 106:16034–16039.
Richards MP, Pettitt PB, Trinkaus E, Smith FH, Paunovic M, Karavanic I (2000) Neanderthal diet at Vindija and Neanderthal predation: the evidence from stable isotopes. Proc. Nat. Acad. Sci. 97:7663–7666.
Richards MP, Pettitt PB, Stiner MC, Trinkaus E (2001) Stable isotope evidence for increasing dietary breadth in the European mid-Upper Paleolithic. Proc. Nat. Acad. Sci. 98:6528–6532.
Richards MP, Harvati K, Grimes V, Smith C, Smith T, Hublin JJ, Karkanas P, Panagopoulou E (2008a) Strontium isotope evidence of Neanderthal mobility at the site of Lakonis, Greece using laser-ablation PIMMS. J. Archaeol. Sci. 35:1251–1256.
Richards MP, Pacher M, Stiller M, Quilès J, Hofreiter M, Constantin S, Zilhão J, Trinkaus E (2008b) Isotopic evidence for omnivory among European cave bears: Late Pleistocene *Ursus spelaeus* from the Peştera cu Oase, Romania. Proc. Nat. Acad. Sci. 105:600–604.
Richards MP, Taylor G, Steele T, McPherron SP, Soressi M, Jaubert J, Orschiedt J, Mallye JB, Rendu W, Hublin JJ (2008c) Isotopic dietary analysis of a Neanderthal and associated fauna from the site of Jonzac (Charente-Maritime), France. J. Hum. Evol. 55:179–185.
Richerson PJ, Boyd R, Bettinger RL (2009) Cultural innovations and demographic change. Hum. Biol. 81:211–235.
Richter D, Rink JW, Schwarcz HP, Julig PJ, Schroeder HB (2001) The Middle to Upper Palaeolithic transition in the Levant and new thermoluminescence dates for a late Mousterian assemblage from Jerf al-Ajla Cave (Syria). Paléorient 27:29–46.
Richter J (2001) For lack of a wise old man? Late Neanderthal land-use patterns in the Altmühl River valley, Bavaria. In: Conard NJ (ed) Settlement Dynamics of the Middle Paleolithic and Middle Stone Age. Kerns Verlag, Tübingen, pp 205–219.
Richter J (2008) Neanderthals in their landscape. In: Demarsin B, Otte M (eds) Neanderthals in Europe. Gallo-Roman Museum, Tongeren, pp 17–32.
Richter J (2009) The role of leaf points in the late Middle Palaeolithic of Germany. Praehistoria 9–10:99–113.
Richter J (2011) When did the Middle Paleolithic begin? In: Conard NJ, Richter J (eds) Neanderthal Lifeways, Subsistence and Technology: One Hundred and Fifty Years of Neanderthal Study. Springer, Dordrecht, pp 7–14.

Rieder H (2001) Erprobung der holzspeere von Schöningen (400 000 Jahre) und Folgerungen daraus. In: Wagner GA, Mania D (eds) Frühe menschen in mitteleuropa - chronologie, kultur, umwelt. Kolloquium Heidelberg 2000, *Homo heidelbergensis* von Mauer e.V., Veröffentlichungen 1. Aachen, Shaker Verlag, pp 91–98.
Rieder H (2003) Nachbau altsteinzeitlicher Speere der groâe Wurf der Frühen Jäger. Biol. Unserer Zeit 33:156–160.
Rieger I (1979) A review of the biology of striped hyaenas, *Hyaena hyaena* (Linne, 1758). Saugetierkundliche Mitteilungen 27:81–95.
Rigaud J-P, Simek J, Thierry G (1995) Mousterian fires from Grotte XVI (Dordogne, France). Antiquity 69:902–912.
Rigaud J-P, Simek JF, Gé T (1996) Structures de combustion du Moustérien de la Grotte XVI à Cénac-et-Saint Julien (Dordogne, France). In: Bar-Yosef O, Cavalli-Sforza LL, March RJ, Piperno M (eds) The Lower and Middle Paleolithic. A.B.A.C.O. Edizioni, Forlì, pp 77–80.
Rightmire GP (1998) Human evolution in the Middle Pleistocene: the role of *H. heidelbergensis*. Evol. Anthropol. 6:218–227.
Rightmire GP (2001) Patterns of hominid evolution and dispersal in the Middle Pleistocene. Quatern. Int. 75:77–84.
Rightmire GP (2007) Later Middle Pleistocene *Homo* In: Henke W, Tattersall I (eds) Handbook of Paleoanthropology. Springer-Verlag, Berlin, pp 1695–1715.
Rink WJ, Schwarcz HP, Ronen A, Tsatskin A (2004) Confirmation of a near 400 ka age for the Yabrudian industry at Tabun Cave, Israel. J. Archaeol. Sci. 31:15–20.
Ritchie EG, Johnson CN (2009) Predator interactions, mesopredator release and biodiversity conservation. Ecol. Lett. 12:982–998.
Rivals F, Kacimi S, Moutoussamy J (2004) Artiodactyls, favourite game of prehistoric hunters at the Caune de l'Arago Cave (Tautavel, France). Opportunistic or selective hunting strategies? Eur. J. Wildl. Res. 50:25–32.
Rivals F, Moncel M-H, Patou-Mathis M (2009) Seasonality and intra-site variation of Neanderthal occupations in the Middle Palaeolithic locality of Payre (Ardèche, France) using dental wear analyses J. Archaeol. Sci. 36:1070–1078
Roach NT, Venkadesan M, Rainbow MJ, Lieberman DE (2013) Elastic energy storage in the shoulder and the evolution of high-speed throwing in *Homo*. Nature 498:483–486.
Roberts DF (1978) Climate and Human Variability, 2nd Edition. Cummings, Menlo Park, CA.
Roberts DF, Bainbridge DR (1963) Nilotic physique. Am. J. Phys. Anthropol. 21:341–370.
Robson SL, Wood B (2008) Hominin life history: reconstruction and evolution. J. Anat. 212:394–425.
Robson SL, van Schaik CP, Hawkes K (2006) The derived features of human life history. In: Hawkes K, Paine R (eds) The Evolution of Human Life History. SAR Press, Santa Fe and Oxford, pp 17–44.
Rode A, Shephard RJ (1971) Cardiorespiratory fitness of an arctic community. J. Appl. Physiol. 31:519–526.
Rode A, Shephard RJ (1994) Prediction of body fat content in an Inuit community. Am. J. Hum. Biol. 6:249–254.
Rode A, Shephard RJ (1995) Basal metabolic rate of Inuit. Am. J. Hum. Biol. 7:723–729.
Rodríguez MÁ, López-Sañudo IL, Hawkins BA (2006) The geographic distribution of mammal body size in Europe. Global Ecol. Biogeogr. 15:173–181.
Roebroeks W (1988) From Find Scatters to Early Hominid Behaviour: A Study of Middle Palaeolithic Riverside Settlements at Maastricht-Belvédère (the Netherlands). Leiden University, Analecta Praehistorica Leidensia 21.
Roebroeks W, Tuffreau A (1999) Palaeoenvironment and settlement patterns of the northwest European Middle Paleolithic. In: Roebroeks W, Gamble C (eds) The Middle Palaeolithic Occupation of Europe. Leiden University Press, Leiden, pp 121–138.
Roebroeks W, van Kolfschoten T (eds) (1994a) The Earliest Occupation of Europe. Leiden University and European Science Foundation, Leiden.

Roebroeks W, van Kolfschoten T (1994b) The earliest occupation of Europe: a short chronology. Antiquity 68:489–503.

Roebroeks W, Villa P (2011) On the earliest evidence for habitual use of fire in Europe. Proc. Nat. Acad. Sci. 108:5209–5214.

Roebroeks W, Conard NJ, van Kolfschoten T (1992) Dense forests, cold steppes, and the Palaeolithic settlement of northern Europe. Curr. Anthropol. 33:551–586.

Roemer GW, Gompper ME, Van Valkenburgh B (2009) The ecological role of the mammalian mesocarnivore. Bioscience 59:165–173.

Rogers AR (1995) Genetic evidence for a Pleistocene population explosion. Evolution 49:608–615.

Rogers AR, Jorde LB (1995) Genetic evidence on modern human origins. Hum. Biol. 67:1–36.

Rogers ES, Black MB (1976) Subsistence strategy in the fish and hare period, northern Ontario: the Weagamow Ojibwa, 1880–1920. J. Anthropol. Res. 32:1–43.

Rohland N, Pollack JL, Nagel D, Beauval C, Airvaux J, Pääbo S, Hofreiter M (2005) The population history of extant and extinct hyenas. Mol. Biol. Evol. 22:2435–2443.

Rolland N (1977) New aspects of Middle Palaeolithic variability in western Europe. Nature 266:251–252.

Rolland N (1981) The interpretation of Middle Paleolithic variability. Man 16:15–42.

Rolland N (1986) Recent findings from La Micoque and other sites in south-western and Mediterranean France: their bearing on the "Tayacian" problem and Middle Palaeolithic emergence. In: Bailey GN, Callow P (eds) Stone Age Prehistory. Studies in Memory of Charles McBurney. Cambridge University Press, Cambridge, pp 121–151.

Rolland N (1987) Comment on "Dietary shifts and the European Upper Paleolithic transition" by Susan Cachel. Curr. Anthropol. 38:594–595.

Rolland N, Dibble HL (1990) A new synthesis of Middle Paleolithic variability. Am. Antiquity 55:480–499.

Roper TJ (1994) The European badger *Meles meles*: food specialist or generalist? J. Zool. 234:437–452.

Rosas A (2001) Occurrence of Neanderthal features in mandibles from the Atapuerca-SH site. Am. J. Phys. Anthropol. 114:74–91.

Rosas A, Bastir M, Martínez-Maza C, García-Tabernero A, Lalueza-Fox C (2006a) Inquiries into Neanderthal craniofacial development and evolution: "accretion" versus "organismic" models. In: Harvati K, Harrison T (eds) Neanderthals Revisited: New Approaches and Perspectives. Springer, Dordrecht, pp 37–69.

Rosas A, Martínez-Maza C, Bastir M, García-Tabernero A, Lalueza-Fox C, Huguet R, Ortiz JE, Julià R, Soler V, de Torres T, Martínez E, Cañaveras JC, Sánchez-Moral S, Cuezva S, Lario J, Santamaría D, de la Rasilla M, Fortea J (2006b) Paleobiology and comparative morphology of a late Neandertal sample from El Sidrón, Asturias, Spain. Proc. Nat. Acad. Sci. 103:19266–19271.

Rosello Izquierdo E, Morales Muniz A (2005) Ictiofaunas Musterienses de la Peninsula Iberica: ¿evidencias de pesca Neandertal? Munibe Antropologia-Arkeologia 57:183–195.

Rosenberg KR (1988) The functional significance of Neandertal pubic length. Curr. Anthropol. 29(4):595–617.

Rosenberg KR (1998) Morphological variations in west Asian postcrania. In: Akazawa T, Aoki K, Bar-Yosef O (eds) Neandertals and Modern Humans in Western Asia. Plenum, New York, pp 367–379.

Rosenzweig ML (1968) The strategy of body size in mammalian carnivores. Am. Midland Nat. 80:299–315.

Roth HL (1890) The Aborigines of Tasmania. Kegan Paul, Trench, Trubner and Co., London.

Rothschild BM, Diedrich CG (2012) Pathologies in the extinct Pleistocene Eurasian steppe lion *Panthera leo spelaea* (Goldfuss, 1810) – results of fights with hyenas, bears and lions and other ecological stresses. Int. J. Paleopath. 2:187–198.

Rots V (2008) Hafting and raw materials from animals. Guide to the identification of hafting traces on stone tools. Anthropozoologica 43:43–66.

Rots V (2009) The functional analysis of the Mousterian and Micoquian assemblages of Sesselfelsgrotte, Germany: aspects of tool use and hafting in the European late Middle Palaeolithic. Quartär 56:37–66.
Rots V (2010) Prehension and Hafting Traces on Flint Tools: A Methodology. Leuven University Press, Leuven.
Rots V (2013) Insights into early Middle Palaeolithic tool use and hafting in western Europe. The functional analysis of level IIa of the early Middle Palaeolithic site of Biache-Saint-Vaast (France). J. Archaeol. Sci. 40:497–506.
Rouadi P, Baroody FM, Abbott D, Naureckas E, Solway J, Naclerio RM (1999) A technique to measure the ability of the human nose to warm and humidify air. J. Appl. Physiol. 87:400–406.
Rougier H, Milota S, Rodrigo R, Gherase M, Sarcina L, Molclovan O, Zilhão J, Constantin S, Franciscus RG, Zollikofer CPE, de Leon MP, Trinkaus E (2007) Pestera cu Oase 2 and the cranial morphology of early modern Europeans. Proc. Nat. Acad. Sci. 104:1165–1170.
Rowley-Conwy P (1984) The laziness of the short-distance hunter: the origins of agriculture in western Denmark. J. Anthropol. Archaeol. 3:300–324.
Rubin CT, Lanyon LE (1984) Dynamic strain similarity in vertebrates: an alternative to allometric limb bone scaling. J. Theor. Biol. 107:321–327.
Ruddiman WF, McIntyre A (1976) North-east Atlantic paleoclimatic changes over the past 600 000 years. Mem. Geol. Soc. Am. 145:111–146.
Rudwick MJS (1964) Inference of function from structure in fossils. Brit. J. Phil. Sci. 15:27–40.
Ruebens K (2006) A typological dilemma: Micoquian elements in continental northwestern Europe during the last glacial cycle (MIS 5d–3). Lithics 27:58–73.
Ruebens K (2009) Bifacial elements in continental northwestern Europe during the last glacial cycle (MIS5d-3): the relationship between Mousterian, Micoquian and 'mixed' assemblages. Papers from the Institute of Archaeology 18:84–100.
Ruebens, K (2013) Regional behavior among late Neanderthal groups in western Europe: a comparative assessment of late Middle Palaeolithic bifacial tool variability. J. Hum. Evol. 65:341–362.
Ruff CB (1991) Climate and body shape in hominid evolution. J. Hum. Evol. 21:81–105.
Ruff CB (1992) Biomechanical analyses of archaeological human material. In: Saunders SR, Katzenburg MA (eds) The Skeletal Biology of Past Peoples: Advances in Research Methods. Alan R. Liss, New York, pp 41–61.
Ruff CB (1993) Climatic adaptation and hominid evolution: the thermoregulatory imperative. Evol. Anthropol. 2:53–59.
Ruff CB (1994) Morphological adaptation to climate in modern and fossil hominids. Yearb. Phys. Anthropol. 37:65–107.
Ruff CB (2000a) Biomechanical analyses of archaeological human skeletals. In: Katzenberg MA, Saunders SR (eds) Biological Anthropology of the Human Skeleton. Alan R. Liss, New York, pp 71–102.
Ruff CB (2000b) Body mass prediction from skeletal frame size in elite athletes. Am. J. Phys. Anthropol. 113:507–517.
Ruff CB, Runestad JA (1992) Primate limb bone structural adaptations. A. Rev. Anthropol. 21:407–433.
Ruff CB, Trinkaus E, Holliday TW (1997) Body mass and encephalization in Pleistocene *Homo*. Nature 387:173–176.
Ruff CB, Larsen CS, Hayes WC (1984) Structural changes in the femur with the transition to agriculture on the Georgia Coast. Am. J. Phys. Anthropol. 64:125–136.
Ruff CB, Scott WW, Liu AY-C (1991) Articular and diaphyseal remodeling of the proximal femur with changes in body mass in adults. Am. J. Phys. Anthropol. 86:397–413.
Ruff CB, Trinkaus E, Walker A, Larsen CS (1993) Postcranial robusticity in *Homo*, I: temporal trends and mechanical interpretation. Am. J. Phys. Anthropol. 91:21–53.

Ruff CB, Niskanen M, Junno J-A, Jamison P (2005) Body mass prediction from stature and bi-iliac breadth in two high latitude populations, with application to earlier higher latitude humans. J. Hum. Evol. 48:381–392.

Ruff CB, Holt B, Trinkaus E (2006a) Who's afraid of the big bad Wolff? "Wolff's law" and bone functional adaptation. Am. J. Phys. Anthropol. 129:484–498.

Ruff CB, Holt BM, Sládek V, Berner M, Murphy WAJ, zur Nedden D, Seidler H, Recheis W (2006b) Body size, body proportions, and mobility in the Tyrolean "Iceman." J. Hum. Evol. 51:91–101.

Russell AP, Bryant HN (2001) Claw retraction and protraction in the Carnivora: the cheetah (*Acinonyx jubatus*) as an atypical felid. J. Zool. 254 67–76.

Russell MD (1985) The supraorbital torus: "a most remarkable peculiarity." Curr. Anthropol. 26:337–360.

Russell MD (1987a) Bone breakage in the Krapina hominid collection. Am. J. Phys. Anthropol. 72:373–379.

Russell MD (1987b) Mortuary practices at the Krapina Neandertal site. Am. J. Phys. Anthropol. 72:381–397.

Sabick MB, Torry MR, Kim Y-K, Hawkins RJ (2004) Humeral torque in professional baseball pitchers. Am. J. Sports Med. 32:892–898

Sacher GA, Staffeldt EF (1974) Relation of gestation time to brain weight for placental mammals: implications for the theory of vertebrate growth. Am. Nat. 108:593–615.

Sahlins MD (1968) Notes on the original affluent society. In: Lee RB, DeVore I (eds) Man the Hunter. Aldine, Chicago, pp 85–89.

Salem PE, Churchill SE (2014) Penetration, tissue damage, and lethality of wood- versus lithic-tipped projectiles. In: Iovita R, Sano K (eds) Multidisciplinary Approaches to the Study of Stone Age Weaponry. Springer, Heidelberg, (in press).

Sandgathe DM (2005) Examining the Levallois Reduction Strategy from a Design Theory Point of View. British Archaeological Reports IS 1417, Oxford.

Sandgathe DM, Hayden B (2003) Did Neanderthals eat inner bark? Antiquity 77:709–718.

Sandgathe DM, Dibble HL, Goldberg P, McPherron SP, Turq A, Niven L, Hodgkins J (2011) On the role of fire in Neandertal adaptations in western Europe: evidence from Pech de l'Azé IV and Roc de Marsal, France. PaleoAnthropol 2011:216–242.

Santa Luca AP (1978) A re-examination of presumed Neandertal-like fossils. J. Hum. Evol. 7:619–636.

Sawyer GJ, Maley B (2005) Neanderthal reconstructed. Anat. Rec. 283B:23–31.

Schaefer O, Hildes JA, Greidanus P, Leung D (1974) Regional sweating in Eskimos and Caucasians. Can. J. Physiol. Pharmacol. 52:960–965.

Schaller GB (1968) Hunting behaviour of the cheetah in the Serengeti National Park, Tanzania. E. Afr. Wildl. J. 6:95–100.

Schaller GB (1972) The Serengeti Lion. University of Chicago Press, Chicago.

Scherer PW, Hahn II, Mozell MM (1989) The biophysics of nasal airflow. Otolaryngol. Clin. N. Am. 22:265–278.

Schillaci MA, Froehlich JW (2001) Nonhuman primate hybridization and the taxonomic status of the Neanderthals. Am. J. Phys. Anthropol. 115:157–166.

Schmidt-Nielsen K (1984) Scaling: Why is Animal Size So Important? Cambridge University Press, Cambridge.

Schmidt JL, Gilbert DL (1978) Big Game of North America. Stackpole Books, Harrisburg.

Schmitt DO, Churchill SE, Hylander WL (2003) Experimental evidence concerning spear use in Neandertals and early modern humans. J. Archaeol. Sci. 30:103–114.

Schoeninger MJ, DeNiro MJ (1984) Nitrogen and carbon isotopic composition of bone collagen from marine and terrestrial animals. Geochim. Cosmochim. Acta 48:625–639.

Schoeninger MJ, Moore K (1992) Bone stable isotope studies in archaeology. J. World Prehist. 6:247–296.

Schofield WN (1985) Predicting basal metabolic rate, new standards and review of previous work. Hum. Nutr.: Clin. Nutr. 39C:5–41.

Scholander PF (1955) Evolution of climatic adaptation in homeotherms. Evolution 9:15–26.

Scholz CA, Johnson TC, Cohen AS, King JW, Peck JA, Overpeck JT, Talbot MR, Brown ET, Kalindekafe L, Amoako PYO, Lyons RP, Shanahan TM, Castañeda IS, Heil CW, Forman SL, McHargue LR, Beuning KR, Gomez J, Pierson J (2007) East African megadroughts between 135 and 75 thousand years ago and bearing on early-modern human origins. Proc. Nat. Acad. Sci. 104:16416–16421.

Schulting RJ, Trinkaus E, Higham T, Hedges R, Richards M, Cardy B (2005) A Mid-Upper Palaeolithic human humerus from Eel Point, South Wales, UK. J. Hum. Evol. 48:493–505.

Schwarcz HP, Simpson JJ, Stringer CB (1998) Neanderthal skeleton from Tabun: U-series data by gamma-ray spectrometry. J. Hum. Evol. 35:635–645.

Schwartz JH, Tattersall I (1996a) Significance of some previously unrecognized apomorphies in the nasal region of *Homo neanderthalensis*. Proc. Nat. Acad. Sci. 93:10852–10854.

Schwartz JH, Tattersall I (1996b) Toward distinguishing *Homo neanderthalensis* from *Homo sapiens*, and vice versa. Anthropologie 34:79–88.

Schwartz JH, Tattersall I, Laitman JT (1999) New thoughts on Neanderthal behavior: evidence from nasal morphology. In: Ullrich H (ed) Hominid Evolution: Lifestyles and Strategies. Edition Archaea, Gelsenkirchen/Schwelm, pp 166–186.

Semal P, Rougier H, Crevecoeur I, Jungels C, Flas D, Hauzeur A, Maureille B, Germonpré M, Bocherens H, Pirson S, Cammaert L, De Clerck N, Hambucken A, Higham T, Toussaint M, van der Plicht J (2009) New data on the late Neandertals: direct dating of the Belgian Spy fossils. Am. J. Phys. Anthropol. 138:421–428.

Senut B (1985) Computerized tomography of a Neandertal humerus from Le Régourdou (Dordogne, France): comparisons with modern man. J. Hum. Evol. 14:717–723.

Serangeli J, Bolus M (2008) Out of Europe - the dispersal of a successful European hominin form. Quartär 55:83–98.

Serre D, Langaney A, Chech M, Teschler-Nicola M, Paunovic M, Mennecier P, Hofreiter M, Possnert G, Pääbo S (2004) No evidence of Neandertal mtDNA contribution to early modern humans. PLoS Biol. 2:313–317.

Shackelford LL (2007) Regional variation in the postcranial robusticity of late Upper Paleolithic humans. Am. J. Phys. Anthropol. 133:655–668.

Shackelford LL, Trinkaus E (2002) Late Pleistocene human femoral diaphyseal curvature. Am. J. Phys. Anthropol. 118:359–370.

Shackleton JC (1987) Oxygen isotopes, ice volume and sea level. Quat. Sci. Rev. 6:183–190.

Shackleton N, Opdyke ND (1973) Oxygen-isotopes and paleomagnetic stratigraphy of Equatorial Pacific core V28–238: oxygen-isotopes, temperatures and ice volumes on a 10^5 and a 10^6 year scale. Quatern. Res. 3:39–55.

Sharpe D (1975) Methods of assessing the primary productivity of regions. In: Lieth H, Whittaker R (eds) Primary Productivity of the Biosphere. Springer-Verlag, New York, pp 147–166.

Shaw CN, Hofmann CL, Petraglia MD, Stock JT, Gottschall JS (2012) Neandertal humeri may reflect adaptation to scraping tasks, but not spear thrusting. PLoS ONE 7:e40349.

Shea BT (1977) Eskimo craniofacial morphology, cold stress and the maxillary sinus. Am. J. Phys. Anthropol. 47:289–300.

Shea J (2005) Bleeding or breeding: Neandertals vs. early modern humans in the Middle Paleolithic Levant. In: Pollock S, Bernbeck R (eds) Archaeologies of the Middle East: Critical Perspectives. Blackwell, Malden, MA, pp 129–151.

Shea J, Davis Z, Brown K (2001) Experimental tests of Middle Palaeolithic spear points using a calibrated crossbow. J. Archaeol. Sci. 28:807–816.

Shea JJ (1988) Spear points from the Middle Paleolithic of the Levant. J. Field Archaeol. 15:441–450.

Shea JJ (1989) A functional study of the lithic industries associated with hominid fossils in the Kebara and Qafzeh Caves, Israel. In: Mellars PA, Stringer CB (eds) The Human Revolution: Behavioural and Biological Perspectives in the Origins of Modern Humans. Princeton University Press, Princeton, NJ, pp 611–625.

Shea JJ (1990) A further note on Mousterian spear points. J. Field Archaeol. 17:111–114.
Shea JJ (1993) Lithic use-wear evidence for hunting by Neandertals and early modern humans from the Levantine Mousterian. In: Peterkin GL, Bricker HM, Mellars PA (eds) Hunting and Animal Exploitation in the Later Palaeolithic and Mesolithic of Europe. Archeological Papers of the American Anthropological Association No.4, pp 189–197.
Shea JJ (1997) Middle Paleolithic spear point technology. In: Knecht H (ed) Projectile Technology. Plenum, New York, pp 79–106.
Shea JJ (1998) Neandertal and early modern human behavioral variability: a regional-scale approach to lithic evidence for hunting in the Levantine Mousterian. Curr. Anthropol. 39:S45-S78.
Shea JJ (2003a) The Middle Paleolithic of the east Mediterranean Levant. J. World Prehist. 17:313–394.
Shea JJ (2003b) Neandertals, competition, and the origin of modern human behavior in the Levant. Evol. Anthropol. 12:173–187.
Shea JJ (2006) The origins of lithic projectile point technology: evidence from Africa, the Levant, and Europe. J. Archaeol. Sci. 33:823–846.
Shea JJ (2008) Transitions or turnovers? Climatically-forced extinctions of *Homo sapiens* and Neanderthals in the east Mediterranean Levant. Quat. Sci. Rev. 27:2253–2270.
Shea JJ (2009) The impact of projectile weaponry on late Pleistocene human evolution. In: Hublin JJ, Richards MP (eds) The Evolution of Hominid Diets: Integrating Approaches to the Study of Palaeolithic Subsistence. Springer, Dordrecht, pp 189–199.
Shea JJ (2013) Stone Tools in the Paleolithic and Neolithic Near East: A Guide. Cambridge University Press, Cambridge.
Shearer G, Kohl DH (1989) Estimates of N_2 fixation in ecosystems: the need for and basis of the ^{15}N natural abundance method. In: Rundel PW, Ehleringer JR, Nagy KA (eds) Stable Isotopes in Ecological Research. Springer-Verlag, New York, pp 342–374.
Shennan S (2001) Demography and cultural innovation: a model and its implications for the emergence of modern human culture. Cambridge Archaeol. J. 11:5–16.
Shephard RJ (1978) Human Physiological Work Capacity. Cambridge University Press, Cambridge.
Shephard RJ (1991) Body Composition in Biological Anthropology. Cambridge University Press, Cambridge.
Shephard RJ, Rode A (1996) The Health Consequences of "Modernization": Evidence from Circumpolar Peoples. Cambridge University Press, Cambridge.
Shephard RJ, Hatcher J, Rode A (1973) On the body composition of the Eskimo. Eur. J. Appl. Physiol. 32:3–15.
Sherry ST, Rogers AR, Harpending H, Soodyall S, Jenkins T, Stoneking M (1994) Mismatch distributions of mtDNA reveal recent human population expansions. Hum. Biol. 66:761–775.
Shetty P (2005) Energy requirements of adults. Pub. Health Nutr. 8:994–1009.
Shichi K, Takahara H, Krivonogov SK, Bezrukova EV, Kashiwaya K, Takehara A, Nakamura T (2009) Late Pleistocene and Holocene vegetation and climate records from Lake Kotokel, central Baikal region. Quatern. Int. 205:98–110.
Shott MJ (1993) Spears, darts, and arrows: Late Woodland hunting techniques in the Upper Ohio Valley. Am. Antiq. 58:425–443.
Siegert MJ (2001) Ice Sheets and Late Quaternary Environmental Change. John Wiley & Sons Ltd, Chichester.
Silberbauer FB (1979) Stable Carbon Isotopes and Prehistoric Diets in the Eastern Cape Province, South Africa. University of Cape Town, Cape Town.
Silva M, Downing JA (1995) The allometric scaling of density and body mass: a nonlinear relationship for terrestrial mammals. Am. Nat. 145:704–727.
Simek J (1987) Spatial order and behavioral change in the French Paleolithic. Antiquity 61: 25–40.

Simek J (1992) Neanderthal cognition and the Middle to Upper Paleolithic transition. In: Bräuer G, Smith FH (eds) Continuity or Replacement? Controversies in *Homo sapiens* Evolution. Balkema, Rotterdam, pp 231–245.
Simmen D, Scherrer JL, Moe K, Heinz B (1999) A dynamic and direct visualization model for the study of nasal airflow. Arch. Otolaryngol. Head Neck Surg. 125:1015–1021.
Sinclair AJ (1975) Long chain polyunsaturated fatty acids in the mammalian brain. Proc. Nutr. Soc. 34:287–291.
Skerry TM, Lanyon LE (1995) Interruption of disuse by short duration walking exercise does not prevent bone loss in the sheep calcaneous. Bone 16:269–274.
Skinner J (1965) The Flake Industries of Southwest Asia: A Typological Study. Columbia University, New York.
Skinner M (1996) Developmental stress in immature hominines from Late Pleistocene Eurasia: evidence from enamel hypoplasia. J. Archaeol. Sci. 23:833–852.
Skinner M (1997) Dental wear in immature Late Pleistocene European hominines. J. Archaeol. Sci. 24:677–700.
Skogland T (1990) Density dependence in a fluctuating wild reindeer herd: maternal vs. offspring effects. Oecologia 84:442–450.
Skrzypek G, Wiśniewski A, Grierson PF (2011) How cold was it for Neanderthals moving to central Europe during warm phases of the last glaciation? Quat. Sci. Rev. 30:481–487.
Slimak L, Giraud Y (2007) Circulations sur plusieurs centaines de kilomètres durant le Paléolithique moyen. Contribution à la connaissance des sociétés néandertaliennes. C. R. Palevol 6:359–368.
Slimak L, Svendsen JI, Mangerud J, Plisson H, Heggen HP, Brugere A, Pavlov PY (2011) Late Mousterian persistence near the Arctic Circle. Science 332:841–845.
Smith EA (1991) Inujjuamiut Foraging Strategies. Aldine de Gruyter, New York.
Smith FA, Lyons SK, Ernest SKM, Jones KE, Kaufman DM, Dayan T, Marquee PA, Brown JH, Haskell JP (2003) Body mass of late Quaternary mammals. Ecology 84:3403.
Smith FH (1976) The Neandertal Remains from Krapina. University of Tennessee, Report of Investigations, Knoxville.
Smith FH (1983) Behavioral interpretation of changes in craniofacial morphology across the archaic/modern *Homo sapiens* transition. In: Trinkaus E (ed) The Mousterian Legacy: Human Biocultural Change in the Upper Pleistocene. British Archaeological Reports IS 164, Oxford, pp 141–163.
Smith FH (1984) Fossil hominids from the Upper Pleistocene of central Europe and the origin of modern Europeans. In: Smith FH, Spencer F (eds) The Origins of Modern Humans: A World Survey of the Fossil Evidence. Alan R. Liss, Inc., New York, pp 137–209.
Smith FH (1985) Continuity and change in the origin of modern *Homo sapiens*. Z. Morphol. Anthropol. 75:197–222.
Smith FH (1991) The Neandertals: evolutionary dead ends or ancestors of modern people? J. Anthropol. Res. 47:219–238.
Smith FH (1994) Samples, species, and speculations in the study of modern human origins. In: Nitecki MH, Nitecki DV (eds) Origins of Anatomically Modern Humans. Plenum, New York, pp 227–249.
Smith FH, Falsetti AB, Donnelly SM (1989) Modern human origins. Yearb. Phys. Anthropol. 32:35–68.
Smith FH, Paquette SP (1989) The adaptive basis of Neandertal facial form, with some thoughts on the nature of modern human origins. In: Trinkaus E (ed) The Emergence of Modern Humans. Cambridge University Press, Cambridge, pp 181–210.
Smith FH, Trinkaus E (1991) Les origines de l'homme moderne en Europe centrale: un cas de continuité. In: Hublin J-J, Tillier A-M (eds) Aux Origines d'Homo sapiens. Presses Universitaires de France, Paris, pp 251–290.
Smith FH, Falsetti AB, Simmons T (1995) Circum-Mediterranean biological connections and the pattern of late Pleistocene human evolution. In: Ullrich H (ed) Man and Environment in the

Palaeolithic. Etudes et Recherches Archéologiques de l'Université de Liège (E.R.A.U.L.), Liège, pp 197–207.
Smith JM, Savage JG (1956) Some locomotory adaptations in mammals. J. Linn. Soc. Zool. 42:602–622.
Smith TM (2013) Teeth and human life-history evolution. A. Rev. Anthropol. 42:191–208.
Smith TM, Tafforeau P, Reid DJ, Grün R, Eggins S, Boutakiout M, Hublin JJ (2007a) Earliest evidence of modern human life history in North African early *Homo sapiens*. Proc. Nat. Acad. Sci. 104:6128–6133
Smith TM, Toussaint M, Reid DJ, Olejniczak AJ, Hublin JJ (2007b) Rapid dental development in a Middle Paleolithic Belgian Neanderthal. Proc. Nat. Acad. Sci. 104:20220–20225
Smith TM, Tafforeau P, Reid DJ, Pouech J, Lazzari V, Zermeno JP, Guatelli-Steinberg D, Olejniczak AJ, Hoffman A, Radovčić J, Makaremi M, Toussaint M, Stringer C, Hublin J-J (2010) Dental evidence for ontogenetic differences between modern humans and Neanderthals. Proc. Nat. Acad. Sci. 49:20923–20928.
Snodgrass JJ, Leonard WR (2009) Neandertal energetics revisited: insights into population dynamics and life history evolution. PaleoAnthropology 2009:220–237.
Snodgrass JJ, Leonard WR, Tarskaia LA, Alekseev VP, Krivoshapkin VG (2005) Basal metabolic rate in the Yakut (Sakha) of Siberia. Am. J. Hum. Biol. 17:155–172.
So JK (1980) Human biological adaptation to arctic and subarctic zones. A. Rev. Anthropol. 9:63–82.
Soffer O (1989a) The Middle to Upper Palaeolithic transition on the Russian Plain. In: Mellars P, Stringer C (eds) The Human Revolution: Behavioural and Biological Perspectives in the Origins of Modern Humans. Princeton University Press, Princeton, NJ, pp 714–742.
Soffer O (1989b) Storage, sedentism and the Eurasian Palaeolithic record. Antiquity 63:719–732.
Soficaru A, Dobos A, Trinkaus E (2006) Early modern humans from the Pestera Muierii, Baia de Fier, Romania. Proc. Nat. Acad. Sci. 103:17196–17201.
Solecki RL (1971) Shanidar: The First Flower People. Knopf, New York.
Solecki RL (1992) More on hafted projectile points in the Mousterian. J. Field Archaeol. 19:207–212.
Solecki RS, Solecki RL (1993) The pointed tools from the Mousterian occupations of Shanidar Cave, northern Iraq. In: Olszewski D, Dibble HL (eds) The Paleolithic Prehistory of the Zagros-Taurus. University of Pennsylvania Museum Press, Philadelphia, pp 119–146.
Sommer R, Benecke N (2005) Late-Pleistocene and early Holocene history of the canid fauna of Europe (Canidae). Mammal. Biol. 70:227–241.
Sommer RS, Benecke N (2006) Late Pleistocene and Holocene development of the felid fauna (Felidae) of Europe: a review. J. Zool. 269:7–19.
Sommerhoff G (1950) Analytical Biology. Oxford University Press, Oxford.
Sørensen B (2011) Demography and the extinction of European Neanderthals. J. Anthropol. Archaeol. 30:17–29.
Sorensen MV, Leonard WR (2001) Neandertal energetics and foraging efficiency. J. Hum. Evol. 40:483–495.
Soressi M, d'Errico F (2007) Pigments, gravures, parures: les comportements symboliques controversés des Néandertaliens. In: Vandermeersch B, Maureille B (eds) Les Néandertaliens. Biologie et Cultures. Éditions du CTHS, Paris, pp 297–309.
Soressi M, Hays M (2003) Manufacture, transport and use of Mousterian bifaces. A case study from the Perigord (France). In: Soressi M, Dibble HL (eds) Multiple Approaches to the Study of Bifacial Technologies. University of Pennsylvania Museum Press, Philadelphia, pp 125–147.
Soressi M, McPherron SP, Lenoir M, Dogandžić T, Goldberg P, Jacobs Z, Maigrot Y, Martisius NL, Miller CE, Rendu W, Richards M, Skinner MM, Steele TE, Talamo S, Texier J-P (2013) Neandertals made the first specialized bone tools in Europe. Proc. Nat. Acad. Sci. 110:14186–14190.
Sotnikova M, Nikolskiy P (2006) Systematic position of the cave lion *Panthera spelaea* (Goldfuss) based on cranial and dental characters. Quatern. Int. 142–143:218–228.

Spencer MA, Demes AB (1993) Biomechanical analysis of masticatory system configuration in Neandertals and Inuits. Am. J. Phys. Anthropol. 91:1–20.
Speth JD (1983) Bison Kills and Bone Counts. University of Chicago Press, Chicago.
Speth JD (1987) Early hominid subsistence strategies in seasonal habitats. J. Archaeol. Sci. 14:13–29.
Speth JD (1989) Early hominid hunting and scavenging: the role of meat as an energy source. J. Hum. Evol. 18:329–343.
Speth JD (2004) News flash: negative evidence convicts Neanderthals of gross mental incompetence. World Archaeol. 36:519–526
Speth JD (2006) Housekeeping, Neandertal-style: hearth placement and midden formation in Kebara Cave (Israel) In: Hovers E, Kuhn SL (eds) Transitions Before the Transition: Evolution and Stability in the Middle Paleolithic and Middle Stone Age. Springer, New York, pp 171–188.
Speth JD (2012) Middle Palaeolithic subsistence in the Near East: zooarchaeological perspectives – past, present and future. Before Farming 2012:1–45.
Speth JD, Speilmann KA (1983) Energy source, protein metabolism, and hunter-gatherer subsistence strategies. J. Anthropol. Archaeol. 2:1–31.
Speth JD, Tchernov E (1998) The role of hunting and scavenging in Neandertal procurement strategies: new evidence from Kebara Cave (Israel). In: Akazawa T, Aoki K, Bar-Yosef O (eds) Neanderthals and Modern Humans in Western Asia. Plenum, New York, pp 223–240.
Speth JD, Tchernov E (2001) Neandertal hunting and meat-processing in the Near East: evidence from Kebara Cave (Israel). In: Stanford CB, Bunn HT (eds) Meat-Eating and Human Evolution. Oxford University Press, Oxford, pp 52–72.
Spiess AE (1979) Reindeer and Caribou Hunters. Academic Press, New York.
Spinapolice EE (2012) Raw material economy in Salento (Apulia, Italy): new perspectives on Neanderthal mobility patterns. J. Archaeol. Sci. 39:680–689.
Spoor F, Hublin JJ, Braunc M, Zonneveldd F (2003) The bony labyrinth of Neanderthals. J. Hum. Evol. 44:141–165.
Stander PE (1992) Foraging dynamics of lions in a semi-arid environment. Can. J. Zool. 70:8–21.
Stanley V (1980) Paleoecology of the arctic-steppe mammoth biome. Curr. Anthropol. 21:663–666.
Steegmann ATJ (1965) A study of relationships between facial cold response and some variables of facial morphology. Am. J. Phys. Anthropol. 23:355–362.
Steegmann ATJ (1967) Frostbite of the human face as a selective force. Hum. Biol. 39:131–144.
Steegmann ATJ (1970) Cold adaptation and the human face. Am. J. Phys. Anthropol. 32:243–250.
Steegmann ATJ (1983) Boreal forest adaptations: the past. In: Steegmann ATJ (ed) Boreal Forest Adaptations: The Northern Algonkians. Plenum, New York, pp 243–267.
Steegmann ATJ, Cerny FJ, Holliday TW (2002) Neandertal cold adaptation: physiological and energetic factors. Am. J. Phys. Anthropol. 14:566–583.
Steele J (1996) On predicting hominid group sizes. In: Steele J, Shennan S (eds) The Archaeology of Human Ancestry: Power, Sex and Tradition. Routledge, London, pp 207–229.
Steele TE (2002) Red Deer: Their Ecology and How They Were Hunted by Late Pleistocene Hominids in Western Europe. Ph.D. Dissertation, Stanford University.
Steele TE (2004) Variation in mortality profiles of red deer (*Cervus elaphus*) in Middle Palaeolithic assemblages from western Europe. Int. J. Osteoarchaeol. 14:307–320.
Steele TE, Klein RG (2005) Mollusk and tortoise size as proxies for stone age population density in South Africa: implications for the evolution of human cultural capacity. Munibe Antropologia-Arkeologia 57:221–237.
Steele TE, Klein RG (2008) Intertidal shellfish use during the Middle and Later Stone Age of South Africa. Archaeofauna 17:63–76.
Steele TE, Klein RG (2009) Late Pleistocene subsistence strategies and resource intensification in Africa. In: Hublin J-J, Richards MP (eds) The Evolution of Hominin Diets: Integrating Approaches to the Study of Palaeolithic Subsistence. Springer, Dordrecht, pp 113–126.

Stefan VH, Trinkaus E (1997) Discrete trait and dental morphometric affinities of the Tabun 2 mandible. J. Hum. Evol. 34:443–468.

Stefan VH, Trinkaus E (1998) La Quina 9 and Neandertal mandibular variability. Bull. Mem. Soc. Anthropol. Paris 10:293–324.

Stefansson V (1913) My Life with the Eskimo. Macmillan, New York.

Stein Mandryk CA (1993) Hunter-gatherer social costs and the nonviability of submarginal environments. J. Anthropol. Res. 49:39–71.

Stenton DR (1991) The adaptive significance of Caribou winter clothing for arctic hunter-gatherers. Inuit Studies 15:3–28.

Stephens DW, Krebs JR (1986) Foraging Theory. Princeton University Press, Princeton, NJ.

Steudel-Numbers KL, Tilkens MJ (2004) The effect of lower limb length on the energetic cost of locomotion: implications for fossil hominins. J. Hum. Evol. 47:95–109.

Steudel-Numbers KL, Weaver TD, Wall-Scheffler CM (2007) The evolution of human running: effects of changes in lower-limb length on locomotor economy. J. Hum. Evol. 53:191–196.

Stewart JR (2004) Neanderthal-modern human competition? A comparison between the mammals associated with Middle and Upper Palaeolithic industries in Europe during OIS 3. Int. J. Osteoarchaeol. 14:178–189.

Stewart JR (2007) Neanderthal extinction as part of the faunal change in Europe during Oxygen Isotope Stage 3. Acta zoologica cracoviensia 50A:93–124.

Stewart JR, Markova A, Musil R, van Kolfschoten T (2003) Neanderthals as part of the broader Late Pleistocene megafaunal extinctions? In: van Andel T, Davies W (eds) Neanderthals and Modern Humans in the European Landscape During the Last Glaciation. McDonald Institute for Archaeological Research, Cambridge, pp 221–231.

Stiner MC (1990a) The Ecology of Choice: Procurement and Transport of Animal Resources by Upper Pleistocene Hominids in West-Central Italy. Ph.D. Dissertation, University of New Mexico.

Stiner MC (1990b) The use of mortality patterns in archaeological studies of hominid predatory adaptations. J. Anthropol. Archaeol. 9:305–351.

Stiner MC (1991a) The faunal remains from Grotta Guattari: a taphonomic perspective. Curr. Anthropol. 32:103–117.

Stiner MC (1991b) Food procurement and transport by human and non-human predators. J. Archaeol. Sci. 18:455–482.

Stiner MC (1991c) An interspecific perspective on the emergence of the modern human predatory niche. In: Stiner MC (ed) Human Predators and Prey Mortality. Westview Press, Boulder, CO, pp 149–185.

Stiner MC (1993) Small animal exploitation and its relation to hunting, scavenging, and gathering in the Italian Mousterian. In: Peterkin GL, Bricker HM, Mellars P (eds) Hunting and Animal Exploitation in the Later Palaeolithic and Mesolithic of Eurasia. Archeological Papers of the American Anthropological Association No.4, pp 107–125.

Stiner MC (1994) Honor Among Thieves: A Zooarchaeological Study of Neandertal Ecology. Princeton University Press, Princeton.

Stiner MC (1998) Comment on "Mousterian large-mammal remains from Kobeh Cave: behavioral implications for Neanderthals and early modern humans" by C.W. Marean and S.Y. Kim. Curr. Anthropol. 39:S98-S103.

Stiner MC (2001) Thirty years on the "broad spectrum revolution" and Paleolithic demography. Proc. Nat. Acad. Sci. 98:6993–6996.

Stiner MC (2004) Comparative ecology and taphonomy of spotted hyenas, humans, and wolves in Pleistocene Italy. Rev. Paléobiol. Genève 23:771–785.

Stiner MC (2006) Middle Paleolithic subsistence ecology in the Mediterranean region. In: Hovers E, Kuhn SL (eds) Transitions Before the Transition: Evolution and Stability in the Middle Paleolithic and Middle Stone Age. Springer, New York, pp 213–231.

Stiner MC, Kuhn SL (1992) Subsistence, technology, and adaptive variation in Middle Paleolithic Italy. Am. Anthropol. 94:306–339.

Stiner MC, Kuhn SL (2006) Changes in the 'connectedness' and resilience of Paleolithic societies in Mediterranean ecosystems. Hum. Ecol. 34:693–712.

Stiner MC, Kuhn SL (2009) Paleolithic diet and the division of labor in Mediterranean Eurasia. In: Hublin JJ, Richards MP (eds) The Evolution of Hominin Diets. Springer, Dordrecht, pp 157–169.

Stiner MC, Kuhn SL, Weiner S, Bar-Yosef O (1995) Differential burning, fragmentation, and preservation of archaeological bone. J. Archaeol. Sci. 22:223–237.

Stiner MC, Munro ND, Surovell TA, Tchernov E, Bar-Yosef O (1999) Paleolithic population growth pulses evidenced by small animal exploitation. Science 283:190–194.

Stiner MC, Munro ND, Surovell TA (2000) The tortoise and the hare: small game use, the broad spectrum revolution, and Paleolithic demography. Curr. Anthropol. 41:39–73.

Stirling I, Oritsland NA (1995) Relationships between estimates of ringed seal (*Phoca hispida*) and polar bear (*Ursus maritimus*) populations in the Canadian Arctic. Can. J. Fish. Aquat. Sci. 52:2594–2612.

Stock J, Pfeiffer S (2001) Linking structural variability in long bone diaphyses to habitual behaviors: foragers from the southern African Later Stone Age and the Andaman Islands. Am. J. Phys. Anthropol. 115:337–348.

Stock JT (2006) Hunter-gatherer postcranial robusticity relative to patterns of mobility, climatic adaptation, and selection for tissue economy. Am. J. Phys. Anthropol. 131:194–204.

Stock JT, Pfeiffer S (2004) Long bone robusticity and subsistence behavior among Later Stone Age foragers of the forest and fynbos biomes of South Africa. J. Archaeol. Sci. 31:999–1013.

Stoner BP, Trinkaus E (1981) Getting a grip on the Neandertals: were they all thumbs? Am. J. Phys. Anthropol. 54:281–282.

Straus LG (1982) Carnivores and cave sites in Cantabrian Spain. J. Anthropol. Res. 38:75–96.

Straus LG (1983) From Mousterian to Magdalenian: cultural evolution viewed from Vasco-Cantabrian Spain and Pyrenean France. In: Trinkaus E (ed) The Mousterian Legacy: Human Biocultural Change in the Upper Pleistocene. British Archaeological Reports IS 164, Oxford, pp 73–111.

Straus LG (1990) The original arms race: Iberian perspectives on the Solutrean phenomenon. In: Kozlowski J (ed) Feuilles de Pierre: Les Industries à Pointes Foliacées du Paléolithique Supérieur Européen. Etudes et Recherches Archéologiques de l'Université de Liège (E.R.A.U.L.), Liège, pp 425–447.

Straus LG (1991) Epipaleolithic and Mesolithic adaptations in Cantabrian Spain and Pyrenean France. J. World Prehist. 5:83–104.

Straus LG (1992) Iberia Before the Iberians: The Stone Age Prehistory of Cantabrian Spain. University of New Mexico Press, Albuquerque.

Straus LG (1993) Upper Palaeolithic hunting tactics and weapons in western Europe. In: Peterkin GL, Bricker HM, Mellars P (eds) Hunting and Animal Exploitation in the Later Palaeolithic and Mesolithic of Eurasia. Archeological Papers of the American Anthropological Association No. 4, pp 83–93.

Straus LG, Clark G (1986) La Riera Cave: Stone Age Hunter-Gatherer Adaptations in Northern Spain. Arizona State University, Tempe, AZ.

Straus LG, Otte M (1995) Stone Age Wallonia (southern Belgium). Curr. Anthropol. 36:851–854.

Strauss RH (1979) The respiratory system. In: Strauss RH (ed) Sports Medicine and Physiology. W.B. Saunders Co., Philadelphia, pp 63–76.

Street M, Terberger T, Orschiedt J (2006) A critical review of the German Paleolithic hominin record. J. Hum. Evol. 51:551–579.

Stringer C (2012) The status of *Homo heidelbergensis* (Schoetensack 1908). Evol. Anthropol. 21:101–107.

Stringer C, Gamble C (1994) In search of the Neanderthals: solving the puzzle of human origins. Cambridge Archaeol. J. 4:95–119.

Stringer C, Pälike H, van Andel TH, Huntley B, Valdes P, Allen JRM (2003) Climatic stress and the extinction of the Neandertals. In: van Andel T, Davies W (eds) Neanderthals and Modern Humans in the European Landscape During the Last Glaciation. McDonald Institute for Archaeological Research, Cambridge, pp 233–240.
Stringer CB (1984) The definition of *Homo erectus* and the existence of the species in Africa and Europe. Cour. Forsch.-Inst. Senckenberg 69:131–143.
Stringer CB (1995) The evolution and distribution of later Pleistocene human populations. In: Vrba ES, Denton GH, Partridge TC, Burckle LH (eds) Paleoclimate and Evolution with Emphasis on Human Origins. Yale University Press, New Haven, pp 524–531.
Stringer CB, Hublin J-J (1999) New age estimates for the Swanscombe hominid, and their significance for human evolution. J. Hum. Evol. 37:873–877.
Stringer CB, Finlayson JC, Barton RNE, Cáceres I, Fernández-Jalvo Y (2008a) Reply to Klein and Steele: Neanderthals and their South African contemporaries. Proc. Nat. Acad. Sci. 105:E116.
Stringer CB, Finlayson JC, Barton RNE, Fernández-Jalvo Y, Cáceres I, Sabin RC, Rhodes EJ, Currant AP, Rodríguez-Vidal J, Giles-Pacheco F, Riquelme-Cantal JA (2008b) Neanderthal exploitation of marine mammals in Gibraltar. Proc. Nat. Acad. Sci. 105:14319–14324.
Stuart CT, Stuart T (2000) Field Guide to the Larger Mammals of Africa. Struik, Cape Town.
Sutcliffe AJ (1985) On the Track of Ice Age Mammals. Harvard University Press, Cambridge.
Suzuki H, Takai F (1970) The Amud Man and his Cave Site. University of Tokyo, Tokyo.
Svoboda J, Ložek V, Vlček E (1996) Hunters between East and West: The Paleolithic of Moravia. Plenum, New York.
Svoboda JA (2005) The Neandertal extinction in eastern central Europe. Quatern. Int. 137:69–75.
Swisher CCI, Curits GH, Jacod T, Getty AG, Suprijo A, Widiasmoro (1994) Age of the earliest known hominids in Java, Indonesia. Science 263:1118–1121.
Tappen NC (1987) Circum-mortem damage to some ancient African hominid crania: a taphonomic and evolutionary essay. Afr. Archaeol. Rev. 5:39–47.
Tappy L (1996) Thermic effect of food and sympathetic nervous system activity in humans. Reprod. Nutr. Dev. 36:391–397.
Tardieu C, Trinkaus E (1994) Early ontogeny of the human femoral bicondylar angle. Am J. Phys. Anthropol. 95:183–195.
Taschini M (1979) L'industrie lithique de Grotta Guattari au Mont Circé (Latium): définition culturelle, typologique et chronologique du Pontinien. Quaternaria 21:179–247.
Tattersall I (1992) Species concepts and species identification in human evolution. J. Hum. Evol. 22:341–349.
Tattersall I, Schwartz JH (2006) The distinctiveness and systematic context of *Homo neanderthalensis*. In: Harvati K, Harrison T (eds) Neanderthals Revisited: New Approaches and Perspectives. Springer, Dordrecht, pp 9–22.
Taylor H, Vazquez-Geffroy M, Samuels S, Taylor D (1999) Continuously recorded suckling behaviour and its effect on lactational amenorrhoea. J. Biosoc. Sci. 31:289–310.
Teitlebaum A, Goldman RF (1972) Increased energy cost with multiple clothing layers. J. Appl. Physiol. 32:743–744.
Tennie C, Call J, Tomasello M (2009) Ratcheting up the ratchet: on the evolution of cumulative culture. Phil. Trans. R. Soc. B 364:2405–2415.
Texier J-P, Porraz G, Parkington J, Rigaud JP, Poppenpoel C, Miller C, Tribolo C, Cartwright C, Coudenneau A, Klein R, Steele T, Verna C (2010) A Howiesons Poort tradition of engraving ostrich eggshell containers dated to 60,000 years ago at Diepkloof Rock Shelter, South Africa. Proc. Nat. Acad. Sci. 107:6180–6185.
Théry I, Gril J, Vernet JL, Meignen L, Maury J (1996) Coal used for fuel at two prehistoric sites in southern France: Les Canalettes (Mousterian) and Les Usclades (Mesolithic). J. Archaeol. Sci. 23:509–512.
Théry-Parisot I, Meignen L (2000) Économie des combustibles (bois et lignite) dans l'abri moustérien des Canalette: de l'expérimentation à la simulation des besoins énergétiques. Gallia Préhist. 42:45–55

Thieme H (1997) Lower Palaeolithic hunting spears from Germany. Nature 385:807–810.
Thieme H (1999) Lower Palaeolithic throwing spears and other wooden implements from Schöningen, Germany. In: Ullrich H (ed) Hominid Evolution: Lifestyles and Strategies. Edition Archaea, Gelsenkirchen/Schwelm, pp 383–395.
Thoma A (1975) Were the Spy fossils evolutionary intermediates between classic Neandertal and modern man? J. Hum. Evol. 4:387–410.
Thompson JL, Nelson AJ (2000) The place of Neandertals in the evolution of hominid patterns of growth and development. J. Hum. Evol. 38:475–495.
Thomson A, Buxton LHD (1923) Man's nasal index in relation to certain climatic conditions. J. Royal Anthropol. Inst. 53:92–122.
Thornthwaite CW (1948) An approach toward a rational classification of climates. Geophys. Rev. 38:55–94.
Tieszen LL (1991) Natural variation in the carbon isotope values of plants: implications for archaeology, ecology and paleoecology. J. Archaeol. Sci. 18:227–248.
Tilkens MJ, Wall-Scheffler C, Weaver TD, Steudel-Numbers K (2007) The effects of body proportions on thermoregulation: an experimental assessment of Allen's rule. J. Hum. Evol. 53:286–291.
Tillier A-M (1983) Le crâne d'enfant d'Engis 2: un exemple de distribution des caractères juvéniles, primitifs et néandertaliens. Bull. Soc. R. Belge Anthropol. Préhist. 34:51–75.
Tillier A-M, Arensburg B, Bruzek J (2008) Identité biologique des artisans moustériens de Kebara (Mont Carmel, Israël): réflexions sur le concept de néanderthalien au Levant méditerranéen. Bull. Mem. Soc. Anthropol. Paris 20:33–58.
Tode A (1954) Mammutjäger vor 100,000 Jahren: Natur und Mensch in Nordwestdeutschland zur letzten Eiszeit, auf Grund der Ausgrabungen bei Salgitter-Lebenstedt. E. Appelhans, Brunswick.
Toigo C, Gaillard J-M, Michallet J (1997) Adult survival pattern of the sexually dimorphic Alpine ibex. Can. J. Zool. 75:75–79.
Tomic-Karovic K (1970) Krapinski Neandertalac i Kanibalizam. In: Malez M (ed) Krapina 1899–1969. Jugoslavenska Akademija Znanosti i Umjetnosti, Zagreb, pp 183–187.
Tompkins RL (1996) Relative dental development of Upper Pleistocene hominids compared to human population variation. Am. J. Phys. Anthropol. 99:103–118.
Toro-Moyano I, Martínez-Navarro B, Agustí J, Souday C, Bermúdez de Castro JM, Martinón-Torres M, Fajardo B, Duval M, Falguères C, Oms O, Parés JM, Anadón P, Julià R, García-Aguilar JM, Moigne AM, Patrocinio Espigares M, Ros-Montoya S, Palmqvist P (2013) The oldest human fossil in Europe, from Orce (Spain). J. Hum. Evol. 65:1–9.
Tournepiche JF, Couture C, Guadelli JL, Michel P (1996) Les restes néandertaliens du repaire d'hyènes de la grotte de Rochelot (Saint-Amant-de-Bonnieure, Charente, France). C. R. Acad. Sci. Paris 322:429–435.
Tremblay A, Poehlman ET, Despres JP, Theriault G, Danforth E, Bouchard C (1997) Endurance training with constant energy intake in identical twins: changes over time in energy expenditure and related hormones. Metab. Clin. Exp. 46:499–503.
Tressler DK (1951) Marine Products of Commerce. Reinhold Publishing Corporation, New York.
Treves A, Naughton-Treves L (1999) Risk and opportunity for humans coexisting with large carnivores. J. Hum. Evol. 36:275–282.
Treves A, Palmqvist P (2007) Reconstructing hominin interactions with mammalian carnivores (6.0–1.8 Ma). In: Gursky SL, Nekaris KAI (eds) Primate Anti-Predator Strategies. Springer, New York, pp 355–381.
Trevors T (2001) Neonatal Morbidity among Macrosomic Infants in the James Bay Cree Population of Northern Quebec. School of Dietetics and Human Nutrition. McGill University, Montreal.
Trinkaus E (1976) The morphology of European and southwest Asian Neandertal pubic bones. Am. J. Phys. Anthropol. 44:95–104.

Trinkaus E (1977) A functional interpretation of the axillary border of the Neanderthal scapula. J. Hum. Evol. 6:231–234.
Trinkaus E (1978) Dental remains from the Shanidar adult Neanderthals. J. Hum. Evol. 7:369–382.
Trinkaus E (1980) Sexual differences in Neandertal limb bones. J. Hum. Evol. 9:377–397.
Trinkaus E (1981) Neanderthal limb proportions and cold adaptation. In: Stringer CB (ed) Aspects of Human Evolution. Taylor and Francis, London, pp 187–224.
Trinkaus E (1982) The Shanidar 3 Neandertal. Am. J. Phys. Anthropol. 57:37–60.
Trinkaus E (1983a) Neandertal postcrania and the adaptive shift to modern humans. In: Trinkaus E (ed) The Mousterian Legacy: Human Biocultural Changes in the Upper Pleistocene. British Archaeological Reports IS 164, Oxford, pp 165–200.
Trinkaus E (1983b) The Shanidar Neandertals. Academic Press, New York.
Trinkaus E (1984) Neanderthal pubic morphology and gestation length. Curr. Anthropol. 25(4):509–514.
Trinkaus E (1985) Cannibalism and burial at Krapina. J. Hum. Evol. 14:203–216.
Trinkaus E (1986) The Neandertals and modern human origins. A. Rev. Anthropol. 15:193–218.
Trinkaus E (1987a) Bodies, brawn, brains and noses: human ancestors and human predation. In: Nitecki MH, Nitecki DV (eds) The Evolution of Human Hunting. Plenum, New York, pp 107–145.
Trinkaus E (1987b) The Neandertal face: evolutionary and functional perspectives on a recent hominid face. J. Hum. Evol. 16:429–443.
Trinkaus E (1988) The evolutionary origins of the Neandertals or why were there Neandertals? In: Trinkaus E (ed) L'Homme de Néandertal: L'Anatomie. Etudes et Recherches Archéologiques de l'Université de Liège (E.R.A.U.L.), Liège, pp 11–29.
Trinkaus E (1989) Neandertal upper limb morphology and manipulation. In: Giacobini G (ed) Hominidae: Proceedings of the 2nd International Congress of Human Paleontology. Jaca Book, Turin, pp 331–338.
Trinkaus E (1993) Femoral neck-shaft angles of the Qafzeh-Skhul early modern humans, and activity levels among immature Near Eastern Paleolithic hominids. J. Hum. Evol. 25:393–416.
Trinkaus E (1995a) Neanderthal mortality patterns. J. Archaeol. Sci. 22:121–142.
Trinkaus E (1995b) Near eastern late archaic humans. Paléorient 21:9–24.
Trinkaus E (1997) Appendicular robusticity and the paleobiology of modern human emergence. Proc. Nat. Acad. Sci. 94:13367–13373.
Trinkaus E (2000) Human patellar articular proportions: recent and Pleistocene patterns. J. Anat. 196:473–483.
Trinkaus E (2005) Anatomical evidence for the antiquity of human footwear use. J. Archaeol. Sci. 32:1515–1526.
Trinkaus E (2006a) The Krapina scapulae. Periodicum Biologorum 108:341–351.
Trinkaus E (2006b) Modern human versus Neandertal evolutionary distinctiveness. Curr. Anthropol. 47:597–620.
Trinkaus E (2007) European early modern humans and the fate of the Neandertals. Proc. Nat. Acad. Sci. 104:7367–7372.
Trinkaus E (2008a) Behavioral implications of the Muierii 1 early modern human scapula. Annu. Roumain Anthropol. 4:27–41.
Trinkaus E (2008b) Kiik-Koba 2 and Neandertal axillary border ontogeny. Anthropol. Sci. 116:231–236.
Trinkaus E (2011a) Late Neandertals and early modern humans in Europe, population dynamics and paleobiology. In: Condemi S, Weniger G-C (eds) Continuity and Discontinuity in the Peopling of Europe: One Hundred and Fifty Years of Neanderthal Study. Springer, Dordrecht, pp 315–329.
Trinkaus E (2011b) Late Pleistocene adult mortality patterns and modern human establishment. Proc. Nat. Acad. Sci. 108:1267–1271.

Trinkaus E (2011c) The postcranial dimensions of the La Chapelle-aux-Saints 1 Neandertal. Am. J. Phys. Anthropol. 145:461–468.

Trinkaus E (2012) Neandertals, early modern humans, and rodeo riders. J. Archaeol. Sci. 39:3691–3693.

Trinkaus E, Biglari F (2006) Middle Paleolithic human remains from Bisitun Cave, Iran. Paléorient 32:105–111.

Trinkaus E, Churchill SE (1988) Neandertal radial tuberosity orientation. Am. J. Phys. Anthropol. 75:15–21.

Trinkaus E, Hilton CE (1996) Neandertal pedal proximal phalanges: diaphyseal loading patterns. J. Hum. Evol. 30:399–425.

Trinkaus E, Ruff CB (1989a) Cross-sectional geometry of Neandertal femoral and tibial diaphyses: implications for locomotion. Am. J. Phys. Anthropol. 78:315–316.

Trinkaus E, Ruff CB (1989b) Diaphyseal cross-sectional morphology and biomechanics of the Fond-de-Foret and the Spy 2 femur and tibia. Anthropol. et Prehist. 100:33–42.

Trinkaus E, Ruff CB (1999) Diaphyseal cross-sectional geometry of Near Eastern Middle Palaeolithic humans: the femur. J. Archaeol. Sci. 26:409–424.

Trinkaus E, Shipman P (1993) Neandertals: images of ourselves. Evol. Anthropol. 1:194–201.

Trinkaus E, Tompkins RL (1990) The Neandertal life cycle: the possibility, probability, and perceptibility of contrasts with recent humans. In: DeRousseau, CJ (ed) Primate Life History and Evolution. Wiley-Liss, Inc., New York, pp 153–180.

Trinkaus E, Zimmerman MR (1982) Trauma among the Shanidar Neandertals. Am. J. Phys. Anthropol. 57:61–76.

Trinkaus E, Churchill SE, Ruff CB (1994) Postcranial robusticity in *Homo*. II: humeral bilateral asymmetry and bone plasticity. Am. J. Phys. Anthropol. 93:1–34.

Trinkaus E, Ruff CB, Churchill SE (1998a) Upper limb versus lower limb loading patterns among Near Eastern Middle Paleolithic hominids. In: Akazawa T, Aoki K, Bar-Yosef O (eds) Neandertals and Modern Humans in Western Asia. Plenum, New York, pp 391–404.

Trinkaus E, Ruff CB, Churchill SE, Vandermeersch B (1998b) Locomotion and body proportions of the Saint-Césaire 1 Châtelperronian Neandertal. Proc. Nat. Acad. Sci. 95:5836–5840.

Trinkaus E, Churchill SE, Ruff CB, Vandermeersch B (1999a) Long bone shaft robusticity and body proportions of the Saint-Césaire 1 Châtelperronian Neandertal. J. Archaeol. Sci. 26:753–773.

Trinkaus E, Ruff CB, Conroy GC (1999b) The anomalous archaic *Homo* femur from Berg Aukas, Namibia: a biomechanical assessment. Am. J. Phys. Anthropol. 110:379–391.

Trinkaus E, Svoboda J, West DL, Sládek V, Hillson SW, Drozdová E, Fisáková M (2000) Human remains from the Moravian Gravettian: morphology and taphonomy of isolated elements from the Dolní Věstonice II site. J. Archaeol. Sci. 27:1115–1132.

Trinkaus E, Milota S, Rodrigo R, Mircea G, Moldovan O (2003) Early modern human cranial remains from the Peştera cu Oase, Romania. J. Hum. Evol. 45:245–253.

Trinkaus E, Biglari F, Mashkour M, Monchot H, Reyss J-L, Rougier H, Heydari S, Abdi K (2007a) Late Pleistocene human remains from Wezmeh Cave, western Iran. Am. J. Phys. Anthropol. 135:371–378.

Trinkaus E, Maki J, Zilhão J (2007b) Middle Paleolithic human remains from the Gruta da Oliveira (Torres Novas), Portugal. Am. J. Phys. Anthropol. 134:263–273.

Trotter M, Gleser GC (1952) Estimation of stature from long bones of American whites and negroes. Am. J. Phys. Anthropol. 10:463–514.

Tuffreau A (1982) The transition Lower/Middle Paleolithic in Northern France. In: Ronen A (ed) The Transition from Lower to Middle Palaeolithic and the Origin of Modern Man. British Archaeological Reports IS 151, Oxford, pp 137–148.

Tuffreau A (1992) Middle Paleolithic settlement in northern France. In: Dibble HL, Mellars P (eds) The Middle Paleolithic: Adaptation, Behavior and Variability. University of Pennsylvania Press, Philadelphia, pp 59–73.

Tullos HS, King JW (1973) Throwing mechanism in sports. Orthop. Clin. North Amer. 4:709–720.

Turner A (1984) Dental sex dimorphism in European lions (*Panthera leo* L.) of the Upper Pleistocene: palaeoecological and palaeoethological implications. Ann. Zool. Fennici 21:1–8.

Turner A (1992) Large carnivores and earliest European hominids: changing determinants of resource availability during the Lower and Middle Pleistocene. J. Hum. Evol. 22:109–126.

Turner A, Antón M (1997) The Big Cats and their Fossil Relatives: An Illustrated Guide to their Evolution and Natural History. Columbia University Press, New York.

Turner C, West RG (1968) The subdivision and zonation of interglacial periods. Eiszeitalter und Gegenwart 19:93–101.

Turner ER (1985) Socio-political organization within the Powhatan chiefdom and the effects of European contact, A.D. 1607–1646. In: Fitzhugh WW (ed) Cultures in Contact: The Impact of European Contacts on Native American Cultural Institutions A.D. 1000–1800. Smithsonian Institution Press, Washington, pp 193–224.

Turq A (1989) Approche technologique et économique du faciès Moustérien de type Quina: étude préliminaire Bull. Soc. Préhist. Française 86:244–256.

Turq A (1992) Raw material and technological studies of the Quina Mousterian in Perigord. In: Dibble HL, Mellars P (eds) The Middle Paleolithic: Adaptations, Behavior and Variability. University of Pennsylvania Museum Press, Philadelphia, pp 75–85.

Tyldesley JA, Bahn P (1983) Use of plants in the European Palaeolithic: a review of the evidence. Quat. Sci. Rev. 2:53–81.

Tyrrell AJ, Chamberlain AT (1998) Non-metric trait evidence for modern human affinities and the distinctiveness of Neanderthals. J. Hum. Evol. 34:549–554.

Ulijaszek SJ (1995) Human Energetics in Biological Anthropology. Cambridge University Press, Cambridge.

Uliyanov YP (1997a) Normogram of nasal aerodynamics. XVI World Congress of Otorhinolarynol. Head Neck Surg., Sydney, Australia, pp 1603–1607.

Uliyanov YP (1997b) Surgical reconstruction of nasal aerodynamics. XVI World Congress of Otorhinolaryngol. Head and Neck Surg., Sydney, Australia, pp 1591–1595.

Ullrich H (1978) Kannibalismus und Leichenzerstuckelung beim Neandertaler von Krapina. In: Malez M (ed) Krapinski Pracovjek i Evolucija Hominida. Jugoslavenska Akademija Znanosti i Umjetnosti, Zagreb, pp 293–318.

Undem BJ, McAlexander M, Hunter DD (1999) Neurobiology of the upper and lower airways. Allergy 54:81–93.

Underdown S (2004) Freezing, fighting and falling: an exploration of trauma causality in the Neanderthals, Fuegians, Eskimo and Aleut. Am. J. Phys. Anthropol. Suppl. 38:198 (abs).

Underdown S (2008) A potential role for transmissible spongiform encephalopathies in Neanderthal extinction. Med. Hypoth. 71:4–7.

US Fish & Wildlife Service (2008) Grey wolf (*Canis lupus*). Department of the Interior, Washington, DC.

Uthmeier T, Kels H, Schirmer W, Böhner U (2011) Neanderthals in the cold: Middle Paleolithic sites from the open-cast mine of Garzweiler, Nordrhein-Westfalen (Germany). In: Conard NJ, Richter J (eds) Neanderthal Lifeways, Subsistence and Technology: One Hundred and Fifty Years of Neanderthal Study. Springer, Dordrecht, pp 25–41.

Valde-Nowak P, Nadachowski A, Wolsan M (1987) Upper Paleolithic boomerang made of a mammoth tusk in south Poland. Nature 329:436–438.

Valensi P (2000) The archaeozoology of Lazaret Cave (Nice, France). In: Burke A (ed) Hunting in the Middle Paleolithic. Int. J. Osteoarchaeol. 10:357–367.

Valensi P, Psathi E (2004) Faunal exploitation during the Middle Palaeolithic in south-eastern France and north-western Italy. Int. J. Osteoarchaeol. 14:256–272.

Valet J-P, Valladas H (2010) The Laschamp-Mono lake geomagnetic events and the extinction of Neanderthal: a causal link or a coincidence? Quat. Sci. Rev. 29:3887–3893.

Valladas H, Joron JL, Valladas G, Arensburg B, Bar-Yosef O, Belfer-Cohen A, Goldberg P, Laville H, Meignen L, Rak Y, Tchernov E, Tillier A-M, Vandermeersch B (1987) Thermoluminescence dates for the Neanderthal burial site at Kebara in Israel. Nature 330:159–160.

Valladas H, Reyss JL, Joron JL, Valladas G, Bar-Yosef O, Vandermeersch B (1988) Thermoluminescence dating of the Mousterian 'Proto-Cro-Magnon' remains from Israel and the origin of modern man. Nature 313:614–616.
Vallverdú J, Allué E, Bischoff JL, Cáceres I, Carbonell E, Cebrià A, García-Antón D, Huguet R, Ibáñez N, Martínez K, Pastó I, Rosell J, Saladié P, Vaquero M (2005) Short human occupations in the Middle Palaeolithic level I of the Abric Romaní rock-shelter (Capellades, Barcelona, Spain). J. Hum. Evol. 48:157–174.
Vallverdú J, Alonso S, Bargalló A, Bartrolí R, Campeny G, Carrancho Á, Expósito I, Fontanals M, Gabucio J, Gómez B, Prats JM, Sañudo P, Solé À, Vilalta J, Carbonell E (2012) Combustion structures of archaeological level O and mousterian activity areas with use of fire at the Abric Romaní rockshelter (NE Iberian Peninsula). Quatern. Int. 247:313–324.
Vallverdú J, Vaquero M, Cáceres I, Allué E, Rosell J, Saladié P, Chacón G, Ollé A, Canals A, Sala R, Courty M, Carbonell E (2010) Sleeping activity area within the site structure of archaic human groups. Curr. Anthropol. 51:137–145.
van Andel T (2003) Glacial environments I: the Weichselian climate in Europe between the end of the OIS-5 Interglacial and the Last Glacial Maximum. In: van Andel T, Davies W (eds) Neanderthals and Modern Humans in the European Landscape during the Last Glaciation. McDonald Institute for Archaeological Research, Cambridge, pp 9–19.
van Andel T, Davies W (eds) (2003) Neanderthals and Modern Humans in the European Landscape during the Last Glaciation. McDonald Institute for Archaeological Research, Cambridge.
van Andel T, Runnels CN (2005) Karstic wetland dwellers of Middle Palaeolithic Epirus, Greece. J. Field Archaeol. 30:367–384.
van Andel TH, Tzedakis C (1996a) European Palaeolithic landscapes: 140,000–30,000 years ago – a summary. In: Bar-Yosef O, Cavalli-Sforza LL, March RJ, Piperno M (eds) The Lower and Middle Palaeolithic. International Union of Prehistoric and Protohistoric Sciences, Forlì, pp 191–204.
van Andel TH, Tzedakis PC (1996b) Palaeolithic landscapes of Europe and environs, 150,000–25,000 years ago: an overview. Quat. Sci. Rev. 15:481–500.
van Andel T, Davies W, Weninger B (2003) The human presence in Europe during the last glacial period I: human migrations and the changing climate. In: van Andel T, Davies W (eds) Neanderthals and Modern Humans in the European Landscape during the Last Glaciation. McDonald Institute for Archaeological Research, Cambridge, pp 31–56.
van de Graaff KM, Fox SI (1986) Concepts of Human Anatomy and Physiology. McGraw-Hill, Boston.
Van der Hammen T, Wijmstra TA, Zagwijn WH (1971) The floral record of the late Cenozoic of Europe. In: Turekian KK (ed) The Late Cenozoic Glacial Ages. Yale University Press, New Haven, pp 391–424.
van der Merwe NJ, Medina E (1991) The canopy effect, carbon isotope ratios and foodwebs in Amazonia. J. Archaeol. Sci. 18:249–259.
van Valkenburgh B (1991) Iterative evolution of hypercarnivory in canids (Mammalia: Carnivora): evolutionary interactions among sympatric predators. Paleobiology 17:340–362.
Van Valkenburgh B, Hertel F (1998) The decline of North American predators during the Late Pleistocene. In: Saunders JJ, Styles BW, Baryshnikov GF (eds) Quaternary Paleozoology in the Northern Hemisphere. Illinois State Museum Scientific Papers Volume XXVII, Springfield, IL, pp 359–375.
Van Valkenburgh B, Koepfli KP (1993) Cranial and dental adaptations to predation in canids. In: Dunstone N, Gorman ML (eds) Mammals as Predators: The Proceedings of a Symposium held by the Zoological Society of London and the Mammal Society, London, 22nd and 23rd November (Symposia of the Zoological Society of London). Clarendon Press, London, pp 15–37.
Van Valkenburgh B, Sacco T (2002) Sexual dimorphism, social behavior, and intrasexual competition in large Pleistocene Carnivorans. J. Vert. Paleontol. 22:164–169.

van Vliet-Lanoë B (1989) Dynamics and extent of the Weichselian permafrost in western Europe (Substage 5e to Stage 1). Quatern. Int. 3–4:109–113.
Vandermeersch B (1981) Les Hommes Fossiles de Qafzeh (Israel). Éditions CNRS, Paris.
Vandermeersch B (1991) La ceinture scapulaire et les membres supérieurs. In: Bar Yosef O, Vandermeersch B (eds) Le Squelette Moustérien de Kébara 2. Éditions C.N.R.S., Paris, p 157 178.
Vandermeersch B, Trinkaus E (1995) The postcranial remains of the Régourdou 1 Neandertal: the shoulder and arm remains. J. Hum. Evol. 28:439–476.
Vaquero M, Chacón G, Fernández C, Martínez K, Rando JM (2001a) Intrasite spatial patterning and transport in the Abric Romaní Middle Palaeolithic site (Capellades, Barcelona, Spain). In: Conard NJ (ed) Settlement Dynamics of the Middle Paleolithic and Middle Stone Age. Kerns Verlag, Tübingen, pp 573–595.
Vaquero M, Vallverdú J, Rosell J, Pastó I, Allué E (2001b) Neandertal behavior at the Middle Palaeolithic site of Abric Romaní, Capellades, Spain. J. Field Archaeol. 28:93–114.
Veicsteinas A (1987) Changes of muscle and fat insulation during exercise in cold water. In: Samueloff S, Yousef MK (eds) Adaptive Physiology to Stressful Environments. CRC Press, Boca Raton, FL, pp 52–57.
Veicsteinas A, Ferretti G, Rennie DW (1982) Superficial shell insulation in resting and exercising men in cold water. J. Appl. Physiol. 52:1557–1564.
Vermeersch PM, Paulissen E, Stokes S, Charlier C, van Peer P, Stringer C, Lindsay W (1998) A Middle Palaeolithic burial of a modern human at Taramsa Hill, Egypt. Antiquity 72:475–484.
Verpoorte A (2006) Neanderthal energetics and spatial behaviour. Before Farming 3:1–6.
Vézina AF (1985) Empirical relationships between predator and prey size among terrestrial vertebrate predators. Oecologia 67:555–565.
Villa P, d'Errico F (2001) Bone and ivory points in the Lower and Middle Paleolithic of Europe. J. Hum. Evol. 41:69–112.
Villa P, Lenoir M (2009) Hunting and hunting weapons of the Lower and Middle Paleolithic of Europe. In: Hublin JJ, Richards MP (eds) The Evolution of Hominid Diets: Integrating Approaches to the Study of Paleolithic Subsistence. Springer, Dordrecht, pp 59–85.
Villa P, Soriano S (2010) Hunting weapons of Neanderthals and early modern humans in South Africa. J. Anthropol. Res. 66:5–38.
Villa P, Boscato P, Ranaldo F, Ronchitelli A (2009a) Stone tools for the hunt: points with impact scars from a Middle Paleolithic site in southern Italy. J. Archaeol. Sci. 36:850–859.
Villa P, Soressi M, Henshilwood CS, Mourre V (2009b) The Still Bay points of Blombos Cave (South Africa). J. Archaeol. Sci. 36:441–460.
Villaverde V, Martínez-Valle R, Guillem PM, Fumanal MP (1996) Mobility and the role of small game in the Middle Paleolithic of the central region of the Spanish Mediterranean: a comparison of Cova Negra with other Paleolithic deposits. In: Carbonell E, Vaquero M (eds) The Last Neandertals, the First Anatomically Modern Humans: A Tale about the Human Diversity. Cultural Change and Human Evolution: The Crisis at 40 KA BP. Universitat Rovira i Virgili, Tarragona, pp 267–288.
Villemeur I (1992) Quelques résultats de l'étude morphologique et mécanique de la main des Néandertaliens. C. R. Acad. Sci. Paris 315:881–884.
Villotte S, Churchill SE, Dutour OJ, Henry-Gambier D (2010) Subsistence activities and the sexual division of labor in the European Upper Paleolithic and Mesolithic: evidence from upper limb enthesopathies. J. Hum. Evol. 59:35–43.
Vinyard PS, Smith FH (2002) Krapina temporal bone morphology: categorical and metric assessment. In: Radovcic J (ed) Proceedings of the Krapina 1899–1999 International Congress, Zagreb.
Vitzthum VJ (2009) The ecology and evolutionary endocrinology of reproduction in the human female. Yearb. Phys. Anthropol. 52:95–136.
Vogel J, Waterbolk H (1972) Groningen radiocarbon dates X. Radiocarbon 14:6–110.

Vogel JC, Waterbolk HT (1963) Groningen radiocarbon dates IV. Radiocarbon 5:163–202.
Vogel S (2001) Prime Mover: A Natural History of Muscle. W.W. Norton & Co., New York.
Vogel S (2005a) Living in a physical world IV. Moving heat around. J. Biosci. 30:449–460.
Vogel S (2005b) Living in a physical world V. Maintaining temperature. J. Biosci. 30:581–590.
Vogel S (2006) Living in a physical world VII. Gravity and life on the ground. J. Biosci. 31:201–214.
Voisin JL (2006) Speciation by distance and temporal overlap: a new approach to understanding Neanderthal evolution. In: Harvati K, Harrison T (eds) Neanderthals Revisited: New Approaches and Perspectives. Springer, Dordrecht, pp 299–314.
Volkov NI, Shirkovets EA, Borilkevich VE (1975) Assessment of aerobic and anaerobic capacity of athletes in treadmill running tests. Eur. J. Appl. Physiol. 34:121–130.
Volpato V, Ruff C, Macchiarelli R, Cowgill LW (2007) Cross-sectional geometry of the Roc de Marsal juvenile Neandertal femur, determined using high-resolution microtomography. Am. J. Phys. Anthropol. Suppl. 44:240 (abs).
Vucetich JA, Peterson RO, Waite TA (2004) Raven scavenging favours group foraging in wolves. Anim. Behav. 67:1117–1126.
Wadley L (2005) Putting ochre to the test: replication studies of adhesives that may have been used for hafting tools in the Middle Stone Age. J. Hum. Evol. 49:587–601.
Waguespack NM (2005) The organization of male and female labor in foraging societies: implications for early Paleoindian archaeology. Am. Anthropol. 107:666–676.
Walker CS, Churchill SE (2014) Territory size in *Canis lupus*: implications for Neandertal mobility. In: Marchi D, Carlson K (eds) Mobility: Interpreting Behavior from Skeletal Adaptations and Environmental Interactions. Springer, Dordrecht (in press).
Walker MJ, Lombardi AV, Zapata J, Trinkaus E (2010) Neandertal mandibles from the Sima de las Palomas del Cabezo Gordo, Murcia, southeastern Spain. Am. J. Phys. Anthropol. 142:261–272.
Walker MJ, Ortega J, López MV, Parmová K, Trinkaus E (2011a) Neandertal postcranial remains from the Sima de las Palomas del Cabezo Gordo, Murcia, southeastern Spain. Am. J. Phys. Anthropol. 144:505–515.
Walker MJ, Ortega J, Parmová K, López MV, Trinkaus E (2011b) Morphology, body proportions, and postcranial hypertrophy of a female Neandertal from the Sima de las Palomas, southeastern Spain. Proc. Nat. Acad. Sci. 108:10087–10091.
Walker R, Gurven M, Hill K, Migliano A, Chagnon N, De Souza R, Djurovic G, Hames R, Hurtado AM, Kaplan H, Kramer K, Oliver WJ, Valeggia C, Yamauchi T (2006) Growth rates and life histories in twenty-two small-scale societies. Am. J. Hum. Biol. 18:295–311.
Wall-Scheffler CM, Geiger K, Steudel-Numbers KL (2007) Infant carrying: the role of increased locomotory costs in early tool development. Am. J. Phys. Anthropol. 133:841–846.
Wallace IJ, Shea JJ (2006) Mobility patterns and core technologies in the Middle Paleolithic of the Levant. J. Archaeol. Sci. 33:1293–1309.
Warren SH (1911) First published report and exhibition of the specimen, May 10th, 1911. Quart. J. Geol. Soc. 67:xcix.
Watson E, Forster P, Richards M, Bandelt HJ (1997) Mitochondrial footprints of human expansions in Africa. Am. J. Hum. Genet. 61:691–704.
Watson JC, Payne RC, Chamberlain AT, Jones RK, Sellers WI (2008) The energetic costs of load-carrying and the evolution of bipedalism. J. Hum. Evol. 54:675–683.
Watts HE, Holekamp KE (2009) Ecological determinants of survival and reproduction in the spotted hyena. J. Mammal. 90:461–471.
Weaver DS (1980) Catastrophe theory and human evolution. J. Anthropol. Res. 36:403–410.
Weaver TD (2003) The shape of the Neandertal femur is primarily the consequence of a hyperpolar body form. Proc. Nat. Acad. Sci. 100:6926–6929.
Weaver TD (2009) The meaning of Neandertal skeletal morphology. Proc. Nat. Acad. Sci. 106:16028–16033.
Weaver TD, Hublin JJ (2009) Neandertal birth canal shape and the evolution of human childbirth. Proc. Nat. Acad. Sci. 106:8151–8156.

Weaver TD, Roseman CC (2005) Ancient DNA, late Neandertal survival, and modern-human-Neandertal genetic admixture. Curr. Anthropol. 46:677–683.
Weaver TD, Steudel-Numbers K (2005) Does climate or mobility explain the differences in body proportions between Neandertals and their Upper Paleolithic successors? Evol. Anthropol. 14:218–223.
Weaver TD, Roseman CC, Stringer CB (2007) Were Neandertal and modern human cranial differences produced by natural selection or genetic drift? J. Hum. Evol. 53:135–145.
Webb P (1951) Air temperatures in the respiratory tracts of resting subjects in cold. J. Appl. Physiol. 4:378–382.
Webb RE (1989) A reassessment of the faunal evidence for Neandertal diet based on some western European collections. In: Patou M, Freeman LG (eds) L'Homme de Néandertal: La Subsistance. Etudes et Recherches Archéologiques de l'Université de Liège (E.R.A.U.L.), Liège, pp 155–178.
Webb TI (1986) Is the vegetation in equilibrium with climate? How to interpret late-Quaternary pollen data. Vegetatio 67:75–91.
Weiner JS (1954) Nose shape and climate. Am. J. Phys. Anthropol. 12:1–4.
Weinstein KJ (2008) Thoracic morphology in Near Eastern Neandertals and early modern humans compared with recent modern humans from high and low altitudes. J. Hum. Evol. 54:287–295.
Wescott DJ (2006) Effect of mobility on femur midshaft external shape and robusticity. Am. J. Phys. Anthropol. 130:201–213.
Weyer ES (1969) The Eskimos: Their Environment and Folkways. Archon, New York.
Whalen RT, Carter DR, Steel CR (1988) Influence of physical activity on the regulation of bone density. J. Biomech. 21:825–837.
Whallon R (2006) Social networks and information: non-"utilitarian" mobility among hunter-gatherers. J. Anthropol. Archaeol. 25:259–270.
Whitaker S (1968) Introduction to Fluid Mechanics. Krieger, Malabar, FL.
White JWC (1993) Climate change: don't touch that dial. Nature 364:186–220.
White MJ (2000) The Clactonian question: on the interpretation of core-and-flake assemblages in the British Lower Paleolithic. J. World Prehist. 14:1–63.
White MJ (2006) Things to do in Doggerland when you're dead: surviving OIS3 at the northwestern-most fringe of Middle Palaeolithic Europe. World Archaeol. 38:547–575.
White MJ, Pettitt P (2012) Ancient digs and modern myths: the age and context of the Kent's Cavern 4 maxilla and the earliest *Homo sapiens* specimens in Europe. Eur. J. Archaeol. 15:392–420.
White R (1992) Beyond art: toward an understanding of the origins of material representation in Europe. A. Rev. Anthropol. 21:537–564.
White TD (1986a) Cannibals at Klasies? Sagittarius 2:6–9.
White TD (1986b) Cut marks on the Bodo cranium: a case of prehistoric defleshing. Am. J. Phys. Anthropol. 69:503–509.
White TD (1992) Prehistoric Cannibalism at Mancos 5MTUMR-2346. Princeton University Press, Princeton.
White TD (2001) Once were cannibals. Sci. Am. August 2001:48–55.
White TD, Toth N (1991) The question of ritual cannibalism at Grotta Guattari. Curr. Anthropol. 32:118–138.
White TD, Asfaw B, DeGusta D, Gilbert H, Richards GD, Suwa G, Howell FC (2003) Pleistocene *Homo sapiens* from Middle Awash, Ethiopia. Nature 423:742–747.
Whittaker RH (1975) Communities and Ecosystems, 2nd Edition. Macmillan, London.
Wiessner P (1983) Style and social information in Kalahari San projectile points. Am. Antiquity 48:253–276.
Wiessner P (2002) Hunting, healing, and *hxaro* exchange. A long-term perspective on !Kung (Ju/'hoansi) large-game hunting. Evol. Hum. Behav. 23:407–436.
Wilkie DR (1960) Man as a source of mechanical power. Ergonomics 3:1–8.
Williams GC (1966) Adaptation and Natural Selection. Princeton University Press, Princeton.

Williams M, Dunkerley D, De Deckker P, Kershaw P, Chappell J (1998) Quaternary Environments, 2nd Edition. Arnold, London.

Wilmsen EN (1978) Seasonal effects of dietary intake on Kalahari San. Fed. Proc. 37:65–72.

Wilson DS (1975) The adequacy of body size as a niche difference. Am. Nat. 109:769–784.

Wilson ML (1993) The "strandloper" concept and its relevance to the study of the past inhabitants of the southern African coastal region. Ann. S. Afr. Mus. 103:293–382.

Winslow CEA, Herrington LP (1949) Temperature and Human Life. Princeton University Press, Princeton, NJ.

Winterhalder B, Smith EA (eds) (1981) Hunter-gatherer Foraging Strategies. University of Chicago Press, Chicago.

Wobst HM (1974) Boundary conditions for Paleolithic social systems: a simulation approach. Am. Antiquity 39:147–178.

Wobst HM (1976) Locational relationships in Paleolithic society. In: Ward RH, Weiss KM (eds) The Demographic Evolution of Human Populations. Academic Press, London, pp 49–58.

Wolff H, Greenwood AD (2010) Did viral disease of humans wipe out Neandertals? Med. Hypoth. 75:99–105.

Wolpoff MH (1968) Climatic influence on skeletal nasal aperture. Am. J. Phys. Anthropol. 29:405–424.

Wolpoff MH (1979) The Krapina dental remains. Am. J. Phys. Anthropol. 50:67–114.

Wolpoff MH (1999) Paleoanthropology, 2nd Edition. McGraw-Hill, Boston.

Wolpoff MH, Caspari R (1996) Why aren't Neandertals modern humans? In: Bar-Yosef O, Cavalli-Sforza LL, March RJ, Piperno M (eds) The Lower and Middle Palaeolithic. Int. Union Prehist. Protohist. Sci., Forli, pp 133–155.

Wolsan M (1993) Évolution des carnivores Quaternaires en Europe centrale dans leur contexte stratigraphique et paléoclimatique. L'Anthropologie 7:203–222.

Wolverton S (2006) Natural-trap ursid mortality and the Kurtén response. J. Hum. Evol. 50:540–551.

Woo SL-Y, Kuei SC, Amiel D, Gomez MA, Hayes WC, White FC, Akeson WH (1981) The effect of prolonged physical training on the properties of long bone: a study of Wolff's Law. J. Bone Joint Surg. 63A:780–787.

Woodroof JG (1967) Tree Nuts: Production, Processing, Products. Avi, Westport, CN.

Wrangham RW, Jones JH, Laden G, Pilbeam D, Conklin-Brittain N (1999) The raw and the stolen: cooking and the ecology of human origins. Curr. Anthropol. 40:567–594.

Wymer J (1982) The Palaeolithic Age. St. Martin's Press, New York.

Wynn T, Coolidge FL (2004) The expert Neandertal mind. J. Hum. Evol. 46:467–487.

Wynn T, Coolidge FL (2010) Beyond symbolism and language: an introduction to Supplement 1, Working Memory. Curr. Anthropol. 51 Suppl. 1:S5–16.

Yamaguchi N, Cooper A, Werdelin L, Macdonald DW (2004) Evolution of the mane and group-living in the lion (*Panthera leo*): a review. J. Zool. 263:329–342

Yellen J, Harpending H (1972) Hunter-gatherer populations and archaeological inference. World Archaeol. 4:244–253.

Yellen JE, Brooks AS, Cornellissen E, Mehlman MJ, Stewart K (1995) A Middle Stone Age worked bone industry from Katanda, Upper Semliki Valley, Zaire. Science 268:553–556.

Yeshurun R, Bar-Oz G, Weinstein-Evron M (2007) Modern hunting behavior in the early Middle Paleolithic: faunal remains from Misliya Cave, Mount Carmel, Israel. J. Hum. Evol. 53:656–677.

Yokley TR (1999) Neandertal Noses: A Descriptive and Comparative Analysis of the Nasal Morphology of the Krapina and Vindija Neandertals. M.A. Thesis, Northern Illinois University.

Yokley TR (2006) The Functional and Adaptive Significance of Anatomical Variation in Recent and Fossil Human Nasal Passages. Ph.D. Dissertation, Duke University.

Yokley TR, Holton NE, Franciscus RG, Churchill SE (2009) The role of body mass in the evolution of the modern human nasofacial skeleton. Paleoanthropology 2009:A39-A40 (abs).

Yom-Tov Y (2003) Body sizes of carnivores commensal with humans have increased over the past 50 years. Func. Ecol. 17:323–327.

Young DD, McCabe TR (1997) Grizzly bear predation rates on caribou calves in northeastern Alaska. J. Wildl. Manage. 61:1056–1066.

Zagwijn WH (1989) Vegetation and climate during the warmer intervals in the Late Pleistocene of western and central Europe. Quatern. Int. 3–4:57–67.

Zarins B, Andrews JR, Carson WGJ (1985) Injuries to the Throwing Arm. W.B. Saunders Company, Philadelphia.

Zilhão J (2001) Middle Paleolithic settlement patterns in Portugal. In: Conard N (ed) Settlement Dynamics of the Middle Paleolithic and Middle Stone Age. Kerns Verlag, Tübingen, pp 597–608.

Zilhão J (2006) Neandertals and moderns mixed, and it matters. Evol. Anthropol. 15:183–195.

Zilhão J, Angelucci DE, Badal-García E, d'Errico F, Daniel F, Dayet L, Douka K, Higham TFG, Martínez-Sánchez MJ, Montes-Bernárdez R, Murcia-Mascarós S, Pérez-Sirvent C, Roldán-García C, Vanhaeren M, Villaverde V, Wood R, Zapata J (2010) Symbolic use of marine shells and mineral pigments by Iberian Neandertals. Proc. Nat. Acad. Sci. 107:7621–7622.

Zilhão J, d'Errico F (1999) The chronology and taphonomy of the earliest Aurignacian and its implications for the understanding of Neandertal extinction. J. World Prehist. 13:1–68.

Zollikofer CPE, Ponce de León MS, Vandermeersch B, Lévêque F (2002) Evidence for interpersonal violence in the St. Césaire Neanderthal. Proc. Nat. Acad. Sci. 99:6444–6448.

Zubrow E (1989) The demographic modelling of Neanderthal extinction. In: Mellars P, Stringer CB (eds) The Human Revolution: Behavioural and Biological Perspectives on the Origins of Modern Humans. Princeton University Press, Princeton, NJ, pp 212–231.

Zumwalt A (2006) The effect of endurance exercise on the morphology of muscle attachment sites. J. Exper. Biol. 209:444–454.

Index

Numbers in italics denote subentries, and not page references.

Accretion hypothesis, 34–36, 37
Ache, 323, 336, 345
Acheulo-Yabrudian, 48, 49, 51
aDNA, 27, 29, 31, 37, 39
Adult mortality, changes in, 358
Aleutian Islanders, 81, 118, 228, 232, 235, 328
Allen's rule, 114, 120, 149
Alloparental care, 315, 341
Allotaxa, 30, 40
Altai Mountains, 10, 27, 29, 113
Altmühlian, 52
Altriciality, 343
Alutiiq, 104
Alyawara, 317
Ameliogenesis, 344
Amudian, 48, 49
Anatomical modernity, 5
Anbarra, 317
Apennine Peninsula, 161, 351
Apex predators, 330
Aquitaine basin, 301, 302, 306, 308, 309, 362
Asia Minor, 158, 171
Asiatic black bear, 255
Attritional mortality profiles, 247
Aurelian European Land Mammal Age, 253
Aurignacian, 48, 53, 285
Australian aborigines, 312

Thin on the Ground: Neandertal Biology, Archeology, and Ecology, First Edition. Steven Emilio Churchill.

Baffinland Inuit, 302
Balkan Peninsula, 161, 351
Bark consumption, 175, 193
Bedding materials, 60, 67, 143, 322
Behavioral dominance, 268, 271–276
Behavioral modernity, 3–4, 5, 6
Bergmann's rule, 102, 114, 149
Bitumen, 59, 68
Blue Babe, 259
Bone modeling/remodeling, 210, 221, 311
Bone needles, 141
Bone points, Middle Paleolithic, 60, 65–66, 297, 362
Bottlenecks, genetic, 28, 350, 356
Brachycephaly, 23
Breast milk, 97, 343, 344
British Isles, 10, 111, 112, 156, 157, 161, 170, 173
Brow ridges, *see* Supraorbital tori
Brown adipose tissue, 139
Brown bear, 255
Bulk harvesting, *see* Resource harvesting
Burden carrying, *see* Load carrying
Burial, 361, 362
Butchery, energetic cost of, 323

Carnivore attacks, on humans, 340
Carnivore extinction, 276
Carnivore kill rates, 330
Carrying capacity, 2, 157, 176, 269, 351, 356, 357, 358
Caucasus Mountains, 158
Cave bear, 204, 253, 255, 271
Cave hyena, 255, 260–262, 281
Cave lion, 255, 258–259, 271, 275, 281
Cephalic index, 130
Chaine opératoire, 338
Character displacement, 265, 266, 268
Charentian Mousterian, 45, 47
  Ferrassie variant, 58
  Quina variant, 56, 58, 67
Châtelperronian, 52, 310, 358, 362
Chyme consumption, 192, 200
Circulating mobility, 290, 292, 298, 300, 309
Clacton spear, *see* Schöningen spears
Classic Neandertals, 30, 38
Climatic stability, 355
Clo, 144, 145
Clothing, 144, 145
Coalitions, 359
Cold-induced vasodilation, 149
Cold-induced thermogenesis, 117, 139

Competition
  exploitative, 2, 265–268
  interference, 2, 268–271
Competition refuge, 264
Competitive exclusion, 265, 358
Completed fertility, 347
Composite technology, *see* Hafting
Conductance, 150
Conductivity, 150
Cooking, 68, 212
Cooperative breeding, 336, 337, 341, 348, 349, 358
Cooperative hunting, 337, 342
Copper Eskimo, 299, 300
Cost of transport, 289, 316, 318, 327
Cree Indians, 92
Crimea, 59, 169, 181, 193, 301, 302, 306, 361
Cro-magnon, 40
Crow Indians, 302
CTE, *see* Cumulative technological evolution
Cultural ratcheting, *see* Cumulative technological evolution
Cumulative technological evolution, 4, 5, 6, 355, 360, 362, 363
Cuon, *see* Dhole

D/O events, 108, 110
Dansgaard/Oeschger oscillations, *see* D/O events
Dasht-e Kavir (Salt Desert), 158
Debitage, 54, 69
Denisovans, 29
Dental calculus, 199
Dental enamel defects, 98
Denticulate Mousterian, 45
Dependency ratio, 306, 339, 340, 341, 348, 358
Dhole, 255, 263–264
Diet breadth, 357, 363
Dire wolf, 263
Dolicocephaly, 23
Durophagy, 261, 262, 263, 265, 267, 276

Ecogeographic variation in body size, 76–78
Effective population size, 28, 350
Effective temperature, 163
"En bombe" cranial vault, 23, 36
"En maison" cranial vault, 23
"En tente" cranial vault, 23
Energetic expense of locomotion, 318
Energy balance, 334
Energy budgets, 6, 288, 289, 317, 321, 322, 330, 333, 334, 342, 343, 346, 348, 354

Eskimo, 104, 120, 129, 130, 132, 145, 293, 299, 302
European jaguar, 253
Eurytopism, 152
Evenki, 94, 96, 127, 140, 346
Exocannibalism, 216
Exogamy, 337
Extinction of Neandertals, 3
  hypotheses for, 2, 6–7
Extractive technology, 53, 58, 327, 342, 350, 363

Facial sinuses, 132
Façonnage, 54, 58
Fasting endurance hypothesis, 101
Fat rendering, 89
Fat sickness, 87
Fatigue damage (bone), 311
Femoral neck-shaft angle, 315
Ferrassie-Quina transition, 58
Fertility, 6, 335, 339, 341, 343, 344, 345, 347, 348, 349, 351, 354, 358, 363
Fire, 68, 142, 145
Fishing, 191
Fission-fusion, 337
Flintknapping (energetic cost of), 323
Forager day trip distances, 308
Forager walking speeds, 308
Foragers, *see* Circulating mobility
Foraging effort, 340, 342
Foraging risk, 338, 341, 359
Four stage model of interglacial development, 160
FOXP2 gene, 29
Fully symbolic sapiens behavior, *see* Behavioral modernity

Galerian European Land Mammal Age, 253
Game driving, 238, 245
Genetic drift, 13, 35, 135
Geophytes, 168, 175, 200, 201, 212, 320
Gestation length, 343
Grandmothering, *see* Alloparental care
Gravettian, 40, 285, 313, 359
Greenland Summit ice cap, 158
G/wi, 299

Habitat partitioning, 277, 279
Hadza, 335, 345
Hafting, 23, 56, 59, 60, 65, 68, 298, 359
Harris lines, 97
Hearths, 68
Heat dumping, 145
Heat exchange, 114, 117
Heinrich events, 108
Hide working, 60, 67, 142, 144, 232, 233, 322, 323
Hiwi, 345
HO mandibular foramen, 50
Homo erectus, 23, 25, 32, 33, 34, 97, 148, 224, 275, 336
Homo heidelbergensis, 7, 10, 14, 33, 34, 36, 37, 42, 340, 343, 346, 351, 353, 359
Homoplasy, 19, 20, 23
Howieson's Poort industry, 360
Humeral bending strength, 229–30
Hunkering down (pseudohibernation), 299, 320, 321
Hunting response, *see* Cold induced vasodilation
Hunting tactics
  ambush, 238
  approach, 240, 241
  disadvantaging, 238
  encounter, 238
  pursuit, 238, 241
Hxaro exchange, 359
Hypercarnivory, 256, 263, 265, 267, 276
Hyperpolar body form, 79, 107, 120, 123
Hypothermia, 146, 148, 150

Iberia, 10, 13, 27, 52, 161, 184, 185, 351, 362
Inactivity (and energy budgets), 321
Infanticide, 349
Interbirth interval, 336, 343, 344, 345, 348
Interbreeding, 10, 28, 29, 30, 39
Intraguild predation, 268, 269, 273, 274
Introgression, *see* Interbreeding
  Inuit, 74, 79, 80, 81, 82, 87, 91, 92, 94, 98, 104, 107, 118, 120, 123, 126, 129, 130, 132, 140, 145, 146, 147, 148, 191, 195, 199, 272, 288, 294, 299, 300, 325
Inuktitut, 104
Inupiaq, 104, 120
Iroquois, 177

Jerzmanowician, 52
Jomon, 228
Ju'/hoansi, 176, 289, 299, 317, 336, 345

Kalaallisut, 104
Keilmessergruppe, *see* Micoquian
Kleptoparasitism, 103, 261, 265, 268, 269, 278, 283, 298
Koniag, 120
Kuzey Anadolu Dağlari highlands, 158

Lactation, 92, 96, 97
Lapps, *see* Sami
Later Stone Age (LSA), 3, 4, 5, 6, 357, 359, 360, 361, 362, 363
LC-PUFA, 88, 89, 98, 187
Lean-to, 143
Lehringen spear, *see* Schöningen spears
Leopard, 255, 259–260
Lesser scimitar cat, 253, 255, 256–258, 276, 281
Levallois Index, 58
Levallois technique, 42, 46, 48, 49, 51, 52, 53, 54, 55, 56, 57, 58, 60, 65, 208
Levant, 37, 38, 42, 48, 50, 51, 53, 158, 161, 174, 181, 192, 217, 295, 301, 362
Levantine Mousterian, 48, 49, 50, 51, 53, 310, 312, 315
  facies, 48–49
Life history theory, 342
Linear enamel hypoplasias, 344
Lissoirs, 60, 67, 362
Load carrying, 296, 312
Local extinction, 153
Logistical collectors, *see* Radiating mobility
Loire valley, 301
Long-chain polyunsaturated fatty acids, *see* LC-PUFA
Lower critical temperature, 115, 128, 129
Lynx, 254–255

Maintainable technology, 297
Male/male competition, 337
Mammoth steppe, 154, 155
Manganese dioxide, 359
Marrow extraction, 89, 184, 214, 215, 216, 265
Mating systems, 337
Matrilocality, 338
Maxillary sinuses, 20, 21
Meat drying index, 294
Meat head hypothesis, 82
Mediterranean (area), 10, 31, 33, 48, 157, 158, 161, 170, 171, 173, 186, 190, 192, 217, 295, 349, 355, 359
Megadroughts, 356
Megaherbivores, 99, 103
Melanocortin 1 receptor (MC1R), 29
Mental eminence, 50
Mesopredators, 330
Metabolic equivalents (MET), 323
Micoquian, 51, 53, 59, 60, 61, 68, 191, 350
Middle Stone Age (MSA), 3, 4, 5, 6, 7, 53, 353, 354, 356, 357, 358, 359, 360, 361, 362, 363
Midfacial prognathism, 14, 20, 23, 35, 130, 330
Minimum ambient temperature, 128, 129
Mobilary art, 359, 362
Mobility zones, 301
Mortality profiles, 348
Mortuary ritual, 215
Mosaicism, 23, 25, 34
Mousterian facies, 42–47
Mousterian of Acheulean Tradition, 43, 48, 58, 67, 350
Mousterian of Micoquian tradition, 52
MSA population growth, 357
MSA site abundance, 356
mtDNA, 27, 28, 29, 31, 36, 338, 345, 356, 357
Mugharan Tradition, *see* Acheulo-Yabrudian
Muscle entheses, 233, 315
Muscle physiological cross-sectional area, 81

Neandertal
  adult mortality, 339, 340, 341, 349
  age at first reproduction, 348
  autapomorphies, 10, 13, 21, 25, 27, 30, 31, 36
  basal metabolic rate (BMR), 84–85, 139
  birth weight, 92
  body mass, 74
  body shape, 25–27
  brain morphology, 23–24
  brain size, 82
  cognition, 336
  core area, 10
  cost of transport, 318, 319
  dental morphology, 25
  diagnostic traits, 14–27
  diet breadth, 216
  divergence time from modern humans, 27
  DNA, 27–29
  femoral curvature, 315
  femoral neck-shaft angle, 315
  femoral shaft cross-sectional shape, 312, 313, 314
  fertility rates, 347
  foraging efficiency, 330–331
  foraging trip distances, 317
  geographic range, 10–11
  gestation length, 343
  growth rates, 347
  humeral shaft cross-sectional shape, 227, 229, 230
  humeral strength asymmetry, 227–229, 230–232, 232–233
  hunting party size, 245
  infant mortality, 348

interbirth interval, 345
life history, 342, 347
lifespan, 348
meat consumption, 165–167, 172–173, 174–175, 211–212
metapopulation size, 351
mortality, 349
muscularity, 27, 80–82, 103, 118, 140, 288
nasal morphology, 19, 21, 134–137, 329–330
neonatal brain size, 94, 343
nuclear genome, 29
phylogeographic structure, 338
population density, 350
respiratory capacity, 327, 329–330
skeletal robusticity, 27, 118, 139, 227, 228, 310
social group size, 338
species status, 30
stature, 72
subpopulations, 28
surface area, 124–127
symplesiomorphies, 5, 14, 23, 25
temporal range, 11–13
thoracic morphology, 25, 327–329
tibial shaft cross-sectional shape, 312, 314
trauma, 245–246, 340, 342
visual cortex, 24
walking speed, 308
weaning age, 344–345
Nepalese porters, 296
Netsilik, 299, 300
Netsilingmiut, 302
Niche divergence, 265, 279
Non-shivering thermogenesis, *see* Cold-induced thermogenesis
Nunamiut, 293, 299, 302, 317

Occlusal fingerprint analysis, 198
Ochre (iron oxide), 359
Olschewian, 52, 362
Opportunistic hunting, 246
Optimal foraging radius, 319
Optimal foraging theory, 319
Osteogenic triggers, 311, 312
Osteoregulation, *see* Osteogenic triggers
Over-harvesting, 355
Oxygen consumption, 329
Oxygen cost of breathing, *see* Work of breathing

Pair-bonding, 337, 359
Palimpsests, 66, 309, 339
Patch choice model, 319
Patrilocality, 338
Percent body fat, 74
Personal adornment, 3, 359, 360, 362
Physical activity level (PAL), 287, 288, 293, 316, 321, 325, 326, 331
Pigments, 359
Piro, 323
Pitjandjara, 317
Platycephaly, 23
Pollen cores
La Grande Pile, 111, 163, 171
Le Bouchet, 160
Marks Tey, 160, 161
Philippi, 160, 161
Ribains, 160
Polygyny, 337
Pontinian, 48
Population bottlenecks, *see* Bottlenecks, genetic
Population sinks, 350
Prey size selection, 242–244
Prey-switching, 321
Projectile weapons, Paleolithic, 66, 357, 358
Pseudomorphs (natural casts), 146
Pume, 317

Quina retouch, 45, 58
Quina-Type Mousterian, *see* Charentian Mousterian

Racloir Index, 67
Radiating mobility, 290, 291, 292, 293, 294, 297, 298, 300, 308, 317
Refugia, 13, 28, 32, 113, 114, 173, 174, 356
Reliable technology, 297
Residence systems, 29, 338
Resins, 59, 68
Resource harvesting, 292, 293, 296
Resting metabolic rate (RMR), 323
Retouch intensity, 57, 67
Retromolar space, 20, 50
Robusticity, 227

Salento peninsula, 301
Sami, 118, 120, 127, 148, 150
San, *see* Ju’/hoansi
Scapular axillary border, 27, 31
Scapular glenoid index, 222
Scavenging, 219
Schöningen spears, 61–65, 241
Selective sweeps, 357
Selk’nam, 144
Sexual dimorphism, 75–76

Sexual division of labor, 232, 322, 324, 327, 342
Shared primitive traits, *see* Neandertal, symplesiomorphies
Shivering, 141
Shock weapons, 70
Siberia, 10, 27, 29, 113, 120, 127, 154, 173, 206, 346, 359
Site density, Middle Paleolithic, 355–356
Sites, early Upper Paleolithic
  Cioclovina, 38
  Eel Point, 285
  Kent's Cavern, 38
  Lagar Velho, 39
  Peştera cu Oase, 38, 204
  Peştera Muierii, 38, 285
Sites, initial Upper Paleolithic
  Cavallo, 38
  Grotte du Renne (Arcy-sur-Cure), 39
  Saint Césaire, 39
Sites, Lower Paleolithic
  Apidima, 34
  Arago, 34, 35, 79
  Barranco León, Orce, 32
  Beroun, 32
  Biache-Saint-Vaast, 36
  Bilzingsleben, 34, 36
  Bodo, 34
  Boxgrove, 34, 79
  Campitello Quarry, 59
  Ceprano, 33, 34
  Chillac, 32
  Clacton-on-Sea, 62
  Cueva Victoria, 32
  Dali, 33, 34
  Dmanisi, 32
  Duero Valley, 32
  Elandsfontein, 34
  Gesher Benot Ya'aqov, 192, 193
  Gran Dolina, Atapuerca, 33, 215
  Jinniushan, 33, 34
  Kabwe, 33, 34
  Korolevo, 32
  Lazaret, 11, 143, 186, 191, 339
  Maba, 34
  Monte Poggiolo, 32
  Montmaurin, 34
  Petralona, 33, 34, 35
  Schöningen, 62, 65
  Sima de los Huesos, Atapuerca, 33, 34, 36, 37, 78, 339, 343, 348
  Sima del Elefante, Atapuerca, 32
  Steinheim, 33, 34, 36
  Stránska Skála, 32
  Swanscombe, 34, 36
  Vallonet, 191
  Ventana Micena, 32
  Vértesszöllös, 34, 36
  Yunxian, 34
  Zuttiyeh, 34
Sites, Middle Paleolithic
  Abri Peyrony, 67
  Abric Romaní, 61, 68, 143, 145, 146, 294, 295, 338, 339
  Achenheim, 154
  Amalda, 186
  Amud, 10, 49, 143
  Arcy-sur-Cure, 197
  Ariendorf, 154
  Barakaevskaïa, 39
  Bau de l'Aubesier, 186
  Baume Flandin, 339
  Baume-Bonne, 186
  Baume-Vallée, 302
  Biache-Saint-Vaast, 11, 36
  Bisitun, 51
  Bolomor, 189
  Breuil, 185, 190
  Buhlen, 339
  Byzovaya, 13, 39, 52
  Cabezo Gordo, 154
  Champ Grand, 301, 338
  Combe Grenal, 46, 61, 152, 215, 247, 294, 362
  Cova Beneito, 188
  Cova Negra, 188
  Cueva Millan, 191
  Cueva Morín, 60
  Dederiyeh Cave, 49
  Denisova, 27, 29, 154, 186
  Douara, 192
  Drachenhöhle, 186
  Ehringsdorf, 11, 36, 154
  El Sidrón, 27, 28, 29, 59, 200, 215, 337, 338, 339, 340, 345, 347
  Esquilleu, 143
  Fate, 154, 275
  Feldhofer, 27, 28, 30, 36, 211
  Figueira Brava, 191
  Figuier Cave, 339
  Fontchevade, 11
  Gabasa Cave, 184
  Gatzarria Cave, 294
  Geula Cave, 49, 50
  Gorham's Cave, 52, 185, 191, 192, 193
  Grotte des Cédres, 186

Grotte des Pêcheurs, 191
Grotte Vaufrey, 247
Grotte XVI, 191
Guattari, 154, 172, 247, 261, 274
Gudenushöhle, 186
Haua Fteah, 190
Hayonim, 49, 188
Hummal, 186
Jonzac, 211, 213, 245
Karabi Tamchin, 302
Kärlich, 61
Kebara, 10, 49, 154, 172, 186, 192, 193, 194, 217, 295
Kiik Koba, 361
Kobeh Cave, 186, 248
Königsaue, 59
Krapina, 31, 36, 38, 61, 77, 98, 214, 215, 274, 315, 340
Kůlna, 39, 302
l'Hortus, 186
l'Observatoire, 186
La Chaise, 11, 36
La Chaise Bourgeois-Delaunay, 11, 36
La Chaise Suard, 11, 36
La Chapelle-aux-Saints, 30, 36
La Cotte de St. Brélade, 154
La Ferrassie, 30, 36, 361, 362
La Folie, 61
La Parte, 114
La Quina, 36, 67, 193, 197, 294, 361
Le Moustier, 36, 294
Le Portel, 184, 275
Le Rond de Saint-Arcons, 302
Lehringen, 62, 63, 69
Les Canalettes, 189, 295
Les Peyrards, 186
Les Pradelles, 211, 215, 293, 295
Lezetxiki, 186
Maastricht-Belvédère, 154, 160, 169
Macassargues, 215
Maggiore, 191
Mamontovaya Kurya, 39
Mas-des-Caves, 192
Mauran, 245, 293
Misliya, 186
Molodova I, 339
Monte Circeo, 30, 36
Monti Lessini, 28, 29, 31, 39
Moscerini, 188, 190, 247
Moula-Guercy, 215
Mujina Pećina, 247
Neumark-Nord, 160, 169
Okladnikov, 27, 113, 154, 186
Orgnac, 191
Payre, 191, 277
Pech de l'Azé, 67, 152
Pêcheurs Cave, 339
Pontnewydd, 11
Qafzeh, 37, 49, 50
Rabutz, 154
Raj Cave, 191, 302, 308
Reilingen, 11, 34, 36
Repolosthohle, 362
Repolust, 186
Roc de Marsal, 361
Rochelot, 274
Saccopastore, 36, 38
Saint-Marcel Cave, 339
Salpêtre de Pompignan, 187, 188, 191
Salzgitter-Lebenstedt, 66, 193, 245, 293, 297
Sant' Agostino, 185, 295
Schweinskopf-Karmelenberg, 154
Scladina, 28, 168, 204, 206, 207, 208, 211, 345
Sclayn Cave, 172
Seßelfelsgrotte, 171
Shanidar, 36, 51, 186, 188, 361
Skhul, 37, 49, 50, 360
Spy, 30, 36, 197, 199
Starosele, 186, 193
Subalyuk, 186, 197
Tabun, 10, 37, 48, 49, 50
Tata, 362
Taubach, 154, 169, 188, 362
Teshik-Tash, 186, 361
Tito Bustillo, 191
Tönchesberg I, 154
Tor Faraj, 296
Trou Magrite, 296
Umm el Tlel, 60
Ust-Kanskaya, 191
Valdegoba, 28
Vanguard Cave, 185, 191
Vaufrey, 191, 302, 309
Veltheim, 154
Vindija, 27, 28, 31, 38, 39, 188, 191, 199, 211, 212, 213, 215
Wallertheim, 172, 186, 293
Wannen, 154
Weimar, 154
Zafarraya, 215
Sites, MSA
Die Kelders, 248
Klasies River Mouth, 247
Tamsara Hill, 37

Snow cover, 67, 97, 102, 284, 293, 320, 321
Social cognition, 336
Social dominance, *see* Behavioral dominance
Social groups, 337
Social network size, 341
Social storage (of calories), 358
Socioecology, 333
South-facing shelters, 143
Spear thrusting, 224, 225, 226, 229, 231, 232, 233, 314
Spears, *see* Schöningen spears
Spearthrowers, 4, 66, 222, 235, 241, 358
Specialized hunting, 246
Specific dynamic action, 87, 323, 325
Specimens
  Amud 1, 31, 37, 195, 198, 199
  Arago XLIV, 79
  Berg Aukas, 354
  Boxgrove 1, 79
  Cova Negra, 275
  Cueva del Camino, 275
  Dederiyeh, 343
  El Sidrón, 198, 200
  Engis 2, 210, 344
  Feldhofer, 210, 314
  Florisbad, 7, 275
  Forbes Quarry 1, 74
  Gibralter 2, 195
  Grotta Breuil, 198
  Grotte de la Tour, 274
  Gruta da Oliveira, 274
  Hoedjiespunt, 275
  Kabwe, 275, 354
  Kalamakia, 274
  Kebara 2, 76, 118, 119, 198, 328, 343
  Krapina, 198
    180, 274
    213, 76
  La Chaise, 197
  La Chapelle aux Saints 1, 118
  La Ferrassie
    1, 76, 124, 125
    2, 77
  Lakonis, 316
  Le Moustier, 199
  Les Pradelles, 202, 208, 209, 275
  Les Rochers-de-Villeneuve, 210, 274, 314
  Lezetxiki 1, 74
  Malarnaud, 195
  Mezmaiskaya, 343
  Monsempron, 199
  Muierii 1, 222
  Okladnikov, 210
  Payre, 202, 208
  Petit-Puymoyen, 197
  Qafzeh, 222
  Regourdou 1, 74
  Rochelot, 197
  Saccopastore 1, 40
  Saint Césaire 1, 197, 210, 212, 246, 310, 314
  Scladina, 202, 206, 207
  Shanidar, 199
    1, 31, 37, 274, 340
    2, 31, 37
    3, 68, 119, 200, 246, 328
    4, 37, 124
    5, 31, 37
  Skhul, 222
  Spy
    1, 199
    2, 124, 199
  Steinheim, 74
  Tabun
    C1, 31, 37, 50, 118, 119, 123, 195, 198, 199, 328, 343, 361
    C2, 50, 195
  Teshik-Tash, 274
  Valdegoba, 275
  Vindija, 197, 210
  Wezmeh Cave, 275
  Zafarraya, 198, 274
  Zeeland Ridges, 202, 210, 212
  Zhoukoudian, 275
  Zuttiyeh, 37
Spotted hyena, *see* Cave hyena
Still Bay industry, 360
Storage pits, 177, 294
Streletskian, 52
Striped hyena, 255, 262
Strontium isotope analysis, 316
Subcutaneous fat, 80, 102, 128, 138, 140
Subsistence tools, *see* Extractive technology
Supraorbital tori, 23, 129
Surgical weapons, 70
Surplus killing, 320
Symbolic behavior, Neandertal, 5, 337, 361, 362
Syngameon, *see* Allotaxa
Syrian Desert, 158
Szeletian, 52

Taubacian, 69
Taurodontism, 25
Taurus Mountains, 158
Technological innovations, Paleolithic, 3–4

Technological stages
  Mode 1, 32
  Mode 2, 42, 60
  Mode 3, 42, 60, 69
  Mode 4, 48, 69
  Mode 5, 69
Technounits, 59
Tents, 143
Terrain-dependent hunting, 238–241
Thermal effect of food, *see* Specific dynamic action
Thermal neutrality, 116, 141, 142, 150
Throwing sticks, Paleolithic, 65
Thyroid hormones, 139
Thyroxine, 117, 139, 140, 149
Tierra del Fuego, 144
Time allocation studies, 322
Total energy expenditure (TEE), 88, 97, 287, 288, 325, 326, 327, 329, 331
Trace element analysis, 344
Trade, 338, 359
Trade-offs (energetic), 334, 335, 348
Tuscany lion, *see* European jaguar
2D:4D ratio, 337
Typical Mousterian, 43

Ulluzian, 52, 362
Ulnar supinator index, 234
Ulnar trochlear notch index, 224
Uniquely derived features, *see* Neandertal, autapomorphies
Upper Paleolithic (UP), 3, 4, 6, 39, 48, 52, 53, 66, 69, 127, 187, 189, 192, 216, 224, 229, 232, 234, 235, 236, 243, 246, 247, 278, 285, 294, 297, 310, 311, 312, 313, 344, 350, 351, 355, 357, 358, 359, 360, 361, 362, 363

Vicarism, 39, 50, 51

Wales, 10
Warming hearths, 143
Weaning, 94, 98, 210, 335, 342, 343, 344, 347
Weaning ages, foragers, 344
Westphalia, 10, 113
Wind chill temperature, 113
Winter mortality (winter-kill), 299, 320
Wolf, 255, 262–263, 281, 304, 305
Wolverine, 254–255
Work of breathing, 329

Ya'mana, 144
Yabrudian, 48, 49
Yup'ik Eskimos, 92
Yupiat, 104

Zagros Mountains, 51, 158
Zagros Mousterian, 48, 51